Crystal Growth –

From Fundamentals to Technology

Edited by

Georg Müller

Crystal Growth Laboratory,
Department of Materials Science, University Erlangen-Nürnberg,
Fraunhofer Institute of Integrated Systems and Device Technology
Erlangen, Germany

Jean-Jacques Métois

Centre de Recherche de la Matière Condensée et des Nanosciences
Campus Luminy
Marseille, France

Peter Rudolph

Institute for Crystal Growth
Berlin, Germany

2004
ELSEVIER

Amsterdam – Boston – Heidelberg – London – New York – Oxford
Paris – San Diego – San Francisco – Singapore – Sydney – Tokyo

ELSEVIER B.V.
Radarweg 29
P.O. Box 211, 1000 AE
Amsterdam, The Netherlands

ELSEVIER Inc.
525 B Street
Suite 1900, San Diego
CA 92101-4495, USA

ELSEVIER Ltd
The Boulevard
Langford Lane, Kidlington,
Oxford OX5 1GB, UK

ELSEVIER Ltd
84 Theobalds Road
London WC1X 8RR
UK

First edition 2004
Reprinted 2005

British Library Cataloguing in Publication Data
A catalogue record is available from the British Library.

Library of Congress Cataloging in Publication Data
A catalog record is available from the Library of Congress.

ISBN: 0-444-51386-8

∞ The paper used in this publication meets the requirements of ANSI/NISO Z39.48-1992 (Permanence of Paper).
Printed in

Preface

Crystals have fascinated mankind for thousands of years – be it snowflakes, minerals or jewels. For at least 50 years crystals have fascinated engineers as the key materials of modern electronics, optoelectronics and other technical fields of application. The formation of crystals in nature, like snowflakes and minerals , as well as the preparation of crystals in laboratories and factories for technical applications is called "crystal growth".

Due to the fact that many of today's technological systems in the fields of information, communication, energy, transportation, medical and safety technologies depend critically on the availability of suitable crystals with tailored properties, their fabrication – crystal growth – has become an important technology.

From this point of view it is evident that teaching crystal growth with the aid of books and schools is of great importance for the further development and dissemination of knowledge and expertise in crystal growth. This idea is one of the reasons why, approximately 50 years ago, a group of scientists working world wide in the field of crystal growth decided to organize themselves within the IOCG and hold International Schools of Crystal Growth. These schools were to be held during the summer, in parallel with their triannual conferences (ICCG). The big success of the previous International Summer Schools of Crystal Growth (ISSCG1 - 11) since is started in the Netherlands in the year 1971, proves the wisdom of the founders of IOCG in this decision.

In 2004, from August 1 to 7, the 12th Int. Summer School of Crystal Growth ISSCG 12 will be held in Berlin (Germany). ISSCG 12 is organized in conjunction with the 14th Int. Conference on Crystal Growth (August 9 - 13 in Grenoble, France) in a close collaboration of the German and French associations of crystal growth, Deutsche Gesellschaft für Kristallwachstum und Kristallzüchtung (DGKK) and the Groupe Francaie de Croissance Cristalline (GFCC).

Concerning the selection of the topics and subjects to be treated in the lectures and seminars we had a rather difficult task due to the duration (one week) of the school. In this limited time frame it is necessary to cover the vast field of crystal growth, from fundamentals to technology to characterization. Another problem is the different interests of the participants and their previous knowledge and experience – ranging from beginners to experts. The latter issue we have addressed by offering introductory seminars on various topics for beginners, one day before the official lecture program starts.

During the school the program is composed of expert lectures each morning and tutorial seminars in the afternoon. The new idea of the tutorial seminars, which are held in parallel for 3 topics, is to offer the participants the possibility of selecting certain topics and to give the lecturers the opportunity to organize the style and the contents according to the level of knowledge of the group of participants.

Nevertheless, the selection of the topics to some degree reflects the background of the editors and we can only apologise for the omission of topics which participants of ISSCG 12 and readers of this book would have wished to have been included – unfortunately no school and no book can be exhaustive.

We would also like to emphasize that the lecturers were solely selected on the basis of their knowledge and expertise in their particular fields and on their ability to contribute within the imposed time frame, but not on their nationality, seniority or other considerations.

This book contains the edited versions of the expert lectures presented during ISSCG 12 in Berlin. In most cases the contents of the manuscripts considerably exceeded the amount of material which could be presented within an one hour lecture. This gives them more the character of a hand book article than just a lecture note. The outline with five chapters is organized according to the lecture program as it was presented during the school, starting with fundamentals (chpt. 1), modeling (chpt. 2), future topics (chpt. 3), crystal growth technology (chpt. 4) and finally crystal defects and characterization (chpt. 5).

As already mentioned, a series of introductory and tutorial seminars were held during the school which are not contained in this book:

- Introduction into Crystal Growth (R. Fornari)
- Crystal Growth Techniques (G. Müller)
- Growth Kinetics (J.-J. Métois)
- Crystal Defects (P. Rudolph)
- Modeling of Crystal Growth (J. Neugebauer and W. Miller)
- Course on Numerical Simulation of Crystal Growth Processes on PCs (T. Jung)
- Growth of Biological Crystals (S. Veesler)
- Phase Diagrams: Analysis and Modeling (M. Mühlberg and D. Klimm)
- MOCVD of Compound Semiconductors (M. Heuken)
- Fundamentals of Epitaxial Growth (A. Pimpinelli)
- Electron Microscopy (A. Albrecht)
- X-ray Methods (H. Leipner)
- Point Defect Analysis (K. Irmscher)

In summary, the aim of the school and this book is to provide the students and readers with advanced knowledge concerning the technologically important subject of inorganic and organic single crystal growth. We hope that this book will be highly valuable for the entire crystal growth community as well as physics, chemistry, materials science engineering, biology, crystallography and remain as an important source for crystal growers, beginners and specialists alike.

July, 2004

The Editors
Georg Müller
Jean-Jacques Métois
Peter Rudolph

Acknowledgement

The editors greatly acknowledge the contributions of all the lecturers of ISSCG 12, especially the authors of this book. In addition we thank the directors of the Institute of Crystal Growth (IKZ, Berlin)), Prof. Roberto Fornari and his predecessor Prof. Winfried Schröder. Both generously supported the preparation and organization of ISSCG 12 in many ways, from personal to financial support!

Also many thanks to our colleagues from the organization team, Sabine Bergmann, Dr. Thorsten Boek, Dr. Anke Lüdge, Dr. Uwe Rehse, Dr. Detlef Schulz, Dr. Jürgen Warneke from the IKZ and Dr. Jochen Friedrich from the Erlangen Crystal Growth Laboratory for their tireless efforts and activities before and during ISSCG 12.

We gratefully acknowledge the support of our esteemed friends and colleagues from the French Association of Crystal Growth (GFCC), Dr. Jean-Jacques Favier, Dr. Stéphane Veesler, Dr. Denis Camel and from the Grenoble team of ICCG 14, Prof. Jacques Villain, Prof. Thierry Duffar and Dr. Peter Wellmann.

Last but not least we acknowledge the permanent support and advice from the President of IOCG, Prof. Robert Sekerka.

Both the ISSCG 12 technical program and the extension of about 60 student participation grants could not be accomplished without major financial contributions from several organizations, public institutions and companies.

The organizers acknowledge the contributions of:

DGKK, ICCG 14/GFCC, European Space Agency., Int. Union of Crystallography, German Association for Crystallography, Crystal Growth Laboratory of FhG Erlangen, InWent GmbH, IKZ-Berlin, DFG.

Companies: Kistler, ECM Infra Fours Physitherm Grenoble, Schunk Kohlenstoff, Linn High Term, Aixtron, Heraeus, Netzsch–Gerätebau, Anteam Industrielektronik, FLP Stähli, Saint-Gobain Winter, Siltronic AG, Freiberger Compound Materials, GTT-Technologies, SiCrystal AG, Röntec, PV Silicon, Matech, Crystec, Sterenat Elektrowärme, Hüttinger Elektronik GmbH.

The preparation of this book was supported by Dr. Jakob Fainberg, Marc Hainke, Stefan Hussy, Dr. Paul Sonda and by the secretary of the Erlangen Crystal Growth Laboratory, Elisabeth Henneberger.

Finally, our thanks to all those from the ELSEVIER publication team who were associated with the production of this book, especially to Jeanette Bakker, John Lardee, Linda Versteeg and Donna Wilson-Wheerd.

The Editors

Georg Müller
Jean- Jacques Métois
Peter Rudolph

Contents

Chapter 2: Modeling

Chapter 3: Crystal Growth Technology

Crystal Growth – from Fundamentals to Technology
G. Müller, J.-J. Métois and P. Rudolph (Editors)

Thermodynamics of modern epitaxial growth processes

Prof. G.B. Stringfellow

College of Engineering, University of Utah, Salt Lake City, Utah 84112, U.S.A.

The purpose of this lecture is to review the basic thermodynamic concepts that lead to a better understanding of: 1) the epitaxial processes used for the growth of semiconductor materials and structures used for photonic and electronic devices and 2) the properties of the materials produced by these techniques.

All epitaxial growth processes are driven by the free energy difference between the nutrient and solid phases. Thus, thermodynamics determines the maximum possible growth rate. For conditions where the rates of interface processes are sufficiently rapid to make mass transport the rate limiting step, a condition that can be approximated as thermodynamic equilibrium exists between the solid and the nutrient right at the interface. Thus, thermodynamics can be used to describe the solid composition in terms of the independent growth parameters. This is true for the near-equilibrium liquid phase epitaxial growth techniques as well as for the vapor phase techniques operating with a very large supersaturation, such as organometallic vapor phase epitaxy (OMVPE) and molecular beam epitaxy (MBE). In addition to determining the overall solid composition of semiconductor alloys, thermodynamics may also determine the stoichiometry of the solid and, consequently, dopant incorporation.

Many important properties of semiconductor materials and alloys produced epitaxially are determined by thermodynamics. Since thermodynamics controls the stoichiometry of the solids produced epitaxially, it controls the densities of native defects as well as deep levels acting as non-radiative recombination centers. The solid composition in alloys is thermodynamically controlled; thus, the most important parameters such as lattice constant and energy band gap are determined by thermodynamic factors. In addition, many semiconductor alloys have miscibility gaps due to the large, positive enthalpy of mixing. Thus, some alloys are not thermodynamically stable. By understanding the epitaxial growth processes, such metastable alloys can often be grown epitaxially. However, the distribution of the constituent atoms is often not random for such alloys. The resulting clustering has deleterious effects on the properties. In addition, under some growth conditions, surface thermodynamic factors drive the system to produce ordered alloys, with a composition modulation with monolayer periodicity along a particular crystallographic direction. Ordering has profound effects on the electrical and optical properties. For instance, the band gap energy can be reduced significantly.

Recently, the role of surface thermodynamics has begun to be recognized. Often, equilibrium is maintained only between the nutrient phase and the surface layers of the epitaxial film. This results in the growth of layers that are metastable in the bulk form. This means that the surface composition and bonding, i.e., the surface reconstruction, can dramatically affect the growth process and the properties of the epitaxial layer. This has led to the use of surfactants to control OMVPE and MBE growth and the resulting epilayer properties. Surfactant effects have been demonstrated on the microstructure of alloys, dopant incorporation, and the growth of low dimensional structures.

1. INTRODUCTION

Today, many semiconductor devices and circuits require epitaxial processes. In Si, this is due to the extremely fine geometries and thin layers required. In the world of compound semiconductors, nearly all devices have always required epitaxy due to the use of alloys, the extremely high quality needed for minority carrier devices, and the fine geometries, especially now, when bandgap engineered structures require quantum wells, wires, and dots.

A number of epitaxial growth techniques has been developed for semiconductors over the last 50 years. The first processes used chlorides for transporting the material for both Si and III/V semiconductors. Liquid phase epitaxy was also extremely useful for the production of layers of high purity III/V semiconductor materials. However, in recent years these techniques have been largely displaced by more flexible techniques for the growth of a large range of materials and special structures, such as organometallic vapor phase epitaxy (OMVPE) [or equivalently metal organic VPE (MOVPE) or MO chemical vapor deposition (MOCVD)], molecular beam epitaxy (MBE), and chemical beam epitaxy (CBE). Each of these techniques will be described very briefly below.

OMVPE has come to be the leading technique for the production of III/V materials, especially for photonic devices. It is particularly successful for devices requiring large areas, such as solar cells, cold cathodes, and light emitting diodes (LEDs). For example, OMVPE is used for commercial scale production of a number of materials and devices including: AlGaInP alloys for visible LEDs and injection lasers[1], the III/V nitrides used for blue and green LEDs and blue injection lasers[2], AlGaInP for the highest efficiency tandem solar cells [3,4], and GaInAsP and AlGaAs structures for infrared lasers and detectors[5,6]. Today, commercially available reactors can be purchased for both laboratory-scale and large production-scale applications from several manufacturers. For in-depth reviews of the OMVPE technique see refs [7-9].

MBE has, for decades, been the leading technique for the production of fine-scale structures. It was the first technique to produce layers showing quantum confinement and has been at the forefront of the development of bandgap engineered structures. It is widely used in production operations, especially for electron switching devices and circuits such as high electron mobility transistors and integrated circuits. Reviews and books are available for in-depth reviews of the technique and applications[10-12].

CBE is essentially a hybrid of OMVPE and MBE. It uses an ultra-high vacuum chamber, as for MBE, but used organometallic or hydride precursors rather than the elemental sources typically used in MBE. This gives certain advantages[13], but this technique remains mainly a laboratory technique which is used less frequently than OMVPE and MBE in production operations.

Each of the epitaxial growth processes is exquisitely complex when viewed in detail at the atomic level. Due to these complexities, even after many thousands of man years of effort we are still nowhere near a complete understanding. For each process, the fundamentals are often broken down into the separate categories of thermodynamics, kinetics, and hydrodynamics and mass transport. In fact, all of these separate factors and processes are tightly coupled.

As a result of the complexity of epitaxy processes, early crystal growth studies were largely empirical, giving epitaxy the appearance of an art. This is partly because of the complex, multicomponent, multiphase systems that are normally of interest and partly because the processes are dynamic. The pursuit of a more perfect understanding of the epitaxial processes has often involved the development of increasingly sophisticated, often *in-situ*, characterization tools. However, with every step in increasing the sophistication of the tools used to study the processes, it becomes clearer that they are even more complex than we had thought. Thus, a complete understanding of any of these processes remains elusive.

In this review, we will concentrate nearly entirely on the thermodynamic aspects of epitaxy. In fact, the epitaxial processes are essentially controlled phase transitions, and, thus, thermodynamic aspects control many of the resulting characteristics of both the processes and the resulting epitaxial layers. Having said that, it must also be acknowledged that the very process of growing an epitaxial layer demonstrates that the system cannot be completely at equilibrium. It is the intentional deviation from equilibrium that gives the driving force for epitaxy. Nevertheless, the extremely slow growth rates and relatively high temperatures typically used for the production of the highly perfect, thin layers we are interested in can give a condition approximating equilibrium between the growing solid and the nutrient phase right at the interface. This guarantees a dominant role for thermodynamics in most cases. As will be seen, the insights gained by thermodynamic analysis are most useful. Thermodynamics defines the maximum growth rate, often determines the composition of the solid being grown including alloy composition, stoichiometry, and dopant incorporation. Thermodynamics also defines the equilibrium microstructure of the solid and so can be used to predict whether the solid alloy will be uniform (random), clustered, or ordered.

2. THERMODYNAMIC DRIVING FORCE FOR EPITAXY

As mentioned above, thermodynamics underlies the epitaxial growth process. At low growth rates and relatively high temperatures, the chemical reaction kinetics play less of a role than in very rapid crystal growth processes. In the limit of infinitely slow growth rates, thermodynamics defines the relationships between the concentrations of the constituents in the vapor and solid phases. Thermodynamics also determines the driving force for any crystal growth process, hence defining the maximum growth rate. To make this section concrete, we will focus on a particular process, namely the OMVPE growth of the III/V semiconductor GaAs.

The equilibrium state for a two phase, $\alpha + \beta$, system is defined in terms of the chemical potentials,

$$\mu_i^{\alpha} = \mu_i^{\beta} \tag{1}$$

where the subscript i indicates the ith component and the superscripts indicate the phase. The chemical potential is usually written in terms of the chemical potential in an arbitrary standard state, denoted by the superscript zero,

$$\mu = \mu^{o} + RT \ln p/p^{o} \tag{2}$$

For an ideal gas mixture,

$$\mu_i = \mu_i^{o} + RT \ln p_i/p_i^{o} \tag{3}$$

where p_i is the partial pressure, equal to the mole fraction x_i multiplied by P, the total pressure, and the standard state is usually pure component i.

For an ideal solid solution, the same expression holds with p_i/p_i^{o} replaced by x_i/x_i^{o}. However, the standard state is pure i, so $x_i^{o} = 1$. The form of eq. (3) is so useful that it is retained even for non-ideal solutions with x_i replaced by the activity, a_i, which may also be considered a product of x_i multiplied by a non-ideality factor, γ_i, the activity coefficient.

For the OMVPE growth of GaAs using trimethylgallium (TMGa) and arsine (AsH_3) the overall reaction is

$$(CH_3)_3Ga(v) + AsH_3(v) = GaAs(s) \tag{4}$$

Assuming the TMGa and AsH_3 to completely decompose in the gas phase to give Ga and As_4, the reaction can be simplified:

$$Ga(v) + 1/4\ As_4(v) = GaAs(s) \tag{5}$$

The equilibrium condition is

$$\mu^v_{Ga} + 1/4\ \mu^v_{As4} = \mu^s_{GaAs} \tag{6}$$

or

$$\mu^{ov}_{Ga} + 1/4\ \mu^{ov}_{As4} + RT \ln p^e_{Ga}(p^e_{As4})^{1/4} = \mu^{os}_{GaAs} + RT \ln a_{GaAs} \tag{7}$$

where the superscript "e" denotes the equilibrium value of partial pressure. Thus,

$$a_{GaAs} / p^e_{Ga}\ (p^e_{As4})^{1/4} = K_{GaAs} \tag{8}$$

where K is the equilibrium constant. This is the basic law of mass action.

When the system is not at equilibrium, the thermodynamic driving force to restore equilibrium is

$$\Delta\mu = \mu^v_{Ga} + 1/4\ \mu^v_{As4} - \mu^s_{GaAs} \tag{9}$$

or

$$\Delta\mu = RT \ln [p_{Ga}\ p_{As4}{}^{1/4} / p^e_{Ga}\ p^e_{As4}{}^{1/4}] \tag{10}$$

This is the driving force for epitaxy. A situation is intentionally created where higher than equilibrium reactant partial pressures drive the system to produce the GaAs solid desired. The maximum quantity of GaAs solid that can be produced is simply the amount (the supersaturation) that would establish equilibrium, and is, thus, fundamentally limited by thermodynamics and the total amount of gas transported through the OMVPE reactor.

For the OMVPE growth of GaAs using arsine and TMGa, the thermodynamic driving force at 1000 K is approximately 80 kcal/mol [14]. The driving forces for the epitaxial growth of GaAs are compared for several epitaxial processes in Fig. 1 [14]. This very high driving force to form solid GaAs is due to the instability of both arsine and TMGa at 1000 K. It will be noted in Fig. 1 that MBE also has a high thermodynamic driving force. In this case, this is due to the instability of elemental Ga and As in the vapor at typical growth temperatures. These high driving forces for formation of the solid have prompted many researchers to dub both OMVPE and MBE as "highly non-equilibrium" growth processes.

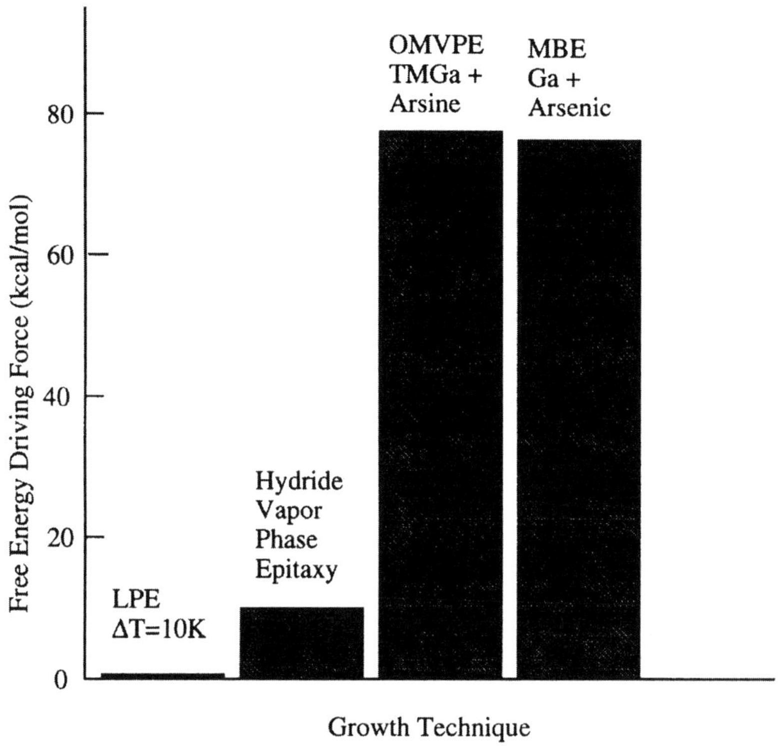

Figure 1. Estimated thermodynamic driving force for several epitaxial growth processes. All calculations are for the growth of GaAs at 1000K. (After Stringfellow [14], reprinted with permission from Elsevier Science.)

This raises the question: how does thermodynamics relate to epitaxial growth for OMVPE and MBE where the driving force is extremely high? The connection between thermodynamics and growth results has long been recognized for liquid phase epitaxy (LPE), where, as seen in Fig. 1, the thermodynamic driving force is relatively small. Here, it seems entirely natural to treat the growth process using thermodynamics and, in fact, phase diagrams are well-known to give vital information about alloy composition, solid stoichiometry, and dopant incorporation. However, even for MBE and OMVPE powerful thermodynamic factors still control much of the growth process. This is because, even for a system with a high supersaturation of the input vapor phase, near equilibrium conditions may prevail near the solid/vapor interface. This means that thermodynamics can provide important information about the growth process and the properties of the resultant materials. However, it may prove necessary to consider the thermodynamic properties of the surface in addition to the vapor and bulk solid phases. Thermodynamic factors largely determine the equilibrium structure of the surface, leading to surface phase diagrams, as discussed below, that give the surface reconstruction as a function of the extensive parameters, such as temperature and the group v partial pressure, as discussed below. Surface reconstruction has profound effects on both the epitaxial growth processes and the properties of the resulting layer.

Ordinarily, in the OMVPE system the growth rate is considerably less than that calculated from thermodynamics. Kinetics, both surface reaction rates (at low temperatures) and diffusion through the gas phase (at higher temperatures), are not rapid enough to allow equilibrium to be established throughout the system at all times. This situation is illustrated by Fig. 2a, where $\Delta\mu$ from eq. (9) is plotted versus reaction coordinate [15]. This allows the schematic representation of the overall, thermodynamic driving force for the growth reaction, represented as $\Delta\mu^*$. The superscript "* " denotes the chemical potential in the input gas phase, where for all reactants $p_i = p^*_i$. The growth rate is proportional to the flux of atoms diffusing through the boundary layer, which is identical to the flux of atoms crossing the interface into the solid. The diagram shows schematically the driving forces necessary to sustain this flux for the diffusion process ($\Delta\mu_D$) and the surface reactions ($\Delta\mu_s$).

Even in cases with a large supersaturation in the input vapor phase, i.e., $\Delta\mu^* >> 0$, near equilibrium conditions may exist at the growing solid surface. This simply requires that the interface kinetics be much more rapid than the diffusion kinetics, i.e., the two processes proceed at the same rate with $\Delta\mu_s << \Delta\mu_D$. This situation, termed diffusion limited growth, is shown schematically in Fig. 2b. Using ordinary growth conditions, with temperatures between approximately 550 and 800 °C, this is the normal situation for the OMVPE growth of GaAs, as deduced from the nearly temperature independent growth rate [16]. For surface kinetically limited processes, the growth rate increases exponentially with increasing temperature [17,18]. This occurs for the OMVPE growth of GaAs at temperatures below approximately 550 °C when TMGa is the Ga precursor, but this temperature depends on the group III precursor used.

In the diffusion limited case, illustrated schematically in Fig. 2b, the interfacial partial pressures, p^i_i, nearly satisfy the equilibrium relationship (assuming the dominant species to be the As tetramer),

$$a_{GaAs} / p^i_{Ga}(p^i_{As4})^{1/4} = K_{GaAs} \tag{11}$$

where a_{GaAs}, the activity of GaAs in the solid, is unity for the growth of pure GaAs. K_{GaAs} is the equilibrium constant at the particular temperature chosen. Since the input vapor is highly supersaturated,

$$p^*_{Ga}(p^*_{As4})^{1/4} >> p^i_{Ga}(p^i_{As4})^{1/4} \tag{12}$$

This is equivalent to stating that $\Delta\mu^* >> 0$. For the typical OMVPE case

$$p^*_{Ga} << 4p^*_{As4} \tag{13}$$

i.e., the V/III ratio is >>1, as will be discussed below. This means that the Ga is nearly depleted at the interface,

$$p^i_{Ga} << p^*_{Ga} \tag{14}$$

while the As_4 partial pressure is hardly diminished,

$$p^{i}_{As4} \approx p^{*}_{As4}, \tag{15}$$

since the same number of As and Ga atoms are removed from the vapor phase to produce GaAs. This situation makes the analysis of growth rate and solid composition particularly simple.

The growth rate is proportional to the flux of Ga and As atoms diffusing through the vapor to the growing interface. For simplicity, this can be analyzed in terms of diffusion through a boundary layer of thickness d. A more complete description is given in reference [19]. The two fluxes are equal, since stoichiometric GaAs is the only product. The flux may be expressed,

$$J = D_{Ga}(p^{*}_{Ga} - p^{i}_{Ga})/RTd \tag{16}$$

where D_{Ga} is the diffusion coefficient of Ga, in whatever form it may appear while diffusing through the boundary layer. In light of eq. (14), the Ga flux and the GaAs growth rate are proportional to p^{*}_{Ga}, as observed experimentally [18]. Equally clear is that the ratio of the concentrations of A and B for alloys with mixing on the group III sublattice, $A_{1-x}B_xC$, will be the same as the ratio p^{*}_{A}/p^{*}_{B}, assuming the diffusion coefficients for the A and B species are nearly the same [20]. This will be discussed in more detail below.

For MBE growth, the situation is quite similar. The growth rate is typically determined by the rate of arrival of Ga atoms at the solid-vapor interface [10]. As is incorporated from the vapor in the amount needed to produce stoichiometric GaAs. Again, at low temperatures, where the group III atoms cannot re-evaporate from the surface, the ratio of group III elements incorporated into a solid alloy with mixing on the group III sublattice is the same as the flux of the group III atoms to the surface. For both OMVPE and MBE, as the temperature is raised to the point that group III atoms can re-evaporate from the surface, thermodynamics begins to control the solid composition [20]. For mixing on the group V sublattice, thermodynamics typically controls the solid composition under typical growth conditions [20].

3. BINARY PHASE DIAGRAMS

The III/V compounds melt congruently, i.e., a liquid and solid having identical compositions are in equilibrium at the melting point, as illustrated in the temperature-composition (T-x) phase diagram for GaAs seen in Fig. 3(a). Thus, the III/V compounds can be grown directly from the melt, a process commonly used to produce large boules of GaAs, GaP, InP, InAs, InSb, and GaSb. This phase diagram shows that only a single III/V compound is stable, with melting point typically well above the melting points of the group III and group V elements. In this diagram the III/V solid appears as an infinitely thin vertical line. This is indicative of the high degree of stoichiometry discussed above. In fact there is a slight, temperature-dependent deviation from stoichiometry, indicated in the inset. The non-stoichiometry is so small that it is extremely difficult to measure, and, in fact, has not been reliably measured for most III/V compounds. However, the deviation from stoichiometry has profound effects on the electrical and optical properties of the material as well as bulk diffusion coefficients, and related to this, the operating lives of many devices.

During the LPE growth of GaAs, the solid is in equilibrium with a liquid phase specified by the liquidus curve shown in Fig. 3(a). At a temperature T_2 the Ga-rich liquid in equilibrium with the GaAs solid has a composition x_2 on the liquidus curve. A horizontal "tie-line" (not shown) joins the liquid and solid compositions in equilibrium. The stoichiometry is controlled by the composition of the nutrient phase. For example, for LPE growth from the Ga-rich liquid, with

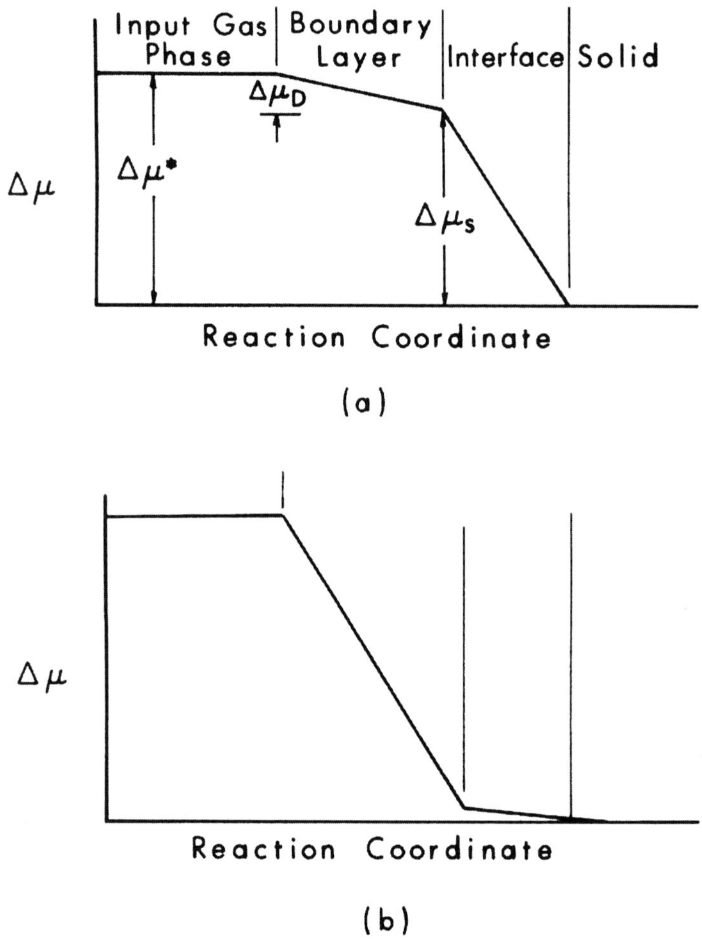

Figure 2. Schematic diagram of chemical potential versus reaction coordinate, showing the drop in chemical potential required for each step in the growth sequence to keep all rates equal: (a) the general case and (b) the case of rapid surface kinetics. (after Stringfellow [15], reprinted from the Journal of Crystal Growth with permission from Elsevier Science.)

composition x_2, the resulting GaAs is on the extreme Ga-rich side of the (very narrow) GaAs phase field. Because of the high As pressures required, GaAs would not typically be grown from the As-rich liquid, with composition x'_2 in this example. The resulting GaAs would be on the extreme As-rich side of the GaAs phase field. The difference in the As/Ga ratio in the solid may be only a few parts per million from one side of the phase field to the other, but this small difference has dramatic effects on the semiconductor properties. Very recently the As-rich GaAs grown at low temperatures by MBE has become of some interest because of its semi-insulating properties. Because of the extremely low temperatures, <200 °C, an As-rich second phase is formed. Thus, growth occurs in the two-phase "GaAs + As" region of Fig. 3(a). It should be noted that the OMVPE growth of GaAs typically produces material near the As-rich side of the solid stability region due to the very high As/Ga ratios at the interface, as discussed below.

In Fig. 3(a) only the condensed phases are shown. However, the system also necessarily contains both Ga and As vapor species. The As (both dimer and tetramer forms) and Ga partial pressures in equilibrium with the condensed phases as a function of composition, at a temperature of approximately 980°C, are shown in Fig 3(b). The stoichiometry of the solid produced at a fixed temperature during LPE growth is not variable. The phase rule specifies that the system is invariant when three phases are in equilibrium in a two component system when the temperature is fixed [21]. However, during vapor phase growth, since the number of phases is reduced by one, the phase rule specifies that one degree of freedom exists even when the temperature is fixed. The stoichiometry can be controlled by either defining the Ga or the As pressure at the solid/vapor interface. Since there is frequently a large excess of the group V

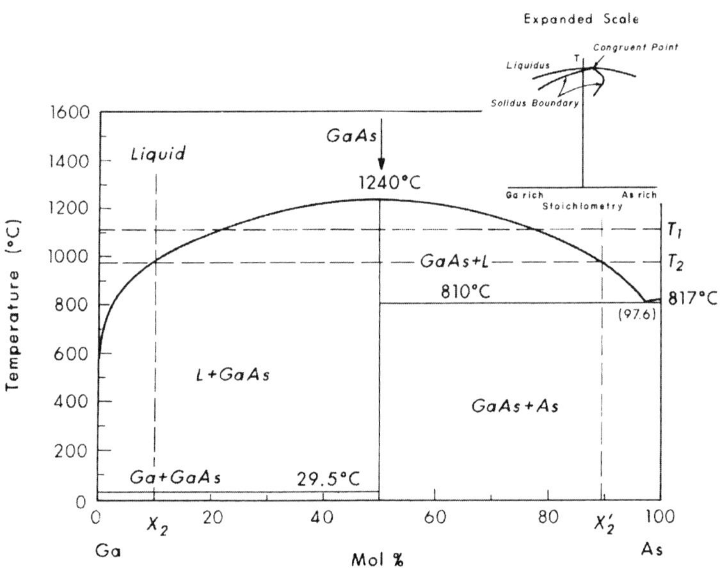

(a)

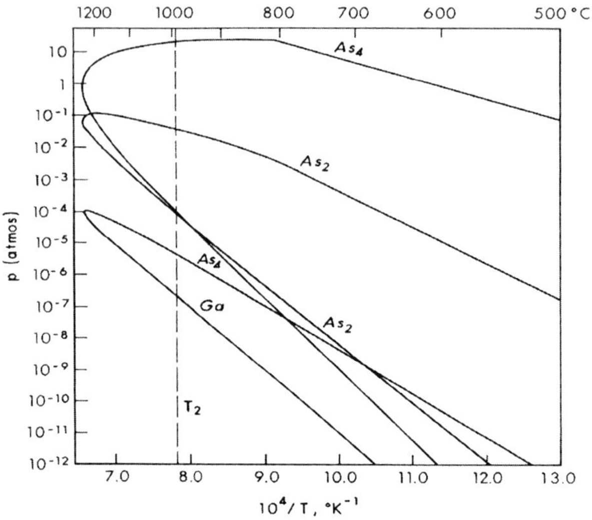

(b)

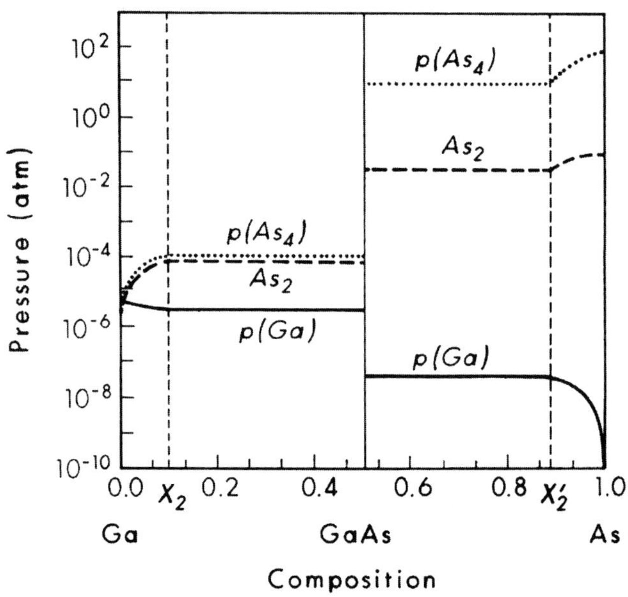

(c)

Figure 3. GaAs phase diagrams: (a) temperature-composition, (b) pressure-temperature, and (c) pressure-composition. The latter is drawn at a temperature T_2 of approximately 980 °C, where the Ga- and As-rich liquidus compositions are x_2 and x_2'. (after Stringfellow [7]).

element during VPE, the stoichiometry is typically controlled by the group V partial pressure input to the system. For the OMVPE growth of GaAs, for example, the As is depleted very little at the interface and the Ga is nearly all used up, as discussed above. In this case the Ga partial pressure at the interface, P^i_{Ga}, is determined by the mass action expression, Eqn (11).

A word of caution is in order here. The literature is filled with papers where the V/III ratio is used improperly to analyze the thermodynamic state of the system, due to a confusion between the input V/III ratio and that at the interface. For example, the former has no obvious connection to the stoichiometry of the solid. A common mistake is to change the "V/III ratio" during growth by changing the input group III flow rate. For conditions where thermodynamic equilibrium is approximately satisfied at the interface, changing the group III flow rate has no effect on the V/III ratio at the interface, the quantity determining the solid stoichiometry. Even more common is a series of experiments where the temperature is varied and the input V/III ratio is held constant. Of course, since the equilibrium constant in Eqn. (11) is strongly dependent on temperature, the V/III ratio at the interface will certainly not be constant.

A slightly different form of the phase diagram can be used to describe the OMVPE growth process, where the independent variables are typically the temperature and the input group III and group V partial pressures. For the growth of GaAs using trimethylgallium (TMGa) and AsH_3 at normal growth temperatures, the precursors can be assumed to be decomposed to elemental Ga and As_4, although incomplete decomposition and the production of As dimers can also be considered, where appropriate. Again, equilibrium is assumed to prevail at the interface, as described above. This results in the phase diagram shown in Fig. 4 [22]. It explains many of the most elementary constraints, first discovered empirically, on the growth of GaAs having a good surface morphology. For input V/III ratios of less than 1 (III/V >1 in Fig. 4) poor surface morphologies are invariably produced. A close examination of the surface for such conditions reveals shiny, hemispherical liquid metal droplets. These frequently reside at the tops of long whiskers protruding from the surface. This is because, as the phase diagram indicates, growth occurs in a two (condensed) phase region. The input vapor is highly supersaturated, i.e., the product of the input As and Ga partial pressures is much larger than a_{GaAs}/K_{GaAs} in Eqn (11). As and Ga are taken from the vapor in a 1:1 ratio to form GaAs. Thus, if the input V/III ratio is not unity, one of the elements is strongly depleted at the growing interface in order to establish equilibrium. When the V/III ratio is less than unity, the group V element is depleted. Because of the low volatility of the group III element, it accumulates on the surface as a second condensed phase. Much better results are obtained for V/III ratios of much greater than unity. It is impossible to form a condensed As phase except for input As partial pressures in excess of the vapor pressure of elemental arsenic, approximately 28 atmospheres at a temperature of 800°C. Thus, the rectangular region labeled GaAs(s) defines the combination of Ga and As input partial pressures yielding single phase GaAs at a particular temperature. Of course, the solid stoichiometry is completely controlled by the input As partial pressure, as described above. For temperatures more likely to be used for OMVPE growth the only change in this diagram would be a reduction in the vapor pressure of pure, elemental As.

As the pressure in the system is reduced into the regime of MBE growth, the free energy of the vapor phase is reduced, which results in major changes in the phase diagram, as discussed in detail by Tsao [12]. For example, the GaAs temperature-composition diagram at a total pressure of 10^{-5} torr is shown in Fig. 5. Under these conditions, the GaAs will sublime before it reaches its melting point. The decreased stability of the liquid phase results in a phase diagram with a fairly extensive region labeled c + v (shaded in Fig. 5) where GaAs is in equilibrium with the As-rich vapor phase. Note that if the stoichiometry of the system is slighty Ga-rich, a liquid phase, essentially pure Ga, is formed. Thus, similar to OMVPE, MBE growth of GaAs is carried out with an excess of As in the vapor phase. The stoichiometry of the solid is then controlled by the As partial pressure at the interface. The pressure-composition phase diagram, Fig. 6, gives additional information defining the range of total pressure leading to successful MBE growth of GaAs. The two phase diagrams define the MBE "window" [12], i.e., the combinations of temperature, total pressure, and V/III ratio leading to growth of a single condensed GaAs phase.

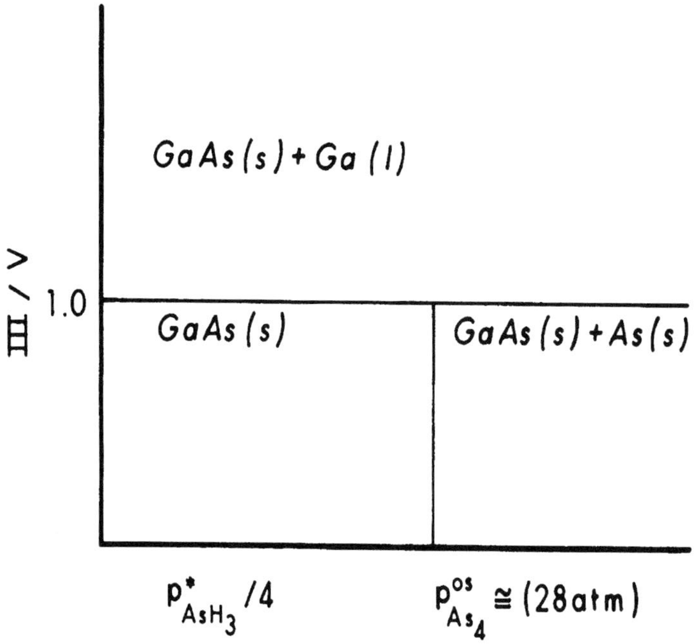

Figure 4. GaAs phase diagram appropriate for OMVPE growth. The regions containing a single solid GaAs phase and the two two-phase regions are shown. (after Stringfellow [22].)

4. SURFACE PHASE DIAGRAMS

The actual surface of a III/V semiconductor is not that expected from a simple termination of the crystalline lattice. This would leave a large concentration of high energy dangling bonds. Thermodynamic factors determine the equilibrium structure of the surface, as for the bulk solid.

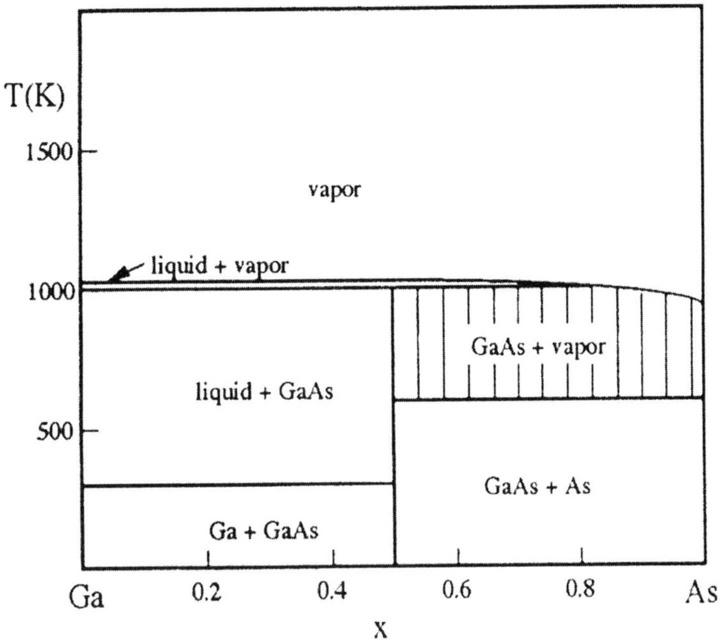

Figure 5. GaAs phase diagram at a total pressure of 10^{-5} torr, applicable to MBE growth. (after Tsao [12].)

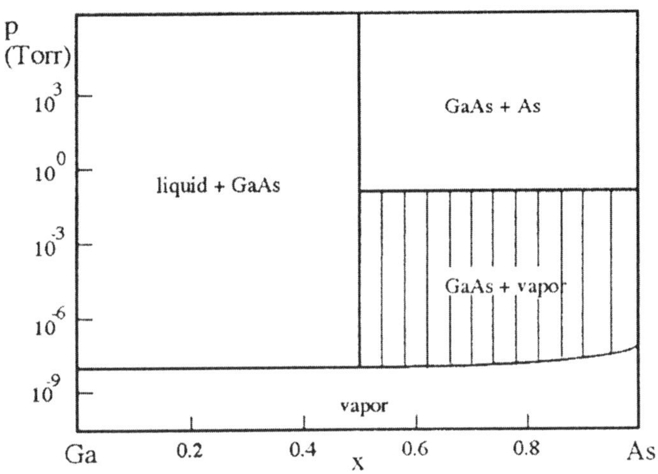

Figure 6. Pressure-composition phase diagram for GaAs under MBE growth conditions at 850 K (after Tsao [12].)

Considering specifically the (001) surface, that most commonly used for the epitaxial growth of III/V semiconductors, termination of the lattice would produce a configuration with two dangling bonds per surface atom, an extremely high energy configuration. Both theoretical calculations [23] and experimental results [24] clearly show that the equilibrium configuration for the As-rich (001) GaAs surface simply involves the joining of pairs of adjacent dangling bonds to form As dimer rows on the surface, similar to the Si dimer rows observed on the reconstructed (001) Si surface [25]. The (2x4) structure has every fourth As dimer missing. The surface atoms also buckle to produce a non-planar arrangement [23]. The reconstructed surface is stable, which gives rise to high surface mobilities for adsorbed atoms, with diffusion lengths reported to be as large as a micron [26]. This must be the key to obtaining the nearly atomically abrupt interfaces reported for OMVPE and MBE growth of quantum well structures widely reported in the literature. Adsorbed atoms that could make two bonds to surface atoms would obviously not be at all mobile. This would lead to statistically rough, three dimensional growth, precluding the possibility of producing quantum wells and other nano-structures.

The surface structures produced during MBE growth of (001) GaAs have been determined by both reflection high energy electron diffraction (RHEED) and scanning tunneling microscopy (STM) [24]. The predominant structure for As-rich conditions is the (2x4) structure, discussed above. For Ga-rich conditions, the (4x2) structure, consisting largely of Ga dimers, is formed. Several other structures are observed, depending primarily on the As/Ga ratio at the surface [27]. Thus, a surface phase diagram can be constructed, as shown in Fig. 7 [28]. Structures similar to those discussed above are also observed for other III/V semiconductors. The major feature appears to be the formation of either group III or group V dimers at the surface to reduce the density of dangling bonds.

The surface structures produced during OMVPE growth are revealed by x-ray scattering [29] and optical probes [28]. The tools most useful for typical laboratories are the optical techniques of reflection-difference spectroscopy (RDS) [28] and surface photo absorption(SPA) [30]. Kamiya et al. [28] used RDS to determine the surface phase diagram, i.e., the reconstruction versus substrate temperature and input AsH_3 flow rate for OMVPE growth. These results are added to the MBE results in Fig. 7. They found that for As-rich conditions, typically used for OMVPE growth, the GaAs surface consists of As dimer rows, producing either the (2x4) structure discussed above or the c(4x4) structure consisting of two layers of As at the surface. Apparently, at a given As flux and growth temperature, the surface structures obtained by MBE and OMVPE are similar, as expected for a thermodynamically controlled phenomenon.

5. SOLUTION THERMODYNAMICS

The relation between the composition of the vapor at the solid/vapor interface and the solid can be approximated very accurately by assuming a condition of thermodynamic equilibrium [31], as described above. On an even finer scale, the same logic can be applied to the surface processes. If the kinetic barrier at the step edge doesn't limit the growth rate, the process is limited by surface diffusion and thermodynamics can be used to describe the surface processes in OMVPE as well as MBE growth [31, 32, 33, 34].

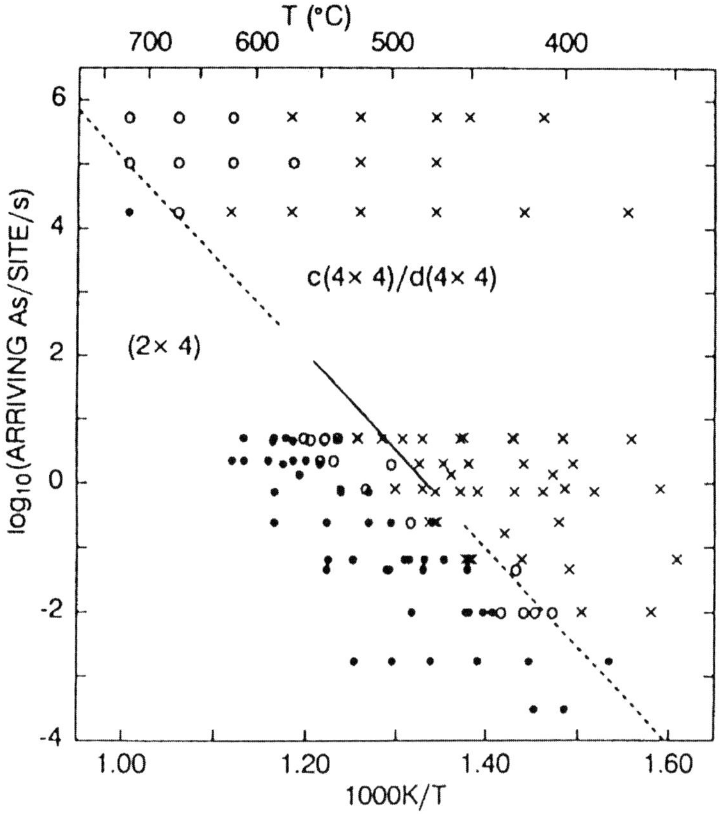

Figure 7. Phase diagram showing the surface reconstruction of (001) GaAs as a function of substrate temperature and the arriving flux of As atoms at the surface. Dots, crosses, and circles represent (2x4), c(4x4), and marginal structures, respectively. (after Kamiya et al [28].)

For example, considering the OMVPE growth of alloys, the equilibrium approximation allows the composition of the solid phase to be accurately calculated as a function of temperature and the composition of the vapor. An important consideration in such calculations is the deviation from ideality of the solid phase. For the III/V alloys, such as $Ga_xIn_{1-x}P$ [35], as well as for the II/VI and IV alloys, for example Ge_xSi_{1-x}, the semi-empirical delta-lattice parameter (DLP) model gives a good representation of the enthalpy of mixing [35]. The valence force field (VFF) model [36] as well as first principles calculation s [37], give accurate estimates of the enthalpy of mixing without adjustable parameters.

The microscopic strain energy associated with stretching and bending the bonds is the most important factor in the thermodynamics of mixing of semiconductor alloys. Thus, the enthalpy of mixing is always positive, and increases with the square of the difference in lattice parameters of the two constituent compounds (or elements for group IV alloys) [35]. This dictates that at equilibrium a random alloy in a certain composition range will decompose into a mixture of two phases when the difference in lattice constants of the component compounds is large, i.e., the phase diagram contains a miscibility gap.

In general, the simpler DLP model can accurately be used for alloy phase diagram calculations. The entropy of mixing for a pseudobinary solution of the type $A_{1-x}B_xC$ is typically simply taken to be the ideal configurational entropy of mixing,

$$\Delta S^M = -R[x \ln x + (1-x) \ln(1-x)] \tag{17}$$

Using these concepts, applied to the solid-vapor equilibria of concern for vapor phase growth, including OMVPE [38] and MBE [39], we can calculate the composition of a multicomponent solid alloy from the temperature and the concentrations of the various components in the vapor phase. Deviations from ideality for the vapor phase are commonly neglected.

The equilibrium conditions for the ternary(or pseudobinary) system may be obtained in exactly the same way as described above for binary systems (Eqn (6)), by equating the chemical potentials of the 2 components in the 2 phases:

$$\mu^v{}_A + \mu^v{}_C = \mu^s{}_{AC} \tag{21a}$$

$$\mu^v{}_B + \mu^v{}_C = \mu^s{}_{BC} \tag{21b}$$

This leads to two mass action expressions, similar to Eqq. (11). As discussed above, equilibrium is assumed to be established at the interface.

As an example of the use of such calculations to understand epitaxial processes, consider the OMVPE growth of $GaAs_{1-x}Sb_x$. The 2 mass action expressions, one for GaAs and one for GaSb, are solved simultaneously with 2 conservation equations, one for solid stoichiometry and one for solid composition [41]. Complete pyrolysis of the source molecules is assumed. The activity coefficients of GaAs and GaSb in the solid are calculated as described above using the DLP model.

The calculation can be performed with no adjustable parameters, yielding solid composition versus vapor composition and substrate temperature during growth. The calculated results are compared with experimental data in Fig. 8 [41]. Several important aspects of VPE are illustrated in this rather complex figure. First, consider the open data points, obtained for an input V/III ratio (the ratio of the input group V to group III molar flow rates) of 2.0. Notice that the calculated curve for V/III = 2.0 fits the data well. The Sb distribution coefficient, defined as kSb = xsSb/xvSb, where xvSb = p*TMSb/(p*TMSb+p*AsH3), is seen to be less than unity. GaAs is more stable than GaSb, thus As is more likely to bond to the Ga on the surface and be incorporated into the solid. The excess Sb evaporates from the surface. An additional important point is that the calculation for a V/III ratio of less than unity yields an antimony distribution coefficient of unity. As discussed above for the case of alloys with mixing on the group III sublattice, when V/III>1, essentially all of the group III elements reaching the interface are incorporated. The case of GaAsSb with mixing on the group V sublattice with V/III ≤ 1 is completely analogous. The establishment of equilibrium at the interface while the input vapor is highly supersaturated requires that the group V elements be virtually exhausted at the interface. A final point relative to Fig. 8 is the presence of a two solid phase region or miscibility gap. Because of the large difference in lattice constant between GaAs and GaSb a miscibility gap exists, as discussed above. However, when the V/III ratio is less than unity, the As and Sb atoms arriving in a random pattern at the surface do not have time to redistribute themselves into GaAs and GaSb rich areas before being covered over by the next layer. Thus, we are able to grow metastable $GaAs_{1-x}Sb_x$ alloys throughout the entire range of solid composition as shown by the solid data points in Fig. 8. Other, even less stable alloys, such as GaPSb and InPSb can also be grown in this way [42].

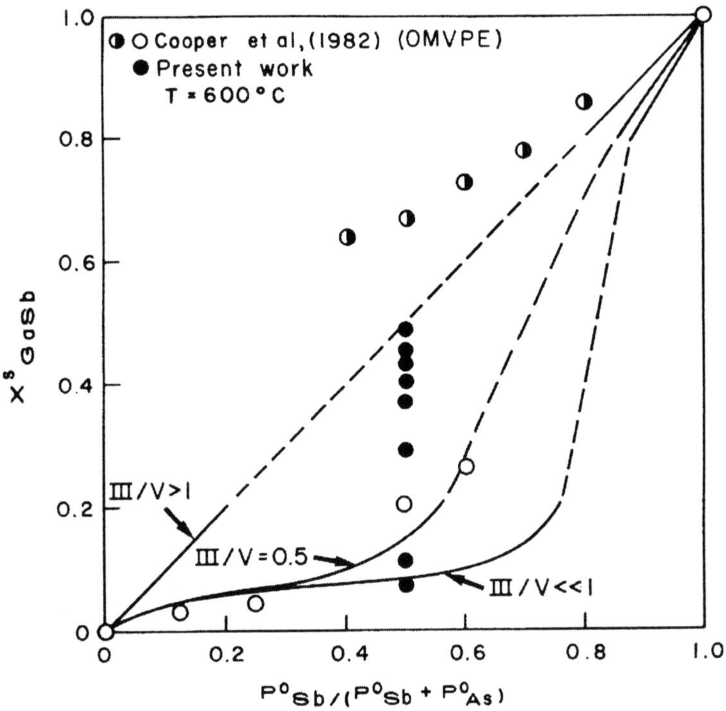

Figure 8. Solid versus vapor composition for the alloy GaAsSb. The data are from the work of Cooper et al [40] and Stringfellow and Cherng [41]. The curves were calculated for various V/III ratios. The broken sections of each curve represent the calculated regions of solid immiscibility. (after Stringfellow and Cherng [41]).

The magnitude of the enthalpy of mixing has significant consequences for OMVPE growth. As mentioned above, it has a direct impact on the composition of the solid alloy grown from a given vapor and, when large, gives rise to a miscibility gap. A dramatic example of this phenomenon involves alloys where N, an extremely small group V element, is used to replace a much larger element such as As or P in an effort to produce 1.3 micron alloys lattice matched to GaAs substrates for photonic devices [43]. For these interesting and increasingly important "dilute N" alloys, the amount of N that can be added, at equilibrium, is limited to values of much less than 1% [36]. Even the important alloy GaInN, used for short wavelength LEDs and lasers, is predicted to have a significant miscibility gap, although the solubility of In in GaN is predicted to be 6% at 850°C [36].

5.1. Surface Thermodynamics

The importance of bulk thermodynamics in determining the solid composition and stoichiometry for semiconductors grown epitaxially, as described above, has been known for decades. In recent years, it has become apparent that the thermodynamics of the surface often plays the dominant role in many aspects of the OMVPE growth process. After all, the growth occurs entirely at the surface. As discussed above, the vapor and surface are nearly in equilibrium for OMVPE and MBE growth of semiconductor materials under most conditions. If the entire system were completely at equilibrium, it would not be important to consider the surface, since the bulk solid would also be in equilibrium with the vapor at the interface. However, the diffusion coefficients in these highly stoichiometric solids, where the vacancy concentrations are extremely low, are so small that the bulk solid is often not able to attain equilibrium with the surface. In this case, the atomic structure and stoichiometry of the solid may be determined by the thermodynamics of the surface phase.

For growth on the common (100) orientation two of the four sp^3 bonds in the bulk are unsatisfied for surface atoms. This would be expected to give very low adatom mobilities, resulting in rough surfaces, as discussed above. This appears to be contrary to experimental observations of large diffusion lengths [44] and the ability to grow quantum well structures with abrupt and smooth interfaces. These are the first indications that each surface atom does not have two dangling bonds. As discussed above, elementary thermodynamic considerations suggest that the surface atoms will rearrange themselves to reduce the surface energy by reducing the number of dangling bonds as well as changing the bonding geometries. First principles calculations have given detailed information about the way in which this might occur [23]. RHEED experiments on semiconductor surfaces have given corresponding experimental information about the surface configuration versus temperature and the group V flux arriving at the surface [45]. The results are found to correspond closely to those obtained using RDS [46]. The surface reconstructions seen for GaAs under conditions used for growth by atmospheric pressure OMVPE are surprisingly similar to those for growth by MBE, as discussed above and seen in Fig. 7.

Typical OMVPE growth conditions, using V/III ratios well in excess of unity, will yield a group V rich surface, producing a (2x4) or (4x4) reconstruction for (001) surfaces, as seen in Fig. 7. Very low group V partial pressures result in the group III rich (4x2) surface. This condition is not normally observed during OMVPE growth where V/III ratios at the interface are essentially always >>1.

A very clear indication of the importance of surface thermodynamics for III/V alloys(as well as Si-Ge and II/VI alloys) is the effect of the surface on the microscopic arrangement of the atoms in the solid, e.g., the production of ordered structures spontaneously during OMVPE growth. Thermodynamic calculations indicate that the (2x4) reconstruction will favor formation of the CuPt structure, with ordering on {111} planes, which is *not* stable in the bulk. Empirically, nearly all III/V alloys are observed to form the CuPt structure during OMPVE growth using typical growth conditions [47]. SPA studies of the effects of temperature, group V flow rate, and substrate misorientation give a clear correlation between the density of [1, $\bar{1}$,0,] P dimers and formation of the CuPt ordered structure during OMVPE growth of GaInP[48].

The thermodynamics at the surface are also expected to have a direct influence on other materials properties, such as dopant incorporation and the concentrations of native defects. The surface reconstruction will also almost certainly affect the surface mobilities of adatoms and the density and nature of steps and kinks on the surface, all of which will affect surface morphologies and defect densities. However, understanding of these phenomena is limited at this time.

5.2. Effect of Surface on Growth Processes

The physical nature of the surface, as described above, has significant consequences for other epitaxial growth phenomena. For example, the surface structure affects adsorption/desorption phenomena. Naturally, the surface reconstruction affects the binding of adatoms at the surface and, hence, the adsorption energy. Thus, it will affect both adsorption/desorption rates as well as heterogeneous reaction rates.

A major consequence of reconstuction on the growth process is the flat interfaces achieved for growth on (001) surfaces. As mentioned above, the adatom mobilities would be extremely small for growth on an unreconstructed (001) surface. Reconstruction dramatically lowers the energy of the surface and thus weakens the bonding to adatoms. This accounts for one extremely significant feature of the OMVPE (and MBE and CBE) growth process, namely that extremely abrupt interfaces can be formed separating layers of different compositions and/or doping levels. Abrupt interfaces are possible only for low energy surfaces, where the adatoms are not strongly bound. The resulting layer-by-layer and step-flow growth processes facilitate the growth of the low dimensionality structures, quantum wells, superlattices, quantum wires, and quantum dots, that form the basis of many of the modern high performance photonic and electronic devices.

Perhaps the most well-defined effect of steps on the growth process relates to ordering, i.e., the spontaneous formation of monolayer superlattice structures during OMVPE growth. This effect has been extensively studied for GaInP alloys grown lattice matched to GaAs substrates. The CuPt structure is formed for most growth conditions. This ordered structure is also formed for most other III/V alloys grown by OMVPE. The driving force for formation of the CuPt structure is related to the thermodynamics of the surface phase. It is driven by the [110] rows of P-P dimers on the (2x4) reconstructed surface. The degree of order is directly related to the density of the [1, $\bar{1}$,0,] P dimers on the surface.

First principles calculations have been used to estimate the relative thermodynamic stabilities of the various ordered structures, including the CuPt structure, in III/V alloys [49]. For the bulk alloys, i.e., ignoring surface effects, the CuPt structure is not found to be stable relative to the disordered alloy. This was initially surprising, since CuPt is nearly the only ordered structure observed experimentally. However, this dilemma is resolved by considering the stabilities of the various ordered structures at the reconstructed surface [50].

For the most commonly observed (2xn), typically (2x4), reconstruction on group V-terminated (001) surfaces, VFF calculations [50] indicate that the B variants of the CuPt structure with ordering on (1, $\bar{1}$,1,) and (1,1, $\bar{1}$,) planes) are the most stable in the layers just beneath the (001) surface. The [110] rows of [1,¯1,0,] oriented group V dimers lead to alternating [110] rows of compressive and tensile strain in the 3rd buried layer. For alloys with mixing on the group III sublattice, such as GaInP, this produces the [110] rows of alternating large and small atoms that comprise the CuPt-B variants. These calculations also predict that the surface structure of alloys with mixing on the group V sublattice, such as GaAsP, will also produce the CuPt-B variants, in agreement with experimental observations [51].

Clearly, the ordering phenomenon is very sensitive to the surface structure. For this reason, ordering is a perfect vehicle for the study of surface processes during epitaxial growth, in general. The occurrence and mechanism of ordering are fascinating materials science problems that reveal much about the important general features of the surface processes occurring during vapor phase epitaxial growth. This paper will concentrate on a review of recent advances in our understanding of the effects of surface processes on ordering.

The flow rate of the group V precursor and growth temperature were found to have a significant effect on the ordering process. The concentration of [-110] P dimers decreases both as the group V partial pressure decreases and as the growth temperature increases. The degree of

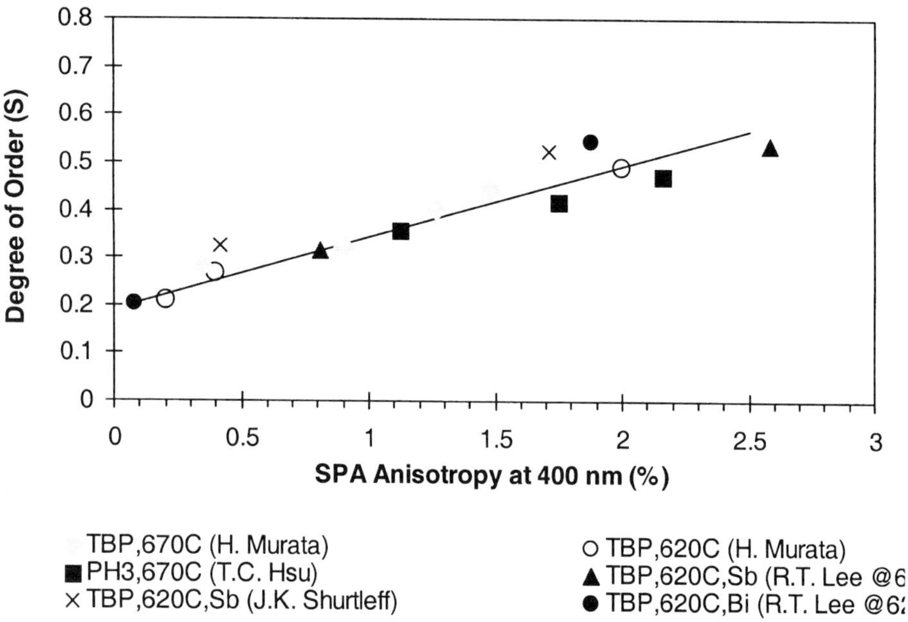

Figure 9. Degree of CuPt order versus SPA difference signal at 400 nm due to P dimers. All the data is for GaInP grown by OMVPE on singular (001) GaAs substrates. (after Lee et a l[53].)

order has been closely correlated with the surface reconstruction [52]. Clearly, the loss of CuPt ordering at low V/III ratios correlates closely with the loss of the (2x4)-like reconstruction. Data from the studies of changing temperature and TBP partial pressure are combined with data for surfactant effects, to be described below, for the plot of the degree of order versus the SPA signal in Fig. 9 [53]. A one-to-one relationship between the degree of order and the concentration of [1,$\bar{1}$,0,] P dimers on the surface is observed for changes in both temperature (620-720°C) and TBP partial pressure (<200 Pa).

6. EFFECTS OF SURFACTANTS

The discussion in the last section indicates that the surface structure can be controlled during growth by the use of temperature and the partial pressure of the P precursor. More recent experiments indicate that this can also be accomplished via the use of surfactants during the epitaxial growth process.

The original use of surfactants during the epitaxial growth (mainly MBE) of semiconductors was for control of the structure and morphology of highly strained layers in both elemental [54] and III/V semiconductors [55]. The addition of dopants during MBE growth was shown to affect surface reconstruction[56]. It also appears to affect the the diffusion of adatoms on the surface[57].The use of isoelectronic surfactants was practically unknown until the recent work described below. This also represented early work on the use of surfactants during OMVPE growth.

7. ANTIMONY

Perhaps the most interesting surfactant, shown to strongly affect the ordering in GaInP, is the isoelectronic dopant Sb. Since it has the same number of valence electrons as P, no significant change in the Fermi level position at the growth temperature is expected. Because it is much larger than P, the solubility is small [58]. Thus, it is likely to accumulate at the surface during growth. The amount of TESb added to the system was sufficient to produce only 10^{18} cm^{-3} ($x = 10^{-4}$) of Sb in the solid [59]. However, the addition of this tiny amount of Sb produces a marked increase in the low temperature PL peak energy [53]. The results indicate that the layers are highly ordered without Sb and essentially disordered when grown with a small amount of Sb present, as indicated in Fig. 10.

The surface photo absorption (SPA) anisotropy intensity at 400 nm decreases markedly versus TESb flow rate during growth. The results indicate a dramatic change in the surface reconstruction induced by the addition of Sb to the system. As discussed above, the intensity of the peak near 400 nm in the SPA anisotropy spectra directly correlates to the concentration of [1,$\bar{1}$,0,] P dimers on the surface.

Comparison of the spectra for the undoped $Ga_{0.52}In_{0.48}P$ layers and layers with Sb clearly indicates that the P dimer concentration on the surface of the samples grown with Sb present has been significantly reduced [53]. This was the first experimental evidence that an isoelectronic dopant, such as Sb, can act as a surfactant to change the surface reconstruction of a III/V semiconductor layer during OMVPE growth.

Since the [1,$\bar{1}$,0,] P dimers produce the surface thermodynamic driving force for formation of the CuPt structure during growth, it is not surprising that the replacement of [1,$\bar{1}$,0,] P dimers by Sb dimers eliminates ordering. The Sb dimer spacing is larger than for P dimers, thus the driving force for ordering is expected to be dramatically reduced. These results confirm the

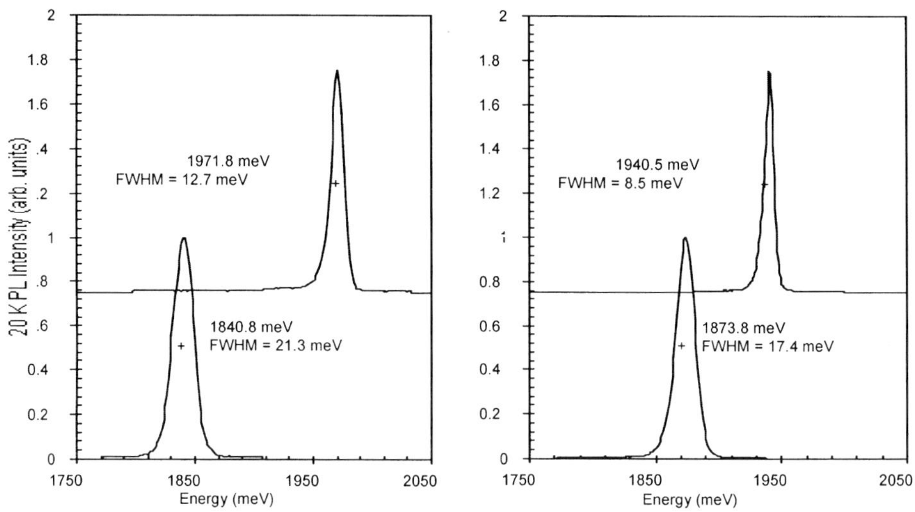

Figure 10. The effect of TESb flow rate on the degree of order and SPA difference signal at 400 nm. (after Lee et al [53].)

observation that the degree of order in $Ga_{0.52}In_{0.48}P$ layers is directly related to the P dimer concentration. In fact, the degree of order versus 400 nm SPA anisotropy signal for the layers grown using Sb fall very near the data discussed above for the effects of temperature and flow rate of the P precursor, as indicated in Figure 9.

The ability to control the ordering and, thus, the bandgap energy of GaInP layers by the addition of Sb during growth suggests the possibility of producing heterostructures and QWs. To grow a D/O heterostructure, the surface of the undoped, ordered $Ga_{0.52}In_{0.48}P$ layer grown first was exposed to tertiary butyl phosphine (TBP) and TESb for five minutes, without growth, to allow Sb to accumulate. The disordered layer, grown with Sb present, followed. Fig. 11 shows the PL spectrum for an order/disorder heterostructures grown by this procedure. A remarkable difference in the bandgap energybetween the two layers of 135 meV was observed. This is the first demonstration of a useful method for the production of atomically engineered structures for advanced electronic and photonic devices. The ability to independently modulate the bandgap and the Fermi level position is the key to the potential usefulness of this technique. The 135 meV bandgap discontinuity is more than 5 kT at room temperature, which should be sufficient for many devices.

These results suggest a new and powerful concept: determination of the surface reconstruction during growth, using a surfactant, as a method of controlling the properties of the resultant epitaxial layer. This has been clearly demonstrated for the control of the bandgap energy in GaInP. However, the implications may be much broader. Surface reconstruction will certainly affect morphology, and hence the ability to grow low dimensional structures. In fact, recent studies have shown that both surface diffusion and attachment kinetics at steps are significantly

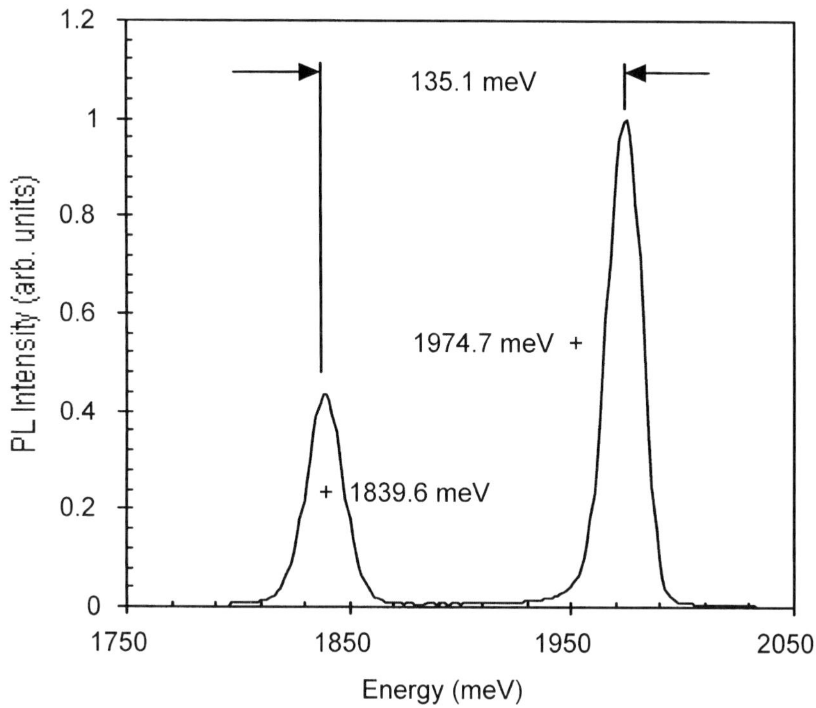

Figure 11. 20-K PL spectrum for a GaInP order/disorder heterostructure grown by OMVPE, with TESb present during growth of the top layer. (after Shurtleff et al [60].)

changed by surfactants such as Sb, Bi, and Te [61]. It would not be surprising to find that the surface reconstruction during growth also controls other important processes such as spinodal decomposition, a process known to occur at the surface during growth. It has also been demonstrated that the surface reconstruction during growth affects the dopant incorporation in GaAs [59]. Thus, surfactants, such as Sb, can be used to control, not only the bandgap energy in alloys, but also the conductivities, mobilities, and minority carrier lifetimes of layers grown by vapor phase processes.

SUMMARY

Thermodynamics plays a central role in determining many aspects of the growth process for many epitaxial techniques such as growth rate and solid composition. It also is the determining factor in many aspects of the extrinsic materials properties associated with the incorporation and distribution of the constituent and dopant elements in the resultant solid epitaxial layer. Only recently have the thermodynamic effects of the surface come to be realized. These factors have been shown to control dopant incorporation and ordering. The use of surfactants to control the structure of the surface during growth is only beginning to be exploited. However, this technique has already proven useful in the control of ordering and dopant incorporation in GaInP grown by OMVPE.

ACKNOWLEDGEMENTS

It is a pleasure to acknowledge the long-term support of the Department of Energy.

REFERENCES

1. C.H. Chen, S.A. Stockman, M.J. Peanasky, and C.P. Kuo, In High Brightness LEDs, Vol. 48 (eds, G. B. Stringfellow and M. G. Craford) Academic Press, Boston,1997.
2. S. Nakamura and G. Fasol. The Blue Laser Diode, Springer-Verlag, Berlin,1997.
3. T. Takamota, E. Ikeda, H. Kurita and M. Ohmori, Appl. Phys. Lett., 70 (1997) 381.
4. K.A. Bertness, S.R. Kurtz, D.J. Friedman, A.E. Kibbler, C. Kramer and J.M. Olson, Appl. Phys. Lett., 65 (1994) 989.
5. N. Grote, In Materials for Optoelectronics(ed, Quillec, M.) Kluwer Academic Publishers, Boston, 1996.
6. M. Razeghi, The MOCVD Challenge, Institute of Physics, Bristol, 1995.
7. G.B. Stringfellow, Organometallic Vapor Phase Epitaxy: Theory and Practice, 2nd Edition, Academic Press, Boston, 1999.
8. D.W. Kisker and T.F. Kuech, Handbook of Crystal Growth, Vol. 3, ed, D.T.J. Hurle, Elsevier, Amsterdam, 1994, pp. 93.
9. G.B. Stringfellow, Materials Science and Engineering (B) 87 (2001) 97.
10. M.A. Herman and H. Sitter, Molecular Beam Epitaxy: Fundamentals and Current Status, Springer-Verlag, Berlin, 1989.
11. C.T. Foxon, Handbook of Crystal Growth, Vol. 3 (ed, D.T.J. Hurle) Elsevier, Amsterdam, 1994, pp. 155.
12. J.Y. Tsao, Materials Fundamentals of Molecular Beam Epitaxy Academic Press, Boston, 1993.
13. H. Hirayama and H. Asahi, Handbook of Crystal Growth, Vol. 3 (ed, D T.J. Hurle) Elsevier, Amsterdam 1994, pp. 185; W.T. Tsang, J. Crystal Growth 120 (1992) 1.
14. G.B. Stringfellow, J. Crystal Growth 115 (1991) 1.
15. G.B. Stringfellow, J. Crystal Growth 68 (1984) 111.
16. GB. Stringfellow, (1999) Organometallic Vapor Phase Epitaxy: Theory and Practice, 2nd Edition, Academic Press, Boston, Section 7.3.

17. D.W. Shaw, Treatise on Solid State Chemistry, Vol 5, ed. N.B. Hannay, Plenum, New York, 1975, p. 283.
18. G.B. Stringfellow, Organometallic Vapor Phase Epitaxy: Theory and Practice, 2nd Edition, Academic Press, Boston, 1999, Chapter 1.
19. K.F. Jensen, Handbook of Crystal Growth, Vol. 3 (ed, D.T.J. Hurle) Elsevier, Amsterdam, 1994, pp. 541.
20. G.B. Stringfellow, (1999) Organometallic Vapor Phase Epitaxy: Theory and Practice, 2nd Edition, Academic Press, Boston, Section 2.4.
21. ibid, Section 2.2.
22. G.B. Stringfellow, in Advanced Crystal Growth, ed. P.M. Dryburgh, B. Cockayne, and K.G. Barraclough, Cambridge: Prentice Hall, 1987, p. 289.
23. D.J. Chadi, J. Vac. Sci. Technol. A, 5 (1987) 834.
24. B.A. Joyce, Advanced Crystal Growth, ed. P.M. Dryburgh, B. Cockayne, and K.G. Barraclough, Cambridge: Prentice Hall, 1987, p. 337-385; M.D. Pashley, J.W. Haberern, W. Friday, J.M. Woodall, and P.D. Kirchner, Phys. Rev. Lett. 60 (1988) 2176; D.K. Biegelsen, R.D. Brignans, J.E. Northrup, and L.E. Swartz, Phys. Rev. B 41 (1990) 5701.
25. M.G. Legally, Y.W. Mo, R. Kariotis, B.S. Swartzentruber, and M.B. Webb, Kinetics of Ordering and Growth at Surfaces, ed. M.G. Lagally, Plenum Press, New York, 1990, p. 145.
26. T. Isu, M. Hata, Y. Morishita, Y. Nomura and Y. Katayama, J. Crystal Growth 115 (1991) 423.
27. C. Deparis and J. Massies, J. Crystal Growth 108 (1991) 157.
28 I. Kamiya, H. Tanaka, D.E. Aspnes, L.T. Florez, L.T.E. Colar, J.P. Harbison and R. Bhat, Appl. Phys. Lett. 60 (1992) 1238.
29. D.W. Kisker, G.B. Stephenson, P.H. Fuoss, F.J. Lamelas, S. Brennan and P. Imperatori, J. Crystal Growth 124 (1992) 1.
30. N. Kobayashi, Y. Kobayashi and K. Uwai, J. Crystal Growth 174 (1997) 544.
31. G.B. Stringfellow, J. Crystal Growth, 68 (1984) 111.
32. T. Nishinaga, T. and T. Suzuki, J. Crystal Growth, 115 (1991) 398.
33. J. Tersoff, M.D. Johnson and B.G. Orr, Phys. Rev. Lett., 78 (1997) 282.
34. J.Y. Tsao, Materials Fundamentals of Molecular Beam Epitaxy, Academic Press, Boston, 1993, Chapter 7.
35. G.B. Stringfellow, J. Crystal Growth, 27 (1974) 21.
36. I.H. Ho and G.B. Stringfellow, Appl. Phys. Lett., 69 (1996) 2701; I.H. Ho and G.B. Stringfellow, Proceedings of MRS, 449 (1997) 871; I.H. Ho and G.B. Stringfellow, J.Crystal Growth, 178 (1997) 1.
37. S.H. Wei and A. Zunger, Phys. Rev. B, 39 (1989) 3279.
38. G.B. Stringfellow, J. Crystal Growth, 62 (1983) 225.
39. A. Koukitu, H. Nakai, T. Suzuki and H. Seki, J. Crystal Growth 84 (1987) 425.
40. C.B. Cooper, R.R. Saxena and M.J. Ludowise, J. Electron. Mater. 11 (1982) 1001.
41. G.B. Stringfellow and M.J. Cherng, J. Crystal Growth 64 (1983) 413.
42. M.J. Jou, and G.B. Stringfellow, J. Crystal Growth 98 (1989) 679.
43. D.J. Friedman, J.F. Geisz, S.R. Kurtz and J.M. Olson, J. Crystal Growth, 195 (1998) 409; S.R. Kurtz, A.A. Allerman, E.D. Jones, J.M. Gee, J.J. Banas and B.E. Hammons, Appl. Phys. Lett., 74 (1999) 729; X. Yang, J.B. Heroux, M.J. Jurkovic and W.I. Wang, Appl. Phys. Lett., 76 (2000) 795; S. Sato and S. Satoh, J. Crystal Growth, 192 (1998) 381.
44. M. Hata, T. Isu, A. Watanabe, Y. Kajikawa and Y. Katayama, J. Crystal Growth, 114 (1991) 203.

45. J.M.V. Hove, P.I. Cohen and C.S. Lent, J. Vacuum Sci. and Technol. A, **1** (1983) 546.
46. D.E. Aspnes, R. Bhat, C. Caneau, E. Colas, L.T. Florez, S. Gregory, J.P. Harbison, I. Kamiya, V.G. Keramidas, M.A. Koza, M.A.A. Pudensi, W.E. Quinn, S.A. Schwarz, M.C. Tamargo and H.J. Tanaka, J. Crystal Growth, 120 (1992) 71.
47. G.B. Stringfellow, Mat. Res. Soc. Symp. Proc., 312 (1993) 35.
48. H. Murata, I. H. Ho, L.C. Su, Y. Hosokawa and G.B. Stringfellow, J. Appl. Phys., 79 (1996) 6895.
49. A. Zunger and S. Mahajan, Handbook on Semiconductors(Ed, T.S. Moss.) Elsevier Science B.V., Amsterdam, 1994, pp. 1399.
50. S.B. Zhang, S. Froyen and A. Zunger, Appl. Phys. Lett. 67 (1995) 3141.
51. G.S. Chen, D.H. Jaw and G.B. Stringfellow, J. Appl. Phys. 69, 4263 (1991).
52. H. Murata, S.H. Lee, I.H. Ho and G.B. Stringfellow, J. Vac. Sci. Technol. B 14 (1996) 3013.
53. R.T. Lee, C.M. Fetzer, J.K. Shurtleff and G.B. Stringfellow, J. Appl. Phys. 87 (2000) 3730.
54. E. Tournie and K.H. Ploog, Thin Solid Films 231 (1993) 43.
55. B.R.A. Neves, M.S. Andrade, W.N. Rodrigues, G.A.M. Safar, M.V.B. Moreira and A.G. de Oliveira, Appl. Phys. Lett. 72 (1998) 1712.
56. H. Oigawa, M. Wassermeier, J. Behrend, L. Daweritz and K.H. Ploog, Surf. Sci. 376 (1997) 185.
57. B. Voightlander, A. Zinner, T. Weber and H.P. Bonzel, Phys. Rev. B 51 (1995) 7583.
58. G.B. Stringfellow, J. Crystal Growth 27 (1974) 21; M.J. Jou, D.H. Jaw, Z.M. Fang, and G. B. Stringfellow, J. Crystal Growth 190 (1990) 208.
59. J.K. Shurtleff, S.W. Jun, and G.B. Stringfellow, Appl. Phys. Lett. 78 (2001) 3038.
60. J.K. Shurtleff, R.T. Lee, C.M. Fetzer and G.B. Stringfellow, Appl. Phys. Lett. 75 (1999) 1914.
61. G.B. Stringfellow, J.K. Shurtleff, R.T. Lee, C.M. Fetzer and S.W. Jun, J. Crystal Growth 221 (2000) 1.

Crystal Growth – from Fundamentals to Technology
G. Müller, J.-J. Métois and P. Rudolph (Editors)

Actual Concepts of Interface Kinetics

Kenneth A. Jackson

Department of Materials Science and Engineering
University of Arizona, 4715 East Fort Lowell Rd., Tucson, AZ 85712, USA

1. GENERAL CONSIDERATIONS

The crystal growth process is usually reversible: the net growth rate is the difference between the arrival rate and departure rate of atoms at the crystal surface At equilibrium these two rates are equal. When both phases are present, the crystal will grow when the interface is below the equilibrium temperature, and it will melt or dissolve when then temperature is above the equilibrium temperature.

1.1. Atoms at an Interface

In general, crystals grow by the addition of atoms or molecules from the mother phase, one at a time. The path of the atoms or molecules between the two phases is reversible, which is known as "microscopic reversibility". The rate at which an atom or molecule joins or leaves the crystal depends on the local environment of the atom or molecule in the growth phase as well as on how many of the nearest neighbor lattice sites are occupied with atoms of the crystal. This determines how tightly the solid atom or molecule is bound to the crystal, and how readily an atom the growth phase can join or leave the crystal.

In growth from the vapor phase or from a solution, it is fairly obvious whether an atom or molecule is part of the crystal or not. Surface atoms or molecules can arrive from the vapor phase or from the solution, move around on the surface by diffusion, and then perhaps leave the surface again. Indeed, if it were not possible to distinguish the atoms of the crystal from those of the neighboring phase, it is difficult to imagine how there could be a nucleation barrier to the formation of new layers, as there usually is in vapor phase or solution growth.

In melt growth the situation is not so obvious. For metals, there is only a few percent difference in density between the melt and the crystal, so it is possible that there could be a gradual transition from the crystal to the melt. But molecular dynamics simulations indicate that atoms which belong to the crystal can be distinguished from those which are part of the melt. The distinction can be made in two ways.

One is using the time trajectory of the atoms. Atoms which are part of the crystal vibrate about a lattice site. Atoms which are part of the liquid migrate randomly. An individual atom at the interface will wander randomly for a while, then perhaps join the crystal where it vibrates about a fixed position, and then perhaps leave the crystal again to wander randomly. There is a clear distinction between these two modes. When an atom joins the crystal, it typically stays there for many vibrational periods, and when it is in the liquid, it typically wanders for many vibrational periods. The vibrational mode can be clearly distinguished from the wandering mode.

The other method of distinguishing between solid and liquid atoms is using the radial distribution function of the atom, which describes the locations of neighboring atoms. The radial distribution function of solid atoms has a distinct minimum which is not present for liquid atoms. That the state of individual atoms at a crystal/melt interface can be distinguished is perhaps a surprising result, since a metal crystal and its melt have such similar properties. It appears that the atoms of the crystal at the interface create a potential well because of their regular positions. Atoms in the liquid can fall into this well, stay there for a while and then hop out again when they have enough energy. The latent heat of the transformation is associated with the potential energy of the atom falling into or hopping out of this energy well.

There has been some discussion in the literature about the importance of the observation in molecular dynamics simulations that there is density variation in a liquid next to a plane wall. The period of the density variations is the atomic diameter, and they extend several atomic diameters into the liquid away from the wall. However, there is no evident dependence of the radial distribution function of the atoms on their distance from the wall. These density variations are apparently due to minor variations in the atom positions near the flat wall. Similar density variations are found in the liquid next to crystal surfaces, and it has been suggested that these density waves influence the crystallization of the liquid atoms. But it is difficult to understand how these minor variations in position of the atoms could influence the energy or momentum distributions of the atoms in such a way as influence the rate at which the atoms join the crystal.

Although atoms are either solid or liquid, there can be a gradual transition from the liquid to the solid in one sense. For the simple metals and some other materials, the interface is rough at the atomic level, as discussed below. There are a few atomic layers at the interface in which there are both solid and liquid atoms. The atoms are either solid or liquid, and these layers contain both.

So in general, we can assume that there are atoms which belong to each phase, and that the transition between the two phases occurs atom by atom or molecule by molecule. This statement is not meant to imply that there is no interaction between adjacent atoms during the crystallization process.

Some crystal interfaces are smooth at the atomic level, and some are rough at the atomic level. The theory of the surface roughening transition is based on the assumption that the atoms of the crystal can be distinguished from the atoms in the phase it is in contact with. It concludes that some interfaces are smooth at the atomic level, and that some are rough at the atomic level, depending in large measure on the difference between the properties of the two phases. If the density of the atoms or molecules is very different in the two phases, as it is for vapor phase growth or solution growth, then the interface is likely to be smooth. If the difference is small, as it is for a metal in contact with its melt, then the interface is likely to be rough.

1.2. General Equation for the Growth Rate of Crystals

The rate of crystallization can be expressed as the product of four terms: a length, a frequency, a term which depends on the structure of the interface, and the free energy difference between the two phases.

$$v = \boldsymbol{a}\, v^{+} \boldsymbol{f}\, u_k \qquad (1)$$

The first term, $\boldsymbol{a}$, is a distance which is related to the atomic or molecular diameter of the growth unit. The second term, v^{+}, is the rate at which atoms join the crystal at active growth sites on the surface. The third term, f, depends on the roughness of the interface, and is the fraction of interface sites which are active growth sites. The difference between the rate at which atoms or molecules join and leave the crystal depends on the local free energy or chemical potential difference between the two phases, which is the driving force for the transformation. The normalized difference between the rate at which atoms or molecules join and leave the crystal is contained in the fourth term u_k:

$$u_k = 1 - \exp\left(\frac{\Delta G}{kT}\right) \tag{2}$$

Here ΔG is the free energy difference between the two phases, k is Boltzmann's constant, and T is the temperature. ΔG can be expressed as an undercooling, as a vapor pressure difference, or as a supersaturation.

The atoms or molecules at the interface join and leave the crystal at rates which depend only on their local environment and on the local departure from equilibrium. Their motion depends on their individual kinetic energies and the local potentials to which they are subject. There is no action at a distance.

There is an ambiguity in defining the area occupied by a growth site on the interface for different orientations of the crystal. We can get around this ambiguity by defining the distance $\boldsymbol{a}$ to be the cube root of the atomic volume, $\boldsymbol{a}^3 = \Omega$, and the area of a growth site on the interface to be $\boldsymbol{a}^2$. Then if an atom is added to each growth site, the interface will advance a distance $\boldsymbol{a}$. This is numerically equivalent to adding a volume Ω to the crystal with each atom, and then defining f to be the density of growth sites on the interface.

Not all of the surface sites on a crystal are active growth sites. The factor f is the probability of finding an active growth site in an area $\boldsymbol{a}^2$ of the interface. Atoms leave and join the crystal surface at many sites, but net growth occurs only at active growth sites. The density of active growth sites can vary widely, and is strongly temperature dependent if the growth depends on the nucleation of new layers. This will be discussed in more detail below. f is typically about ¼ for a rough surface, but can be very small on a smooth surface

The rate at which atoms or molecules join the crystal at active growth sites, ν^+, depends on the phase into which the crystal is growing, and on the mobility of growth units in that phase.

The classical theory for vapor growth was developed by Knudsen [1] and for melt growth by Wilson [2] and Frenkel [3]. In these models, the density of growth sites, f, was not discussed. It was assumed implicitly that atoms or molecules could join a crystal at any surface site. But these models contain the essence of the physical description of the rate at which atoms or molecules can join the crystal at active growth sites, ν^+.

Below is a brief review of two models for growth on smooth surfaces: surface nucleation, and growth aided by screw dislocations. These are the two classical models for f, the fraction of interface sites which are active growth sites. These models apply only to growth on smooth surfaces, and their applicability depends on how smooth the surface is.

Reliable information about the distribution of active growth sites on a surface can be obtained from Monte Carlo computer simulations, and detailed information about growth rates for specific materials and orientations requires molecular dynamics modeling.

1.3. Entropy Change on Crystallization

A crystal invariably has a lower enthalpy than its melt. The difference is latent heat. The atoms of a crystal must have sufficient energy to leave the crystal and join the higher enthalpy growth phase.

The rate at which atoms leave the crystal to return to the growth phase at active growth sites must contain this energy difference as a Boltzmann factor:

$$\nu^- = \nu_0^- \exp\left(-\frac{L}{kT}\right) \quad (3)$$

At the melting point, T_M, the rate at which atoms join the crystal must contain a similar factor:

$$\nu^+ = \nu_0^- \exp\left(-\frac{L}{kT_M}\right) \quad (4)$$

This exponential factor is implicit in ν^+ in Eqn. (1), so that the difference between ν^+ and ν^- is normalized in u_K. $L/T_M = \Delta S$ is the entropy difference between the two phases. This entropy term compensates for the increase in energy when atoms leave the crystal. Its magnitude depends on the difference in order between the crystal and the growth phase.

An atom or molecule in the crystal must increase its energy to leave the crystal, but it is relatively easy for it to find a place where it can join the growth phase. An atom or molecule in the growth phase decreases its energy on joining the crystal, but it is geometrically more difficult for it to join the crystal, because it can only do so at a lattice site. This represents and increase in its state of order. These two effects, one energetic, and the other geometrical, compensate to make the net rate of transition equal to zero at equilibrium.

For melt growth of metals, $\Delta S/k$ is typically about one, so this factor is about 1/3. For small organic molecules, $\Delta S/k$ is typically about 6. The difference in order between the crystal and the melt should account for one unit of this, as for a metal. The other five units come from the rotational disorder of the melt. The molecule must have the correct orientation in order to join the crystal. This suggests that less than 1% of the molecules are in the right orientation to join the crystal.

1.4. Early Models for Melt Growth

The first analysis for the growth rate of a crystal from a liquid was published by Wilson [2] in 1900. He related the rate at which atoms join the crystal using the diffusion coefficient of the liquid. About 30 years later, Frenkel [3] expressed the growth rate in terms of the viscosity of the liquid. The two treatments are equivalent since the diffusion coefficient of a liquid, D, is related to its viscosity, η, by the Stokes-Einstein relationship:

$$D = \frac{kT}{3\pi\eta a} \quad (5)$$

In this equation, a is the diameter of the diffusing atom or molecule.

It is found experimentally that the diffusion coefficient and the reciprocal of the viscosity have the same temperature dependence. Experiments on glass-forming systems, where it is relatively easy to measure the growth rate as a function of temperature, indicate that the same activation energy also applies for crystal growth.

These early expressions for melt growth assumed that all surface sites are active growth sites. The modern version of these equations includes the factor f, as well as the entropy factor of Eqn. (4).

In addition, the diffusion jump distance in the liquid is not the atomic diameter as was assumed in this early model. An atom in the liquid diffuses by a series of small motions of length Λ, which is the average distance which an atom moves during a diffusive motion.

In some cases, a rearrangement of the configuration and positions of other atoms in the liquid is required in order for the an atom or molecule to join the crystal. In these cases, it is the average time required for an atom to move a diffusion jump distance, Λ, which determines the rate at which an atom or molecule can join the crystal. An atom which moves a distance Λ to join a crystal, advances the interface by an atomic diameter. Taking this into account, and adding the factor f as well as entropy term from Eqn. (4) and expressing the growth rate in terms of the diffusion coefficient in the liquid, results in the modern version of the Wilson-Frenkel expression for the growth rate:

$$v = \frac{6\,a\,D}{\Lambda^2} f \exp\left(-\frac{\Delta S}{k}\right)\left[1 - \exp\frac{\Delta G}{kT}\right] \tag{6}$$

This is the appropriate expression for the crystallization rate for many materials, where the mobility of the atoms limits the growth rate. For glass forming materials, the growth rate can be measured rather readily. In these materials, at large undercooling, the term in the square bracket changes slowly with temperature, so the temperature dependence of the growth rate is dominated by the temperature dependence of the molecular mobility. This equation also agrees with the results of molecular dynamics simulations of the crystallization of silicon, as discussed below.

1.5. Growth Rate from the Melt

Metal crystals grow very rapidly at small very small undercoolings, and so growth usually occurs with the interface close to the equilibrium temperature. In this case, the term in the square brackets can be approximated to give a linear relationship between the growth rate and the interface undercooling, which is often written in the simple form:

$$v = \mu\,\Delta T \tag{7}$$

here $\Delta T = T_M - T$, and μ is called the kinetic coefficient.

The growth rate of metals is usually limited by diffusive processes. For pure metals, the growth rate is limited by heat flow, that is, by how fast the latent heat can diffuse away from the interface. For alloys, compositional diffusion is also important. For a metal growing from the melt, μ is typically about 0.5 m/s.K.

It is important to remember that growth processes take place only at the interface. The atoms at the interface are subject to the conditions at the interface: the local temperature, the local composition, the local curvature of the interface. The thermal fields and compositional fields in the sample as a whole influence the local conditions at the interface, but the kinetic processes at the interface on the atomic scale depend only on the local conditions there. The atoms at the interface respond only to their immediate environment.

1.6. Nucleation of Layers

When the surface of a crystal is essentially atomically flat, with only a few adatoms in the next atomic layer, there is a kinetic barrier to the formation of new layers surface. On the surface of a defect free crystal, new layers must form by a nucleation process. Applying standard nucleation theory to the formation of new layers gives a rate of nucleation which can be written as:

$$I = I_0 \exp\left(-\frac{T^*}{\Delta T}\right) \tag{8}$$

Here T* depends on the edge free energy of the surface step, and ΔT is the undercooling. On a smooth surface, the density of active growth sites, f in Eqn. (1), depends on the density of growth sites provided by the nucleation process.

1.7. Growth on Screw Dislocations

Charles Frank (see [4]) suggested that defects in the crystal structure could help to form new layers, so that a nucleation process was not necessary for growth. He pointed out that a screw dislocation which ended at a surface could provide a continuous step on a surface for growth. He suggested that the surface step would wind up into a spiral, with the spacing between the steps on the spiral given by the critical radius, which is proportional to the undercooling. This means that the factor f is proportional to ΔT, and since u_K is also proportional to ΔT at small undercoolings, Frank predicted that the growth rate of a crystal should be proportional to ΔT^2.

Dislocations are not necessary for crystal growth: large dislocation-free crystals of silicon are grown every day. However, growth on defects can be important when there is a barrier to the formation of new layers. When the rate at which new layers nucleate is very slow, then growth can occur at the growth sites provided by defects. A growth rate which is proportional to ΔT^2, as Frank predicted, is often observed at small undercoolings on surfaces where there is a nucleation barrier to growth. At larger undercoolings, the nucleation rate increases so that the nucleation of new layers takes over the growth process, and the growth rate increases dramatically.

2. MOLECULAR DYNAMICS SIMULATIONS OF CRYSTAL GROWTH

Molecular dynamics simulations have provided detailed information about the process of crystal growth at the atomic level.

2.1. Crystallization from the Melt

The diffusion coefficients in molecular dynamics (MD) simulations of the liquid phase have been determined using both the Lennard-Jones (LJ) potential, which is valid for argon and works reasonably well for many metals, and for the Stillinger-Weber (SW) potential, which is valid for silicon. The diffusion processes in the liquid for both LJ and the SW simulations are very similar. The activation energies for diffusion scale with the melting points. Liquid silicon is metallic, and the atoms in all these liquids behave essentially as spheres. Although both the LJ and SW liquids are very similar, the crystals which form are quite different.

For the LJ potential, the force which any two atoms exert on each other depends only on the distance between them, and not on the location of any other atoms. This is called a pair potential. With this potential, the atoms crystallize with a face centered cubic structure, which is a close-packed structure.

The Stillinger-Weber [5] potential is a three-body potential developed especially for silicon. Each atom in crystalline silicon has four nearest neighbors, and the angle between the bonds to the nearest neighbors is important. And so the potential depends on not only the distance from each neighboring atom, but also on the angle between each pair of neighboring atoms. The potential provides a good approximation for the structure and properties of liquid silicon, and, below the melting point, a crystal forms with the diamond cubic structure, as the potential was designed to do.

The crystallization behavior of the two is quite different.

MD simulations using the SW potential [6] are shown in Fig. 1. The solid line is the Wilson-Frenkel expression, Eqn. (6), fitted to the data using the activation energy derived from the diffusion data for liquid silicon. This correctly describes the computer simulation data. This curve is similar to the experimental data for the crystallization of SiO_2 and GeO_2.

MD simulations using the LJ potential [7] are shown in Fig. 2. The open circles are the simulations results. The solid line is Eqn. (6), using the activation energy for liquid diffusion from simulation. It is obvious that the data are not described by this equation. The growth is not thermally activated.

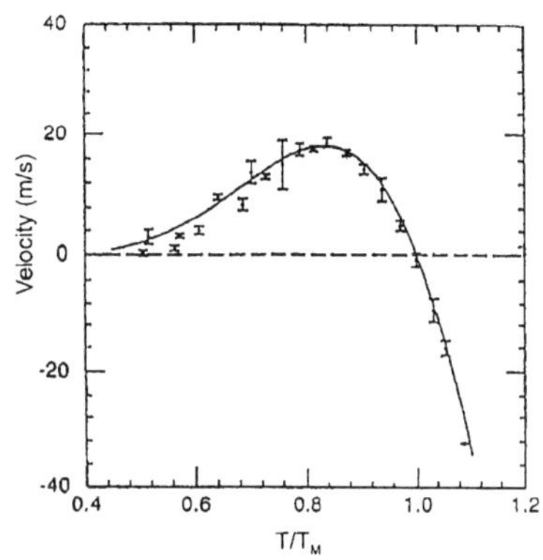

Figure 1. Molecular dynamics growth rates for the (100) face of silicon, using the Stillinger-Weber potential. The solid line is Eqn. (6).

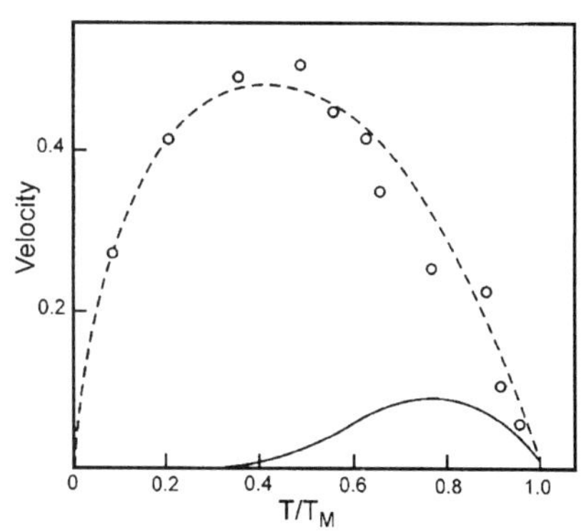

Figure 2. Molecular dynamics growth rates for the (100) face of an fcc crystal, using the Lennard-Jones potential. The solid line is Eqn. (6), the dashed line is Eqn. (10).

This result suggests that an atom can move a small distance to join the crystal without interference from other atoms. In this case, it is the mean free time, which is the time required for the atom to traverse its mean free path, λ, which determines the growth rate.

An atom can traverse its mean free path to join the crystal at the average thermal velocity, $(3kT/m)^{1/2}$. v^+ in Eqn. (1) is then the reciprocal of the mean free time between collisions in the liquid,

$$v^{+} = 1/\tau = \bar{v}/\lambda = \sqrt{3kT/m}/\lambda \tag{9}$$

The growth rate can be written:

$$v = \frac{a}{\lambda}\sqrt{\frac{3kT}{m}}\exp\left(-\frac{\Delta S}{k}\right)\left[1-\exp\left(-\frac{L\Delta T}{kT_M T}\right)\right]f \tag{10}$$

The dashed line through the data was calculated using Eqn. (10), with $\lambda = 0.1a$, and $f = 0.25$. This implies that the atoms run unobstructed from their positions in the liquid at the interface into the neighboring solid site at the interface, with the average thermal velocity of the atoms in the liquid. This growth rate does not depend on the rate at which atoms can move around in the liquid by diffusion. Unlike the diffusion jump distance, Λ, the mean free path, λ, is not strongly temperature dependent. This is the appropriate expression for the solidification rate of metal crystals, and also for the crystallization of the inert gases.

There are potential wells formed at the interface by the regular positions of the solid atoms at the interface. In the simulations, a liquid atom falls into one of these wells, stays there for a while, vibrating about an average position, and then hops to and moves randomly in the liquid for a while until it falls back into the same, or a different, well.

For silicon, and for molecular materials in general, there is an activation energy associated with rearrangement of the liquid structure. This activation energy determines the temperature dependence of the diffusion coefficient in the liquid, the viscosity of the liquid, and the crystallization rate. This is unlike the results from simulations with the LJ potential, where the atoms are essentially spheres. In this case, the activation energy for liquid diffusion and viscosity are the same, but the crystallization process is not thermally activated.

These results suggest that materials can be divided into two groups based on their crystallization behavior:

(i) One group contains materials for which no rearrangement of the liquid structure is required for any individual atom to join the crystal. The crystallization is not thermally activated. These materials can crystallize at very low temperatures, and they do not form glasses.

(ii) The other group contains materials where there must be some structural rearrangement of the liquid around an atom or molecule before it can join the crystal. The structural rearrangement process is thermally activated with the same activation energy as liquid diffusion and viscosity. These materials will not crystallize at very low temperatures, so that they can, in principle, be made into glasses.

In the discussion above, the number of active surface sites, f, has been assumed to be constant. This is usually the case for metals and the inert gases. However, for many materials, surface nucleation is required for growth, in which case f is strongly temperature dependent.

3. THE KOSSEL-STRANSKI MODEL

Stranski assumed that the strength of binding of an atom to the surface depends on its number of nearest neighbors. He applied this to a simple cubic crystal, which is also known as a Kossel crystal [8], and so this is now known as the Kossel-Stranski model. This is, of course, a simple approximation to how the binding depends on the number of neighbors, but this model does contain the essence of an explanation of why there is a strong anisotropy in the growth rate of some crystals. This model is still the basis for modern statistical mechanical models of crystal growth.

3.1. Bonding at an Interface

Stranski [9] suggested that the reason crystals grow with a crystalline shape is because there is a barrier to the formation of new layers on the closest packed faces. He suggested that the barrier exists because the rate at which individual atoms leave the surface of a crystal depends on how tightly they are bound to the surface. In this model, the binding of an atom to the surface of a crystal depends on how many nearest neighbor bonds it has. An isolated atom sitting on a flat surface has only one bond to the crystal, and so will leave relatively rapidly. An atom at a kink in a step has three bonds to the crystal, and so will leave more slowly. An atom which is part of the flat face of the crystal has five bonds to the crystal, and will leave it much more slowly. The surface of a simple cubic crystal where the surface atoms have various numbers of nearest neighbors in the crystal is illustrated in Fig. 3.

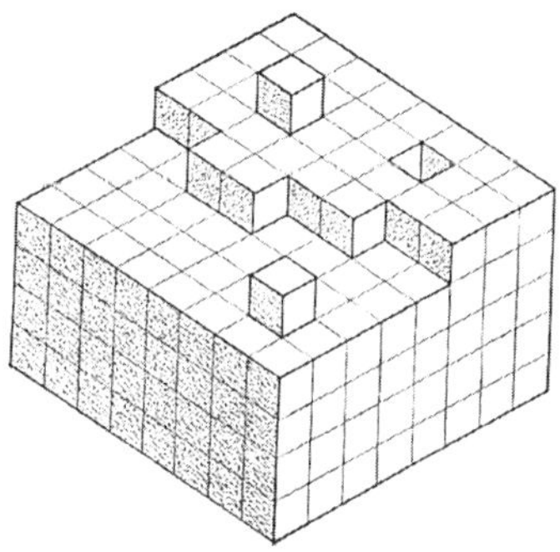

Figure 3. Configurations on the surface of a simple cubic crystal.

The rate at which atoms join the crystal is assumed to be independent of the local configuration, but the rate at which an atom leaves the crystal depends on how many of its nearest neighbor sites are occupied by atoms of the crystal. The normalized net rate at which atoms join and leave the crystal at a site with n nearest neighbors is:

$$u_k^n = 1 - \exp\left(\frac{L}{kT_M} - \frac{n\phi}{kT}\right) \tag{11}$$

Here ϕ is the bond energy, given by $\phi = 2L/Z$, where Z is the number of nearest neighbor sites in the crystal. n is the actual number of nearest neighbors of the atom at the surface of the crystal. Because each bond is shared between two atoms in the bulk of the crystal, the binding energy per atom, L is equal to $Z\phi/2$.

An important concept introduced by Stranski is the repeatable step site, which is also called a "kink site" because it is the site at a kink in a surface step. It is a site, as can be seen in Fig. 3, where an atom has half of its nearest neighbors, $n = Z/2$. An atom in the kink site breaks these Z/2 bonds when it leaves the surface, so the energy for it to leave is $Z\phi/2 = L$, which is the latent heat per atom. When an atom joins or leaves a kink site, the repeatable step site moves along the step, so the free energy associated with the surface configuration does not change. Except for edge effects, the entire crystal can be built up by the motion of these kink sites.

At equilibrium, the rate of arrival of atoms at a kink site is the same as the rate of departure from the kink site, as can be seen from the Eqn. (11) by substituting $n = Z/2$, so that $n\phi = Z\phi/2 = L$. The normalized rate of departure of an atom from a kink site as given by Eqn. (11) is identical to normalized average rate of departure of atoms from the crystal as in Eqn. (2).

In the general expression for the growth rate, Eqn. (1), f is the fraction of surface sites which are active growth sites. All active growth sites are kink sites. But on very rough surfaces, not all of the kink sites are active growth sites. For growth on a rough crystal surface, the factor f is more or less a constant, approximately ¼, a value which depends on the configuration of a rough surface. A surface cannot be constructed which has only repeatable step sites, and computer modeling suggests that the value of ¼ for f is about the best that can be done. The surface can be made rougher, but the effective number of growth sites does not increase much.

If we knew the site distribution function for the surface, that is, the probability of an atom on the surface having **n** nearest neighbors, we could replace the terms f and u_K in Eqn. (1) with a sum of the probability for each type of site, times the net rate of addition of atoms at that type as in Eqn.(11). The site distribution function can be estimated at equilibrium, but it is best determined by computer simulation.

3.2. Surface Roughness

It turns out that the mathematics of the Kossel-Stranki model is identical to the mathematics of the model which was originally developed by Ising [10] for the ferromagnetic Curie point, see for example [11]. The same mathematical model also applies to ordering in an alloy. In the Ising model, there is one interaction energy if the neighboring atom has the same spin, and a different interaction energy if it does not. For crystal growth, an atom has one interaction energy with an occupied neighboring site, and a different interaction energy with an empty neighboring site. The same mathematics applies to each of these cases, and so they are said to belong to the same Ising Universality Class. The Ising model is simple in principle, but becomes very complex in detail. It has been extensively studied in statistical mechanics.

Monte Carlo computer simulations can be readily performed on the Ising system [12]. And then the cluster distributions and the other behavior obtained from the computer simulations can scaled and used to describe the cluster distributions and the behavior of real systems. In applying the Ising model to crystal growth, it is assumed that there is a lattice in both phases, in order to enable the analysis, but that does not affect the universality of the result. The distribution of cluster sizes is the same for each of the systems in the Ising universality class, independent of whether or not there is a lattice. But the cluster distributions can be determined most readily by doing a calculation or a computer simulation using a lattice.

At first it seems strange that the cluster distributions should be so universal. But consider standard nucleation theory. The presence or absence of a lattice is not mentioned. The resulting equations for the cluster distributions are applied to the nucleation of a crystal in a liquid, a precipitate in a crystal, or a liquid in a vapor, etc. Similarly, the cluster distributions derived by analysis on a lattice can be applied universally to all the members of the Ising universality class.

It is worth emphasizing that the critical temperature of a member of a universality class cannot be predicted from an analysis of the cooperative behavior, because in the past it has been assumed that the critical point for the surface roughening temperature in the Ising model can be used to predict the surface roughening transition for real crystals. The Ising model describes the behavior of the system in the vicinity of the critical point, which is similar for all members of the same universality class. But, for example, the Curie temperature of a ferromagnetic material cannot be predicted from the Ising model. Nor can the roughening transition of the surface of any particular crystal be predicted from the Kossel-Stranski model.

The Ising model contains nucleation behavior as one limit. For example, in a magnetized ferromagnet at a low temperature, all the spins are aligned. In a demagnetized sample, all the spins are locally aligned, but there are magnetic domains of the opposite alignment, so there is no net magnetization of the sample. If a magnetic field is applied to a magnetized sample in a direction which requires the spins to flip, then there will be a nucleation barrier to the formation of the new magnetic domains. At a low temperature, there will be a critical size for new domains, which depends on the strength of the applied magnetic field. The formation of domains of reverse magnetization can be described by the standard equations for nucleation. But at temperatures approaching the Curie temperature, many flipped spins occur spontaneously, and small sub-critical regions of flipped spins will increase in size and number. These will interact and merge. Close to the curie temperature, standard nucleation theory does not apply to the formation of a domain of reverse magnetization, because the density of flipped spins is much too high. But the Ising model does incorporate this behavior, even though it is so difficult to describe analytically.

3.3. Monte Carlo Simulations of Crystallization

In these simulations, which are based on the Kossel-Stranski model, atoms arrive randomly at sites on the surface at some specified rate, and atoms leave the surface at a rate which is determined by a probability factor which depends on the number of adjacent crystalline atoms. If the departure probability is bigger than a random number, then the atom leaves, and if not, it does not. The arrival probability, P^+, and departure probability, P_n^-, can be written as:

$$P^+ = \nu^+$$
$$P_n^- = \nu^+ \exp\left(\frac{L}{kT_M} - \frac{n\phi}{kT}\right) \tag{12}$$

These are similar to Eqn. (11), but in the computer the arrival rate at a repeatable step, ν^+, must be set small enough so that the departure rate for a site with n = 1, P_1^-, is less that one. The simulation is conducted by atoms individually joining and leaving the crystal at each site on the interface with these probabilities. n is the actual number of nearest neighbors of the surface atom. At a repeatable step site where n = Z/2, the arrival rate is equal to the departure rate at equilibrium. In the computer, atoms join and leave the crystal according to these simple rules and so the computer simulation provides a statistical analysis of this model. The statistics are never exact, since the simulated systems are finite in size and simulation time. But the atoms form clusters and nucleate new layers, and so on, just as we suppose happens in the real world. When the bond energy is comparable to kT then there are many adsorbed atoms on the surface, there is no nucleation barrier to the formation of new layers. When the bond energy is large compared to kT, a single atom on the surface has a high probability of leaving the surface, and so there are few on the surface. In this regime, the formation of nuclei on the surface is difficult. Many atoms have to get together on the surface to form a cluster of critical size and the growth of new layers involves the lateral spreading of these clusters. In this regime, growth occurs by nucleation and the motion of steps on the surface.

3.4. Equilibrium Surface Structure

The density of growth sites on the surface of a crystal depends above all on the roughness of the crystal surface. This is an intrinsic property of the surface at equilibrium. It depends on the nature of the two phases separated by the interface and on the atomic structure of the interface.

A simple two-dimensional model for the equilibrium structure of a crystal surface or interface [13] identifies an important parameter, which is known as the Jackson α-factor, on which the surface roughness depends:

$$\alpha = \frac{L}{kT_M}\frac{\eta_l}{Z} \tag{13}$$

Here η_l is the number of nearest neighbor sites in the substrate layer and in the surface layer respectively, and Z is the total number of nearest neighbors. The two-dimensional mean field model predicts a singularity in the surface roughness at a critical value of α, $\alpha_C = 2$. This value for the surface roughening transition is in very good agreement with the observed behavior of melt-growing crystals. That is, for surfaces with a small α-factor, the surface is rough and the crystals can grow readily without surface nucleation. Surfaces with a large α-factor are smooth, and there is a nucleation barrier to the growth of each layer.

The α-factor has two components, one of which depends on the change in entropy of the transformation, $\Delta S = L/T_M$, and the other, η_l/Z, on the geometry of the crystal face. The factor η_l/Z is largest for the closest-packed face and smaller for the other faces of the same crystal.

For materials with a small entropy change on melting, all the crystal faces will be rough. These materials will exhibit relatively isotropic growth, and will grow rapidly at small interface undercoolings. For materials with a large entropy change on melting, the closest packed faces will be smooth and so there is a nucleation barrier to growth on those faces.

3.5. Monte Carlo Computer Simulation Results

The above discussion outlines the basic story of surface roughening, but there are many more details that have been worked out primarily using Monte Carlo computer modeling, but also mathematical analyses of approximations to the Ising model. A full analytical treatment of a multi-layer interface has been done for one special case, but there is no general solution. Multilevel cluster expansion models have been used to demonstrate that there is an analytical singularity in the thermodynamic properties of the surface of an Ising crystal at the surface roughening transition, but it is a very weak singularity. The surface roughening transition in the three-dimensional Ising Model is at $\alpha_C = 3.2$, rather that at $\alpha_C = 2$, as in the two-dimensional Bragg-Williams model. The surface tension and most of the other properties of the surface are continuous through this transition[14]. Fig. 4 is a plot of the surface roughness [15], which is defined as the number of unsatisfied bonds on the surface.

The surface roughness increases as the alpha factor decreases, but it is difficult to locate the surface roughening transition from this plot. The surface roughness is similar for each of the models: the three dimensional Bragg-Williams (0th order cluster expansion), the 1st order cluster expansion, and for Monte Carlo modeling (three dimensional Ising model), even though the analytical singularity which defines the surface roughening transition is different for each of these models. 8exp(-α) provides a reasonable approximation for the surface roughness. As the surface becomes rougher, the roughness is spread over more atomic layers. But in all the models, nothing much happens to the surface roughness at the roughening transition. The surface roughness increases continuously through the roughening transition.

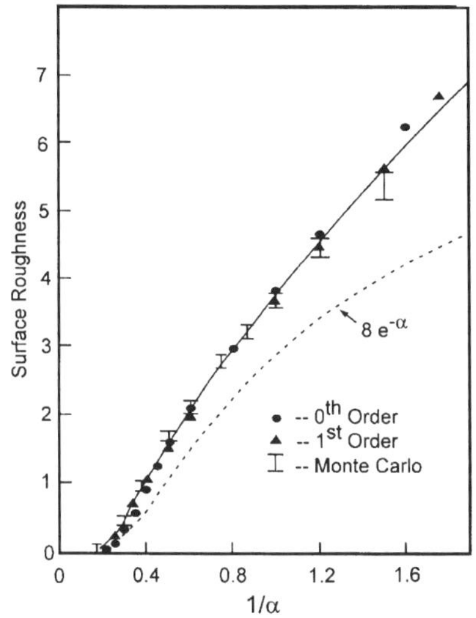

Figure 4. Surface roughness versus $1/\alpha$.

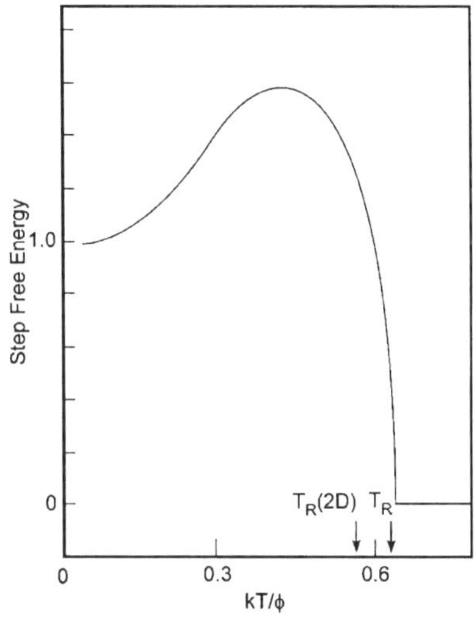

Figure 5. Step free energy versus kT/ϕ.

This is not true for the free energy of a step on the surface. The free energy of a step decreases as the surface roughening transition is approached, and it goes to zero at the surface roughening transition [16] as shown in Fig. 5.

Far above the surface roughening transition, the surface is smooth, although steps on the surface may be jagged. As the surface roughening transition is approached, the steps become increasingly jagged. There is a major contribution to the free energy of the steps from the entropy associated with this jaggedness. When the surface is rough, the steps are lost in the roughness, as illustrated in Fig. 6.

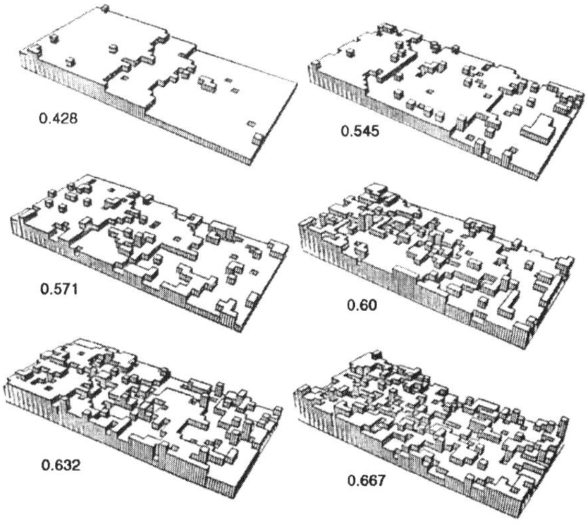

Figure 6. Configurations of surfaces from Monte Carlo simulations. The configuration on the lower right is a rough surface, below the surface roughening transition. The other configurations are for surfaces progressively farther above the roughening transition [16]. The parameter is kT/ϕ.

It has been shown that above the surface roughening transition, the interface is like a fluid/fluid interface [17]; the interface ignores the crystal lattice. On the rough side of the surface roughening transition, there is no difficulty in forming new layers. There is no barrier to continuous growth, and so the growth rate is linear with undercooling. On the smooth side of the surface roughening transition, steps on the surface have a finite free energy, and so the growth rate depends on the rate of nucleation of new layers. This is illustrated in Fig. 7, where growth rates from Monte Carlo computer simulation studies are shown. The curve labeled 1.08 is just on the rough side of the transition, and the growth rate is linear with undercooling. The other curves are for smooth surfaces. They come into the origin with zero slope [18], and farther from the roughening transition the curves are flatter and more "nucleation-like".

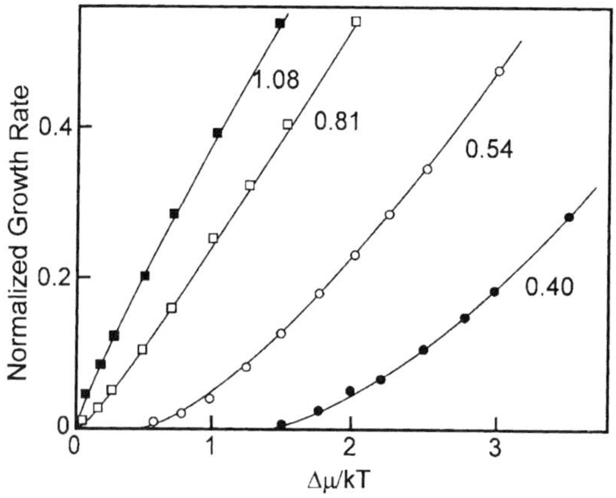

Figure 7. Normalized growth rate versus undercooling above and below the roughening transition from Monte Carlo computer simulations. The parameter is α_C/α.

As suggested by Fig. 7, it has been shown, using linear response theory, that the growth rate on rough interfaces is linear with undercooling, whereas the growth rate on smooth surfaces approaches equilibrium with zero slope. And so there is a discontinuity in the kinetic properties of the interface at the roughening transition.

The free energy of a step depends on the proximity of the surface roughening transition, and so the difficulty in nucleating new layers increases with the distance from the roughening transition. The growth rate for crystals with large entropies of transformation is highly anisotropic.

3.6. Simulations of Silicon Growth

As an example of growth anisotropy, Monte Carlo simulations have been carried out for silicon, with the roughening transition in the simulations scaled to the roughening transition for silicon [19]. The entropy change for melt growth of silicon, $\Delta S/k = 3.6$, so the α-factor for (111) is 2.7, and for (100) is 1.8, and it is lower for all the other faces. The crystallization rate for the (100) face, as determined by both experiment and MD simulations, is approximately $v = 0.12\ \Delta T$ m/sec. In Monte Carlo simulations, growth rate on the (111) face is limited by the rate of nucleation of new layers, whereas the growth rate on the (100) face is much more rapid, and linear with undercooling, as illustrated in Fig. 8.

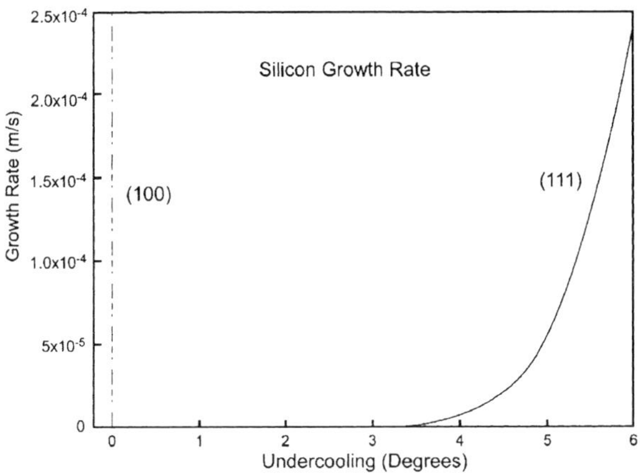

Figure 8. Growth rate versus undercooling for silicon (100) and (111) .

The growth rates in the figure were scaled to v = 0.12 ΔT for the (100) orientation. The difference between the growth behavior for the two faces comes from the Monte Carlo modeling.

Typical Czochralski growth rates for silicon are between $5x10^{-5}$ and 10^{-4} m/sec. At these growth rates the undercooling on the (100) face, as well as for all the other growth orientations except for (111), is less than a milli-degree. So these interfaces will be essentially at the melting point isotherm. The undercooling where the steps are nucleating on the (111) face is about 5 degrees. This is in accord with experiment and explains the large facet observed on silicon (111) during growth. The undercooling on the two faces for the same growth rate differs by about four orders of magnitude.

By fitting the growth rate on the (111) face to a nucleation model, the edge free energy of the steps on the (111) face is only about 10% of the free energy of the (111) face.

3.7. Kinetic Roughening

Experimentally, the growth rate is one of the most anisotropic properties of a crystal. As can be seen in Fig. 7 or 8, at small undercoolings there can be very large anisotropies in the growth rate. At small undercoolings, some faces grow and others do not: the difference in growth rates can be orders of magnitude. On very slowly growing faces, defects can significantly promote growth. However at large undercoolings, when the growth rate lifts off the horizontal axis in Fig. 7, even for the slow-growing directions, and the anisotropy in the growth rate decreases dramatically. In this regime, a surface which was smooth at equilibrium becomes rough [20]. This has been termed kinetic roughening, which is illustrated in Fig. 9. This results in a transition in the growth morphology from faceted or spiky crystals at smaller undercoolings to isotropic growth at large undercoolings.

The surface roughening transition plays the same role in the deposition of thin films as in growth from the melt. At high substrate temperatures, the equilibrium configuration of the surface is rough. At lower substrate temperatures, the surface becomes smooth, and deposition proceeds by the nucleation and spreading of layers. At very low temperatures, the rate of at which atom leave the surface decreases, and a significant fraction of the incident atoms stick. The critical nucleus size decreases, and the surface becomes kinetically rough.

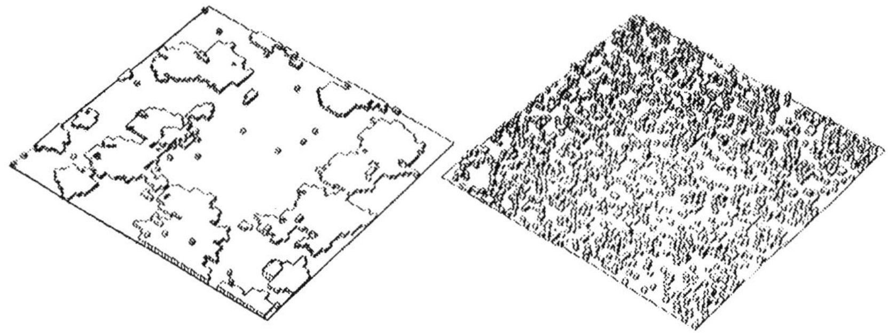

Figure 9. Kinetic roughening for growth at $\alpha_C/\alpha = 0.4$ as in Fig. 7.
Left: $\Delta\mu/kT = 1.5$; right: $\Delta\mu/kT = 3.5$.

4. THE FLUCTUATION DISSIPATION THEOREM

The measurement of fluctuations in the interface position at equilibrium can be used to determine both the kinetic coefficient and the surface tension [21, 22]. This correlation makes use of the Onsager fluctuation dissipation theorem [23], which states that: *"The rate at which a system, which has been displaced from equilibrium, returns to equilibrium is the same as the rate at which fluctuations in the system decay at equilibrium"*

4.1. Interface Fluctuations

Application of this theorem to crystal growth is illustrated by data for the time dependence of the number of atoms in the crystal at equilibrium in a Monte Carlo simulation [24] shown in Fig. 10.

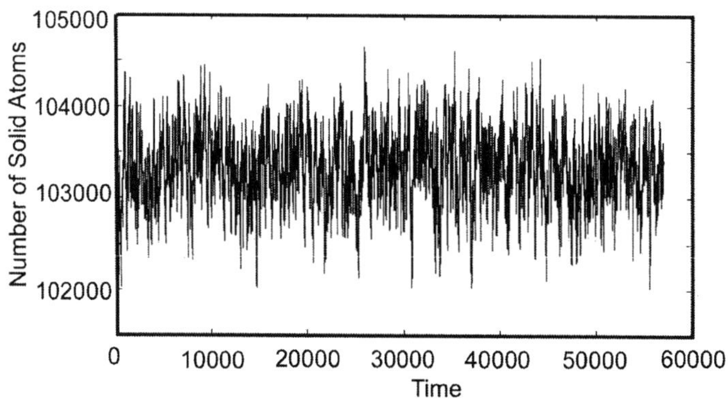

Figure 10. The number of atoms in the solid in a Monte Carlo simulation of a crystal at equilibrium with its melt fluctuates in time. These data were taken with an imposed temperature gradient which provided a restoring force on the interface position.

The rate at which fluctuations decay in data such as Fig. 10 can be obtained using a time correlation function:

$$C(t) = \frac{< \Delta N_S(t) \cdot \Delta N_S(0) >}{< \Delta N_S^2(t) >} \tag{14}$$

Here $N_S(t)$ is the number of solid atoms at any given time, t. There is an average number of solid atoms, $<N_S>$, and the number of solid atoms fluctuates about this average value. The instantaneous departure from this average value is $\Delta N_S(t) = N_S(t) - <N_S>$. The carets indicate average values.

The time correlation function is evaluated by designating some arbitrary time to be t = 0. The number of solid atoms at that time is recorded, and then the number of solid atoms is recorded as a function of time after this initial time. The product $\Delta N_S(t) \cdot \Delta N_S(0)$ can be calculated for each subsequent time, t. Then another time is picked as t = 0, and the process is repeated. Then the values of $\Delta N_S(t) \cdot \Delta N_S(0)$ for each starting time are averaged. If $\Delta N(0)$ is positive, then at short times, $\Delta N_S(t)$ is also likely to be positive, and if $\Delta N_S(0)$ is negative, then at short times, $\Delta N_S(t)$ is also likely to be negative, but the product $\Delta N_S(t) \cdot \Delta N_S(0)$ will be positive in either case.

For t = 0, the time correlation function, C(0) = 1. At some long time later, the number of atoms in the solid will be randomly greater than or less than the average, and so the product $\Delta N_S(t) \cdot \Delta N_S(0)$ will be randomly positive and negative, with an average value of 0. So C(t) = 0 for large t. The time correlation function should have the form:

$$C(t) = \exp(-t/\tau) \tag{15}$$

If C(t) decays exponentially as indicated, then the time constant, τ, can be determined from the data. The time correlation function is shown in Fig. 11 for the data in Fig. 10. The slope of the line is $1/\tau$.

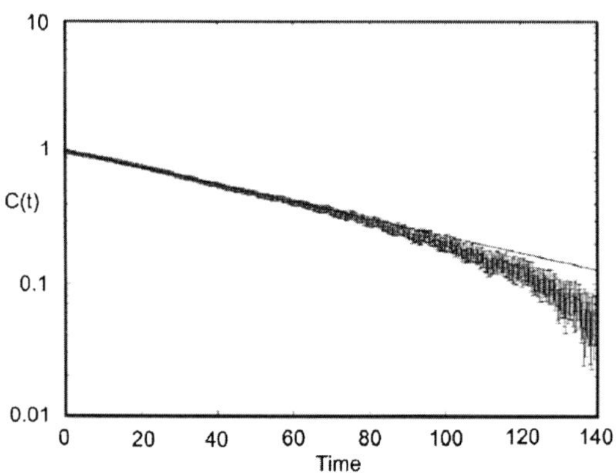

Figure 11. Time correlation function, C(t), for the data shown at left. The line is an exponential fit to the data for short times.

The fluctuations in the number of solid atoms in Fig. 10 seems random, but there is a time constant buried in the fluctuations which can be extracted using time correlations, as shown in Fig. 11.

In order for this scheme to work, there must be a fixed average number of solid atoms, and the number of solid atoms fluctuates about this average. If the interface was unconstrained and maintained at the melting point isotherm, the number of solid atoms would vary randomly. The data in Fig. 10 were obtained with an imposed temperature gradient, G = dT/dz. On average, there was solid on one side of the melting point isotherm and liquid on the other side. The gradient provides a restoring force which tends to return the interface to the melting point isotherm if it is displaced. The location of the melting point isotherm defines the equilibrium position of the interface, which in turn defines the average number of solid atoms.

$$T = T_M + Gz \tag{16}$$

We could equally well have a constant volume system, and use the pressure generated by the volume change to define the equilibrium number of solid atoms. Or we could have an adiabatic system, and use the temperature change caused by the latent heat to establish the equilibrium number of solid atoms.

If the growth rate of the crystal depends linearly on the undercooling, we can write, as in Eqn. (7):

$$v = \mu \Delta T \tag{17}$$

where μ is the kinetic coefficient

If the interface is displaced in a temperature gradient from the melting point isotherm by adding $\Delta N_S(0)$ atoms to the solid, the interface will, on average, be above the melting point by an amount which depends on the temperature gradient. Melting will proceed at a rate which depends on local interface temperature and the kinetic coefficient. It can be shown that the time dependence for a displaced interface to return to its equilibrium position in the temperature gradient, where $\langle \Delta N_S \rangle = 0$, is given by:

$$\Delta N_S(t) = \Delta N_S(0) \exp(-\mu G t) \tag{18}$$

The rate of melting decreases as interface approaches the melting point isotherm, as indicated by Eqn. (18). Similarly, if atoms are removed from the solid, the crystal will grow until the interface returns to the equilibrium isotherm.

Onsager's Fluctuation Dissipation Theorem states that rate of decay of fluctuations of the interface should be the same as the rate at which the displaced interface returns to the melting point isotherm. The value of time constant, τ, which is buried in the fluctuations in Fig. 10, depends on the magnitude of the restoring force on the interface, which is provided by the temperature gradient. It should be the same as the time constant, $1/\mu G$, in Eqn. (18), for return of a displaced interface to its equilibrium position. And so, the kinetic coefficient, μ, can be derived from a time correlation analysis of the fluctuations in the simulations, and is given by:

$$\mu = \frac{1}{\tau G} \tag{19}$$

If the crystal growth rate does not depend linearly on the growth rate, then this analysis will not work.

The fluctuations in pressure in a constant volume system could be monitored to give the number of solid atoms, or the temperature in an adiabatic system could be monitored to give the number of solid atoms. In any case, the total number of atoms must be large enough so that statistically significant fluctuations are observed, and small enough so that the fluctuations are not averaged out.

4.2. Determination of the Kinetic Coefficient from Fluctuations

Fig. 12 shows the kinetic coefficient as determined from Monte Carlo simulations of the (100) face of a simple cubic crystal.

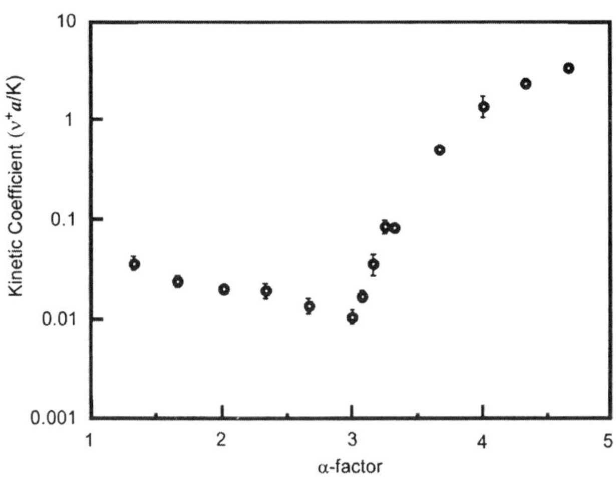

Figure 12. Kinetic coefficient as a function of the α-factor from Monte Carlo simulations of the (100) face of a simple cubic crystal [24].

The interface was at equilibrium in a temperature gradient. The interface fluctuations were analyzed using time correlations, and the kinetic coefficient was determined from Eqn. (19).

Below the surface roughening transition, the growth rate is linear with undercooling, and so a valid kinetic coefficient is obtained. The growth rate is not linear with undercooling above the surface roughening transition, and so the data do not correspond to a kinetic coefficient. Instead, the time correlation function is picking up very rapid changes adatom density on the smooth surface.

The theoretical surface roughening transition for this interface is at $\alpha = 3.2$. There is a sharp change in the kinetic coefficient at the surface roughening transition.

As illustrated in Figs. 4 and 6, many of the properties of the surface are continuous through the surface roughening transition. It is difficult to locate the transition in these figures. The surface tension is also continuous through the transition. But the kinetics of the interface motion are discontinuous at the surface roughening transition as demonstrated dramatically by Fig. 12.

The break in the data in Fig. 12 corresponds to the transition from the region where the growth rate is linear with undercooling in Fig. 7, to where the growth rate approaches the melting point with zero slope.

5. NON-EQUILIBRIUM SEGREGATION IN BINARY SYSTEMS

When the rate of advance of the interface becomes comparable to the rate at which atoms can move by diffusion the quasi-equilibrium treatment based on thermodynamics is modified by kinetic effects.

5.1. Experimental Observations

The first extensive observations of non-equilibrium crystallization were made by Duwez [25] using a technique he called splat quenching. A molten drop of the alloy was propelled onto a curved copper sheet. The copper sheet was held at an angle so that the droplet spread out along it into a thin layer, typically a few microns thick. The samples solidified very quickly. The samples were analyzed using X-rays and transmission electron microscopy (TEM). It was found that many metastable solid solutions could be formed. That is, the composition of the solid which formed was not the equilibrium solubility as reported on the phase diagram. Instead, solids containing much more of the second component than the equilibrium value were obtained. In some systems, instead of a two phase region as in the phase diagram, single phase solid solutions were obtained for all compositions of the alloy.

The alloys which form metallic glasses are usually a mixture of metallic elements and semi-metals or semiconductor elements, which crystallize into a complicated crystal structure. And so these materials have large entropies of fusion. Since these alloys crystallize slowly, they can be quenched into a glassy state where the atom mobility is very small. For pure metals, the crystallization rate is so fast that this cannot happen.

The composition of the solid which forms during crystallization is given by the distribution coefficient, also known as the **k**-value, which is defined as $\mathbf{k} \equiv C_B^S / C_B^L$, where C_B^S and C_B^L are, respectively, the concentrations of the solid and liquid at the interface.

Quantitative data are available for the dependence of the **k**-value on growth rate during the very rapid recrystallization of laser-melted, ion-implanted silicon [26, 27]. These data were obtained by first ion implanting a dopant into a silicon wafer. Then a very short duration, high power laser pulse, a centimeter or so in diameter, was directed at the surface. The total energy in the pulse is enough to melt a one to two micron thick layer of the surface of the wafer. The wafer was at room temperature, so the bulk of the wafer acts as an efficient heat sink. The liquid layer recrystallizes in a microsecond or so, with a growth rates in the range of meters per second. Typical data for the observed concentration profiles are shown in Fig. 13.

The as-implanted dopant distribution is shown, together with the final distribution of the dopant after laser melting of the surface. If the equilibrium distribution coefficient, 7×10^{-4}, had applied, all of the dopant would have been pushed to the surface by the crystallization process. The data can be fitted only by using a **k**-value of 0.1, which is over 100 times the equilibrium value. Similar data have been collected for aluminum alloys.

The growth rate dependence of the **k**-value is also responsible for the so-called "facet effect" observed during the slow growth of semiconductor crystals. An increased incorporation of most dopants at a faceted region of the interface during growth at normal laboratory or production growth rates is observed.

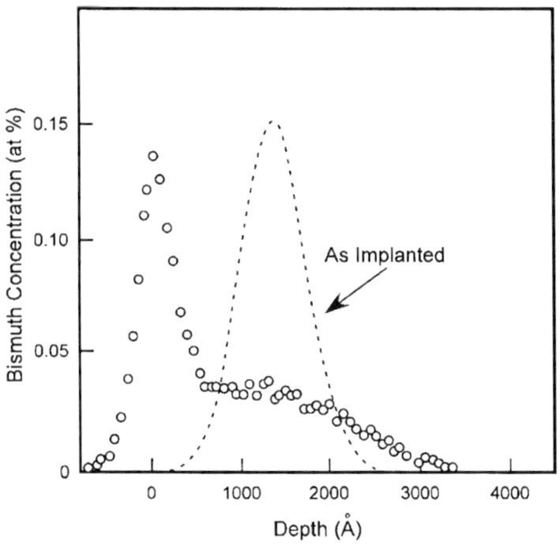

Figure 13. The as-implanted distribution of bismuth in silicon, and the distribution after a surface layer of the crystal was melted and rapidly recrystallized. After crystallization, there is a high concentration of bismuth at the surface, and RBS instrument broadening spreads out this distribution so that some of the bismuth appears to be outside the surface.

5.2. Monte Carlo Computer Modeling

Computer modeling has reproduced all of the main features of these observations, including the orientation dependence of the **k**-value. In addition the modeling has provided a definitive explanation for the orientation dependence. This modeling does not rely on any special properties of the atoms, it assumes the same interactions between atoms which are responsible for the equilibrium properties of the alloy. The modeling does incorporate the non-equilibrium effects which occur when the rate of advance of the interface becomes comparable to the rate at which atoms can move by diffusion.

Simulations of "diffusionless" transformations have been carried out, where the transformation takes place by the motion of the interface, but the atoms do not move: their position is fixed on lattice sites. These simulations correspond, for example, to a "shear" or martensite-type transformation where the interface moves very rapidly, at rates approaching the speed of sound. The atoms do have sufficient time only to shift their positions a small amount to conform to the new structure as the transformation front passes. Growth and melting of the alloy in these simulations was observed below and above the $\mathbf{T_0}$ line, respectively, where $\mathbf{T_0}$ is the locus of the temperatures on a phase diagram where the free energy of the solid alloy is equal to the free energy of the liquid alloy with the same composition. This is clearly the expected behavior for a diffusionless transformation since the kinetics of a diffusionless transformation should depend on the difference between the free energies of the two phases, rather than on the difference between the chemical potentials in the two phases of the species present. For the diffusionless case, freezing or melting should depend on which phase has the lower free energy, and this occurs above and below $\mathbf{T_0}$, as is observed in the simulations.

Data from Monte Carlo simulations of alloys which have been accumulated for a variety of growth temperatures, growth rates and diffusion coefficients, both above and below the roughening transition [28 - 31] are presented in Fig. 14. All of these data fall on a single curve when they are plotted against the dimensionless parameter β, as in Fig. 14.

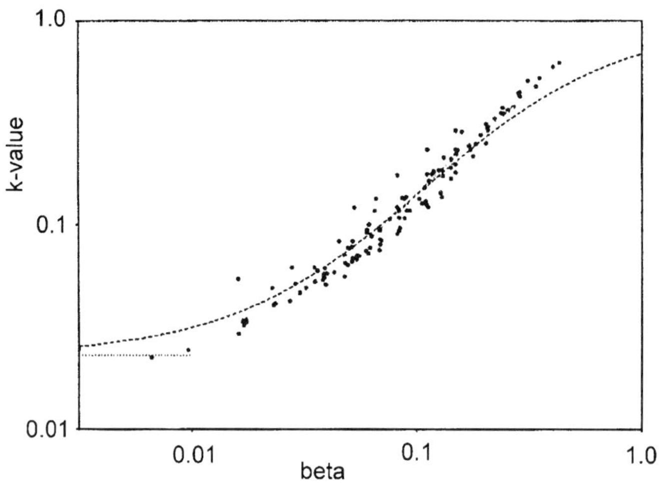

Figure 14. The distribution coefficient (k-value) data from Monte Carlo computer simulations for a wide variety of growth conditions, plotted against β (Eqn. 20).

The dimensionless parameter β which makes all the simulation data superpose is:

$$\beta = \sqrt{\frac{a \mathrm{v} \mathrm{u}_\mathrm{k}}{\mathrm{D}}} \tag{20}$$

here ***a*** is the lattice dimension, v is the growth rate, u_k is the normalized chemical potential difference driving the growth process, and D is the diffusion coefficient in the liquid.

β was first identified by Temkin [32] in his analytical modeling of alloy crystallization. He suggested that β depends on the time it takes a fluctuating interface to pass an atom, compared to the diffusion jump time.

Near equilibrium, where both v and u_k are small, β is small, and the **k**-value approaches the equilibrium value. When diffusion is relatively slow, and the growth rate is relatively large, β becomes large, and the **k**-value approaches one.

An alternate explanation is based on whether a B atom at the interface is likely to be engulfed during the time it spends at the interface. How far the interface moves during the average time that a liquid B atom spends at the interface can be compared with the distance that an atom can diffuse during that time. If the interface moves farther than the B atom can diffuse, the probability that the B atom will be incorporated into the crystal will be greatly increased.

In the general form for the growth rate of a crystal, Eqn. (1), the rate at which a liquid atom at the interface joins the crystal is $\nu^+ \boldsymbol{f}$, which is the probability that the atom is at an active growth site, times the rate it joins the crystal when it is at an active growth site. The average time, τ_i that a liquid atom spends at the interface before it joins the crystal is the reciprocal of this rate, $1/\nu^+ \boldsymbol{f}$.

The distance which the interface advances during this time is $l_v = v\tau_i = v/v^+f$. The distance which an atom can move by diffusion during this time is $l_D = \sqrt{D\tau_i} = \sqrt{D/v^+f}$. The ratio of these two distances is:

$$\frac{l_v}{l_D} = \frac{v}{v^+f} \Big/ \sqrt{\frac{D}{v^+f}} = \sqrt{\frac{avu_k}{D}} \tag{21}$$

which is β, as in Eqn. (20).

Near equilibrium, β is small. For very fast growth or very slow diffusion, β approaches infinity.

5.3. Analytical Model

For near-equilibrium growth, the growth rate for the **i**th component of an alloy can be written as [28, 31]:

$$v_i = av^+f[C_i^L - C_i^S \exp(\frac{\Delta\mu_i^0}{kT})] \tag{22}$$

Where the term in the square brackets becomes $\mathbf{u_k}$ for each species. The chemical potential of species **i** in each phase can be written as: $\mu_i = \mu^0_i + kT \ln C_i$. In Eqn. (22), $\Delta\mu^0_i$ is the difference between the μ^0_i in the two phases. The overall growth rate is given by the sum of the v_i. This formulation is valid when the atoms in the liquid move around rapidly compared to the rate at which the interface passes. Each growth site then samples the average composition of the liquid. Under these conditions, each atom effectively acts independently on being incorporated into the crystal, as is assumed implicitly in Eqn. (22). As a result, this formulation predicts that the distribution coefficient does not change significantly with growth rate, and so it does not include the phenomena known as solute trapping.

When the growth rate becomes comparable to the rate at which atoms can move around in the parent phase, the equations must be modified because the atoms can no longer act independently. Eqn. (22) can be modified to take into account these interactions by introducing a parameter *P*, which is related to β. *P* is zero when β is zero, and *P* is one when β is infinite.

$$1 - P = \frac{1}{1 + A\beta} \tag{23}$$

Here A is a constant. When *P* = 0, the atoms act independently; when *P* = 1, all species present in the liquid at the interface are incorporated at the same rate. In the modified formulation, Eqn. (22) becomes:

$$v_i = av^+f[C_i^L - C_i^S \exp(\frac{(1-P)\Delta\mu_i^0 + P\Delta F^0}{kT})] \tag{24}$$

where ΔF^0 is the free energy difference between the crystal and the liquid.

For small v, when near equilibrium conditions prevail at the interface, P is small, and Eqn. (24) reduces to the quasi-equilibrium equation for the growth rate, Eqn. (22). When β is very large, that is when the diffusion rate is small compared to the growth rate, P approaches 1. The growth rate of all the species depends of the free energies of the alloy in the two phases, rather than on their individual chemical potentials. Because the atoms cannot move as the interface passes, they must enter the crystal cooperatively. The composition of the solid will be the same as the composition of the liquid.

A simple approximate expression for the distribution coefficient for a small concentration of a second component, B, can be obtained from Eqn. (24) by noting that the equilibrium **k**-value, $\mathbf{k_e}$ is given by:

$$\mathbf{k_e} = \exp\left(-\frac{\Delta\mu_B^0}{kT}\right) \tag{25}$$

Since $v_B/\boldsymbol{a}v^+\boldsymbol{f}$ is usually small, and ΔF^0 is small compared to $\Delta\mu_i^0$ for small concentrations of B, we have:

$$k \approx k_e^{1-P} = k_e^{1/(1+A\beta)} \tag{26}$$

For near equilibrium growth, β is small, so $\mathbf{k} \approx \mathbf{k_e}$. For large β, **k** approaches 1.

For growth on a rough interface, the growth rate, v, is proportional to u_k. In this case, β is proportional to v, and so Eqn. (26) can be written as:

$$k \approx k_e^{1/(1+A'v)} \tag{27}$$

where A' is a constant which can be determined from the constant A.

5.4. Comparison with Experiment

The critical experiments to explore the solute trapping phenomenon were performed by laser melting of a thin layer on the surface of a silicon single crystal which had been ion implanted with a dopant as described above. In Fig. 15, experimental data are presented for the growth rate dependence of the **k**-value for silicon implanted with bismuth [30 -37], tin [38] and germanium [39].

Monte Carlo computer simulations have been performed using the crystal structure of silicon, inputting the equilibrium **k**-values for silicon doped with germanium, tin and bismuth. Data from these simulations are also presented in Fig. 15. The agreement between Eqn. (26) and the simulation results is quite good. There is also quite good agreement with the experimental data for germanium, and there is good agreement with some of the experimental data for bismuth. There is not good agreement with some of the experimental data for bismuth, or with the experimental data for tin, which are from the same laboratory.

It is well known that the **k**-value of dopants in silicon, as well as in compound semiconductors, depends on the orientation of the growth front. In order to investigate this, an extensive set of simulations were carried out in three dimensions using the crystal structure of silicon, with various orientations of the interface. The growth rates for the simulations using the crystal structure of silicon were correlated to experimental growth rates as outlined above.

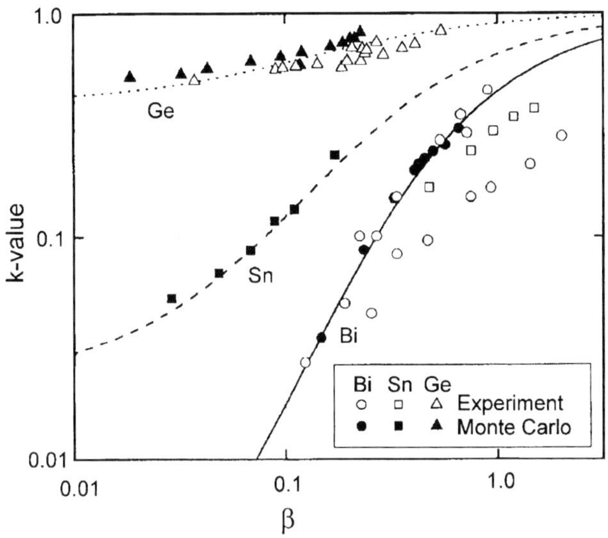

Figure 15. Simulation data and experimental data for the **k**-value for silicon doped with germanium, tin and bismuth versus the parameter β (Eqn. 20). The lines are Eqn. (26) using A = 8, and with $\mathbf{k}_e$ equal to 0.4 for germanium, 0.023 for tin, and 0.0007 for bismuth.

The data agree well with experimental measurements of the orientation dependence of bismuth incorporation into laser-melted silicon wafers [40]. The **k**-value does not depend on orientation when plotted as a function of β rather than as a function of the growth rate. The magnitude of u_k, the undercooling required for growth, is quite different for on-facet and off-facet growth at the same growth rate. This difference accounts for the anisotropy in the incorporation of dopants on different orientations during growth at the same rate. In physical terms, this means that in order to keep up with the growth rate on a rough surface, the steps on a smooth surface must move very rapidly, and this increases the incorporation of the dopant on the facet.

Eqn. (26) is also in reasonable agreement with experimental results for the growth rate dependence of **k**-value for aluminum alloys and for a nickel alloy, all using the same value of the constant A.

Eqn. (24) can be used to calculate kinetic phase diagrams, such as the illustration in Fig. 16. As the growth rate increases, the solidus and liquidus lines both collapse towards the $\mathbf{T}_0$ line given by the equality of the free energies of the two phases.

For growth under conditions where these non-equilibrium effects are present, the kinetic phase diagram and the associated **k**-value, rather than the equilibrium values, should be used to describe the growth conditions and segregation effects.

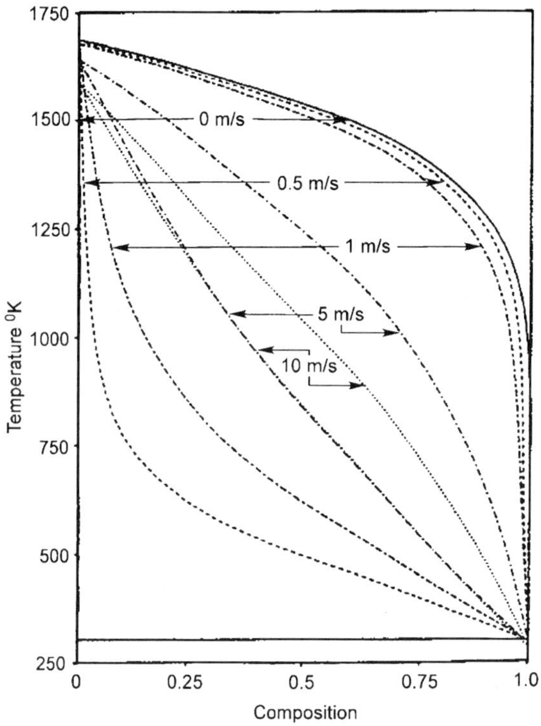

Figure 16. Kinetic phase diagram for various growth rates calculated from Eqns. (24). The equilibrium solidus is indistinguishable from the left axis.

ACKNOWLEDGEMENT

The author would like to thank NASA for financial support, and for NASA's continuing strong support of crystal growth studies.

REFERENCES

1. M. Knudsen, Ann. Phys. 34 (1909) 593.
2. H. A. Wilson. Phil. Mag. 50 (1900) 238.
3. J. Frenkel, Physik Z. der Sowjet Union 1 (1932) 498.
4. W. K. Burton, N. Cabrerra, F. C. Frank, Phil. Trans. Roy. Soc. A243 (1951) 299.
5 F. H. Stillinger, T. Weber, Phys. Rev. B 31 (1985) 5262.
6. M. H. Grabow, G. H. Gilmer, A. F, Bakker. Mat. Res. Soc. Symp. Proc. 141 (1989) 349.
7. J. Q. Broughton, G. H. Gilmer, K. A. Jackson, Phys. Rev. Lett. 49 (1982) 1496.
8. W. Kossel, Nachr. Ges. Wiss. Göttingen (1927) 135.

9. I. N. Stranski, Z. Phys. Chem. (Leipzig) 136 (1928) 259.
10. E. Ising, Z. fur Physik 31, (1925) 253.
11. D. Chandler, Introduction to Modern Statistical Mechanics, Oxford University Press, (1987) Ch. 6 and 7.
12. G. H. Gilmer, K. A. Jackson, 1976 Crystal Growth and Materials, (Ed.: E. Kaldis, H. J. Scheel), North-Holland (1977) 80.
13. K. A. Jackson, Liquid Metals and Solidification ASM Cleveland (1958) 174; Growth and Perfection of Crystals, (Ed.: Doremus, Roberts, D. Turnbull), Wiley (1958) 319.
14. J. D. Weeks, G. H. Gilmer, Adv. Chem. Phys. 40 (1979) 157.
15. H. J. Leamy, K. A. Jackson, J. Appl. Phys., 42 (1971) 2121.
16. H. J. Leamy, G. H. Gilmer, K. A. Jackson, Surface Physics of Materials, (Ed.: J. B. Blakeley), Academic Press (1975) 121.
17. J. D. Weeks, S. T. Chui, Phys. Rev. B14 (1976) 4978.
18. H. J. Leamy, G. H. Gilmer, J. Cryst. Growth 24/25 (1974) 499.
19. K. M. Beatty, K. A. Jackson, J. Crystal Growth 211 (2000) 13.
20. G. H. Gilmer, K. A. Jackson, 1976 Crystal Growth and Materials, (Ed.: E. Kaldis, H. J. Scheel), North-Holland (1977) 80.
21. W. J. Briels, H. L. Tepper, Phys. Rev. Lett. 79 (1997) 5074.
22. J. J. Hoyt, B. Sadigh, M. Asta, S. M. Foiles, Acta Mater. 47 (1999) 3181.
23. L. E. Reichl, A Modern Course in Statistical Mechanics, University of Texas Press (1991) 545-560.
24. D. N. Bentz, K. A. Jackson, J. Cryst. Growth, in course of publication.
25. P. Duwez, ASM Trans. Quarterly 60 (1967) 606; Prog. in Solid State Chem. 3 (1967) 377; P. Duwez, Ed., Metallic Glasses, ASM (1978).
26. P. Baeri, J. M. Poate, S. U. Campisano, G. Foti, E. Rimini, A. G. Cullis, Appl. Phys. Lett. 37 (1980) 912.
27. C. W. White, S. R. Wilson, B. R. Appleton, F. W. Young, Jr., J. Appl. Phys. 51 (1980) 738.
28. K. A. Jackson, Interface Science 10 (2000) 159.
29. K. A. Jackson, G. H. Gilmer, D. E. Temkin, Phys. Rev. Lett. 75 (1995) 2530.
30. K. A. Jackson, G. H. Gilmer, D. E. Temkin, K. M. Beatty, J. Crystal Growth 163 (1969) 461.
31. K. A. Jackson, K. M. Beatty, J. Crystal Growth, to be published.
32. D. E. Temkin, Sov. Phys. Cryst. 17 (1972) 405.
33. C. W. White, S. R. Wilson, B. R. Appleton, F. W. Young, Jr., J. Appl. Phys. 51 (1980) 738.
34. C. W. White, B. R. Appleton, B. Stritzker, D. M. Zehner, S. R. Wilson, Mat. Res. Soc Symp. Proc. 1 (1981) 59
35. P. Baeri, G. Foti, J. M. Poate, S. U. Campisano, A. G. Cullis, Appl. Phys. Lett. 38 (1981) 800.
36. P. Baeri, G. Foti, J. M. Poate, S. U. Campisano, E. Rimini, A. G. Cullis, Mat. Res. Soc Symp. Proc. 1 (1981) 67.
37. M. J. Aziz, J. Y. Tsao, M. O. Thompson, P. S. Peercy, C. W. White, W. H. Christie, Mat. Res. Soc Symp. Proc. 35 (1985) 153.
38. D. E. Hoglund, M. J. Aziz, S. R. Stiffler, M. O. Thompson, J. Y. Tsao, P. S. Peercy, J. Cryst. Growth 109 (1991)107.
39. D. P. Brunco, M. O. Thompson, D. E. Hogland, M. J. Aziz, H. J. Grossman, J. Appl. Phys. 78 (1995) 1575 .
40. K. M. Beatty, K. A. Jackson, J. Crystal Growth 174 (1997) 28.

Crystal Growth – from Fundamentals to Technology
G. Müller, J.-J. Métois and P. Rudolph (Editors)

Theory of Crystal Growth Morphology

Robert F. Sekerka[a*]

[a]University Professor of Physics and Mathematics,
Carnegie Mellon University, Pittsburgh, PA 15213, USA

Crystal growth morphology results from an interplay of crystallographic anisotropy and growth kinetics, the latter consisting of interfacial processes as well as long-range transport. A baseline for crystal morphology is the equilibrium shape that results from minimizing the anisotropic surface free energy of a crystal under the constraint of constant volume. This equilibrium shape is given by the classical Wulff construction but can also be represented by an analytical formula based on the $\boldsymbol{\xi}$-vector formalism of Hoffman and Cahn. We give analytic criteria for missing orientations (sharp corners or edges) on the equilibrium shape, both in two dimensions (classical) and in three dimensions (new). Crystals that grow under the control of interfacial kinetic processes (limiting case of fast long-range transport) tend asymptotically toward a "kinetic Wulff shape," the analog of the Wulff shape, except it is based on the anisotropic interfacial kinetic coefficient rather than the anisotropic surface free energy. If long-range transport were not an issue, a crystal would presumably nucleate with its equilibrium shape and then proceed to evolve toward its "kinetic Wulff shape," ultimately becoming bounded by surfaces of the more slowly growing orientations, as described by Frank. But long-range transport of heat or solute is important during at least some stage of a crystal growth process. We would therefore expect crystal morphology to be influenced by the shape and anisotropic properties of the container in which a crystal is grown. But the situation is vastly more complicated because the transport processes themselves are unstable. This leads to shape instabilities on the scale of the geometric mean of a transport length (typically a diffusivity divided by the growth speed) and a capillary length (of the order of atomic dimensions). The resulting shapes can be cellular or dendritic, but can also exhibit corners and facets related to the underlying crystallographic anisotropy. Such complex crystal morphologies are very difficult to model because of the need to track a sharp crystal-nutrient interface in detail. Within the last decade, however, powerful phase field models have been used to treat simultaneously all of the above phenomena. Phase field models are based on a diffuse (rather than a sharp) crystal-nutrient interface and incorporate an order parameter to indicate the phase. Evolution of the phase field and the transport fields (temperature, composition) is based on coupled partial differential equations that can be derived from an entropy functional. Computed morphologies can exhibit cells, dendrites and facets and the geometry of the isotherms and isoconcentrates can also be determined.

---

*Thanks are due to the Division of Materials Research of the National Science Foundation for the support of this research over a period of several decades.

1. INTRODUCTION

The quintessential notion of a crystal conjures up the image of a transparent, translucent, or opaque object, possibly of alluring color, having smooth surfaces and sparkling facets. The external shape of such a crystal constitutes its so-called morphology. Many gemstones acquire their morphology by cutting and polishing, but these are artificial morphologies which will not be dealt with in this paper. Here, we shall be concerned with the natural morphologies of crystals that arise as a consequence of the conditions of their growth, possibly from a flux, a solution, a vapor or a melt. To focus our ideas, we shall concentrate on crystal morphologies that result from growth from the melt.

The morphology of grown crystals arises in part from crystalline anisotropy, the fact that crystallographic properties depend on orientation. Generally speaking, both bulk and surface properties of crystals are anisotropic, but surface anisotropies are generally more important. That is because bulk properties such as thermal conductivity $\check{\mathbf{k}}$ and chemical diffusivity $\check{\mathbf{D}}$ are second rank tensors, which for the very important class of cubic crystals are isotropic.[2] But surface properties, such as the surface free energy γ or the interface kinetic coefficient μ (see Eq. (21) below) are anisotropic even for cubic crystals, and the latter can be strongly anisotropic, especially for materials having large entropies of fusion, as characterized by Jackson's α factor. [1]

We shall see that the anisotropy of γ gives rise to the concept of an equilibrium shape, known as the Wulff shape; whereas, the anisotropy of μ gives rise to an analogous growth shape that we shall call the "kinetic Wulff shape." These shapes can possess corners and facets and are usually quite different from one another. If long-range transport of energy and/or mass were extremely fast during a crystal growth process, the growth would be controlled by interfacial processes and, in that case, a crystal would nucleate in its equilibrium (Wulff) shape and grow asymptotically toward its kinetic Wulff shape. But in many growth processes, long-range transport is important, or even dominant, and so growth morphologies can be much more intricate. In fact, we shall see that the diffusive nature of long-range transport processes gives rise to instabilities, known as morphological instabilities, and that these instabilities lead to shapes that can have smooth but highly convoluted surfaces, for example, dendrites. Such shapes are quite difficult to compute from classical sharp interface models in which the interface that separates bulk phases has zero thickness, but can now be computed from diffuse interface models, such as the phase field model. Examples will be presented in the final section.

2. EQUILIBRIUM AND KINETIC WULFF SHAPES

We begin by examining the equilibrium and growth shapes of a crystal in contact with its melt. Analogous shapes exist for a crystal in contact with a vapor or a solution, provided that the appropriate values of γ and μ are employed.

[2]For crystals of lower symmetry, $\check{\mathbf{k}}$ and $\check{\mathbf{D}}$ would be anisotropic; for instance, for ice, which is hexagonal, they would have axial symmetry, isotropic in the basal plane and different along the c-axis. Elastic moduli are fourth rank tensors and would be anisotropic even for cubic crystals.

2.1. Equilibrium shape

Under isothermal conditions, a crystal in equilibrium with its melt takes on a so-called equilibrium shape in order to minimize its interfacial free energy

$$\Omega^{\mathrm{XS}} = \int_A \gamma(\hat{\mathbf{n}})\, dA, \tag{1}$$

subject to the constraint of constant crystal volume. The quantity $\gamma(\hat{\mathbf{n}})$ is the anisotropic interfacial free energy (excess Kramers potential) per unit area, $\hat{\mathbf{n}}$ is a unit vector perpendicular to the interface, and the integral is over the interface (surface) of the entire equilibrium shape. $\gamma(\hat{\mathbf{n}})$ is often represented graphically by means of a γ-plot, a polar plot of γ versus the spherical polar angles θ and φ of $\hat{\mathbf{n}}$. In two dimensions, γ is a function of only one angle, θ, measured from some appropriate crystallographic direction, and its γ-plot can be represented by a planar diagram, such as shown in Figure 1a.

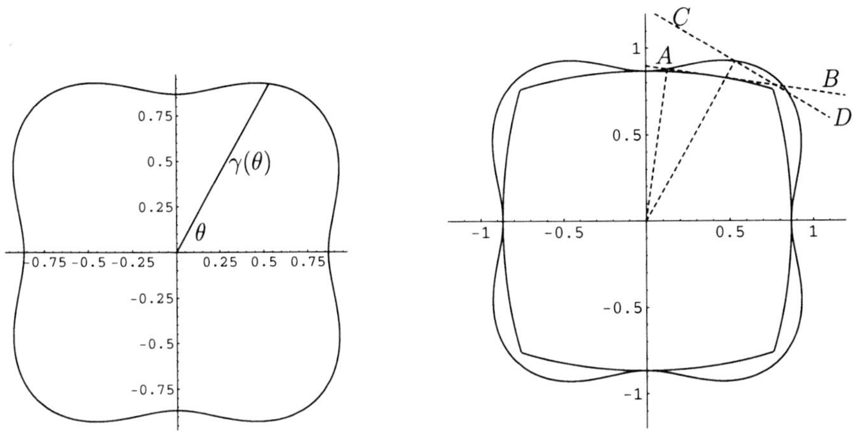

(a) γ-plot (b) Wulff construction

Figure 1. γ-plot (a) and Wulff construction (b) for a two-dimensional crystal having square symmetry, illustrated for $\gamma/\gamma_0 = 1 - (1/7.5)\cos(4\theta)$. The length of the radius of the γ-plot for any value of θ is $\gamma(\theta)$. For the Wulff construction, one draws perpendiculars to each such radius where that radius intersects the γ-plot and then selects the shape that can be reached from the origin without crossing any such perpendiculars (inner convex hull). Two such perpendiculars are shown, AB which contributes to the equilibrium shape and CD which lies outside the equilibrium shape. This particular equilibrium shape has missing orientations at its four corners and nearly flat sides.

The equilibrium shape that corresponds to a given γ-plot is given by a well-known geometrical construction due to Wulff [2]. According to the Wulff construction, the equilibrium shape is the inner convex hull bounded by planes (Wulff planes) drawn perpendicular to each $\hat{\mathbf{n}}$ at a distance $\gamma(\hat{\mathbf{n}})$ from the origin. In two dimensions, one draws perpendicular lines, instead of planes, as illustrated in Figure 1b. Since it is possible for planes corresponding to a given orientation to be farther from the origin than planes from some other orientation, certain orientations on the equilibrium shape may be missing. In such cases, the equilibrium shape has edges and/or sharp corners, and the remainder of the shape is made up of smooth curved surfaces or planar facets. For facets to occur, the γ-plot must have sufficiently deep cusps, which strictly speaking could only be the case at the absolute zero of temperature (because of entropic contributions to γ), but practically speaking, there can be nearly flat faces at finite temperatures.

2.1.1. Herring sphere

An elegant criterion for missing orientations to occur was given by Herring [3,4] and is related to the so-called Herring sphere construction. For a given orientation $\hat{\mathbf{n}}$, Herring considered a sphere that is tangent to the γ plot at $\gamma(\hat{\mathbf{n}})$ and that passes through the origin. By using the geometrical theorem that an angle inscribed in a semicircle is a right angle, Herring showed that if the orientation $\hat{\mathbf{n}}$ appears on the equilibrium shape, it appears at an orientation $\hat{\mathbf{N}}$ that points outward from the origin along the diameter of that sphere. He then recognized that if such a sphere were totally inside the γ-plot, then that orientation would appear on the equilibrium shape; otherwise, if some part of the γ-plot were to intrude inside the Herring sphere, its Wulff plane would "cut off" the orientation corresponding to the point of tangency, and that orientation would not appear. This is illustrated in two dimensions in Figure 2a. This powerful criterion not only gives rise to a visual picture but can be applied independently of any analytical representation of $\gamma(\hat{\mathbf{n}})$.

Rather than deal with a tangent sphere, one can develop a related criterion by means of the well-known transformation of inversion (with respect to the origin). Instead of the γ-plot, one considers a polar plot of $1/\gamma$ versus $\hat{\mathbf{n}}$. Then one recognizes that a sphere passing through the origin inverts to a corresponding plane. In particular, a sphere that is tangent to the γ-plot, such as a Herring sphere, inverts to a plane that is tangent to the $1/\gamma$-plot. If a Herring sphere is inside the γ-plot, its corresponding plane will be outside the $1/\gamma$-plot. Moreover, if some portion of the γ-plot is inside a Herring sphere, the corresponding portion of the $1/\gamma$-plot will be outside the corresponding plane. This is illustrated in two dimensions in Figure 2b.

Thus, if the $1/\gamma$ - plot is convex, then all of its tangent planes will lie outside, and all orientations will appear on the equilibrium shape. Otherwise, there will be missing orientations. In fact, if the $1/\gamma$ - plot is not convex, we can make a convex body from it by adding tangent planes. The portions of the $1/\gamma$ - plot between these tangent planes and the origin will correspond to missing orientations, and all other orientations will appear. Thus, not only do we know that orientations are missing, but precisely which ones. This is illustrated in two dimensions in Figure 2b in which the tangent line AB subtends missing orientations; since there is fourfold symmetry, three other such tangent lines would need to be drawn to obtain a convex curve.

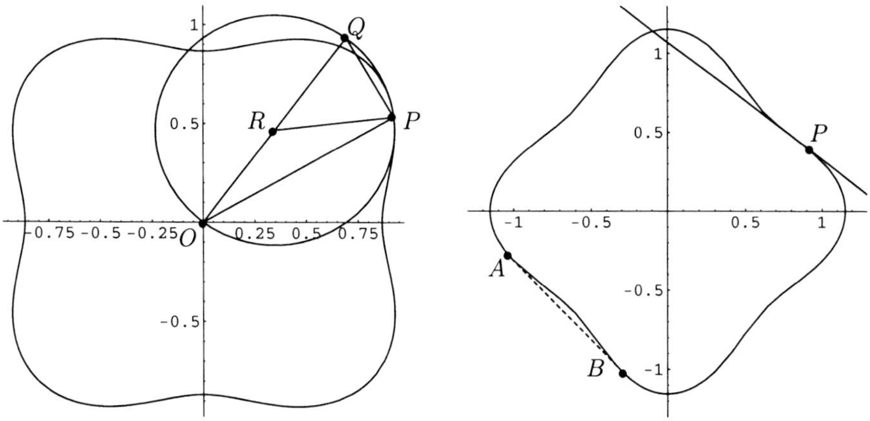

(a) Herring's tangent circle construction (b) $1/\gamma$ - plot and tangent lines

Figure 2. Illustration of the Herring sphere construction in two dimensions. (a) Draw a circle centered at R that passes though the origin and is tangent to the γ-plot at P. If the orientation at the point of tangency were to appear on the equilibrium shape, it would do so at point Q because OPQ is a right angle. If the circle is intersected by the γ-plot (as it is for P) the orientation corresponding to P will be missing on the equilibrium shape. (b) If any portion of the $1/\gamma$-plot lies outside one of its tangent planes, this is equivalent to the γ-plot lying inside the corresponding tangent sphere. Thus, if the $1/\gamma$-plot is not convex, there will be missing orientations on the equilibrium shape, for example, those that lie on part of the $1/\gamma$-plot subtended by the common tangent line AB.

2.1.2. Analytical criteria for missing orientations

The convexity of $1/\gamma$ is easy to quantify in two dimensions, where γ depends only on a single angle θ. The curvature of a polar plot $r = f(\theta)$ is

$$c = \frac{r^2 + 2r_\theta^2 - r r_{\theta\theta}}{[r^2 + r_\theta^2]^{3/2}} \tag{2}$$

where a subscript θ denotes differentiation with respect to θ. For $r = f(\theta) = 1/\gamma(\theta)$, Eq. (2) becomes

$$c = \frac{\gamma + \gamma_{\theta\theta}}{[1 + (\gamma_\theta/\gamma)^2]^{3/2}}. \tag{3}$$

Since the denominator in Eq. (3) is positive, convexity is lost whenever

$$\gamma + \gamma_{\theta\theta} < 0. \tag{4}$$

In three dimensions, the analytical criterion for missing orientations is much more complicated because $\gamma(\theta, \varphi)$ depends on both the polar angle θ and the azimuthal angle φ. This problem can be made tractable by using the $\boldsymbol{\xi}$-vector formalism of Hoffman and

Cahn [5,6]. One first extends the function $\gamma(\hat{\mathbf{n}})$ to a three dimensional vector space, $\mathbf{P}$, by defining

$$\tilde{\gamma}(\mathbf{P}) := P\,\gamma(\hat{\mathbf{n}}) \tag{5}$$

where $\hat{\mathbf{n}} = \mathbf{P}/P$ and $P = |\mathbf{P}|$. Then the $\boldsymbol{\xi}$-vector is defined by

$$\boldsymbol{\xi}(\hat{\mathbf{n}}) := \tilde{\nabla}\,\tilde{\gamma}(\mathbf{P}) = \gamma(\hat{\mathbf{n}})\,\hat{\mathbf{n}} + P\tilde{\nabla}\gamma(\hat{\mathbf{n}}) \tag{6}$$

where $\tilde{\nabla} = \partial/\partial\mathbf{P}$ is the gradient operator in $\mathbf{P}$ space. From its definition in Eq. (5), $\tilde{\gamma}(\mathbf{P})$ is a homogeneous function of degree one in the components P_ν of $\mathbf{P}$, so $\boldsymbol{\xi}$ is a homogeneous function of degree zero in P_ν and therefore depends only on $\hat{\mathbf{n}}$. It follows that $\gamma = \boldsymbol{\xi}\cdot\hat{\mathbf{n}}$, so $\hat{\mathbf{n}}\cdot d\boldsymbol{\xi} = 0$, and $d\gamma = \boldsymbol{\xi}\cdot d\hat{\mathbf{n}}$.

In terms of $\boldsymbol{\xi}$, the anisotropic Gibbs-Thomson equation for a crystal in equilibrium with its melt is

$$T = T_M - \frac{T_M}{L_V}\nabla_S\cdot\boldsymbol{\xi} \tag{7}$$

where T is the temperature, T_M is the melting point, L_V is the latent heat of fusion per unit volume of solid, and ∇_S is the surface divergence operator. Since $\nabla_S\cdot\mathbf{r} = 2$ where $\mathbf{r}$ is the position vector, the equilibrium shape is given in parametric form by

$$\mathbf{r}(\hat{\mathbf{n}}) = \mathcal{L}\boldsymbol{\xi}(\hat{\mathbf{n}}) \tag{8}$$

where $\mathcal{L} = 2T_M/[(T_M - T)L_V]$ is a scale factor. Eq. (8) states that the equilibrium shape is similar in shape to the $\boldsymbol{\xi}$-plot, which is a polar plot of $\boldsymbol{\xi}$ as a function of orientation $\hat{\mathbf{n}}$.

In two dimensions, $\boldsymbol{\xi}$ takes the simple form

$$\xi_x = \gamma\cos\theta - (d\gamma/d\theta)\sin\theta \tag{9}$$
$$\xi_y = \gamma\sin\theta + (d\gamma/d\theta)\cos\theta \tag{10}$$

where the subscripts on ξ designate cartesian components. Figure 3 shows a plot of $\boldsymbol{\xi}$ for $\gamma/\gamma_0 = 1 - (1/7.5)\cos(4\theta)$. The plot has "ears" that contain the missing orientations, but these ears have to be truncated to obtain the equilibrium shape, which is convex. [7]

It is easy to establish the relationship between the $\boldsymbol{\xi}$-plot and the $1/\gamma$-plot. The unit normal $\hat{\mathbf{N}}$ to the $1/\gamma$-plot, which is the surface $r - 1/\gamma = 0$, lies along $\nabla(r - 1/\gamma) = \hat{\mathbf{r}} + (1/\gamma^2)\nabla\gamma$, where $\hat{\mathbf{r}}$ is a unit vector along $\mathbf{r}$, so[3]

$$\hat{\mathbf{N}} = \frac{\gamma\hat{\mathbf{r}} + (1/\gamma)\nabla\gamma}{|\gamma\hat{\mathbf{r}} + (1/\gamma)\nabla\gamma|} = \frac{\gamma\hat{\mathbf{r}} + r\nabla\gamma}{|\gamma\hat{\mathbf{r}} + r\nabla\gamma|} = \frac{\boldsymbol{\xi}}{\xi}, \tag{11}$$

where ξ is the magnitude of $\boldsymbol{\xi}$. According to Eq. (11), which we now rewrite in the form

$$\hat{\mathbf{N}}(\hat{\mathbf{n}}) = \frac{\boldsymbol{\xi}(\hat{\mathbf{n}})}{\xi(\hat{\mathbf{n}})}, \tag{12}$$

the normal to the $1/\gamma$-plot lies along $\boldsymbol{\xi}(\hat{\mathbf{n}})$. For cases in which the $1/\gamma$-plot is not convex, we know that the equilibrium shape will have missing orientations; the corresponding

[3]In making the identification with $\boldsymbol{\xi}$, we note that $\mathbf{r}$ plays the role of $\mathbf{P}$ and ∇ plays the role of $\tilde{\nabla}$.

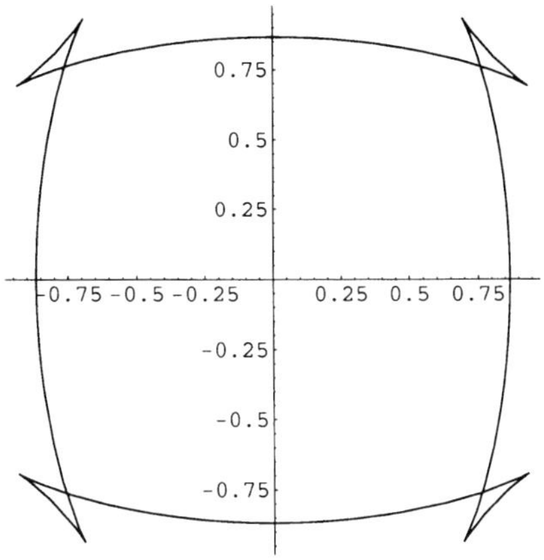

Figure 3. $\boldsymbol{\xi}$-plot corresponding to a two-dimensional γ-plot of the form $\gamma/\gamma_0 = 1 - (1/7.5)\cos(4\theta)$. The ears must be truncated to obtain the equilibrium shape, which is convex. The missing orientations at the corners of the equilibrium shape are those that are present on the ears.

$\boldsymbol{\xi}$-plot will have ears that have to be truncated to obtain the equilibrium shape, which is known to be convex. These ears correspond to $\boldsymbol{\xi}$ having multiple values for some range of $\hat{\mathbf{n}}$, in fact, precisely that range of $\hat{\mathbf{n}}$ for which portions of the $1/\gamma$-plot are nearer to the origin than the tangent planes needed to convexify it.

One can use Eq. (12) to obtain an explicit analytical criterion for the occurrence of missing orientations on the equilibrium shape. This may be done by recognizing that convexity of the $1/\gamma$-plot is lost whenever at least one of its principle curvatures K_1 or K_2 (or possibly both) changes sign, so their product K_1K_2, which is known as the Gaussian curvature, passes through zero. It is well known from differential geometry that the Gaussian curvature is an invariant. Thus for a surface $\mathbf{r} = \mathbf{r}(u, v) = \hat{\mathbf{N}}(\hat{\mathbf{n}})/\gamma(\hat{\mathbf{n}})$ where u and v are parameters on which $\hat{\mathbf{n}}$ depends, it can be shown that K_1K_2 can be expressed as a determinant of a matrix that connects $\hat{\mathbf{N}}_u$ and $\hat{\mathbf{N}}_v$ to $\mathbf{r}_u$ and $\mathbf{r}_v$, where the subscripts u and v indicate partial derivatives, as follows:

$$K_1K_2 = PS - QR \tag{13}$$

where

$$\begin{pmatrix} \hat{\mathbf{N}}_u \\ \hat{\mathbf{N}}_v \end{pmatrix} = \begin{pmatrix} P & R \\ Q & S \end{pmatrix} \begin{pmatrix} \mathbf{r}_u \\ \mathbf{r}_v \end{pmatrix}. \tag{14}$$

By inserting the explicit form of $\mathbf{r}$, one can convert this determinant to a scalar triple product involving $\boldsymbol{\xi}$ and its derivatives, namely

$$K_1K_2 = \gamma^2 \frac{\boldsymbol{\xi} \cdot \boldsymbol{\xi}_u \times \boldsymbol{\xi}_v}{\xi|\boldsymbol{\xi}_u \times \boldsymbol{\xi}_v|}. \tag{15}$$

It follows that the equilibrium shape will be on the verge of having missing orientations whenever $\boldsymbol{\xi} \cdot \boldsymbol{\xi}_u \times \boldsymbol{\xi}_v = 0$. In order to proceed further, we employ spherical coordinates and choose $u = \theta$ and $v = \varphi$. Then

$$\hat{\mathbf{n}} = \sin\theta\cos\varphi\,\hat{\mathbf{i}} + \sin\theta\sin\varphi\,\hat{\mathbf{j}} + \cos\theta\,\hat{\mathbf{k}} \tag{16}$$

where $\hat{\mathbf{i}}$, $\hat{\mathbf{j}}$ and $\hat{\mathbf{k}}$ are cartesian unit vectors. The $\boldsymbol{\xi}$-vector is given by

$$\boldsymbol{\xi} = \gamma\,\hat{\mathbf{n}} + \gamma_\theta\,\hat{\boldsymbol{\theta}} + (\gamma_\varphi/\sin\theta)\,\hat{\boldsymbol{\varphi}} \tag{17}$$

where $\hat{\mathbf{n}}$, $\hat{\boldsymbol{\theta}}$ and $\hat{\boldsymbol{\varphi}}$ are unit vectors of the spherical coordinate system. The scalar triple product can be evaluated by straightforward but tedious algebra. The final result is[4]

$$K_1K_2 = \frac{\gamma^4}{\xi^4\sin^2\theta}\left[(\gamma+\gamma_{\theta\theta})(\gamma\sin^2\theta + \gamma_{\varphi\varphi} + \gamma_\theta\sin\theta\cos\theta) - (\gamma_{\theta\varphi} - \gamma_\varphi\cot\theta)^2\right]. \tag{18}$$

The factor of $\sin^2\theta$ in the denominator is related to the fact that φ is an azimuthal angle and has a scale factor proportional to $\sin\theta$, but this factor is cancelled by terms in the numerator as $\theta \to 0$ and poses no problem. For isotropy one obtains $K_1K_2 = \gamma^2$, as expected, since the $1/\gamma$-plot is just a sphere of radius $1/\gamma$.

2.1.3. Illustration for cubic symmetry

We illustrate the use of Eq. (18) for the case in which γ has cubic symmetry of the form

$$\gamma = \gamma_0[1 + aQ_4(\hat{\mathbf{n}})] \tag{19}$$

where γ_0 and a are constants and

$$Q_4(\hat{\mathbf{n}}) \;=\; n_x^4 + n_y^4 + n_z^4 = \sin^4\theta(\cos^4\varphi + \sin^4\varphi) + \cos^4\theta. \tag{20}$$

For positive values of a, convexity of the $1/\gamma$-plot will be lost whenever $a > 1/3$, and ears will appear on the $\boldsymbol{\xi}$-plot, as illustrated in Figure 4 for $a = 1.0$. As a increases from 1/3, these ears increase in size. The range of missing orientations increases as a increases and dihedral ears that we call "flaps" reach the $\langle 110\rangle$ orientations for $a > 2/3$. After truncation of these ears and/or flaps, the remaining figure is the equilibrium shape, which tends toward an octahedron for very large a. For negative values of a, as a decreases from zero, missing orientations will begin to occur at the $\langle 110\rangle$ directions when $a < -2/9 = -0.22222$. This is illustrated in Figure 5 for $a = -0.2$, for which all orientations appear on the equilibrium shape, and $a = -0.5$ at which the $\boldsymbol{\xi}$-plot has large flaps that join the $\langle 111\rangle$ directions. The equilibrium shape tends toward a cube as a tends to negative values of large magnitude.

---

[4]I first derived the expression in brackets in this formula in 1992 and presented it (complete with a misprint on the slide) at my Frank Prize lecture in San Diego. The scale factor $\gamma^4/(\xi^4\sin^2\theta)$ and the detailed relationship to other criteria are the result of subsequent work, in the process of being published.

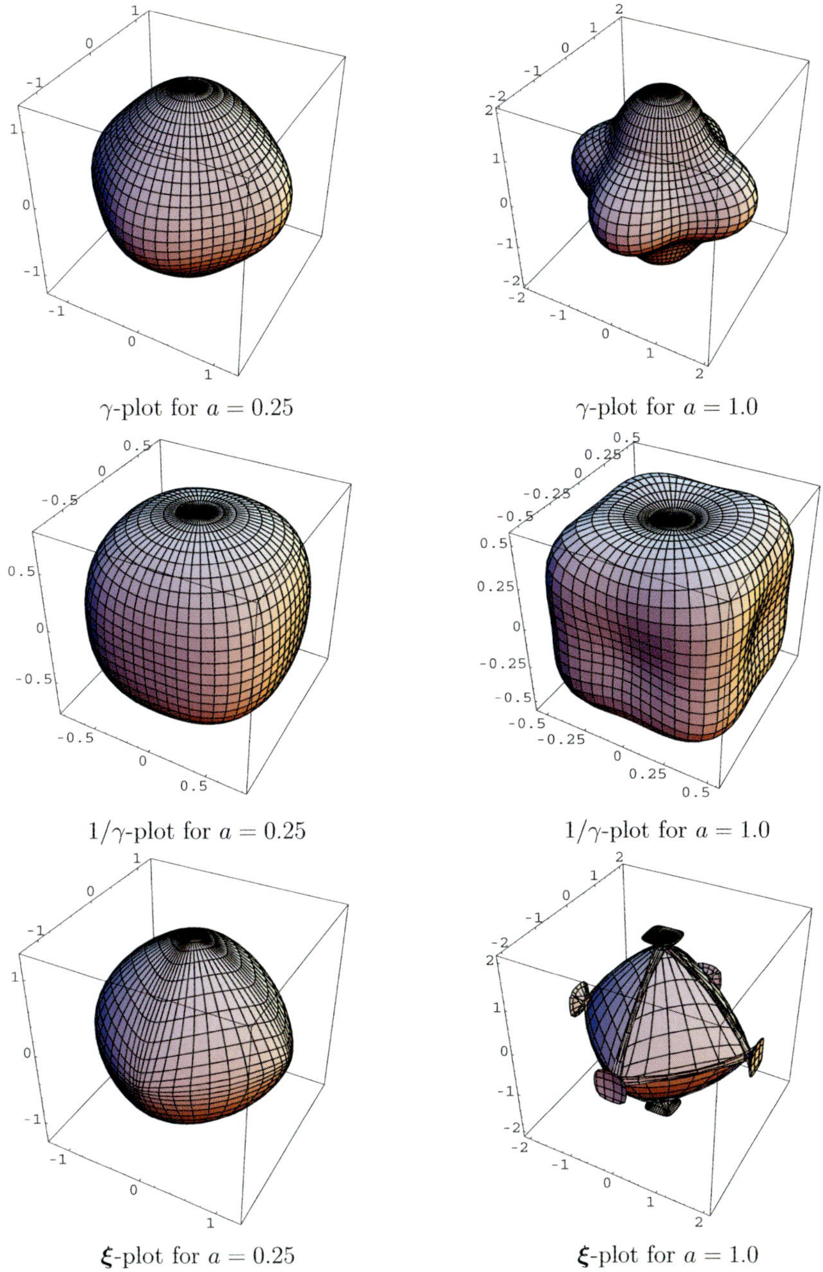

Figure 4. γ-plots, $1/\gamma$-plots and $\boldsymbol{\xi}$-plots for $\gamma = \gamma_0[1+aQ_4(\hat{\mathbf{n}})]$ for positive values of a. For $a = 0.25$, all orientations appear on the equilibrium shape. For $a = 1.0$, the $1/\gamma$-plot is concave and the $\boldsymbol{\xi}$ plot has ears and flaps that must be truncated to give the equilibrium shape, which resembles an octahedron with curved faces.

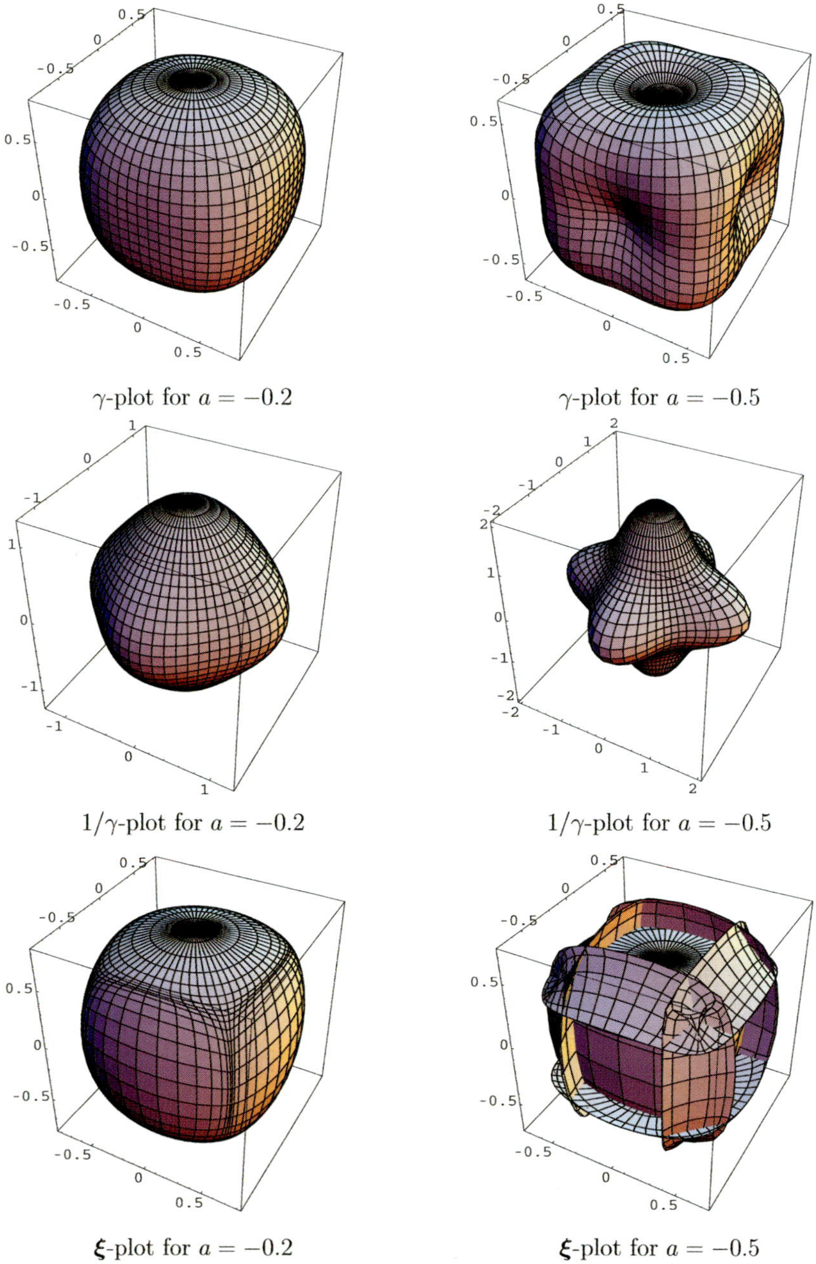

Figure 5. γ-plots, $1/\gamma$-plots and $\boldsymbol{\xi}$-plots for $\gamma = \gamma_0[1 + aQ_4(\hat{\mathbf{n}})]$ for negative values of a. For $a = -0.2$, all orientations appear on the equilibrium shape. For $a = -0.5$, the $1/\gamma$-plot is concave and the $\boldsymbol{\xi}$-plot has ears and flaps that must be truncated to obtain the equilibrium shape, which resembles a cube with curved faces.

2.2. Kinetic Wulff shape

Next, we consider the shape of a crystal that grows in the limiting case for which growth is controlled locally by anisotropic interface kinetics. In this case, heat flow is so fast that the entire system, both solid and liquid, is practically at a uniform temperature T_I. Moreover, relatively speaking, the growth must be so slow that the difference in temperature needed to carry off the latent heat is negligible. Then we adopt a growth law of the form

$$U = \mu(\hat{\mathbf{n}})\Delta T \tag{21}$$

where U is the local normal growth speed, the interface undercooling $\Delta T := T_M - T_I$ is a constant, and we have exhibited the dependence of the kinetic coefficient on interface orientation, characterized by its unit normal vector $\hat{\mathbf{n}}$.[5] The growth law represented by Eq. (21) leads to an exact solution for any initial shape, obtained by updating the initial shape in time by following trajectories of constant orientation, rather than following the movement of each element along its local normal. This method is based on Chernov's [10] use of the method of characteristic curves of partial differential equations, but has also been related to a physical model (so-called kinematic wave theory) by Frank [11]. In terms of the unit vectors $\hat{\mathbf{n}}$, $\hat{\boldsymbol{\theta}}$ and $\hat{\boldsymbol{\varphi}}$ of a spherical coordinate system, for which $U(\theta, \varphi) = \mu(\theta, \varphi)\Delta T$, the velocity along a trajectory of constant orientation is given by

$$\mathbf{V} = U\hat{\mathbf{n}} + \frac{\partial U}{\partial \theta}\hat{\boldsymbol{\theta}} + \frac{1}{\sin\theta}\frac{\partial U}{\partial \varphi}\hat{\boldsymbol{\varphi}}. \tag{22}$$

But since Eq. (22) is independent of time, these trajectories are straight lines.

The directions of these straight lines can be determined geometrically as follows: A polar plot of the reciprocal of U can be represented in spherical coordinates by the equation

$$f(r, \theta, \varphi) := r - \frac{1}{U(\theta, \varphi)} = 0. \tag{23}$$

Its normal is therefore along

$$\nabla f = \hat{\mathbf{r}} + \frac{1}{r}\frac{1}{U^2}\frac{\partial U}{\partial \theta}\hat{\boldsymbol{\theta}} + \frac{1}{r\sin\theta}\frac{1}{U^2}\frac{\partial U}{\partial \varphi}\hat{\boldsymbol{\varphi}} = \frac{\mathbf{V}}{U}, \tag{24}$$

where we have identified $\hat{\mathbf{r}} = \hat{\mathbf{n}}$ and used Eq. (23) to eliminate r.

Examination of Eqs. (22-24) shows that they are analogous to those that arise in the process of finding the equilibrium shape of a particle of fixed volume for anisotropic surface tension, as discussed in the previous subsection. They define a "Wulff shape" that is self-similar to the convex hull of a polar plot of the $\boldsymbol{\xi}$-vector given by Eq. (17). Furthermore, we saw that the direction of $\boldsymbol{\xi}$, as a function of θ and φ, is along the normal of a polar plot of $1/\gamma(\theta, \varphi)$. Therefore, U is analogous to γ and $\mathbf{V}$ is analogous to $\boldsymbol{\xi}$. Thus, when Eq. (22) applies, a crystal grows such that trajectories of surface elements having constant

[5]In writing Eq. (21), we have omitted any correction of the local interface temperature for capillarity (see Eqs. (27) and (28)). As shown by Angenent and Gurtin [8] and illustrated by Yokoyama and Sekerka [9], this is valid except in the vicinity of regions of high curvature and only affects the growth shape locally. More general kinetic laws can be handled by allowing the kinetic coefficient to depend explicitly on T, i.e., $\mu(\hat{\mathbf{n}}, T)$, but we illustrate the concepts here for the simple form of Eq. (21).

orientation will move along straight lines until a shape, analogous to the Wulff shape, is approached asymptotically. This shape is often referred to as the "kinetic Wulff shape," since it is also the Wulff shape for the kinetic coefficient $\mu(\theta, \varphi)$ which is proportional to $U(\theta, \varphi)$. Since the kinetic Wulff shape can have missing orientations and corners, which occur where the $1/\mu(\theta, \varphi)$ plot is concave, the faster growing orientations "grow out" and eventually cease to exist, leaving the crystal shape to be bounded by its more slowly growing orientations. If the kinetic Wulff shape is faceted, the resulting crystal shape is faceted, in agreement with our common impression of a crystal morphology.

3. LONG-RANGE TRANSPORT

In reality, of course, growth limited by the control of local interface kinetics is an idealization, and one must treat long-range transport and concomitant non-isothermal temperature fields. We illustrate this for solidification of a single component crystal from its pure melt, for the simple case in which the density is independent of phase and uniform throughout the system. We proscribe fluid convection, so transport of heat is purely diffusive.[6] The temperature T_L in the liquid (melt) is governed by the equation

$$\nabla^2 T_L = \frac{1}{\kappa_L} \frac{\partial T_L}{\partial t} \tag{25}$$

where $\nabla^2 = \partial^2/\partial x^2 + \partial^2/\partial y^2 + \partial^2/\partial z^2$ is the Laplacian in cartesian coordinates, t is the time and κ_L is the thermal diffusivity, assumed to be constant. Similarly in the solid (crystal) we have

$$\nabla^2 T_S = \frac{1}{\kappa_S} \frac{\partial T_S}{\partial t}. \tag{26}$$

For some crystals, the thermal diffusivity is anisotropic, but for cubic crystals it is still isotropic, so we treat this still realistic case of isotropy for simplicity.

For a solidification problem, solutions to Eq. (25) and Eq. (26) must be joined at the solid-liquid interface, a free boundary of unknown shape and location that is assumed to be a surface of zero thickness where the temperature has the value T_I. One might first assume the temperature of the interface to be the thermodynamic melting point T_M, but a better approximation would be the equilibrium temperature T_E which corrects the melting temperature for the local interface mean[7] curvature K. For isotropic[8] solid-liquid surface tension γ, this correction for capillarity is given by the Gibbs-Thomson equation

$$T_E = T_M - T_M \frac{\gamma}{L_V} K \tag{27}$$

where L_V is the latent heat of fusion per unit volume. But the interface temperature may differ from T_E due to interface motion. This effect can be represented by a kinetic

[6]Of course, fluid convection cannot be proscribed in actuality, and must either be treated or virtually eliminated, e.g., by doing experiments in microgravity.
[7]The mean curvature $K = 1/R_1 + 1/R_2$ where R_1 and R_2 are the principal radii of curvature, signed positive for a spherical crystal.
[8]For anisotropic surface tension, the result is more complicated and involves derivatives of γ, the so-called Herring equation [3,4], equivalent to Eq. (7).

law in which the normal growth speed U is the product of a kinetic coefficient μ and the interface undercooling $T_M - T_I$:

$$U = \mu(T_E - T_I). \tag{28}$$

In general, μ can depend strongly on crystallographic orientation as well as temperature, the latter leading to a nonlinear dependence on interface undercooling $T_E - T_I$. By combining Eq. (27) and Eq. (28) we obtain

$$T_I = T_M - T_M \frac{\gamma}{L_V} K - \frac{U}{\mu}. \tag{29}$$

For very rapid interface kinetics, $\mu \to \infty$ and $T_I \to T_E$, a condition known as local equilibrium. At the moving solid-liquid interface, energy must be conserved, which leads to the additional boundary condition

$$L_V U = (k_S \nabla T_S - k_L \nabla T_L) \cdot \hat{\mathbf{n}} \tag{30}$$

where k_S and k_L are thermal conductivities of solid and liquid and $\hat{\mathbf{n}}$ is the unit outward normal to the crystal.

Eq. (25) and Eq. (26), together with initial conditions, far-field boundary conditions, and the interfacial boundary conditions, Eq. (29) and Eq. (30), constitute a free boundary problem for the shape and location of the crystal-melt interface, and hence for crystal morphology. This is a difficult problem to solve. Analytical solutions are known only for cases in which the interface is assumed to be isothermal, which leads to bodies having the shapes of quadric surfaces (ellipsoids, paraboloids and hyperboloids) [12,13]. Numerical solutions can be obtained, but usually with great difficulty. All of this is complicated, however, by the fact that many solutions to this problem are unstable. Such instability is suggested by directional solidification experiments in which the solid-liquid interface is observed sometimes to be cellular, rather than planar, as well as free growth into supercooled liquid that can result in dendritic forms. This possibility of morphological instability was studied thermodynamically by Tiller, Jackson, Rutter and Chalmers [14] and dynamically by Mullins and Sekerka [15,16] and is taken up in the next subsection.

3.1. Morphological stability

The phenomenon of morphological stability is concerned with the spontaneous change (instability) of shape (morphology) of a surface or interface. More specifically, in the context of crystal growth from the melt, it is concerned with the instability in shape of the crystal-melt interface, which is often called the solid-liquid interface, even though that term might be more appropriate for growth of crystals from liquid solutions.

In order for the question of morphological instability to be well set, it is essential to ask it with respect to a well-defined base state, in which the crystal-melt interface evolves in time according to a well-defined growth law. This growth law ordinarily comes from a solution to the appropriate free-boundary problem (sometimes idealized) in the sense that all transport equations (say, for heat and mass transport) and boundary conditions are satisfied. Then this well-defined base state (sometimes called the unperturbed solution) is tested for stability by introducing a shape perturbation, solving the resulting perturbed problem, and deducing from the solution whether the perturbation will grow or decay in time. If the perturbation is very small, the problem can be linearized in its amplitude; this leads to linear stability theory, which addresses the question of local instability in the sense that only neighboring solutions to the unperturbed solution are tested. If the perturbation is large, the question of global stability may be addressed in principle, but the mathematical analysis of such nonlinear situations is not very tractable, so our knowledge in this area is quite meager.

Another way of viewing the phenomenon of morphological instability relates to a lack of uniqueness to the solution of the growth problem. Given that a base state (unperturbed solution) exists, one might inquire if there are other possible solutions that also obey the transport equations and boundary conditions. Under some growth conditions (choice of material, initial conditions, boundary conditions) the unperturbed solution might be unique, in which case it will be both locally and globally stable. Under other growth conditions, there might be a lack of uniqueness, and other neighboring solutions or solutions that are not neighboring could exist, enabling the possibilities of local and global instability, respectively.

In particular, as some parameter that characterizes the growth conditions is changed, the system might change from a realm in which there is a unique solution to a realm in which multiple solutions exist. Sometimes, when this parameter exceeds some critical value, the solution can bifurcate (split into two branches). It might also split into more than two branches, although the word "bifurcation" rather than "polyfurcation" is still usually used when this occurs. Alternatively, additional solutions can come into existence but appear not to be connected continuously to the unperturbed solution as the "bifurcation parameter" is changed. In any case, instabilities and bifurcations are intimately related, the former being associated with the means of transition from a given solution branch to another branch under conditions for which a bifurcation exists.

We shall begin by treating in detail a linear stability analysis of the very simple case of directional solidification at constant velocity of a pure, single component crystal with an initially planar interface. We will give sufficient detail to allow the reader to follow the calculation of the dispersion relation that determines the conditions for morphological stability of this system. We will then generalize this treatment to include a second component, a solute, but with little detail, emphasizing the results and a comparison with the thermodynamically-based constitutional supercooling criterion. Then we will examine briefly base states with non-planar geometries and provide an introduction to nonlinearities.

3.1.1. Directional solidification, single component

We develop in detail a linear morphological stability analysis for directional solidification of a single component system. Solidification (crystallization) takes place by means of heat conduction in the solid (crystal) and the liquid (melt). We treat the case in which the base state (the unperturbed solution that will be tested for stability) consists of unidirectional solidification by means of motion at constant velocity, V, of a planar solid-liquid interface. This is intended to model a crystal growth configuration in which the crystal is moved relative to sources and sinks of heat in such a way as to keep the heat flow macroscopically one-dimensional. For simplicity of analysis, we proscribe convection in the melt and assume that the temperatures T_S and T_L in the solid and liquid, respectively, obey Laplace's equation

$$\nabla^2 T_{S,L} = 0. \tag{31}$$

This is known as the quasi-steady-state approximation and can be justified [15] near the onset of instability on the basis that (1) the latent heat of fusion, L_V, per unit volume is large compared to the specific heat per unit volume, C_V, times a typical temperature difference in the system, ΔT, and (2) the thermal length, $\alpha_{S,L}/V$ (where $\alpha_{S,L}$ is the smaller of the thermal diffusivity of solid or liquid) is large compared to the largest cross sectional dimension of the system (which will also be the largest wavelength of a perturbation of interest, see below). Under these conditions, the unperturbed solution (which we denote by a zero superscript in parentheses) can be written in the form

$$T_S^{(0)} = T_M + G_S z, \quad z < 0; \tag{32}$$

$$T_L^{(0)} = T_M + G_L z, \quad z > 0 \tag{33}$$

where G_S and G_L are (constant) temperature gradients in the solid and liquid, respectively, T_M is the melting temperature, and z measures distance into the melt from the solid-liquid interface, which is located in the plane $z = 0$ in a coordinate system that moves (uniformly) along with it in the positive z direction. The expressions in Eqs. (32-33) satisfy Eqs. (31), and the gradients G_S and G_L are manifestations of the sources and sinks of heat mentioned earlier; they enable us to account for the far field boundary conditions without dwelling on details. From the principle of conservation of energy (this is a special case of Eq. (30) above) the latent heat of fusion must be carried off by conduction into the solid and the liquid, so G_S and G_L cannot be selected independently of the growth velocity, but must obey the equation

$$L_V V = k_S G_S - k_L G_L \tag{34}$$

where k_S and k_L are the respective thermal conductivities of solid and liquid.

We next reconsider the same problem but with a perturbed interface of the form $z = h(x,t)$ where x is measured (along the unperturbed interface) perpendicular to z, and t is time. (We could consider a three-dimensional problem in which the interface shape also depends on y, but this introduces no essential generalities for a linear stability analysis, so we treat the two-dimensional case for simplicity.) Since the interface is no longer planar, its equilibrium temperature depends on its curvature according to the Gibbs-Thomson equation

$$T_I = T_M - T_M \Gamma K \tag{35}$$

where Γ is a capillary length equal to γ/L_V, where γ is the crystal-melt interfacial free energy (assumed isotropic here for simplicity) and K is the interface curvature. Provided that $h(x,t)$ is small compared to the wavelength, λ, of relevant perturbations, the problem can be linearized in $h(x,t)$. Then we have a principle of superposition, so we can study, without loss of generality, one Fourier component, which amounts to taking

$$h(x,t) = \delta(t)\cos(\omega x) \tag{36}$$

where δ is the amplitude of a perturbation of wavelength $\lambda = 2\pi/\omega$. Consequently, the curvature can be approximated by

$$K \approx -\partial^2 h/\partial x^2 = \delta(t)\,\omega^2 \cos(\omega x) \tag{37}$$

so Eq. (35) becomes

$$T_I(x,t) = T_M - T_M \Gamma \delta(t) \omega^2 \cos(\omega x). \tag{38}$$

Therefore, on the solid-liquid interface,

$$T_S(x, h(x,t), t) = T_L(x, h(x,t), t) = T_I(x,t). \tag{39}$$

The general form for the conservation of energy at the interface is given by Eq. (30), which to first order h can be written

$$L_V \left(V + \frac{\partial h}{\partial t} \right) = k_S \frac{\partial T_S}{\partial z} - k_L \frac{\partial T_L}{\partial z}. \tag{40}$$

We are now in a position to solve the perturbed problem. We let

$$T_S = T_S^{(0)} + T_S^{(1)}, \quad T_L = T_L^{(0)} + T_L^{(1)} \tag{41}$$

where the quantities with superscript (1) are the perturbed temperature fields, which are small corrections, of order h (or equivalently δ) to the unperturbed fields. Since Eqs. (31) are linear, they are also satisfied by the unperturbed fields, and we can take trial solutions in the form

$$T_S^{(1)} = A_S \exp(\omega z) \cos(\omega x), \quad T_L^{(1)} = A_L \exp(-\omega z) \cos(\omega x) \tag{42}$$

where we have chosen solutions that decay as $z \to \mp\infty$, respectively. Here, A_S and A_L are quantities (independent of x and z but weakly dependent on t) that must be determined by the boundary conditions represented by Eqs. (39). In determining these quantities, it is crucial that each term in the boundary conditions be expanded consistently to first order in δ. For instance,

$$\begin{aligned} T_S(x, h(x,t), t) &\approx T_S^{(0)}(0) + h(x,t) \left(\frac{dT_S^{(0}}{dz} \right)_{z=0} + T_S^{(1)}(x,0,t) \\ &= T_M + \delta(t)\, G_S \cos(\omega x) + A_S \cos(\omega x). \end{aligned} \tag{43}$$

In this manner, we find

$$A_S = -\delta(t) \left(G_s + T_M \Gamma \omega^2 \right), \quad A_L = -\delta(t) \left(G_L + T_M \Gamma \omega^2 \right) \tag{44}$$

which is consistent with our assumption that A_S and A_L are of order δ. We can now substitute into Eq. (40), noting, for instance, that

$$\begin{aligned} \left(\frac{\partial T_S}{\partial z} \right)_{z=h(x,t)} &\approx \left(\frac{\partial T_S^{(0)}}{\partial z} \right)_{z=0} + h(x,t) \left(\frac{\partial^2 T_S^{(0)}}{\partial z^2} \right)_{z=0} + \left(\frac{\partial T_S^{(1)}}{\partial z} \right)_{z=0} \\ &= G_S + \omega A_S \cos(\omega x), \end{aligned} \tag{45}$$

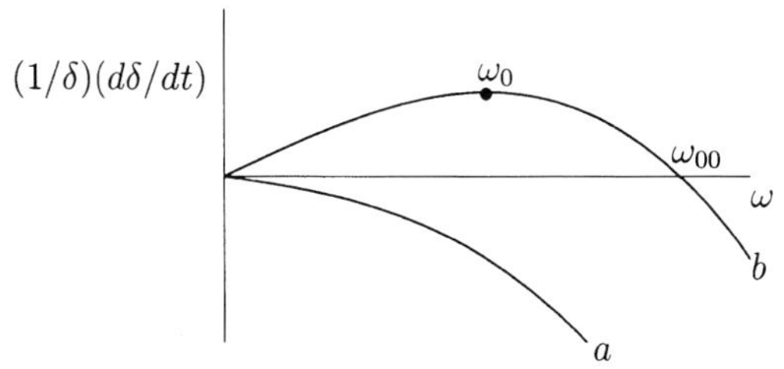

Figure 6. Sketch of Eq. (46) as a function of ω. If $(1/\delta)(d\delta/dt)$ is positive for any value of ω, the interface is unstable. Curve a is for $G^* > 0$, stability, while curve b is for $G^* < 0$, instability. The marginal wavenumber ω_{00} is given by Eq. (49) and the wavenumber ω_0 of the fastest growing perturbation is given by Eq. (55).

to obtain the following differential equation for the perturbation amplitude:

$$\frac{1}{\delta}\frac{d\delta}{dt} = \frac{(k_S + k_L)}{L_V}\omega\left[-T_M\Gamma\omega^2 - G^*\right] \tag{46}$$

where the conductivity weighted average temperature gradient

$$G^* = \frac{k_S G_S + k_L G_L}{k_S + k_L}. \tag{47}$$

Eq. (46) is the main result of our calculation because it determines whether there will be a relative increase (instability) or decrease (stability) in the magnitude of the perturbation amplitude with time. Its right hand side is sketched in Figure 6 as a function of ω. If it is positive for any value of ω, the interface is unstable. The term containing Γ is always negative and represents the stabilizing effect of capillarity (effect of crystal-melt interfacial free energy). The term containing G^* is stabilizing if G^* is positive and destabilizing if G^* is negative, so the interface will be unstable whenever

$$G^* < 0, \quad \text{(instability)}. \tag{48}$$

Under conditions of instability, it follows from Eq. (46) that there is instability for Fourier components that satisfy

$$0 \le \omega \le \omega_{00} = \left[-G^*/\left(T_M\Gamma\right)\right]^{1/2}, \tag{49}$$

or in terms of wavelength,

$$\infty \ge \lambda \ge \lambda_{00} = 2\pi\left[\left(T_M\Gamma\right)/\left(-G^*\right)\right]^{1/2}. \tag{50}$$

For all practical purposes, an instability with wavelength greater than the maximum cross-sectional dimension of the crystal, which dimension we shall denote by H, will not be observable, so Eq. (49) and Eq. (50) should be replaced by

$$2\pi/H \leq \omega \leq \omega_{00} = \left[-G^{*}/\left(T_M \Gamma\right)\right]^{1/2}, \tag{51}$$

and

$$H \geq \lambda \geq \lambda_{00} = 2\pi \left[\left(T_M \Gamma\right)/\left(-G^{*}\right)\right]^{1/2}, \tag{52}$$

which means that a finite negative value of G^{*} is actually needed to obtain an observable instability.

Eq. (46) can be integrated with respect to time to obtain

$$\delta = \delta_0 \exp[\sigma(\omega)t], \tag{53}$$

where $\sigma(\omega)$ is an abbreviation for the right hand side of Eq. (46), i.e.,

$$\sigma(\omega) := \frac{(k_S + k_L)}{L_V}\omega\left[-T_M \Gamma \omega^2 - G^{*}\right], \tag{54}$$

which is known as a dispersion relation, since it relates the exponential rate of increase of a perturbation to its wavenumber, ω. By differentiation of σ with respect to ω, it is found that its maximum value (fastest growing perturbation) occurs for

$$\omega_0 = \omega_{00}/\sqrt{3}, \qquad \lambda_0 = \lambda_{00}\sqrt{3}. \tag{55}$$

For $G^{*} \approx -10$ K/cm, $T_M \approx 1000$ K and $\Gamma \approx 10^{-8}$ cm, we have $\lambda_0 \approx 0.1$ cm, which is probably much smaller than typical values of H for crystals of interest. The corresponding value of σ is about 0.2 s^{-1} for metallic systems, so the perturbations develop in a few seconds, which is small compared to the typical time needed to grow a crystal.

With the aid of Eq. (34) and Eq. (36), the instability criterion represented by Eq. (48) can be rewritten in the form

$$\frac{L_V V + 2k_L G_L}{k_S + k_L} < 0, \quad \text{(instability)} \tag{56}$$

which requires $G_L < 0$, i.e., the interface can only be unstable if it grows into sufficiently supercooled liquid. This is hardly surprising, but only an approximate criterion for several reasons. First, Eq. (52) requires that

$$H \geq 2\pi \left[\left(T_M \Gamma\right)/\left(-G^{*}\right)\right]^{1/2} \tag{57}$$

in order that there be an observable range of unstable wavelengths. Second, this analysis is based on the use of Laplace's equation, according to the quasi-steady-state approximation mentioned above. A more compete analysis based on fully time-dependent equations for heat flow in the moving frame of reference is beyond the scope of this article, but leads, even for equal thermal properties (conductivity k and diffusivity α) in solid and liquid, to a dispersion relation that is much more complicated than Eq. (54). This dispersion relation, given by Eq. (10) of reference [17], yields a quadratic equation for σ that can even have complex roots, although these represent solutions that oscillate as they damp in time. Allowing for moving reference frame effects but assuming a steady state in the moving frame of reference affords some simplification, leading to the criterion [17]

$$G^* < \frac{L_V V}{4k}\left(-3\epsilon^{1/3} + \epsilon\right), \quad \text{(instability)} \tag{58}$$

where

$$\epsilon = \frac{T_M \Gamma k V}{L_V \alpha^2} \tag{59}$$

is a small parameter that characterizes the degree to which the system is stabilized by capillarity. Therefore, according to either Eq. (57) or Eq. (58), a finite negative value of G^* is actually required for instability.

3.1.2. Directional solidification, binary alloy

We next analyze the stability of a planar interface for the problem discussed above with the important modification that solidification takes place from a binary alloy melt. For the sake of simplicity, we treat the case of a dilute alloy melt of solute concentration c that has a phase diagram with straight liquidus and solidus lines, the former of slope m (negative if the solute depresses the freezing point) and the latter of slope m/k, where k is the distribution coefficient ($k < 1$ for $m < 0$ and $k > 1$ for $m > 0$). According to this sign convention, the quantity $m(k-1)$ is always positive. The distribution coefficient k is the ratio, c_S/c , of the concentrations of solute in solid and liquid in local equilibrium at the crystal-melt interface, and we take it to be constant for simplicity.

In order to account for the presence of solute, we will need to modify some of the equations in the previous section and add some new equations. The temperature fields will still be assumed to obey Laplace's equation but the solute field in the liquid will be governed by

$$\nabla^2 c + \frac{V}{D}\frac{\partial c}{\partial z} = 0 \tag{60}$$

which is still within the quasi-steady-state approximation but contains an additional term that arises because of our use of a moving frame of reference. This is an important term because the length D/V is not usually much larger than H and λ, a consequence of the fact that D is typically three to four orders of magnitude smaller than the thermal diffusivities, α_S and α_L. On the other hand, the solute diffusivity of the solid is typically several orders of magnitude smaller than D, so we ignore diffusion in the solid, assuming that whatever is frozen in from the liquid will remain intact. This is why nonuniform solute concentrations in the liquid, related to interface instability, will lead to permanent solute segregation in the crystal.

Eq. (35) must be modified to account for the presence of solute in the melt, resulting in

$$T_I = T_M + mc - T_M \Gamma K \tag{61}$$

through which the temperature and solute fields become coupled. Eq. (30) still applies to guarantee conservation of energy at the interface, but solute must also be conserved at the interface, resulting in

$$(1-k)c\,\mathbf{v}\cdot\hat{\mathbf{n}} = (-D\nabla c)\cdot\hat{\mathbf{n}} \tag{62}$$

where the quantity $(1-k)c$ on the left hand side represents the jump, $c - c_S$, in concentration at the interface. Finally, the solute is assumed to approach a bulk concentration C_∞ as $z \to \infty$.

The unperturbed temperature and concentration fields that characterize the base state (which will be tested for instability) are

$$T_S^{(0)} = T_0 + G_S z, \quad z < 0; \tag{63}$$

$$T_L^{(0)} = T_0 + G_L z, \quad z > 0 \tag{64}$$

and

$$c^{(0)} = C_\infty + C_\infty \left(\frac{1}{k} - 1\right) \exp\left(-Vz/D\right), \tag{65}$$

where

$$T_0 = T_M + mC_\infty/k. \tag{66}$$

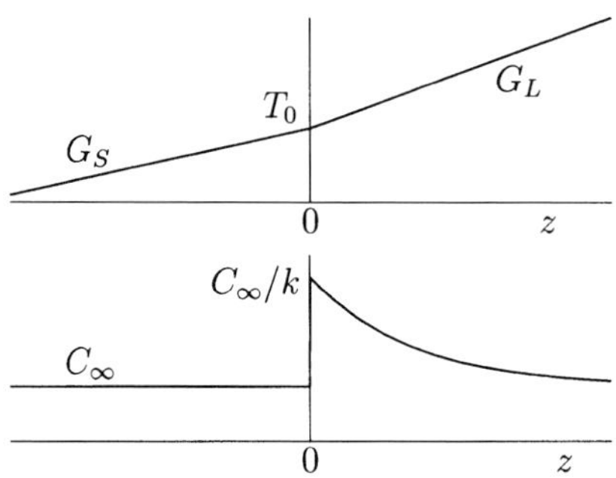

Figure 7. Unperturbed temperature and concentration fields as a function of distance, z, from the interface for steady state unidirectional solidification, as described by Eqs. (63-65). The temperature fields are linear with gradients G_S and G_L, respectively, in solid and liquid. The concentration field decays exponentially to C_∞ with decay length D/V from its value C_∞/k at the interface.

Note that Eqs. (63-64) resemble Eqs. (32-33) except that the interface temperature has been shifted due to the presence of solute, which has concentration C_∞/k at the interface and decays exponentially to C_∞ with decay length D/V. These unperturbed fields are sketched in Figure 7. Eq. (34) still applies, relating the thermal gradients to the growth velocity.

The perturbed fields can be expressed in the form

$$T_S^{(1)} = B_S \exp(\omega z)\cos(\omega x), \quad T_L^{(1)} = B_L \exp(-\omega z)\cos(\omega x) \tag{67}$$

and

$$c^{(1)} = B \exp(-\omega^* z)\cos(\omega x) \tag{68}$$

where

$$\omega^* = \frac{V}{2D} + \left[\omega^2 + \left(\frac{V}{2D}\right)^2\right]^{1/2}. \tag{69}$$

The quantities B_S, B_L and B are then found by satisfying all boundary conditions, leading to a differential equation for δ of the form [16]

$$\sigma = \frac{1}{\delta}\frac{d\delta}{dt} = \frac{2\bar{k}}{L_V}\frac{\omega\left[-T_M\Gamma\omega^2 - G^* + mG_C F_1(\omega)\right]}{\left[1 + (2\bar{k}/L_V V) mG_C F_2(\omega)\right]} \tag{70}$$

where $\bar{k} := (k_S + k_L)/2$ is the average thermal conductivity and

$$G_C = \frac{C_\infty (k-1) V}{Dk} \tag{71}$$

is the gradient of the unperturbed concentration field at the interface (the derivative of Eq. (65) with respect to z, evaluated at $z = 0$),

$$F_1(\omega) = \frac{\omega^* - (V/D)}{\omega^* - (V/D)(1-k)} \tag{72}$$

and

$$F_2(\omega) = \frac{\omega}{\omega^* - (V/D)(1-k)}. \tag{73}$$

Comparison of Eq. (70) with Eq. (46) and Eq. (54) shows that they differ only with respect to the two terms containing G_C. These terms would vanish for a single component for which $C_\infty = 0$. Because of our conventions regarding m and k, the quantity mG_C is always positive. Recalling the definition of ω^* from Eq. (69), we see that the functions $F_1(\omega)$ and $F_2(\omega)$ are also positive for all values of ω. Hence, the denominator of Eq. (70) is always positive, and it remains only to study the sign of the quantity in square brackets in the numerator to test for instability.

If we could be sure that $V/D \ll \omega$, then we would have $F_1(w) \approx 1$ and consideration of the sign of the numerator of Eq. (70) would be similar to that encountered in analyzing Eq. (46), leading to

$$G^* < mG_C, \quad \text{(instability, modified CS)} \tag{74}$$

in place of Eq. (48). We see that the presence of a solute has a destabilizing effect, since instability can now take place even for a positive temperature gradient. Thus, *supercooling*, as a criterion of instability, becomes replaced by *constitutional supercooling* (CS) for alloys, a term coined by Tiller, Rutter, Jackson and Chalmers [14] to denote supercooling with respect to the alloy constitution (phase) diagram. Eq. (74) is known as the modified constitutional supercooling (modified CS) criterion, and can be compared with the CS criterion by using Eq. (71) to rewrite it in the form

$$\frac{G^*}{V} < \frac{mC_\infty(k-1)}{Dk}, \quad \text{(instability, modified CS)} \tag{75}$$

which looks just like the CS criterion except that the CS criterion has G_L in place of G^*. This can lead to significantly different results, since by using the definition of G^*, Eq. (75) can be written in the form

$$\frac{G_L}{V} < -\frac{L_V}{2k_L} + \left(\frac{k_S + k_L}{2k_L}\right)\frac{mC_\infty(k-1)}{DK}, \quad \text{(instability, modified CS)}. \tag{76}$$

Figure 8 shows that a plot of G_L/V versus C_∞ along the stability-instability demar-

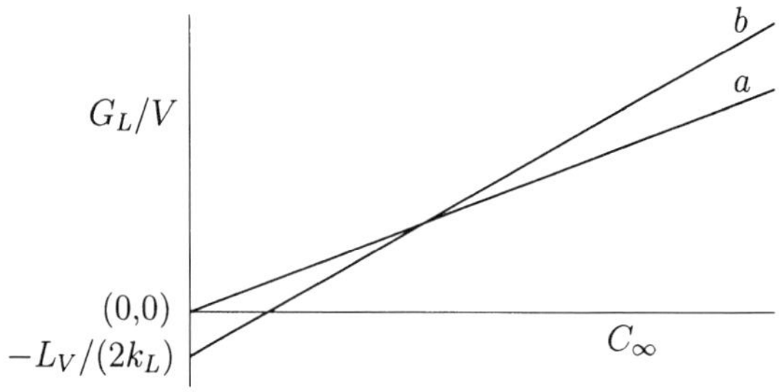

Figure 8. Comparison of the constitutional supercooling (CS) criterion (line a) with the modified CS criterion (line b) corresponding to Eq. (76), illustrated for $k_S = 2k_L$. Because of a finite latent heat, the modified CS line does not pass through the origin.

cation results in a straight line with a finite intercept and a slope that is different by a factor of $(k_S + k_L)/2k_L$, compared to CS. For small V, large G_L and values of D that are often uncertain by about a factor of two, these differences could go undetected; however, accurate measurement of liquid diffusivities in microgravity would surely reveal them. We see, therefore, that the dynamical theory of morphological stability is in qualitative agreement with CS for small V but can differ quantitatively depending on the relative thermal conductivities of solid and liquid and the latent heat.

We now return to analyze the full dispersion relation, Eq. (70), the numerator of which is sketched as a function of ω in Figure 9. In general, instability first takes place at a value of ω that can be comparable to V/D, so the approximation $F_1(\omega) \approx 1$, which leads to the modified CS criterion, is not always warranted. In fact, instability first takes place at a critical value of ω corresponding to a tangency condition at which both the numerator and its derivative with respect to ω vanish simultaneously. This leads [18] to a cubic equation that must be solved to find the critical value of ω which must then be substituted into $F_1(\omega)$ to calculate the stability criterion. It turns out that this cubic equation depends on two dimensionless parameters, the distribution coefficient k and the parameter

$$A = k\frac{\Gamma V}{D}\frac{T_M V}{mG_C D} = \frac{k^2}{(1-k)}\frac{\Gamma V}{D}\frac{T_M}{(-m)c_\infty}. \tag{77}$$

Consequently, the results can be written in the form

$$G^* < mG_C\mathcal{S}(A,k), \quad \text{(instability)} \tag{78}$$

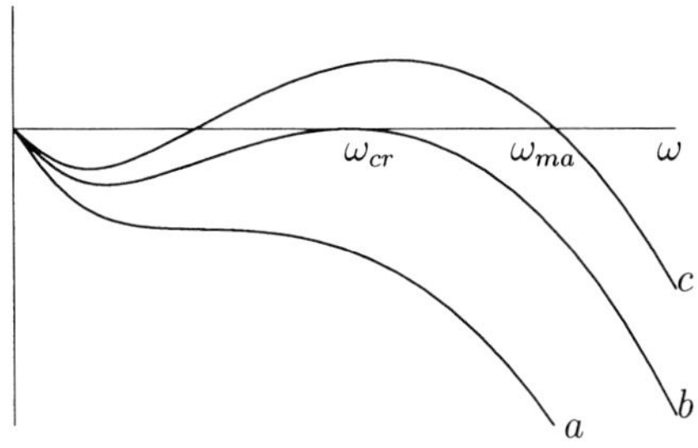

Figure 9. Sketch of the numerator of Eq. (70) versus ω for stability (curve a), marginal stability (curve b) and instability (curve c). The condition of marginal stability corresponds to a double root at ω_{cr} at which the function and its derivative with respect to ω vanish simultaneously. The value ω_{ma} is the wavenumber for marginal stability, above which there is stabilization by capillarity.

where the function $\mathcal{S}(A,k)$ varies from 1 to 0 as A varies from 0 to 1. The function $\mathcal{S}(A,k)$ has been calculated and is shown graphically in references [17] and [18]. It gives rise to additional stabilization of the system due to capillarity. See Figure 10 below for a plot of $\mathcal{S}(A,1/2)$ which is typical for other values of k as well. Except at very high growth speeds, $A \ll 1$, due to the smallness of Γ, and so $S \approx 1$, leading to the modified CS criterion, Eq. (74).

A special case arises for $k = 1/2$ because the cubic equation becomes quite simple; we can use this case for illustrative purposes since it displays all of the important features of the general case. Thus for $k = 1/2$, the critical value of ω and the corresponding critical wavelength are

$$\omega_{cr} = \frac{V}{2D}\frac{\left(1 - A^{2/3}\right)^{1/2}}{A^{1/3}}, \quad \lambda_{cr} = 4\pi\frac{D}{V}\frac{A^{1/3}}{\left(1 - A^{2/3}\right)^{1/2}} = \frac{\lambda_0}{A^{2/3}(1 - A^{2/3})^{1/2}}, \tag{79}$$

where $\lambda_0 = 2\pi\gamma T_M/(-mC_\infty)$, and

$$\mathcal{S}\left(A, \frac{1}{2}\right) = 1 - \frac{3}{2}A^{1/2} + \frac{1}{2}A. \tag{80}$$

Figure 10 shows the critical wavelength λ_{cr} for instability and the function $\mathcal{S}$ as functions of $\log_{10} A$ for $k = 1/2$.

For an alloy of fixed composition, we can examine the effect of changing V. For small V, A will be small and

$$\lambda_{cr} = 4\pi\frac{D}{V}A^{1/3} \propto \frac{1}{V^{2/3}}, \quad \mathcal{S} \approx 1, \tag{81}$$

whereas for large V, A is no longer small, and as $A \to 1$, we find

$$\lambda_{cr} = 4\pi\frac{D}{V}\frac{1}{(1 - A)^{2/3}} \to \infty, \quad \mathcal{S} \approx 0. \tag{82}$$

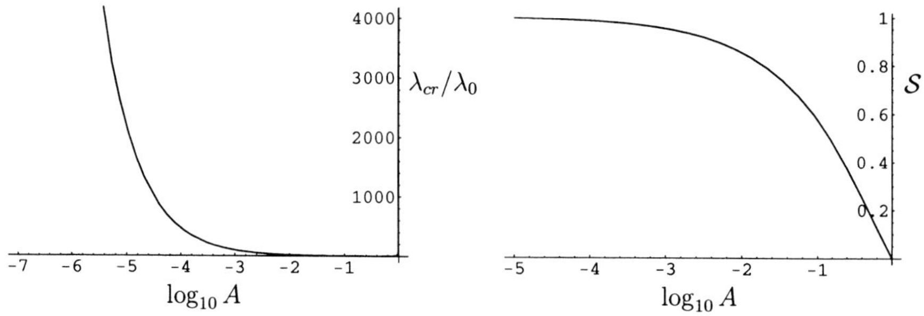

Figure 10. Plots of the critical wavelength λ_{cr} for instability from Eq. (79) and the function $\mathcal{S}$ from Eq. (80) as functions of $\log_{10} A$ for $k = 1/2$. For $A \to 1$, $\lambda_{cr} \to \infty$ according to Eq. (82), but this occurs over such a narrow range of A that it is invisible on the graph.

Thus as $A \to 1$, the instability criterion given by Eq. (78) tends to that for a single component, Eq. (48); i.e., the destabilizing effect of solute is completely negated by capillarity. For $A > 1$, analysis [18] shows that Eq. (48) becomes the criterion for instability. Thus, $A \geq 1$ was originally called absolute stability [16] because the analysis was done for positive G^*, whereas in reality, it is a condition for which the destabilizing effect of solute is completely negated and the criterion for instability becomes actual supercooling (rather than some kind of constitutionally related supercooling).

The overall picture at fixed positive G^* is therefore the following, as illustrated in Figure 11: At low V, the modified CS criterion, Eq. (75), applies and increasing C_∞ makes the system more unstable. As V increases, capillary stabilization comes into play, and stabilization becomes possible for higher values of C_∞, eventually corresponding to the criterion $A = 1$ at large V (where C_∞ is proportional to V along the stability - instability demarcation). Thus, for a fixed value of C_∞ above the minimum of a curve in Figure 11, the interface first becomes unstable with increasing V and then restabilizes for sufficiently large V. Below such a minimum, the interface is stable for all V.

3.1.3. Non-planar base states

It is possible to carry out, at least within the quasi-steady-state approximation, morphological stability analyses for base states in which the solid-liquid interface is non-planar and moves at variable velocity. Easy cases to treat are spheres [15,19] and circular cylinders [20,21] because these are shapes of uniform mean curvature ($2/R$ and $1/R$, respectively, where R is the radius). Thus, the Gibbs-Thomson equation can be satisfied by a uniform shift of the interface temperature (which, however, depends on time, because R is a function of t). We note the following differences in analyzing the morphological instability of a sphere growing into a pure supercooled melt:

- The base state depends on time. According to the quasi-steady-state approximation, an unperturbed sphere of a single component, growing from its nucleation radius in a supercooled melt at temperature T_∞, attains a maximum growth velocity at twice this radius [22] and settles asymptotically into a behavior in which R is proportional

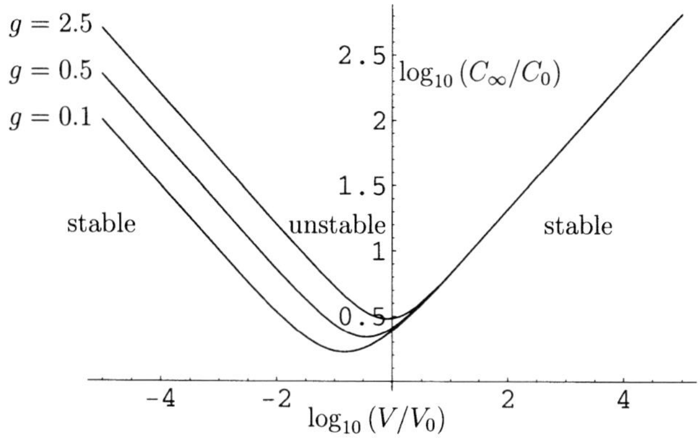

Figure 11. Log-log plot of the critical concentration, C_∞, versus V at three values of G_L for $k = 1/2$, $C_0 = L_V D/[(k_S + k_L)(-m)]$ and $V_0 = 2L_V D^2/[T_M\Gamma(k_S + k_L)]$. At low V, the modified CS criterion, Eq. (74), applies and the curves depend on G_L through the parameter $g := k_L(k_S+k_L)T_M\Gamma G_L/(DL_V)^2$. As V increases, capillary stabilization comes into play, and stabilization becomes possible for higher values of C_∞. For a fixed value of C_∞ above the minimum of a curve, the interface first becomes unstable with increasing V and then restabilizes for sufficiently large V.

to $t^{1/2}$. This behavior is caused by a reduction in the effective supercooling to $T_M - 2T_M\Gamma/R - T_\infty$ from the nominal supercooling $T_M - T_\infty$.

- The perturbed sphere can be studied by means of eigenfunctions (known as spherical harmonics) of the angular part of the Laplacian operator, rather than the cosines used in the planar case [15]. In other words, the geometry dictates the eigenfunctions.

- As the sphere grows, it becomes unstable with respect to a given eigenfunction at a critical radius. This resembles a nucleation phenomenon, but it can be understood by comparison with the planar case as a weakening of the stabilizing influence of capillarity, relative to the destabilizing effect of a negative temperature gradient, as growth proceeds.

- As the sphere grows, it becomes successively unstable to harmonics of higher index (the higher the index, the more nodes). The lower order harmonics become unstable at a sphere of radius that is typically 10-20 times the nucleation radius.

- Within the quasi-steady-state approximation, the time development of the perturbations is algebraic, rather than exponential as in the planar case, because of the dependence on time of the underlying base state [19].

In polar coordinates, a perturbed shape can be represented in the form

$$r = R + \delta\, Y_{\ell m}(\theta, \varphi) \tag{83}$$

where $R(t)$ is the time-dependent radius of the unperturbed sphere, $\delta(t)$ is the time-dependent amplitude of a perturbation, and $Y_{\ell m}(\theta, \varphi)$ is a spherical harmonic of order ℓ, m, where $\ell = 1, 2, 3 \cdots$ and m is an integer in the range $-\ell \leq m \leq \ell$. Local equilibrium ($\mu \to \infty$) at the solid-liquid interface is assumed, so Eq. (35) takes the form

$$T_I = T_M[1 - \Gamma K] = T_M \left[1 - \frac{2\Gamma}{R} - \frac{\Gamma\delta}{R^2}(\ell - 1)(\ell + 2)\, Y_{\ell m}(\theta, \varphi)\right] \tag{84}$$

where $\Gamma = \gamma / L_V$ is a capillary length and higher order terms in δ/R have been neglected. In Eq. (84), the term $2\Gamma/R$ comes from the unperturbed sphere and the term in $Y_{\ell m}(\theta, \varphi)$ comes from the perturbation.

The analysis proceeds by solving for the temperature fields in solid and liquid to first order in δ. If the interface were an isotherm, the isotherms of these fields would become distorted near the perturbation into shapes that resemble the perturbation. This would tend to enhance the growth of the perturbation. But this distortion of the isotherms is mitigated by the fact that the interface is not an isotherm, as represented by the term containing $Y_{\ell m}(\theta, \varphi)$ in Eq. (84), and this leads to stabilization. Detailed analysis of the sign of the quantity $(1/\delta)d\delta/dt$ leads to the conclusion that the sphere is unstable whenever $\ell > 1$ and

$$R > \frac{R^*}{2}\left[(\ell + 1)(\ell + 2) + \ell(\ell + 2)\frac{k_S}{k_L} + 2\right] \tag{85}$$

where $R^* := 2\Gamma T_M/(T_M - T_\infty)$ is the nucleation radius, in which T_∞ is the far field temperature. Thus, the sphere becomes unstable to an ellipsoidal shape ($\ell = 2$) whenever $R/R^* > 7 + 4k_S/k_L$ and to more undulating shapes at larger values of R, corresponding to larger values of ℓ. If it were not for capillarity, i.e., if $\Gamma = 0$, the sphere would be unstable at all sizes to perturbations of all wavelengths.

The criterion for instability can be written in an alternative way in terms of the magnitude $-G_L = [T_M(1 - R^*/R) - T_\infty]/R$ of the (negative) temperature gradient at the solid-liquid interface of the unperturbed sphere. Thus, instability occurs whenever

$$-G_L > \frac{T_M\Gamma}{R^2}\left[(\ell + 1)(\ell + 2) + \ell(\ell + 2)\frac{k_S}{k_L}\right]. \tag{86}$$

Eq. (86) supports the interpretation that growth into a supercooled melt ($-G_L > 0$) is destabilizing while capillarity (term in Γ) is stabilizing. For large ℓ, one can interpret $\lambda \sim 2\pi R/\ell$ as the wavelength of a perturbation, in which case Eq. (86) at marginal stability (replace $>$ by $=$) and for $k_S = k_L$ yields

$$\lambda \sim 2\pi \left[\frac{2T_M\Gamma}{|G_L|}\right]^{1/2}. \tag{87}$$

Thus the scale of the instability is the geometric mean of a capillary length Γ and a thermal length $T_M/|G_L|$. Another form of Eq. (87) can be obtained in terms of the growth velocity $V = |G_L|k_L/L_V$ and the capillary length $d_0 := \Gamma T_M c_V/L_V$ where c_V is the heat capacity per unit volume. The result is

$$\lambda \sim 2\pi \left[2d_0\frac{\kappa_L}{V}\right]^{1/2} \tag{88}$$

which is essentially the geometric mean of the capillary length d_0 and the thermal length κ_L/V. This is a general characteristic of morphological instability phenomena, independent of the shape of the unperturbed body. Thus Langer and Müller-Krumbhaar [23] first proposed that a dendrite tip radius ρ should be about equal to λ, which leads to

$$\sigma^* := \frac{2d_0\kappa_L}{\rho^2 V} = \frac{1}{(2\pi)^2} \approx 0.025. \tag{89}$$

It is amazing that Eq. (89) is in pretty good agreement [24] with experiment, although the value of the numerical constant is surely fortuitous. It turns out that the scaling suggested by Eq. (89) is essentially correct, but the value of σ^* depends on a more delicate analysis, such as that provided by microscopic solvability theory [25–27].

Morphological stability theory is very general, and can also be extended to include departures from local equilibrium (interface kinetics) as well as anisotropy. For a comprehensive review, see the article by Coriell and McFadden [28].

3.1.4. Nonlinearities

We next turn briefly to nonlinearities. For conditions of instability, linear stability theory predicts that certain perturbations will grow. Returning to Eq. (70) as a basis for discussion, exponential growth (see Eq. (53)) of a sinusoidal perturbation of amplitude $\delta(t)$ will soon result in a situation in which $|\delta(t)/\lambda| \sim 1$, even though $|\delta(0)/\lambda| \ll 1$. Thus, previously neglected terms that are higher order in $\delta(t)/\lambda$ will become important. These nonlinearities will cause the Fourier modes to couple, and the time evolution of the interface will become quite different from that predicted by linear stability theory.

One possibility would be for the interface to restabilize but with a nonplanar periodic shape. In two dimensions, such shapes would be called "bands" or "roll cells," whereas in three dimensions, shapes known as "cells" or "nodes," that can sometimes form hexagonal arrays, are possible. If the interface restabilizes with a nonplanar shape, steady state crystal growth will still be possible, but for alloys there will be solute segregation (microsegregation) in the direction perpendicular to the growth direction. Alternatively, the system might not restabilize into a nonplanar steady state, but continue to branch, forming a dendritic (treelike) structure that might display some characteristic length scales (say, primary, secondary and tertiary arm spacings) but still have some stochastic characteristics.

To gain a little more insight into nonlinearities, we discuss briefly weakly nonlinear stability theories, first introduced to this problem by Wollkind and Segel [29]. Such theories are supposed to be valid near marginal stability, at which only a single Fourier component becomes unstable. The basic idea is that only the fundamental component is important (although it couples with itself through its harmonics) and δ is still small enough to enable a series expansion consisting of a few terms. Thus, Eq. (70) is replaced by an equation of the form

$$\frac{d\delta}{dt} = a\delta + b\delta^3 \tag{90}$$

which is known as a Landau equation. The parameters a and b are calculated by expanding all equations and boundary conditions to third order in δ, but the details are beyond the scope of this article. (The quadratic term in δ is missing because the problem has a

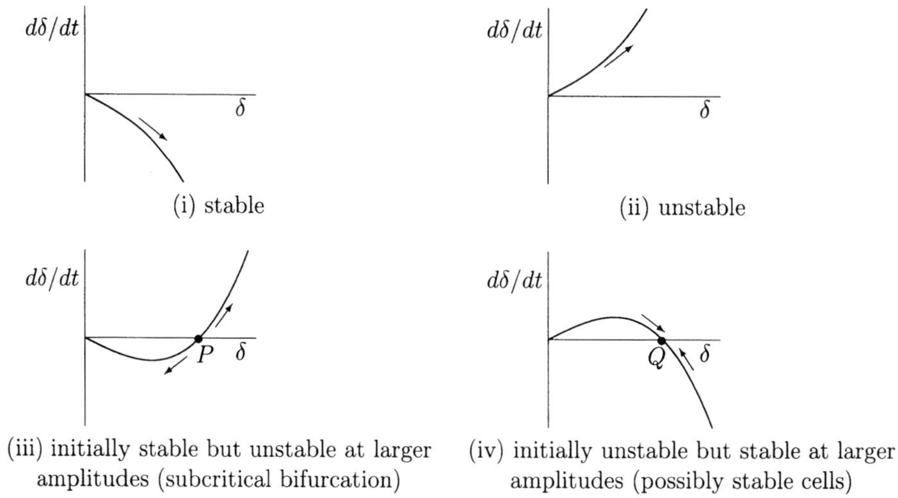

Figure 12. Plots of $d\delta/dt = a\delta + b\delta^3$ versus δ for the four cases discussed in the text. (i) $a < 0,\ b < 0$; (ii) $a > 0,\ b > 0$; (iii) $a < 0,\ b > 0$; (iv) $a > 0,\ b < 0$. The arrows along the curves point in the direction of increasing time. The state corresponding to point P is unstable whereas the state corresponding to point Q is stable against changes in the amplitude δ, so Q represents a possibly stable cellular steady state.

translational symmetry such that $x \to x + \pi$ is the same as $\delta \to -\delta$.) If $b = 0$ in Eq. (90), the result is Eq. (70) with the parameter a in place of σ. Thus $a < 0$ corresponds to linear stability and $a > 0$ corresponds to linear instability. Since b can have either sign, there are four possibilities as follows:

(i) $a < 0,\ b < 0$; the system is stable for all δ.

(ii) $a > 0,\ b > 0$; the system is unstable for all δ.

(iii) $a < 0,\ b > 0$; the system is stable for small δ but becomes unstable for sufficiently large δ. Thus, there is a threshold value of δ for instability, and if this threshold is exceeded, instability can take place prior to conditions for linear instability. This is called a subcritical bifurcation.

(iv) $a > 0,\ b < 0$; the system is unstable for small δ but becomes stable for sufficiently large δ. There is no threshold value for instability, which first takes place under conditions for linear instability. This is called a supercritical bifurcation.

The cases (i-iv) above are illustrated in Figure 12 on plots of $d\delta/dt$ versus δ. The arrows on the curves are drawn in the direction of increasing time. We see that cases (iii) and (iv) admit the possibility that $d\delta/dt = 0$ for a finite value of δ, namely

$$\delta_{nl} = (-a/b)^{1/2} \tag{91}$$

which represents a nonplanar steady state. In case (iii), subcritical bifurcation, this nonplanar steady state is unstable (the arrows lead away from it) whereas in case (iv), it is stable with respect to changes in δ (the arrows lead toward it). Thus, for case (iv), this non-planar steady state can represent a cellular interface of sinusoidal shape, to which the system might restabilize subsequent to the morphological instability of the planar interface.

Of course Eq. (90) represents only the first two terms of an expansion, so a proper description of cellular interfaces requires handling the full nonlinear problem. Cells that show distinct deviations from sinusoidal shapes, along with their concomitant solute segregation, have been calculated by numerical methods by McFadden and Coriell [30]. Moreover, deep cells have been calculated by Ungar and Brown [31] by special mapping techniques that allow the deep grooves between the cells to be treated differently than the regions near the cell tips. These results display important phenomena such as period doubling and joining up of branches that emanate from the planar interface solution with different wavelengths. Moreover, for cases in which a weakly nonlinear expansion such as Eq. (90) would predict a subcritical bifurcation and an unstable nonplanar solution, a fully nonlinear analysis often enables the calculation of stable cells of larger amplitude.

4. PHASE FIELD MODEL

Morphological stability analysis shows us that computations of crystal morphology require the solution of a more complex free boundary problem in which the effects of capillarity must be included. Neglecting these effects gives rise to solutions for idealized shapes that are unstable on all length scales of continuum models. Adding to this the fact that the surface tension is actually anisotropic and that anisotropic interface kinetics can give rise to shapes related to this anisotropy as well, we were faced with a formidable free boundary problem. This provided motivation for the phase field model in which all of these effects could be incorporated in a more tractable way.

4.1. Basis of the model

In the phase field model [32–34], the sharp interface is replaced by a diffuse interface and an auxiliary parameter φ, the phase field, is introduced to indicate the phase. The quantity φ is a continuous variable that takes on constant values in the bulk phases, say 0 in the solid and 1 in the liquid, and increases from 0 to 1 over a thin layer, the diffuse interface. A partial differential equation is formulated to govern the time evolution of φ. It incorporates the interfacial physics of the problem in such a way that the diffuse interface has an excess energy, which gives rise, for a sufficiently thin interface, to a surface tension γ. Bending of the diffuse interface automatically introduces capillarity, Eq. (27). A diffusivity related to the time evolution of φ gives rise to a linear kinetic law, Eq. (21). Both the surface tension and the kinetic coefficient can be made to be anisotropic. The partial differential equation for φ is coupled to other equations that determine the relevant fields that govern transport, temperature in the case of energy transport and composition in the case of solute transport.

We indicate briefly the general procedure for constructing the phase field equations for solidification of a single component from its pure melt. For simplicity of presentation, we assume that all quantities are isotropic, that the density is uniform in solid and liquid, and that there is no convection in the liquid. We postulate that the internal energy $\mathcal{U}$ and the entropy $\mathcal{S}$ in any subvolume $\mathcal{V}$ of our system are given by

$$\mathcal{U} := \int_{\mathcal{V}} [u + \frac{1}{2}\varepsilon_u^2|\nabla\varphi|^2]\, d^3x \tag{92}$$

and

$$\mathcal{S} := \int_{\mathcal{V}} [s(u,\varphi) - \frac{1}{2}\varepsilon_s^2|\nabla\varphi|^2]\, d^3x \tag{93}$$

where $u(\mathbf{r}, t)$ is the local density of internal energy, $\varphi(\mathbf{r}, t)$ is the phase field, $\mathbf{r}$ is the position vector, t is time, and ε_u^2 and ε_s^2 are constants. We regard these expressions to be *functionals* of $u(\mathbf{r}, t)$ and $\varphi(\mathbf{r}, t)$; in other words, $\mathcal{U}$ and $\mathcal{S}$ depend on functions, rather than just variables. The quantities u and s are internal energy and entropy densities that pertain to a homogeneous phase having a uniform value of φ. The terms involving $|\nabla\varphi|^2$ are corrections that are only important in the diffuse interface where φ changes from its value $\varphi = 0$ in bulk solid to its value $\varphi = 1$ in bulk liquid. The term $\frac{1}{2}\varepsilon_u^2|\nabla\varphi|^2$ is called a gradient energy and $-\frac{1}{2}\varepsilon_s^2|\nabla\varphi|^2$ is called a gradient entropy. Together they give rise to a gradient free energy, such as used in Cahn-Hilliard theory [35].

Dynamical equations are based on the functionals given by Eq. (92) and Eq. (93) and the concepts of local energy conservation and local entropy production. Since energy is conserved,

$$\dot{U}_{\text{prod}} := \frac{d}{dt}\mathcal{U} + \int_{\mathcal{A}} [\mathbf{q} - \varepsilon_u^2\dot{\varphi}\nabla\varphi \cdot \hat{\mathbf{n}}] d^2x = 0. \tag{94}$$

The rate of entropy production is

$$\dot{S}_{\text{prod}} := \frac{d}{dt}\mathcal{S} + \int_{\mathcal{A}} [\frac{\mathbf{q}}{T} + \varepsilon_s^2\dot{\varphi}\nabla\varphi \cdot \hat{\mathbf{n}}] d^2x \geq 0. \tag{95}$$

Here, $\mathcal{A}$ is the area surrounding the arbitrary subvolume $\mathcal{V}$, $\hat{\mathbf{n}}$ is its unit outward normal, and a dot above a variable denotes partial differentiation with respect to time. The vector $\mathbf{q}$ is the classical heat flux and $\mathbf{q}/T$ is the classical entropy flux. The additional fluxes in the area integrals are nonclassical fluxes associated with the gradient energy and gradient entropy corrections. These nonclassical fluxes arise whenever elements of the diffuse interface enter or leave a control volume, as discussed by Wang et al. [36]. From Eq. (94) we obtain

$$\int_{\mathcal{V}} [\dot{u} + \nabla \cdot \mathbf{q} - \varepsilon_u^2\dot{\varphi}\nabla^2\varphi]\, d^3x = 0. \tag{96}$$

Since Eq. (96) holds in every arbitrary subvolume, the integrand itself must vanish and we obtain

$$\dot{u} + \nabla \cdot \mathbf{q} - \varepsilon_u^2\dot{\varphi}\nabla^2\varphi = 0. \tag{97}$$

From Eq. (95) we obtain

$$\int_{\mathcal{V}} \left[\mathbf{q} \cdot \nabla \left(\frac{1}{T}\right)\right] d^3x + \int_{\mathcal{V}} \left[\left(\frac{\partial s}{\partial \varphi}\right)_u + \frac{\varepsilon_f^2}{T}\nabla^2\varphi\right] \dot{\varphi}\, d^3x \geq 0, \tag{98}$$

where $\varepsilon_f^2 = \varepsilon_u^2 + T\varepsilon_s^2$. Eq. (98) can be satisfied for every subvolume $\mathcal{V}$ by assuming linear constitutive laws of the form

$$\mathbf{q} = M_u \nabla \left(\frac{1}{T}\right) = -k\nabla T \tag{99}$$

where $M_u > 0$ and $k = M_u/T^2$ is the thermal conductivity, and

$$\tau\dot{\varphi} = \left(\frac{\partial s}{\partial \varphi}\right)_u + \frac{\varepsilon_f^2}{T}\nabla^2\varphi \tag{100}$$

where τ is a positive time constant that is related to the interface kinetic coefficient in Eq. (108).

Eq. (100) is the equation for the time evolution of the phase field. Substitution of Eq. (99) into Eq. (97) leads to a compatible energy equation, essentially an equation for time evolution of the temperature. This becomes clear once explicit functions for u and s in terms of independent variables T, φ are specified. For example, we could take an internal energy density of the form

$$u(T,\varphi) = u_0 + c_V(T - T_M) + L_V p(\varphi) + \frac{W_u}{2} g(\varphi) \tag{101}$$

where u_0 is a constant, c_V is a constant heat capacity per unit volume, L_V is a constant latent heat per unit volume, $g(\varphi) = \varphi^2(1-\varphi)^2$ is a double well potential, W_u is a constant strength parameter for the double well, and $p(\varphi) = \varphi^3(10 - 15\varphi + 6\varphi^2)$ is a smooth function of φ that increases monotonically from $p(0) = 0$ to $p(1) = 1$. We could also take a Helmholtz free energy density $f = u - Ts$ of the form

$$f(T,\varphi) = u_0 - Ts_0 + c_V(T - T_M) - c_V T \ln(T/T_M) + L_V (1 - T/T_M)\, p(\varphi) + \frac{W_f}{2} g(\varphi) \tag{102}$$

where s_0 is a constant and $W_f = W_u + TW_s$, where W_s is a constant. Then, by means of a thermodynamic equation, we deduce that

$$\left(\frac{\partial s}{\partial \varphi}\right)_u = -\frac{1}{T}\left(\frac{\partial f}{\partial \varphi}\right)_T = L_V\left(\frac{1}{T_M} - \frac{1}{T}\right) p'(\varphi) - \frac{W_f}{2} g'(\varphi), \tag{103}$$

where the primes on p and g denote differentiation. We therefore obtain the phase field equations

$$c_V\dot{T} + L_V\dot{p}(\varphi) = k\nabla^2 T + \varepsilon_u^2\dot{\varphi}\nabla^2\varphi - \frac{W_u}{2}\dot{g}(\varphi) \tag{104}$$

and

$$\tau\dot{\varphi} = \frac{\varepsilon_f^2}{T}\nabla^2\varphi + L_V\left(\frac{1}{T_M} - \frac{1}{T}\right) p'(\varphi) - \frac{W_f}{2} g'(\varphi). \tag{105}$$

The term containing L_V in Eq. (104) gives rise to latent heat evolution at the interface and incorporates the boundary condition Eq. (30) of the sharp interface model. The term containing L_V in Eq. (105) provides a bias to the double well potential represented by $g(\varphi)$ and this causes the crystal to melt or grow, depending on the sign of $T - T_M$. By means of asymptotic analysis [37] in the limit of a very thin interface, one can show that the interface thickness (as φ varies from 0.05 to 0.95) is about 6ℓ where

$$\ell = \frac{\varepsilon_f}{\sqrt{W_f}}. \tag{106}$$

The surface tension is given by

$$\gamma = \frac{W_f \ell}{6} = \frac{\varepsilon_f \sqrt{W_f}}{6} \tag{107}$$

and the kinetic coefficient by

$$\mu = \frac{6 L_V \ell}{T_M^2 \tau}. \tag{108}$$

Eqs. (106-108) can be used to relate the parameters of the model to physical properties and to the thickness of the diffuse interface, a computational parameter. Asymptotic analysis for a thicker interface [38] leads to a somewhat different relationship of model parameters to physical properties. Anisotropy in γ and μ can be introduced by allowing ε_f and τ to depend on $\hat{\mathbf{N}} = \nabla\varphi/|\nabla\varphi|$, which plays the role of a unit normal vector in the interfacial region. In this case, Eq. (105) must be replaced by a more complicated equation that contains derivatives of ε_f. Details and examples are presented in [39–42], where these anisotropies are related to the $\boldsymbol{\xi}$-vector formalism of Hoffman and Cahn.

5. DISCUSSION AND CONCLUSIONS

We conclude with a list of remarks and generalizations, including a few examples of computed results.

- Since the phase field model incorporates all boundary conditions at the crystal-melt interface, including anisotropy, the morphologies computed from it display most of the morphological features observed experimentally. This is evident in Figure 13, courtesy of Wheeler, Murray and Glicksman [43], for solidification of ice in two dimensions at low supercoolings. Cleaving of dendrites, sidebranching and coarsening are evident.

- Convection in the melt can be very important and have profound effects on morphological stability phenomena [44]. In general, morphological stability phenomena display directionality that depends on the details of the flow pattern [45,46]. Such flows also lead to perturbations that travel along the flow as they grow. An extensive review of the effects of flow on morphological stability has been given by Davis [47].

- Anisotropies of bulk (e.g., thermal conductivity [48]) and surface (e.g., crystal-melt interfacial free energy [49]) properties lead to changes in the stability-instability threshold as well as to preferred directions, as would be expected, and these can influence the geometry of cells and their concomitant solute segregation [50]. Recent phase field modeling of Uehara and Sekerka [51] and Debierre et al. [52] exhibits faceting and missing orientations on growing crystals and dendrites.

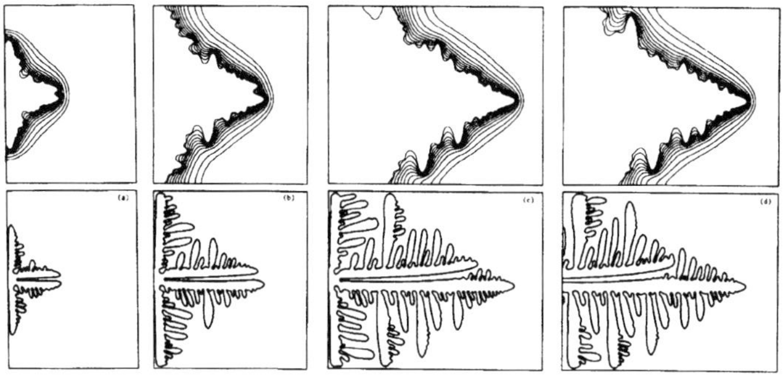

Figure 13. Temperature (top) and interface (bottom) of a dendrite computed from the phase field model in two dimensions by Wheeler, Murray and Glicksman [43]. Cleaving, sidebranching and coarsening are evident. Note in the lower right frame how one of the secondary branches cuts off a primary branch.

- Departures from local equilibrium at the solid-liquid interface (growth kinetics) can be large and highly anisotropic, and can affect the resulting instabilities and subsequent patterns, especially for rapid solidification [53]. Moreover, nonequilibrium solute segregation (solute trapping) can have large effects on morphological stability at high growth velocities [54].

- The phase field model also applies to the analogous problem of isothermal precipitation. It has been generalized by Wheeler, Boettinger and McFadden [55–58] and others [59–61] to apply to the solidification of alloys, in which case there are coupled partial differential equations for time evolution of the phase field, the temperature and the composition. Computations based on the alloy phase field model have led to a much better understanding of solute segregation and pattern selection at cellular interfaces and during dendritic growth [62–66]. Figure 14 from the work of Bi [67] shows solute segregation and the trapping of liquid droplets in cell grooves during the directional solidification of a binary alloy. Figure 15 from the work of George and Warren [68] shows dendrite morphology and solute segregation in three dimensions for solidification of an alloy and interfacial free energy having cubic symmetry of the form of Eq. (19). The influence of crystallography on the side fins of the six main ⟨100⟩ dendrites is truly spectacular.

- The phase field model has also been extended to include hydrodynamics, both for pure materials and for alloys [69–73]. Computations including hydrodynamics are difficult, but results are beginning to emerge [74,75]. Hydrodynamics has also been added to the phase field model by means of hybrid methods by Tönhardt and Amberg [76–78] and Beckermann et al. [79] and has led to somewhat more tractable models [80] for computing solidification microstructures.

- Solution adaptive grids have been used to facilitate phase field modeling in two

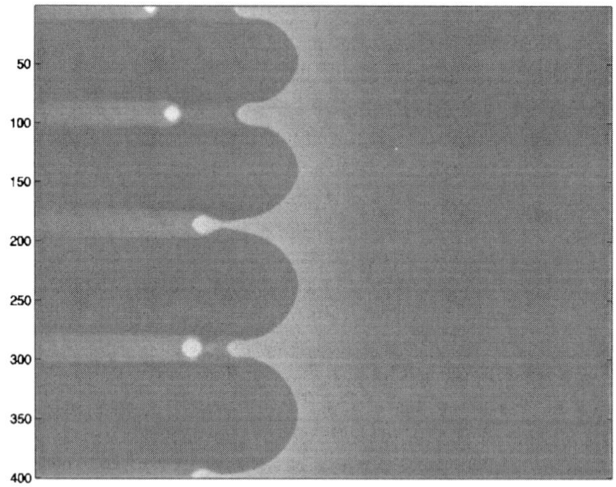

Figure 14. Cellular interface computed by Bi [67] from the phase field model for directional solidification of a binary alloy. The light regions show solute ($k < 1$) segregation in the liquid and in the cell grooves. A secondary instability in the cell grooves leads to the encapsulation of liquid droplets.

Figure 15. Three dimensional alloy dendrite computed by W. George and J. Warren [68] of NIST using the phase field model with interfacial free energy having cubic symmetry of the form given by Eq. (19). Note the six main branches in the $\langle 100 \rangle$ directions. The red color indicates regions rich in a solute having $k < 1$ that is rejected on freezing.

dimensions [81,82]. They are especially useful in helping to resolve dendrite sidebranching with computational efficiency. Solution adaptive grids are practically mandatory for modeling dendrites in three dimensions and were first used for this purpose for the sharp interface model [83]. Computations using the phase field model have also been extended to three dimensions, both by using adaptive grids [84,85] and by means of hybrid methods involving random walk algorithms [86]. Computations based on these models have allowed for simulations that can be used to study dendrite sidebranching and to test three-dimensional predictions of microscopic solvability theory [87].

- It has been demonstrated that the phase field model gives rise to phenomena such as solute trapping and solute drag [55,88–93] as well as other effects related to departures from local equilibrium at a sharp solid-liquid interface. In order to eliminate these non-equilibrium phenomena and compare with well-known results for the case of local equilibrium, "corrections" based on thin interface asymptotics have been formulated [94–96].

- Kobayashi, Warren and Carter [97] have generalized the phase field model by including a complex order parameter that enables the orientation of a grain to be tracked. This allows one to model grain growth and grain rotation in solid-solid transformations. Moreover, Kassner et al. [98] have included the effects of strain for a solid in contact with a melt. Finally, Gránásy, Börzsönyi and Pusztai [99] have used a diffuse interface model to compute nucleation and have incorporated this with the phase field model for growth to simulate nucleation and growth of polycrystalline aggregates.

- Today, the phase field model is the model of choice for computation of complex interface morphologies that result subsequent to morphological instability. It has already enhanced our theoretical understanding of the origin and complexity of these morphologies. Improvements in the model itself and computational techniques to solve it continue to be developed. Ultimately, one would like to incorporate the phase field model in codes to enable realistic simulations of crystal growth processes that can serve as guidance for the design of engineering systems to improve crystal yield and quality. Very recently, Amberg [100] developed a semi-sharp phase field method that has great promise because of its ability to produce results that agree with sharp interface models but without severe restrictions on computational parameters.

Acknowledgment

The author is grateful to the Division of Materials Research of the National Science Foundation for financial support over a period of several decades when much of this research was conducted. Thanks are also expressed to S. R. Coriell and C. L. C. Sekerka for critiquing drafts of this manuscript. This manuscript is dedicated to two people, Dr. D.T.J. Hurle who heard my 1992 Palm Springs lecture on morphological stability and inspired me to write it all down, and to my late father John J. Sekerka who attended my Frank Prize lecture in San Diego but will not be around when I give this lecture in Berlin.

REFERENCES

1. K.A. Jackson, D.R. Uhlmann and J.D. Hunt, J. Crystal Growth 1 (1967) 1.
2. G. Wulff, Z. Krystallogr. 34 (1901) 449.
3. C. Herring, Surface Tension as a Motivation for Sintering, in: The physics of powder metallurgy, a symposium held at Bayside, L.I. New York, August 24-26, 1949, W.E. Kingston (ed.), McGraw-Hill, New York, 1951, p. 143.
4. C. Herring, The Use of Classical Macroscopic Concepts in Surface-Energy Problems, in: Structure and properties of solid surfaces, a conference arranged by the National Research Council, Lake Geneva, WI, September 1952, R. Gomer and C.S. Smith (eds.), U. Chicago Press, Chicago, 1953, p. 5.
5. D.W. Hoffman and J.W. Cahn, Surface Science 31 (1972) 368.
6. J.W. Cahn and D.W. Hoffman, Acta Met. 22 (1974) 1205.
7. W.W. Mullins, J. Math. Phys. 3 (1962) 754.
8. S. Angenent and M.E. Gurtin, Archive for Rational Mechanics and Analysis 108 (1989) 323.
9. E. Yokoyama and R.F. Sekerka, J. Crystal Growth 125 (1992) 389.
10. A.A. Chernov, Soviet Physics Crystallography 8 (1964) 401.
11. F.C. Frank, On the Kinematic Theory of Crystal Growth and Dissolution Processes, in: Growth and Perfection in Crystals, R.H. Doremus, B.W. Roberts, and D. Turnbull (eds.), John Wiley & Sons Inc., New York, 1958, p. 411.
12. R.F. Sekerka and S-L Wang, Moving Phase Boundary Problems, in: Lectures on the Theory of Phase Transformations, second edition, H.I. Aaronson (ed.) TMS, Warrendale, 2000, p. 231
13. R.F. Sekerka, Morphology: From Sharp Interface to Phase Field Models, in: Fifty Years of Progress in Crystal Growth, R. Feigelson (ed.), to be published by Elsevier.
14. W.A. Tiller, J.W. Rutter, K.A. Jackson and B. Chalmers, Acta Met. 1 (1953) 428.
15. W.W. Mullins and R.F. Sekerka, J. Applied Physics 34 (1963) 323.
16. W.W. Mullins and R.F. Sekerka, J. Applied Physics 35 (1964) 444.
17. R.F. Sekerka, Phase Interfaces: Morphological Stability, in: Encyclopedia of Materials Science & Engineering, M.B. Bever (ed.), Pergamon, Oxford, 1986, p. 3486.
18. R.F. Sekerka, J. of Applied Physics 36 (1965) 264.
19. J.W. Cahn, in: Crystal Growth, H.S. Peiser (ed.), Pergamon, Oxford, 1967, p. 681.
20. S.R. Coriell and R.L. Parker, J. Applied Physics 36 (1965) 632.
21. S.R. Coriell and S.C. Hardy, J. Research National Bureau of Standards 73A (1969) 65.
22. R.F. Sekerka, Melt Growth, in: Proceedings International School of Crystallography, 7th Course: Interfacial Aspects of Phase Transformation, Erice-Traponi, Sicily, B. Mutafschiev (ed.), D. Reidel Publishing Co., Dordrecht, 1982, p. 489.
23. J.S. Langer and H. Müller-Krumbhaar, J. Crystal Growth 42 (1977) 11.
24. J.S. Langer, R.F. Sekerka and T. Fujioka, J. Crystal Growth 44 (1978) 414.
25. A. Barbieri and J.S. Langer, Phys. Rev. A39 (1989) 5314.
26. M. Ben Amar and P. Pelcé, Phys. Rev. A39 (1989) 4263.
27. Y. Pomeau and M. Ben Amar, Dendritic Growth and Related Topics, in: Solids Far From Equilibrium, C. Godrèche (ed.), Cambridge University Press, Cambridge, 1992,

p. 365.
28. S.R. Coriell and G.B. McFadden, Morphological Stability, in: Handbook of Crystal Growth 1B, Transport and Stability, D.T.J. Hurle (ed.), North-Holland, Amsterdam, 1993, p. 785.
29. D.J. Wollkind and L.A. Segel, Proc. Roy. Soc. 268 (1970) 351.
30. G.B. McFadden and S.R. Coriell, Physica 12D (1984) 253.
31. L.H. Ungar and R.A. Brown, Phys. Rev. B31 (1985) 5931.
32. J.S. Langer, unpublished notes, August 1978.
33. J.B. Collins and H. Levine, Phys. Rev. B31 (1985) 6119.
34. J.S. Langer, Models of Pattern Formation in First-Order Phase Transitions, in: Directions in Condensed Matter Physics, G. Grinstein and G. Mazenko (eds.), World Scientific, Singapore, 1986, p. 165.
35. J.W. Cahn and J.E. Hilliard, J. Chem. Phys. 28 (1958) 258.
36. S-L. Wang, R.F. Sekerka, A.A. Wheeler, B.T. Murray, S.R. Coriell, R.J. Braun and G.B. McFadden, Physica D69 (1993) 189.
37. G.B. McFadden, A.A. Wheeler and D.M. Anderson, Physica D144 (2000) 154.
38. A. Karma and W.-J. Rappel, Phys. Rev. E53 (1996) 3017.
39. G.B. McFadden, A.A. Wheeler, R.J. Braun, S.R. Coriell and R.F. Sekerka, Phys. Rev. E48 (1993) 2016.
40. A.A. Wheeler and G.B. McFadden, Eur. J. Appl. Math 7 (1996) 367.
41. A.A. Wheeler and G.B. McFadden, Proc. Roy. Soc. London A 453 (1997) 1611.
42. R.F. Sekerka, Fundamentals of phase field theory, in: Advances in Crystal Growth Research, K. Sato, Y. Furukawa and K. Nakajima (eds.), Elsevier, Amsterdam, 2001, p. 21.
43. A. Wheeler, B. Murray and M.E. Glicksman, J. Crystal Growth 154 (1995) 386.
44. S.R. Coriell and R.F. Sekerka, J. Physico-Chem. Hydrodyn. 2 (1981) 281.
45. R.T. Delves, Theory of Interface Stability in: Crystal Growth, B. Pamplin (ed.), Pergamon, Oxford, 1974, p. 40.
46. S.R. Coriell, G.B. McFadden, R.F. Boisvert and R.F. Sekerka, J. Crystal Growth 69 (1984) 15.
47. S.H. Davis, Effects of Flow on Morphological Stability, in: Handbook of Crystal Growth 1B, Transport and Stability, D.T.J. Hurle (ed.), North-Holland, Amsterdam, 1993, p. 859.
48. S.R. Coriell, G.B. McFadden, and R.F. Sekerka, J. Crystal Growth 100 (1990) 459.
49. S.R. Coriell and R.F. Sekerka, J. Crystal Growth 34 (1976) 157.
50. G.B. McFadden, S.R. Coriell and R.F. Sekerka, J. Crystal Growth 91 (1988) 180.
51. T. Uehara and R.F. Sekerka, J. Crystal Growth 254 (2003) 251.
52. J-M. Debierre, A. Karma, F. Celestini and R. Guérin, Phys. Rev. E68 (2003) 041604.
53. S.R. Coriell and R.F. Sekerka, Interface Stability During Rapid Solidification, in: Rapid Solidification Processing: Principles and Technologies 2, R. Mehrabian, B.H. Kear and M. Cohen (eds.), Claitor's, Baton Rouge, 1980, p. 35.
54. S.R. Coriell and R.F. Sekerka, J. Crystal Growth 61 (1983) 499.
55. A.A. Wheeler, W.J. Boettinger and G.B. McFadden, Phys. Rev. A45 (1992) 7424.
56. A.A. Wheeler, W.J. Boettinger, Towards a Phase Field Model for Phase Transitions in Binary Alloys, in: On the Evolution of Phase Boundaries, M.E. Gurtin and G.B. Mc-

Fadden (eds.), The IMA Volumes in Mathematics and Its Applications 43, Springer-Verlag, Berlin, 1992, p. 127.
57. A.A. Wheeler, W.J. Boettinger and G.B. McFadden, Phys. Rev. E47 (1993) 1893.
58. W.J. Boettinger, A.A. Wheeler, B.T. Murray, G.B. McFadden and R. Kobayashi, A phase-field, diffuse interface solidification model for pure metals and binary alloys, in: Modeling of Coarsening and Grain Growth, S.P. Marsh and C.S. Pande (eds.), TMS, Warrendale, 1993, p. 45.
59. Z. Bi and R.F. Sekerka, Physica A261 (1998) 95.
60. Ch. Charach and P.C. Fife, SIAM J. Appl. Math. 58 (1998) 1826.
61. Ch. Charach and P.C. Fife, Open Systems, Information Dynamics 5 (1998) 99.
62. J.A. Warren and W.J. Boettinger, Acta Metall. Mater. 43 (1995) 689.
63. M. Conti, Phys. Rev. E 56 (1997) 3197.
64. J.A. Warren and W.J. Boettinger, Prediction of dendritic microstructure patterns using a diffuse interface phase field model, in: Modeling of Casting, Welding and Advanced Solidification Processes VII, M. Cross, J. Campbell (eds.), TMS, Warrendale, 1995, p. 601.
65. J.A. Warren and W.J. Boettinger, Numerical Simulation of Dendritic Alloy Solidification using a phase field model, in: Solidification Processing 1997, J. Beach and H. Jones (eds.), Department of Engineering Materials, University of Sheffeld, UK, 1997, p. 422.
66. Z. Bi and R.F. Sekerka, J. Crystal Growth 237 (2002) 138.
67. Z. Bi, Directional Solidification of a Binary Alloy Using the Phase Field Model, Doctoral Thesis, Carnegie Mellon University, Pittsburgh 2001.
68. W. George and J.A. Warren, Parallel 3D Dendritic Growth Simulator using the Phase-Field Method, J. Comp. Phys., 177 (2002) 264.
69. M.E. Gurtin, D. Polignone and J. Viñals, Math. Models Methods Appl. Sci. 6 (1996) 815.
70. D.M. Anderson and G.B. McFadden, Phys. Fluids 9 (1997) 1870.
71. D.M. Anderson, G.B. McFadden and A.A. Wheeler, Ann. Rev. Fluid Mech. 30 (1998) 139.
72. D.M. Anderson, G.B. McFadden and A.A. Wheeler, Physica D135 (2000) 175.
73. R.F. Sekerka and Z. Bi, Phase Field Model of Multicomponent Alloy with Hydrodynamics, in: Interfaces for the Twenty-First Century, M.K. Smith, M.J. Mixis, G.B. McFadden, G.P. Neitzel and D.R. Canright (eds.), Imperial College Press, London, 2001, p. 147.
74. D.M. Anderson, G.B. McFadden and A.A. Wheeler, Physica D2711 (2001) 1.
75. D.M. Anderson, G.B. McFadden and A.A. Wheeler, A Phase-Field Model of Solidification with Convection: Numerical Simulations, in: Interfaces for the Twenty-First Century, M.K. Smith, M.J. Mixis, G.B. McFadden, G.P. Neitzel and D.R. Canright (eds.), Imperial College Press, London, 2001, p. 131.
76. R. Tönhardt and G. Amberg, J. Crystal Growth 194 (1998) 406.
77. R. Tönhardt and, G. Amberg, J. Crystal Growth 213 (2000) 161.
78. R. Tönhardt and G. Amberg, Phys. Rev. E62 (2000) 828.
79. C. Beckermann, H.J. Diepers, I. Steinbach, A. Karma and X. Tong, J. Comp. Phys. 154 (1999) 468.

80. W.J. Boettinger, J.A. Warren, C. Beckermann and A. Karma, Ann. Rev. Mater. Res. 32 (2002) 163.
81. R.A. Braun and B.T. Murray, J. Crystal Growth 174 (1997) 41.
82. N. Provatas, N. Goldenfeld and J. Dantzig, J. Comput. Phys. 148 (1999) 265.
83. A. Schmidt, J. Comput. Phys. 125 (1996) 293.
84. N. Provatas, N. Goldenfeld and J. Dantzig, Phys. Rev. Letters 80 (1998) 3308.
85. J-H Jeong, N. Goldenfeld and J.A. Dantzig, Phys. Rev. E64 (2001) 41602.
86. A. Karma and M. Plapp, Phys. Rev. Letters 84 (2000) 1740.
87. A. Karma, Y.H. Lee and M. Plapp, Phys. Rev. E61 (2000) 3996.
88. W.J. Boettinger, A.A. Wheeler, B.T. Murray and G.B. McFadden, Material Science and Engineering, A178 (1994) 217.
89. M. Conti, Phys. Rev. E55 (1997) 701.
90. M. Conti, Phys. Rev. E55 (1997) 765.
91. N.A. Ahmad, A.A. Wheeler, W.J. Boettinger and G.B. McFadden, Phys. Rev. E58 (1998) 3436.
92. Ch. Charach, C.K. Chen and P.C. Fife, J. Stat. Phys. 95 (1999) 1141.
93. Ch. Charach and P.C. Fife, J. Crystal Growth 198/199 (1999) 1267.
94. G.B. McFadden, A.A. Wheeler and D.M. Anderson, Physica D144 (2000) 154.
95. A. Karma, Phys. Rev. Letters 87 (2001) 115701.
96. B. Echebarria, R. Folch, A. Karma and M. Plapp, submitted to Phys. Rev. E.
97. R. Kobayashi, J.A. Warren and W.C. Carter, Physics D140 (2000) 141.
98. K. Kassner, C. Misbah, J. Müller, J. Kappey and P. Kohlert, J. Crystal Growth 225 (2001) 289.
99. L. Gránásy, T. Börzsönyi and T. Pusztai, J. Crystal Growth 237-239 (2002) 1813.
100. G. Amberg, A semi-sharp phase field method for quantitative phase change simulations, to appear in Phys. Rev. Letters.

Crystal Growth – from Fundamentals to Technology
G. Müller, J.-J. Métois and P. Rudolph (Editors)

Crystallization Physics in Biomacromolecular Solutions

A.A. Chernov[a], P.N. Segre[b], A.M. Holmes[c]

[a]BAE SYSTEMS, NASA Marshall Space Flight Center, Mail Code SD46, MSFC, AL 35812

[b]NASA Marshall Space Flight Center, Mail Code SD46, MSFC, AL 35812

[c]University of Alabama, Huntsville, NASA Marshall Space Flight Center, Mail Code SD46, MSFC, AL 35812

These lecture notes are complementary to, and partly overlapping with, the reviews [1, 2] on which the lecture by A.A. Chernov is based. State of the art in biomacromolecular crystallization physics is reviewed with an emphasis on the unresolved problems. Some of the problems are still open for inorganics as well. The lecture starts with an elementary introduction for newcomers on biological macromolecules with an emphasis on proteins (Sec. 1). Sec. 2 briefly summarizes crystallization techniques - very simple in terms of physics, but complicated chemically. Biomacromolecules are overwhelmingly crystallized from aqueous solutions. Their solubility decreases when the precipitant – an organic salt, polyethelene glycol or a buffer that changes pH – is added to the biomacromolecular solution.

Sec. 3 is devoted to nucleation rate and time lag. Results following from application of the classical "development" technique based on the temperature dependence of protein solubility present the macroscopic approach to the problem. Light scattering from supersaturated solutions provides data closer to the molecular level. Sec. 4, on crystal growth and perfection, summarizes the kinetic coefficients for crystal growth as measured by AFM. This section is also illustrated with examples indicating that regular polyhedral morphology associated with the layer growth may be replaced by skeleton and dendritic growth modes following from the thermodynamic or kinetic roughening transitions. Impurity distribution is addressed as a factor contributing, directly or indirectly, to crystal imperfection (Sec. 5), inducing irregular molecular orientations within the lattice and conformational changes within the molecules.

1. BIOMACROMOLECULE – STRUCTURE AND FUNCTION

Living organisms contain 70% of water, 15% of proteins, 7% of nucleic acids, 5% of polysacharids, lipids and precursors, 1% of small molecules and 1% of inorganic ions [3]. The biomacromolecules crucial for life – proteins and nucleic acids, 22% total – are produced, perform their functions and degrade in these aqueous solutions. They are the material components of the chemical life machinery. Each molecule has its own very specific function in the long and interlacing chains of chemical reactions ("signaling pathways") occurring in living cells and intercellular fluids. These chains form the "flowchart" of the life machinery.

The very narrow specificity is not the only general feature typical of the macromolecules.

The other is the composition similarity between macromolecules performing similar functions in different classes of the living nature. For instance, the proteins responsible for various functions in, say, food digestion, breathing or bone formation and dissolution in humans, and animals (and, for some functions, even in plants) are only several percent different in their chemical composition and, thus, folding and spacial structure. This similarity allows classification and a chance to discover still hidden general laws controlling structure and function of the biomolecules (their "periodic table"). On the way to this discovery, modern molecular biology and biochemistry tries to understand numerous specific elements of the chemical machinery (~ 30,000 in human genome), i.e. specific biomolecules and their interactions with their closest neighbors in the reaction chains.

As in organic chemistry of small molecules, the macromolecule has properties determined by their three-dimensional (3D) structure. This structure directly and unambiguously follows from the molecule composition. Each biomolecule is a specifically folded chain (or several chains) built of 20 amino acids, if this is a protein molecule, or of 4 nucleotides if this is a nucleic acid molecule. The amino acid is small organic molecule with a molecular weight of about 100 *Da* (Daltons). Each amino acid includes the carboxyl ($COOH^-$) and amino ($NH_2{}^+$) groups and a specific side chain. When the carboxyl belonging to one amino acid reacts with the amino group of the other, they form the strong covalent-ionic peptide bond $CONH_2$, releasing the H_2O. The protein chain is thus the polypeptide chain.

Among the 20 amino acids, 5 are acidic (2) or basic (3), i.e. their side chains easily lose either the proton H^+ or the OH^- group and become charged. Therefore, they are highly hydrophilic. Less hydrophilic are the 7 amino acids with polar side chains. The remaining 8 amino acids expose hydrocarbon groups and are hydrophobic. Therefore, in a physiological solution, pH ~7, the protein chain folds to form the molecular globulae, with the hydrophilic and polar amino acids sitting on the surface and hiding the hydrophobic amino acids in the inner core. The probability for each amino acid to be located on an "average" macromolecular surface is estimated statistically from the already found 3D structures [4]. Thus, from the amino acid content in the chain of interest, one may make a first guess on solubility.

The amino acid sequence in the protein chain, tens to thousands of amino acids long, fully determines the way this chain is folded in solution, resulting in the functional relatively compact globular molecule. A de-folded molecule is not functional. That is why the 3D atomic structure is the background for structural biology. The amino acid or nucleotide sequence is now routinely determined by chemical analyses (e.g. subsequent hydrolysis). However, this sequence does not yet allow to predict the 3D folding – the folding problem, despite a long, intensive effort, is not solved.

Some structural folding elements - α, helicies, β, sheets - may be guessed from the amino acid sequence in the polypeptide chain, but it doesn't help much in predicting the full 3D structure with the required high precision. The reason for such a requirement is that a chemical reaction between two molecules occurs if the atoms on the surfaces of the both fit within a part of an angstrom, i.e. with the accuracy at which any chemical reaction starts. This, for instance, is required when a sugar molecule (the substrate[1]) approaches and is trapped by an active site on the lysozyme molecule (the enzyme). As soon as the sugar is trapped, the electron density within the whole enzyme molecule is redistributed, causing the enzyme molecule to deform and cleave the sugar. Another example illustrating the demand to know precise atomic positions is

---

[1] Substrate is the molecule acted upon by the enzyme.

drug design. Presently, the absolute majority of drugs work by entering and blocking the active site on the target protein molecule, thus stopping malfunction of this protein. If atomic positions within the active site are known, computer search for the relevant small molecule to block the protein activity becomes possible.

Nuclear magnetic resonance of protons and other species composing biomacromolecules in solution allows to solve the 3D atomic structure of relatively small biomolecules, less than about a few tens of thousands of *kDa* in molecular weight. However, the overwhelming majority of the 3D biomacromolecular structures, ~15,000 in the Protein Data Bank, have been found deciphering, by x-ray diffraction, the crystals' structure built of these biomacromolecules. Each crystallographic study of a new molecule needs the corresponding crystal to be grown. The crystals must allow high structural atomic resolution. This means that sufficiently intensive (above the noise level) diffraction reflections from all the lattice planes, including those low density planes that are separated by less than ~2Å, or even 1Å, should be recorded. The theoretical resolution is half the wavelength of the x-ray beam, i.e. ~0.5Å. The best resolution obtained so far is ~0.8Å. For comparison, various biomolecules have an effective diameter, and thus the lattice spacing, within the approximate range of 20-3000Å corresponding to a molecular weight of $10^4 - 3 \cdot 10^6$ *Da*. Thus, the resolution down to 1Å means the rather tough requirement to the crystal perfection, which is the major challenge for crystal growth.

The structural resolution depends, to some extent (~20%), on the diffraction equipment and is usually better when intensive synchrotron radiation is used. The time required today to solve the protein structures varies within several weeks, while finding solution composition allowing growth of perfect crystal takes months or years. That is why crystallization is a bottleneck for biomolecular crystallography and, among other issues, of structural biology.

The issue of protein crystallization is covered in several books [5, 6], reviews [2, 7] in the recent special issue of the Journal of Structural Biology [8] and in Proceedings of nine International Conferences on Crystallization of Biological Macromolecules published in the Journal of Crystal Growth [9-15] and Acta Crystallographica [16, 17].

2. THE TECHNIQUES

The strategy of biocrystallization comes from the demand to crystallize hundreds of thousands of different molecules operating in living organisms with no use of the crystals themselves, except for obtaining the macromolecular structure, and from major difficulties associated with biological and biochemical procedures to obtain, in pure form, the molecules of a protein, nucleic acid or a combination of both. These molecules are produced only in milligram scale amounts or less. Typical solubility varies around the mg/ml range so that 20-100 μl volumes of solution are usually employed. The major practical approach is to add precipitant that decreases solubility of the protein to be crystallized, to the protein solution. Precipitants are usually taken from commercially available crystallization kits containing ammonium, sodium, potassium, cadmium sulfates, chlorides, phosphates, nitrates, citrates, tartrates and the other salts that decrease solubility and induce either crystallization or the non-useful amorphous precipitation. The polyethelene glycol (PEG), neutral polymer of molecular weights in the range of 400-8000 *Da* is also a very effective precipitant. Organic solvents, like ethanol, acetone, dioxane, etc., are used sometimes as well.

A buffer solution of small molecules, also in the mM concentration range, is another mandatory ingredient that maintains constant pH during crystallization (usually ±2 units around pH7). Protein solubility often does not depend significantly on temperature, though many examples of regular or retrograde solubility are known. The temperature dependence, however, is not known *a priori* so that it is usually tried if the salt or the pH change-induced precipitation is not successful in obtaining high quality crystals. Though more than 15,000 proteins have been successfully crystallized and their molecular structure solved, phase diagrams are hardly known for more than ten of the proteins. After all, the diagrams are different for different precipitants.

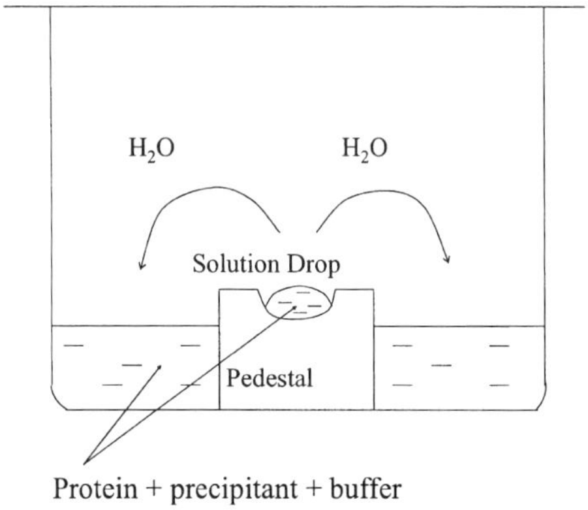

Figure 1. Sitting Drop Technique.

The four major techniques usually used are: 1) Sitting or hanging drop; 2) Batch; 3) Dialysis; and 4) Crystallization in gel. 1. In the sitting or hanging drop technique (Figure 1), a solution droplet with added buffer and precipitant is exposed, in a sealed volume, to the air in contact with a large reservoir containing the same precipitant at a higher concentration than that in the droplet. The equilibrium water vapor pressure above solution is lower when the solute concentration is higher. Therefore, water from the droplet evaporates and diffuses through the air to the reservoir. As a result, all concentrations in the droplet increase, inducing crystallization. In the hanging drops version, the droplet is hanging on the lid. 2. In the batch technique, protein solution with a buffer is placed in a small well and then the precipitant is pipetted into the well. 3. Dialysis is implemented in tubes, separated by a membrane into two compartments. The membrane allows only the small precipitant ions or molecules (e.g., PEG) to go through it into the macromolecule compartment, where it causes crystallization. In all these techniques, numerous protein-precipitant-buffer concentration combinations are used to screen the best conditions by trial and error. The batch technique fits best to build robots to screen the numerous conditions. 4. In the gel technique, warm liquid agarose or silica gel is added to the protein solution in a tube or a wide capillary (~1mm – 1cm in diameter or more). The mixture is tempered until gellation occurs. Then, liquid precipitant solution is poured above the gel. The slow diffusion of ions or small molecules (like PEG) through the gel results in the precipitant gradient in the capillary with gel and protein. Such a gradient allows simultaneous implementation of a wide range of conditions within one tube or capillary. The gel technique is used in vessels possessing various geometries to control precipitant (and protein) diffusion [18].

3. NUCLEATION

3.1. Making solution supersaturated

In the widely used batch method, the protein and precipitant solutions are mechanically mixed by injection of one into the reservoir filled with the other.

Figure 2 exemplifies ferritin crystals precipitated by $CdSO_4$. The images show crystals grown in two identical wells, from identical solution. These photos were taken several days after pipetting the solutions together. The disparity between the wells in the crystallization pattern (crystal size and quantity) is a typical characteristic of the lack of reproducibility with this technique. This is the mixing that that seems to cause irreproducibility when supersaturation is significant.

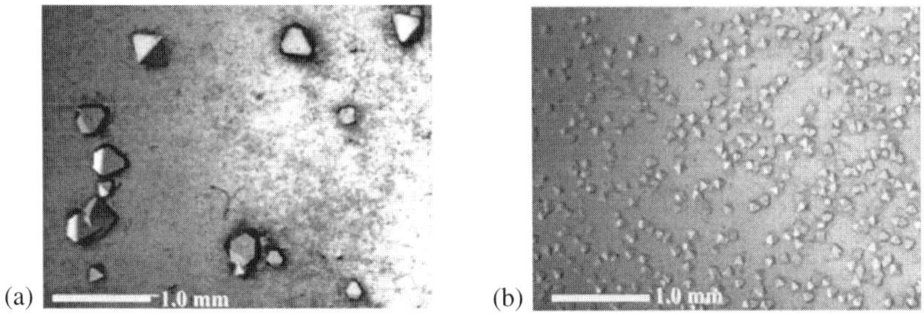

Figure 2. The samples (both at the same magnification) were prepared by pipet dispensing and vortex mixing the precipitating solution ($CdSO_4$) with protein solution (ferritin) in one vial and then "immediately" aliquoted into separate wells. The difference in the ferritin crystals' precipitation pattern shows a lack of reproducibility using this method.

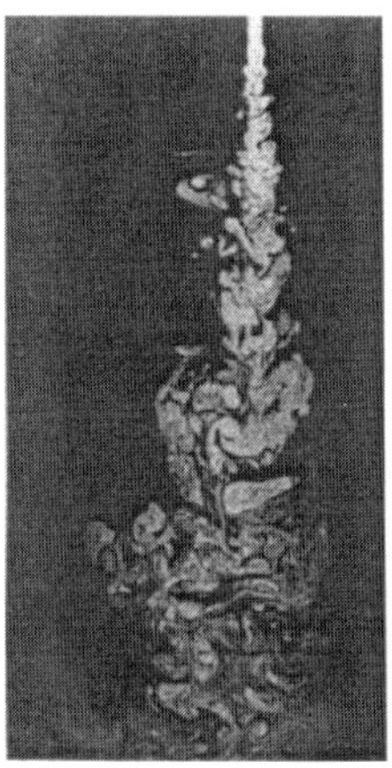

Figure 3. Turbulence induced by injection of a fluid showing laser-induced flourescence, $Re = 2.5 \cdot 10^3$ for the jet stream.

The reason for the difference between Figs 2 a and b is suggested by Figure 3 [19], since the Reynolds number (*Re*) for precipitant injection may be also close to 10^3 [1]. Figure 3 illustrates the turbulence when one liquid (white in Fig 3) is injected into the reservoir filled with the other (black in Fig 3). The Reynolds number for pipetting depends on the individual mode of pipetting - pipette depth, angle, technician, etc. At immediate contact between the protein and precipitant, the supersaturation highly exceeds the final pre-calculated composition that assumes homogeneous mixing between the two liquids. The supersaturation in the case of biomacromolecules is often so high that invisible nuclei may appear before the homogeneous mixture is reached. Automated pipetting is of some help in practice, but it does not provide an insight into the processes. This example illustrates the importance of reproducibility in mixing for protein crystallization, though the problem is not solved yet. A similar mixing problem exists in industrial crystallization of conventional materials from solutions.

3.2. Nucleation rate

A systematic analysis of nucleation was carried out recently [20] with lysozyme, the most studied model protein (M = 14.3 kDa, effective molecular diamter ≅ 2nm). Lysozyme solubility steeply increases with temperature: dissolution enthalpy ΔH = 15.6 kcal/mol [21]. This dependence allows avoidance of mixing in order to achieve sufficiently high supersaturation and to apply the classical "development" technique [22].

In this technique, a liquid volume is exposed to the supercooling sufficient for nucleation to occur during the "exposure" time. During this time, the as yet invisible crystal embryos are supposed to form. The temperature is then increased to the level within the metastable zone, where new nuclei do not form, but extant crystal embryos may grow to visible size. In the lysozyme studies [20], the number of crystallites observed in the microscope, N, increased linearly with the exposure time, τ. The slope of the line provides the homogeneous nucleation rate, J, while the number of the heterogeneously nucleated crystals is given by the segment on the ordinate (Fig 4, line 1). The nucleation rate, J, follows the classical exponential dependence on the reciprocal squared supersaturation:

$$J\,(\,1/\mathrm{cm}^3\ \mathrm{s}) = B\exp(A/kT\sigma^2). \tag{1}$$

This rate exponentially depends on the nucleation work A/σ^2 normalized by the thermal energy, kT, and $\sigma \equiv \Delta\mu/kT$, where $\Delta\mu$ is the difference between chemical potentials of the old

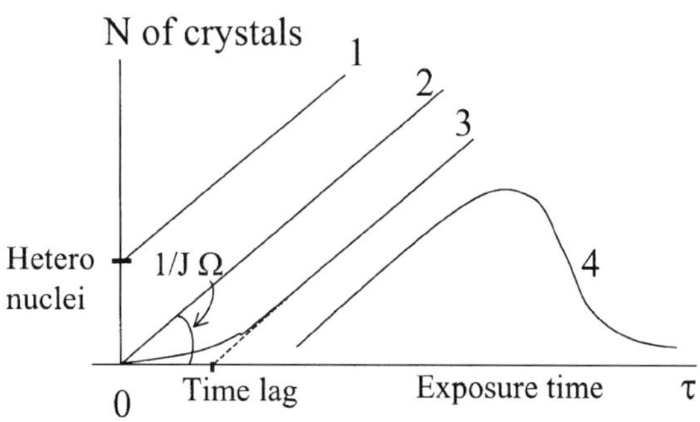

Figure 4. The number N of crystals appeared at the nucleation rate $J(\sigma)$ in the solution volume Ω exposed for the time τ to a high supersaturation σ and then to a lower supersaturation, allowing only the crystal growth with no new nucleation. The intersection between line 1 and the N-axis provides the number of heterogeneous nuclei. The line 2 characterizes homogenous nucleation. The line 3 is typical of the system with slow pre-nucleation processes where the time lag is essential; line 4 also shows the maximum, suggesting a ripening during the sample exposure at high supersaturation. The slope of lines 1, 2 and 3 allows the determination of the nucleation rate, J, and provides the expectation time $1/J\Omega$ for the crystal to appear.

and the new phases and the constant A includes the interfacial energy, γ = 0.68 erg/cm^2 for lysozyme [20], between the crystal and solution. The pre-exponential factor, B, is proportional to the surface area, $4\pi(2\omega\gamma/\Delta\mu)^2$, of the (spherical) critical cluster where ω is the specific molecular volume. By adding a new molecule, the cluster is transformed into a supercritical nucleus that overcomes the Gibbs free energy barrier, A/σ^2. Then, the cluster may grow "sliding down" the potential energy hill. The pre-exponential factor B was not addressed in [20], although an earlier study [23] found this factor to linearly increase with lysozyme concentration c in solution, $B{\sim}c$, rather than $B{\sim}c^2$, as predicted by the widely used theory that ignores the partition function of the pre-critical clusters, i.e. their translational and vibrational degrees of freedom. These findings [24, 25] (mainly for nucleation in a vapor) are still being debated [26].

3.3. Time lag

3.3.1. Macroscopic observations

The time lag is defined as the time required for the system to reach an equilibrium size distribution of pre-critical clusters, after the onset of supersaturation. If the time lag is measurable, the $N(\tau)$ curve at a given σ takes the form shown schematically by line 3 in Figure4. Then, the time lag is measured as the segments on the τ-axis cut by the extension of the linear $N(\tau)$ dependence [22, 26, 27]. The time lag is usually considered to be long (many hours) in viscous glassy materials [28] , but is negligibly short in melts and solutions of small molecules. A modified "development" technique applied to lysozyme [29] revealed that the number of macrocrystals, N, that appeared by the exposure time, τ, is a linear function of this time at $\tau \leq 100$ min. However, the $N(\tau)$ dependencies in these experiments measured at different supersaturations are similar to line 4 in Fig 4 - they intersect the τ-axis at different distances from the origin $\tau = 0$, i.e. the time lag reaches minutes.

The time lag should not be confused with expectation time, $1/J\Omega$, to form a stable supercritical nucleus, i.e. a crystal appearing in the solution of a known total volume Ω. The expectation time is just the reciprocal of the slope of the $N(\tau)$ curve normalized by the volume Ω. Both the expectation time and the time lag are of kinetic origin. The expectation time is the average time before a viable nucleus appears in the equilibrium ensemble of molecular aggregates. This time is taken by fluctuations (random attachments and detachments of molecules to and from molecular aggregates) to form one nucleus, which size exceeds critical cluster size. In the simplest isotropic approximation, radius of this cluster is $2\omega\alpha/\Delta\mu$, while the number of such viable aggregates appearing per unit volume per unit time, J, is given by eq.(1). These fluctuations are assumed to occur in the molecular ensemble within which the equilibrium with respect to reversible formation of dimers, trimers, and higher, though subcritical, oligomers, had already been established after the onset of supersaturation, e.g. via temperature decrease. However, establishment of this equilibrium subcritical size distribution of oligomers also takes some time after the onset of a given supersaturation. This is the latter time, which is the time lag. In other words, time lag is the period during which the system climbs up the potential ladder to the step where all the subcritical aggregates exist in equilibrium quantities. The expectation time is required for the system to climb up the next, supercritical, step. The third way to understand the difference between the expectation time and the time lag is to note that a reverse to the not supersaturated solution within the time lag period practically excludes formation of a nucleus. Such a reverse within the expectation time period still allows one to get a nucleus, though with a lower probability.

Equilibrium cluster size distribution is achieved when the largest critical clusters containing n_c molecules appear in the solution. Association of the n_c molecules occurs via reversible coagulation of single molecules into dimers, trimers and larger oligomers. The reversibility means that the oligomer clusters may lose molecules and decay, i.e. the number of molecules (monomers) in each cluster fluctuates. The random attachment and detachment process means diffusion of a point, representing a number of molecules in a cluster, along the axis of cluster size, n. The Gibbs free energy rises as the cluster size is going up because, at $n < n_c$, the cluster surface energy increases with the cluster size, but is not yet compensated by gain in the bulk free energy. Therefore, the diffusivity along the n-axis is the function of n. This diffusivity is also proportional to the attachment frequency, ν, to one molecular site on the cluster. Therefore, ignoring complications associated with the diffusivity dependence on n, the time lag is $\sim n_c^2/\nu$, i.e. is decreasing together with n_c^2 when the supersaturation increases.

In the experiments [29], the time lag varies from several minutes at the highest $\sigma = \Delta\mu/kT = 2.6$ through ~60 minutes at the lowest one, $\Delta\mu/kT = 2$. Although the analysis of this finding is missing in the ref [29], the time lag, indeed, may be expected to be noticeable in the macromolecular solutions because of the extremely low frequency ν at which the molecules join the crystalline lattice and clusters. However, because of the limited studies accomplished so far with proteins, the time lag has not yet been discriminated from the expectation time needed to observe viable crystal nuclei.

3.3.2. Light Scattering.

Because of their large size (2-20nm), the biomacromolecules are good light scatterers. This is a big advantage in the study of aggregation processes monitoring cluster development from single molecules to large clusters. Both are detected by dynamic light scattering (DLS) as a single peak on the scattering intensity vs. the scatterer size dependence. In the DLS instrument, an essentially parallel light beam, ~100μm in cross-section, is directed to a ~ 1cm wide cuvette filled with solution to be analyzed. The light is scattered by density fluctuations in the liquid and is registered at different angles (often at the 90° angle). A fluctuation – a small region in the solution within which the scatterer density spontaneously increases – exists during decay time, inversely proportional to the scatterer (macromolecule) diffusivity in the solution. The decay time is provided by the temporal self correlation function of the scattered signal. From this decay time, the scatterer diffusivity is determined. This diffusivity obeys the Einstein relation and is inversely proportional to the scatterer effective radius (radii), which is the subject of analysis. The result is recorded as the scattering intensity of a function of the scatterer radius. If several scatterers are present, the instrument provides a spectrum of several peaks although relative amounts of various clusters is not just proportional to intensities of the corresponding peaks, because the scattering intensity is proportional to both the scatterer volume and its number density in solution.

In the light scattering experiments with trypsin (M = 95kDa) and thaumatin (M = 22kDa) [30], where the supersaturation was unknown, the first supercritical nuclei appeared after several hours. A similar delay time was observed earlier [31], when the authors monitored the major light scattering peak, corresponding the scatterer size varying from the monomer to the larger supercritical clusters. This peak moved, fluctuating right and left, corresponding to the molecular attachments of detachments of the, probably, monomer molecules to and from the clusters. The average cluster size increased not more than twice. After the ~ 4 hour fluctuation time, a steep, continuous increase of the cluster size was observed. The fluctuation period includes both the time lag and the expectation time for nucleation, while steep continuous cluster size increase occurs for the supercritical clusters, probably crystallites. However, discrimination between the time lag and expectation time has not yet been made.

Figs. 5, 6 and 7 illustrate cluster size evolution in different protein systems. Figure 5 shows a typical light scattering recording of the developing catalase nanoclusters. To avoid contaminants that may result in undesired heterogeneous nucleation, the protein solution is initially highly purified by high performance liquid chromatography (HPLC) into its monomeric state, as demonstrated by the very narrow size distribution in the inset. The monomer diameter is ~8 nm. Protein crystallization is initiated by adding a buffer solution to the protein monomer solution, which addition, changing pH, brings the system into a supersaturated state. The main figure shows the transition over time from pure monomers to growing clusters, which are probably crystals. Significantly, there is a crossover or transition point in the data at a time ~60 min. We interpret this as the time for the sub-critical cluster size distribution to form, also referred to as the lag time. For times shorter than the time lag, monomers are seen to slowly aggregate together climbing up the potential barrier. As soon as these aggregates reach a critical size, R_{crit}~15 nm, for crystal nucleation, crystal growth proceeds four times faster. The slope of the $R(\tau)$ dependence, together with the information on the cluster number density, includes data on the attachment frequency of the molecules to the crystallites. Measuring the slope at different temperatures (in a narrow range compatible with the protein integrity) may shed a light on the so far unknown activation energy for molecular attachments.

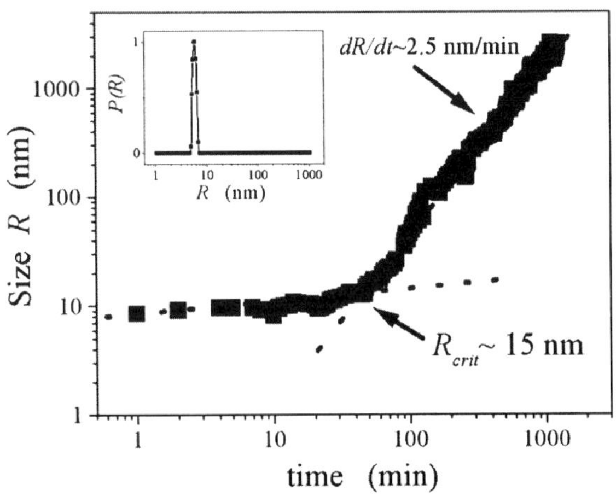

Figure 5: Growth of catalase crystals from a highly purified solution as determined by DLS. Inset: The size disribution of the solution before initiating crystallizing conditions. Main Figure: Radius R vs. time, τ. Crystal growth occurs after an induction, or lag time of ~50 minutes.

One of the major concerns in getting reproducible results in protein crystals studies is the purity (i.e. purely monomeric, no aggregates or fragments) of the starting protein solution.

DLS methods can readily quantify sample purity (along with the HPLC), as they are particularly sensitive to even very small amounts of aggregates. Aggregates of monomers or even small amounts of dust or dirt, can potentially serve as sources of heterogeneous nucleation of crystals.

To illustrate this we show in Figure 6 a similar study to that in Figure 5, but with a less pure

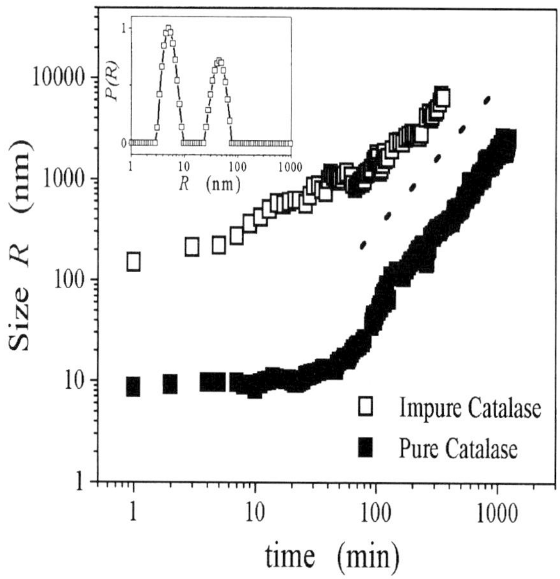

Figure 6: Comparison of the growth of catalase crystals from impure and highly purified solutions. Inset: The (unnormalized) size distribution *P(R)* of the impure solution before initiating crystallizing conditions. The number ratio of aggregates to monomers is extremely small, N_{agg}/N_{mon}~10^{-5}. Main Figure: Radius R vs. time τ. The crystallization process changes dramatically in the presence of small numbers of impurities.

starting solution. We achieved this by heating a purified solution for 1 hour at a high temperature (45^{0}C) that caused stable small amorphous aggregates to form. Upon cooling back to the growth temperature of 23 ^{0}C, we have a size distribution containing R~8 nm monomers as well as R~50 nm aggregates, as shown in the inset of Figure 6. A buffer solution was then added to this modified protein solution to initiate crystallization. The crystal formation and growth in this modified solution was significantly different than that from the purified solution. Large aggregates were present at the earliest times and an unambiguous determination of a lag time (if any) was not possible. Moreover, at any given time the clusters were always much larger in the modified than in the purified samples. These results show that small amounts of aggregates present in the starting solution can serve as nucleation centers, inducing heterogeneous nucleation, which yields significantly different results than the homogenous nucleation.

While the results from purified catalase show a distinct time lag and/or long expectation time before the appearance of critical and supercritical clusters, other proteins can show quite different behavior. In Figure 7 we show DLS results of the crystal size vs. time for the protein
Figure 7: Growth of apoferritin crystals from a purified solution as determined by dynamic

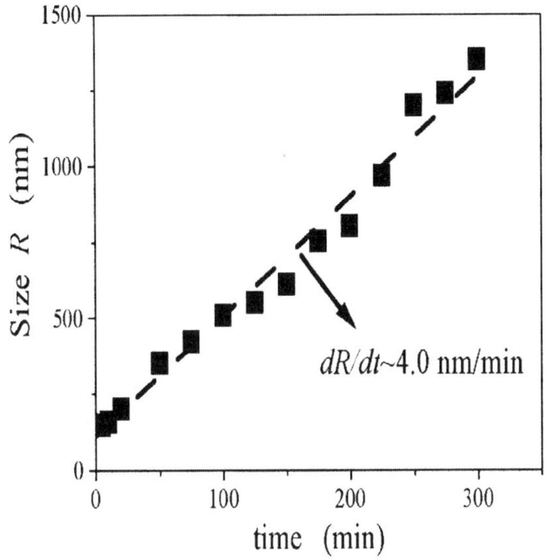

light scattering. Crystal growth occurs without any measurable time lag.

apoferritin. In marked contrast to catalase: growth proceeds right after onset of supersaturation. We note that apoferritin also has a much larger step kinetic coefficient for crystal growth than catalase, $\beta_{stApo} = 6.0 \cdot 10^{-4}$cm/s vs. $\beta_{stCat} = 3.2 \cdot 10^{-5}$cm/s [2] (see Sec. 4.1 for definition of the β_{st}). We would expect the induction time to vary with the step kinetic coefficient because, ultimately, this time is related to the attachment rate of monomers [22, 32, 33].

3.4. Processes in the cluster-solution mixture

According to ref [29], the $N(\tau)$ dependencies for each supersaturation σ employed reached a maximum at $\tau \cong 120$ min and then, within another ~100 min, decreased, tending to reach a saturation value several times lower than the maximum. This dependence is schematically shown for one supersaturation in Figure4, line 4. This maximum was interpreted by the authors as the result of the Ostwald ripening. This ripening is known to be driven by the difference in solubility between small and larger crystallites, and is related to the crystal surface energy (the

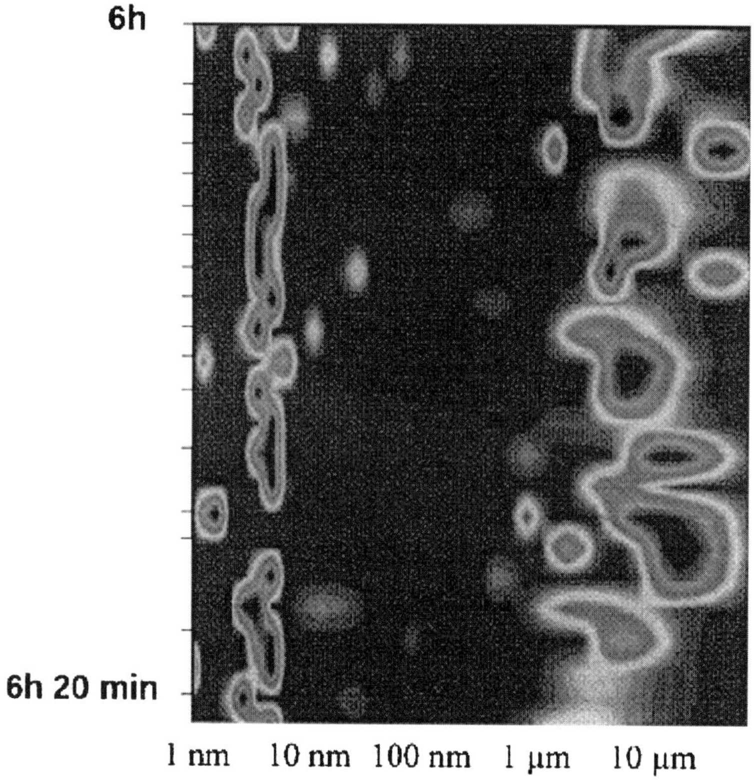

Figure 8. Dynamic light scattering intensity from trypsin solution taken at different times after onset of supersaturation. This time is plotted along the ordinate. Grade of white in each horizontal cross-section corresponds to the scattering intensity induced by the cluster of the size plotted along abscissa [30].

Gibbs-Thomson effect). It is difficult, however, to expect such a ripening to occur in solution at the "development" supersaturation at which all the crystals are supposed to grow, though we do not have an alternative explanation.

Interestingly, after the appearance of clusters, the light scattering data on trypsin and thaumatin [30] revealed two or three peaks (besides the close-to-monomer peak) from clusters ranging from 30 nm to 10 μm. This is unlike the situation in the catalase and ferritin solutions where only one peak corresponding to the aggregates larger than the monomer exists. In the trypsin and thaumatin solutions, these peaks presenting larger aggregates appeared and disappeared within the time range of ~ 10 minutes to an hour (Figure 8). This figure is superposition of the aggregate spectra taken at subsequent times: in each cross-section of the picture by a horizontal line (corresponding to a given moment of time), intensity of the white is proportional to the density of aggregates which size is plotted along this abscissa (in logarithmic scale). The closed white loops in Figure 8, i.e. the aggregates of a size mainly between 1 and 10 μm, exist within several minutes each (the whole scale on the ordinate is ~ 20 min). On the other hand, the average cluster growth velocity estimated from the whole ~7 hour long experiment was small, $\sim 10^{-8}$cm/s, and cannot account for such a rapid variation in the measured aggregate sizes. Unless this loop pattern is a noise in the light scattering system, the rapid reversible changes in the size distribution might be associated with macroscopic cluster size inhomogenieties and preferential association of these different clusters in "clouds" comparable in size with the probe beam cross-section and slowly moving within the liquid sample. The cluster size distribution, per se, may come from variation of time when the super-critical nuclei first appear, cluster coalescence or ripening mentioned above, or other interaction between the clusters, e.g. via their diffusion fields within the clouds. The presence of different cluster populations in solution may induce different populations in the ultimate nucleation output seen, for example, in Figure2.

It has recently been discussed that in a solution of small molecules [34] and biomacromolecules [35, 36], nucleation might occur via the formation of liquid-like aggregates, which later spontaneously acquire crystalline structure. This process, if it indeed occurs, may modify the cluster size distribution and assist coalescence.

These experiments suggest that an evolution in the nanocrystalline ensemble, rather than just independent creation and growth of each cluster, may occur after the onset of supersaturation. We can use an analogy with, e.g. deeply undercooled gallium melt (~0.5 of its melting point, 302K). In this case, five polymorph modifications have been observed as the final product of solidification [22].

4. CRYSTAL GROWTH

4.1. Crystal growth kinetics

Crystal growth kinetics is determined by the interface incorporation rate and the bulk transport rate of monomers in the solution to the crystal face. Under typical protein crystallization conditions, the crystal faces are below the thermodynamic surface roughening transition (Sec.4.2), i.e. they are not disordered by thermal fluctuations, but remain molecularly smooth and facetted (Figs. 9, 10 a). Consequently, these crystals grow by subsequent deposition of layers (tangential to the surface), typically one lattice spacing thick. Incorporation of molecules at the edges of the layers, the steps, is characterized by the step kinetic coefficient, β_{st}, defined via the step propagation velocity, v_{st} as [22]:

$$\beta_{st} = v_{st}/\omega(c-c_e) \qquad (2)$$

Here, ω is the molecular volume, c_e is the equilibrium concentration of solution and c is the solution concentration immediately at the step. The kinetic coefficients have been measured by atomic force microscopy for ferritin (6.10^{-4}cm/s), catalase ($3.2.10^{-5}$cm/s), lysozyme (5.10^{-5} cm/s), and thaumatin (2.10^{-4}cm/s) [2]. For comparison, the step kinetic coefficients for KDP, ADP, $BaNO_3$, K-Al-alums are within the (4-12) $\cdot 10^{-2}$ cm/s range. The step kinetic coefficient is directly related to the frequency of molecular attachments onto the growing clusters. It should also be related to the time needed for molecules to initially cluster together to form a critical crystal nuclei, though this comparison has not been made so far. An evident obstacle on the way to this comparison is the unknown roughness of the steps and the clusters and the crystal surfaces.

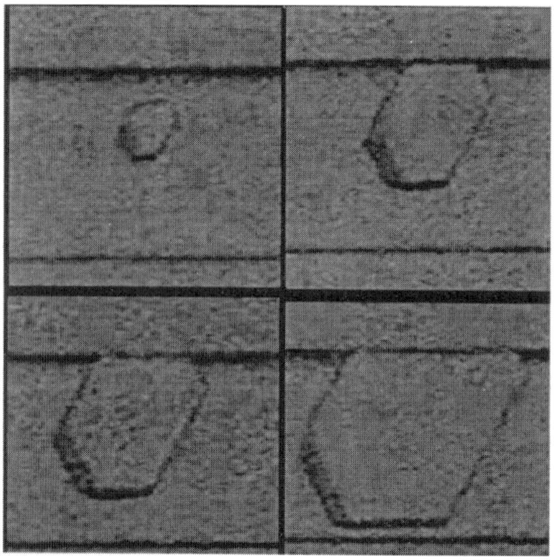

Figure 9. A glucose isomerase crystal, nucleated and growing in solution flowing through the channel 70μm wide. Photos provided by NASA MSFC LOCAD facility (unpublished data, L. Monaco, A.

The macroscopic growth rate of a crystal face (perpendicular to the surface) is given by $V = p\, v_{st} = \beta_{st}\, p\, \omega(c-c_e) \equiv \beta\omega(c-c_e)$, where p is the density of steps on the crystal face (number of steps crossing the 1 cm long normal to these steps), normalized by the step height (cm) and β is the kinetic coefficient for the face. The step density is determined by either 2-D nucleation or by screw dislocation and may depend on supersaturation. On the (110) face of lysozyme, p = $1.3.10^{-2}$ (and is independent of supersaturation), so that $\beta \approx 10^{-6}$cm/s [37]. Typical length, δ, determining the bulk diffusion rate, D/δ, may be taken to be equal to either the crystal size in a stagnant solution or to the smaller diffusion boundary layer thickness if convection is present. Thus, for a crystal of an effective radius $\sim 10^{-1}$ cm and $D = 10^{-6}$ cm^2/s (lysozyme), $\beta\delta/D \sim 10^{-1} \ll 1$, i.e. the growth rate is controlled by surface incorporation rate. Analysis of lysozyme crystal growth in space suggests that bulk transport may be essential also for the ~5 mm size ysozyme crystals [38]. Growth of ferritin crystals seems to be controlled by bulk transport in a wider range of crystal size [2, 39]. However, the macroscopic growth velocities were never measured under controlled solution flow conditions or with the supersaturation known right at the growing interface. These measurements are becoming available while making use of microfluidic devices engraved in a chip [40]. The microfluidic device uses the planar glass chips (~2 cm x 4 cm x 0.2 cm) that include several wells (containing up to 100 μl solution each) connected by a network of microchannels ~50 x 100 μm in the cross-section. The networks and active well locations vary from chip to chip. The chip is covered by a pressure lid, hermetically plugging each well into the pressure pipeline connected to the computer controlled pumps. The latter allow maintenance of desirable pressure in each well. The computer program actuates the prescribed set of pressures desired to a sensitivity of +/- 5.0000 psi, allowing the flow to achieve a steady state in the channels.

Figure 9 shows the glucose isomerase crystal growing in the microfluidic channel in a flowing solution. This crystal grows extraordinarily fast – it covers the 70 μm channel cross-section in less than a minute! No systematic growth rate measurements at different supersaturations have been done yet for this protein. The step kinetic coefficients are missing as well.

4.2. Facetting.

Protein crystals are typically facetted, as it is clear from octahedral ferritin crystal shown in Figure 10a. However, the facetted crystals (Figure 10a) may acquire a rounded shape, if thermal energy, kT, is close to the interfacial energy per molecule. If so, the thermal fluctuations destroy perfect molecular packing on the crystal face, and the so-called thermodynamic roughening transition [22]occurs. That may happen with a change in the solution composition and the corresponding decrease in the crystal - solution interfacial free energy. The rough interface contains kinks (the half-crystal molecular configurations where new molecules join the lattice) at the density comparable to the number density of the molecules on the surface. Therefore, the rough, disordered surfaces do not need generation of the new layers (steps), and grow much faster than the smooth, flat, crystal faces. The rough surfaces easily loose their stability during growth, and the crystal acquires dendritic shape shown in Fig 10c. The facetted crystal may also lose its morphological stability (Fig 10b), but the steps on the surfaces and terraces between the steps still exist and, therefore, the sharp corners are preserved. If a smooth interface is at conditions close to the thermodynamic

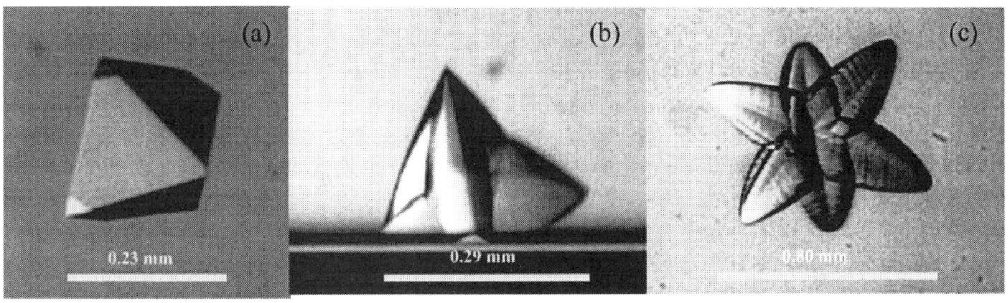

Figure 10. Ferritin crystals growing at different solution compositions and supersaturations acquire regular polyhedral shape (a), skeleton shape at the beginning of morphological instability below the surface roughening transition (b), dendritic shape from morphological instability above surface roughening transition (c).

roughening transition, then roughening may be initiated by an increase of the supersaturation. This is the *kinetic* roughening transition. Neither thermodynamic nor kinetic roughening on protein crystal surfaces has been studied so far.

5. BIOCRYSTAL PERFECTION

5.1. Types of defects

Biocrystals contain the same types of lattice defects as those present in crystals of small molecules. A major factor specific of biomacromolecules is that many of them are prone to conformational changes. Both the conventional lattice defects and the conformational disorder compromise the crystal quality and x-ray diffraction resolution. However, the role that the conventional lattice defects and the molecular conformations play in the ultimate x-ray diffraction quality of the crystals has not yet been systematically addressed. We may guess that inhomogeneous internal stress in a biomacromolecular crystal lattice induces conformational changes and misorientation of the macromolecules, also inhomogeneous, throughout the crystal.

Protein crystals grown in microgravity have shown, in about 20% of the experiments, a higher perfection and larger size than the crystals grown on the ground in presence of solutal convection. However, the reasons and conditions why and when such an improvement may happen are still unknown. A viable hypothesis is that crystals may grow trapping less of the ever present impurities in the stagnant solution [41] and, hence, have a higher perfection.

5.2. Trapping of impurities

It has been found that ferritin crystals grown in reduced gravity contain fewer impurities, the ferritin dimers and some others, than their terrestrial counterparts [41]. This finding may be understood as follows. At the initial stage of growth, a newly created crystal is enriched with impurities that are present in the mother liquor, if the distribution coefficient, k, characterizing the impurity trapping

$$k \equiv c_{is}/c_{il} \tag{3}$$

exceeds unity, $k > 1$. Here, c_{is} and c_{il} (cm^{-3}) are the impurity (i) concentrations in the solid, s, (crystal) and in the liquid, l, (solution). The c_{il} is defined as the concentration of impurities in the solution at the interface of the growing crystal. Then, the impurity balance at this interface, $z = 0$, is

$$- D_i\, \partial c_{il}/\partial z = (c_{il} - c_{is})V = (1\text{-}k)c_{il}V \tag{4}$$

where D_i is the diffusivity of impurities and V (cm/s) is the crystal growth velocity. Evidently, if $k > 1$, then the impurity amount per unit volume in the crystal exceeds that in the solution. The enrichment of the crystal with impurities occurs at the expense of the surrounding solution. Therefore, if the solution is stagnant, the crystal must be surrounded by a zone depleted of impurities. As crystallization proceeds, growth occurs in a solution that is becoming purer and purer. Each impurity molecule trapped distorts the lattice, causing internal stress, mosaicity and possible short range disorder in molecular conformations, thus compromising the crystal diffraction quality [42, 43]. The overall diffusion purification of a crystal occurs, of course, at the expense of the crystal core, which is enriched by the impurity incorporation during the early nucleation and growth stages. X-ray diffraction intensity is proportional to the diffracting volume. Therefore, unless the small impurity enriched crystal core does not induce mosaic blocks all over the crystal, it will essentially not affect the crystal quality of a large single crystal. The impurity depletion zone is a function of the radius of the growing crystal and, therefore, a smaller crystal has a narrower depletion zone. If convection is present in solution, the impurity depletion zone is destroyed and new impurity amounts are delivered continuously to the growing surface by the liquid flow. Therefore, in the presence of convection, the crystal engulfs more impurity and should be of lower quality. To check the hypothesis, it is therefore of paramount importance to measure the distribution coefficients as determined by interfacial processes rather than by bulk transport. So far, only a slight uncontrolled stirring or natural convection was used.

From the practical standpoint, it is more convenient to measure the distribution coefficient counting the impurity concentration not with respect to a volume, but to the concentration c_p of the crystallizing protein in the crystal and in solution [44]:

$$K \equiv (c_i/c_p)_{solid} / (c_i/c_p)_{liquid} = k(c_{pl}/c_{ps}). \tag{5}$$

Since the protein solubility usually does not exceed several percent or less, $c_{pl}/c_{ps} << 1$ and $k >> K$.

The concentration of the impurity trapped by the crystals was measured by dissolving the crystals in pure solvent, and analyzing the solution by high performance liquid chromatography (HPLC). For the ferritin dimers, in the ferritin crystal, $K = 4$, $k \approx 2000$. This can be compared with dimer incorporation into ferritin crystals, grown in space, $K = 1.4$ [41]. In other words, the resistance to diffusion in stagnant solution results in the lower effective distribution coefficient that relates the impurity concentration in the crystal to its average concentration in the solution bulk rather than to the concentration immediately at the growing interface [39, 45].

The dimensionless parameter discriminating the impurity trapping controlled by the bulk transport from the trapping controlled by the processes on the growing interface is $(k\text{-}1)V\delta/D_i$, where δ is the typical length that determines bulk transport rate. In a stagnant solution, δ is of the order of the crystal size, ~ 0.1 – 5mm, whereas in the presence of convection, δ is the thickness of the diffusion boundary layer, depending on the flow rate and geometry, but is less than the crystal size. For the ferritin monomer (a sphere 13nm in diameter), diffusivity is $3.2.10^{-7}cm^2/s$ [2], so that the dimer diffusivity may be estimated from Einstein diffusivity equation to be $D_i \sim 2.10^{-7}cm^2/s$. Therefore, at $V \sim 10^{-7}cm/s$, in a stagnant solution, $\delta \sim 10^{-2}cm$ and $k \sim 2.10^3$, one has $(k\text{-}1)\delta V/D_i \approx 20 > 1$, which means the trapping process falls within the realm of diffusion limited incorporation.

6. CONCLUSIONS

Biomacromolecular crystals are being grown overwhelmingly from aqueous solution with supersaturation induced by adding solution of a precipitant (mainly inorganic salts or polyetheleneglycol) to the protein solution.

Pipetting the precipitant solution into precipitant solution, or *vice versa,* is associated with turbulent mixing, which may be a source of irreproducibility in the number and size of the spontaneously nucleated and then growing crystals in the batch technique. This irreproducibility should be expected at high supersaturations when the time required for complete mixing noticeably exceeds the time required for the supercritical cluster to appear.

The several nm-scale size of the protein molecules allows to directly monitor molecular aggregation and crystal nucleation processes in supersaturated solutions by dynamic light scattering. In different systems, such monitoring suggests nucleation to occur with and without time lag. However, rigorous experimental discrimination between the nucleation time lag and just the expectation time for a nuclei to appear is missing. The large size of biomacromolecules and slow growth of the crystals opened the way to directly visualize growth processes on molecular level by atomic force microscopy. The AFM studies revealed molecular attachments and detachments, trapping of impurities and creation of defects. Overall, biomacromolecular crystallization follows the same laws of crystallization of small molecules. However, the 10-

100 times larger size of biomacromolecules leads to the much lower kinetic coefficient for proteins. On the other hand, protein solutions allow growth of rather perfect, though small, single crystals at supersaturations up to several hundred percents, vs. several percent for conventional solution growth. Though the interface incorporation processes are typically very slow, some of the biomacromolecular crystals experience bulk transport control growth so that solution flow influences the overall growth kinetics. Biomacromolecular crystals may have both smooth and rough interfaces, though the thermodynamic and kinetic roughening transitions have not yet been addressed.

Creation of defects in biomacromolecular crystals and the growth conditions-to-crystal perfection link remains a big challenge of essential significance from both the fundamental standpoint of self-assembly in biosystems and practical importance of biomacromolecular crystal growth. The latter is indispensable to reveal three dimensional structures of biomolecules by x-ray diffraction from crystals and, making use of these structures, to understand molecular functions in living nature on an atomic/ionic level. Also, computer based drug design is impossible without the atomic-resolution structure of the biomolecules.

REFERENCES

1. A.A. Chernov, J. Struct. Biol. 142 (2003) 3.
2. P.G. Vekilov and A.A. Chernov, Sol. St. Phys. 57 (2002) 1.
3. D. Voet and J.G. Voet, Biochemistry, John Wiley, New York, 1995.
4. C.H. Schein, Biotechnology. 8 (1990) 308.
5. A. McPherson, Crystallization of biological macromolecules, Cold Spring Harbor Laboratory Press, Cold Spring Harbor New York, 1999.
6. A. Ducruix and R. Giege, eds. Crystallization of Nucleaic Acids and Proteins. A Practical Approach. IRL Press, Oxford, 1992.
7. A.A. Chernov and H. Komatsu, in Science and Technology of Crystal Growth, Kluwer Academic, Dordrecht, 1995.
8. Macromolecular crystallization in the structural genomics era. J. Struct. Biol. Vol. 142, 2003.
9. J. Crystal Growth. 76 (1986).
10. J. Crystal Growth. 90 (1988).
11. J. Crystal Growth. 110 (1991).
12. J Crystal Growth. 122 (1992).
13. J Crystal Growth. 168 (1996).
14. J Crystal Growth. 196 (1999).
15. J Crystal Growth. 232 (2002).
16. Acta. Cryst. D. 50 (1994).
17. Acta. Cryst. D. 58 (2002).
18. J.M. Garcia-Ruiz and A. Moreno, J. Crystal Growth. 178 (1997) 393.
19. P.E. Dimotakis, J. Fluid Mech. 409 (2000) 69.
20. O. Galkin and P.G. Vekilov, J. Phys. Chem. 103 (1999) 10965.
21. C.A. Schall, J.S. Riley, E. Li, E. Arnold, and J.M. Wiencek, J. Crystal Growth. 165 (3) (1996) 299.

22. A.A. Chernov, Modern Crystallography III, Crystal Growth,. Springer Series Solid State. Vol. 36, Springer, Berlin, 1984.
23. P.G. Vekilov, L.A. Monaco, B.R. Thomas, V. Stojanoff, and F. Rosenberger, Acta Crystallogr. Section D. 52 (1996) 785.
24. J. Lothe and G.M. Pound, J. Chem. Phys. 36 (1962) 2080.
25. J.L. Katz, The Journal of Chemical Physics. 52 (9) (1970) 4733.
26. D. Kashchiev, Nucleation, Butterworth/Heinemann, Oxford, 2000.
27. S. Toshev, A. Milchev, and S. Stoyanov, J. Crystal Growth. 13/14 (1972) 123.
28. I. Gutsow and J. Schmelzer, The vitreous state, Springer, Berlin, 1995.
29. A. Penkova, N. Chayen, E. Saridakis, and C. Nanev, Acta Cryst. D. 58 (2002) 1606.
30. E. Saradakis, K. Dierks, A. Moreno, M.W.M. Dieckmann, and N. Chayen, Acta Cryst. D. 58 (2002) 1957.
31. A.J. Malkin and A. McPherson, J. Crystal Growth. 128 (1993) 1232.
32. Y. He, B.J. Ackerson, W.v. Megen, S.M. Underwood, and K. Schatzel, Phys. Rev. E. 54 (1996) 5286.
33. K. Schatzel and B.J. Ackerson, Phys. Rev. Lett. 68 (1992) 337.
34. B. Garetz, J. Matic, and A. Meyerson, Phys. Rev. Lett. 89 (2002) 175501.
35. P.R. ten Wolde and D. Frenkel, Science. 277 (1997) 1975.
36. O. Galkin and P.G. Vekilov, Proc. Natl. Acad. Sci. USA. 97(12) (2000) 6277.
37. P.G. Vekilov, in Studies and Concepts in Crystal Growth, Pergamon, Oxford, 1993.
38. F. Otalora, J.M. Garcia-Ruiz, L. Carotenuto, D. Castagnalo, M.L. Novella, and A.A. Chernov, Acta Cryst. D. 58(10) (2002) 1681.
39. C.P. Lee and A.A. Chernov, J. Crystal Growth. 240 (2002) 531.
40. S. Spearing, S.Y. Son, J. Allen, and L.A. Monaco in First International Conference on Microchannels and Minichannels. Rochester, New York, 2003.
41. B.R. Thomas, A.A. Chernov, P.G. Vekilov, and D.C. Carter, J. Crystal Growth. 211 (2000) 149.
42. S.-T. Yau, B.R. Thomas, and P.G. Vekilov, Phys. Rev. Lett. 85 (2000) 353.
43. A.A. Chernov, Physics Reports. 288 (1997) 61.
44. B.R. Thomas and A.A. Chernov, J. Crystal Growth. 232 (2001) 237.
45. H. Lin, F. Rosenberger, J.I.D. Alexander, and A. Nadarajah, J. Crystal Growth. 151 (1995) 153.

Crystal Growth – from Fundamentals to Technology
G. Müller, J.-J. Métois and P. Rudolph (Editors)

Dendritic Crystal Growth in Microgravity

M.E. Glicksman
Materials Science & Engineering Department,
Rensselaer Polytechnic Institute, Troy, New York 12180.3590 USA

Microgravity is the unique environment created during space flight in free fall along a low-Earth orbit (LEO). Although the precise acceleration conditions achieved on a space craft during free fall in LEO depend upon a number of factors, some of which will be discussed in these notes, one generally can achieve a quasi-steady acceleration of ca. $10^{-6}g_0$, where g_0 denotes the average acceleration of gravity experienced on the surface of the Earth at sea level, $\approx 9.8\text{m}\cdot\text{s}^{-2}$. In addition, if one attempts crystal growth in microgravity, a spectrum of "g-jitter" is unavoidably encountered, which is random, oscillatory accelerations and vibrations. G-jitter, which can influence some crystal growth processes, arises from the presence of rotating equipment, such as pumps and fans, and from the movement of astronauts who might be on board. The main focus of these notes will be to review the scientific accomplishments of three space flight crystal growth experiments carried out on board the space orbiter/shuttle *Columbia* in 1994, 1996, and 1997. Each experimental hardware system was a primary component of the United States Microgravity Payload Missions (USMP 2, 3, and 4). These experiments addressed fundamental questions concerning dendritic crystal growth, and were named by the National Aeronautics and Space Administration (NASA) as the *Isothermal Dendritic Growth Experiment* (IDGE).

IDGE-1 and IDGE-2 were experimental hardware systems dedicated to testing scientifically the relevant transport and interfacial physics theories describing the steady-state dendritic crystal growth of pure succinonitrile (SCN), which is a body-centered cubic (BCC) crystal with extremely low anisotropy of the crystal-melt interfacial energy. IDGE-3 was a follow-on experiment that used an advanced hardware system to test similarly the steady-state crystal growth of pure pivalic acid (PVA), which is a face-centered cubic (FCC) crystal with a relatively large anisotropy of the crystal-melt interfacial energy. IDGE-3 also accommodated on-board video recording and down-linked telemetry of dendritic growth and mushy-zone melting processes, some of which will be presented during the lecture sequence. The results from these microgravity space-flight experiments, each conducted during a typical 2-week orbital mission, built a solid experimental foundation that tested prevailing transport theories of dendritic crystal growth. In addition, space experiments confirmed the broad validity of so-called dendritic "scaling laws," which will be developed fully in these notes. Finally, the three space flight experiments created an archive that is proving useful for "benchmark" testing of modern dynamical theories of dendritic growth, including advanced computer simulation techniques based on methods such as phase field and level sets.

1. HISTORY AND BACKGROUND

1.1. Approach

Websters International Dictionary defines the word "dendrite" as 1) 'a branching figure resembling a tree produced on or in a mineral; 2) a crystallizing arborescent form.' Indeed, in keeping with these definitions, snow flakes and frost patterns are among the most obvious examples of naturally formed dendritic crystals, the occurrence of which is ubiquitous. The fundamental solidification process to be considered here is dendritic growth. Dendrites (tree-like crystals, from the Greek word $\delta\epsilon\nu\delta\rho o\nu$) are now known to represent the evolved microstructure of an unstable solid-melt interface [1]. Dendrites, in fact, are the most common form of crystal growth encountered when metals and alloys solidify under low thermal gradients. From a commercial standpoint, dendrites also invariably constitute the crystal form encountered in the manufacture of alloy castings, primary metal ingots, and industrial weldments. Except in the restrictive cases of controlled growth of bulk single crystals, where dendritic morphologies are purposely avoided to prevent micro-chemical segregation, dendrites virtually always appear when supercooled melts and solutions are crystallized.

Aside from its underlying technical importance and applications in engineering, geology, and biology, dendritic growth also represents a fascinating category of self-organizing pattern formation phenomena, which in recent years has become a deeply researched subject within the broader field of non-linear dynamics. In fact, the current interest among condensed matter physicists in studying dendritic growth is directly attributable to the simplicity of dendritic growth as a non-linear phase transformation, and to the extraordinary richness of dendritic behavior observed in experiments and predicted from relatively simple theory.

The purpose of these notes is to present a didactic view of dendritic crystal growth, and, thereby, provide a logical guide to the reader of the key concepts underlying this common form of crystal growth. We provide neither an exhaustive bibliography nor a complete history of this complex subject, but rather more of a focussed overview of the major phenomena that form the framework of our current understanding of dendritic crystal growth.

1.2. Steady-state characteristics of dendrites

In most cases, the formation of dendritic crystals involves the coupling of two different processes: 1) the steady-state propagation of the tip region, accounting for the formation of the main, or primary stem, and 2) the time-dependent crystallization of the secondary and tertiary side branches. These processes working together establish the most obvious length scales of a dendrite. Figure 1 shows three dendrites typical of pure substances growing from their supercooled melts: On the left is a snow flake crystal, which is a flattened H_2O dendrite formed in air supersaturated with water. In the middle is a dendrite of a body-centered cubic crystal, succinonitrile (SCN) $[CN - (CH_2)_2 - CN]$, typical of a crystal with nearly isotropic interfacial energy. On the right of Figure 1 is a dendrite of a face-centered cubic crystal, pivalic acid (PVA), $[(CH_3)_3 - C - COOH]$, typical of a non-faceting cubic crystal with strongly anisotropic interfacial energy. A fully developed SCN dendrite consists of a smooth paraboloidal shaped tip region that is nearly a perfect body of revolution, behind which occurs a trailing periodic "wake" of branches that spreads

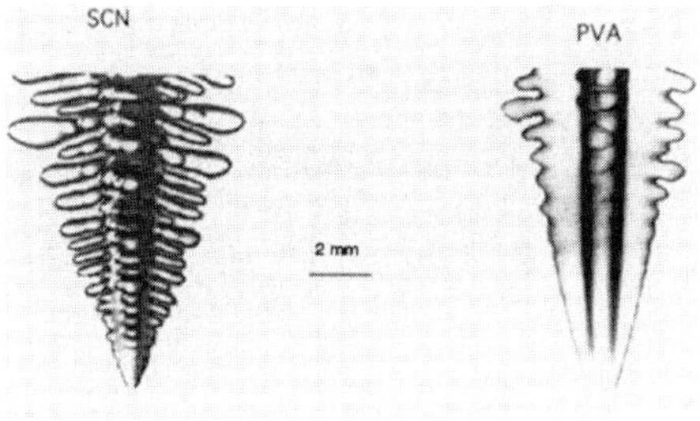

Figure 1. **Left**: Common water/ice snow flake formed from supersaturated moist air. Ice, a rhombohedral crystal, grows easily in its six-fold prismatic directions ⟨0010⟩ in the plane of the micrograph, but thickens with difficulty along its C-axis [0001]. Dendritic patterns in ice are always flattened, and "flake-like". The kinetics of ice dendrites in supercooled pure water were studied by Fujioka and Sekerka [2]. **Middle**: SCN dendrite growing from its supercooled melt. Its tip form is paraboloidal, and is nearly a body of revolution, as expected for a cubic substance with nearly isotropic interfacial energy. The primary growth directions are the six cube-edge directions, ⟨100⟩. Strong local thermal gradients surrounding the tip cause the growth direction of the side branches to depart from the ⟨100⟩, which is the growth direction expected from considerations of crystallographic symmetry. **Right**: PVA dendrite growing from its supercooled melt. Its tip form is distinctly cruciform in cross-section, and the side branches extend close to the expected [010] growth directions. These growth features are typical of crystals with more anisotropic interfacial energy.

away from the primary stem with an opening angle from about 60° to 90°, depending on the crystal's interfacial energy anisotropy. In a yet larger field of view (not displayed), the growth angle of the side branches for SCN eventually approaches exactly 90° with respect to its [100] tip growth direction, because of the underlying cubic crystallographic symmetry. One must remember that a dendritic crystal is still a single crystal with a complex three-dimensional morphology. The PVA dendrite shown on the right-hand side of Figure 1, by contrast, grows with a distinctive cruciform tip, reminiscent of a "Phillips-head" screwdriver tip, that exhibits prominent bulges or branching sheets in the two orthogonal cube-edge directions around the ⟨100⟩ primary growth axis. The side branches in PVA and other anisotropic crystals tend to initiate with a longer delay from the tip, and then extend outward almost perfectly aligned in the expected ⟨100⟩ directions.

1.3. Time-dependent aspects of dendrites

Curiously, until recently, the time-dependent aspects of dendritic growth were virtually ignored. Indeed, most theories were limited to mathematical descriptions of branchless, or so-called "needle crystals" growing at steady-state. As much as dendrites represent the end-state of morphologically unstable crystal growth, it is somewhat surprising that their steady-state characteristics were studied first. As illustrated below in Figure 2, secondary and, eventually, tertiary branches form during dendritic growth. These time-dependent features are critically important to the properties of crystals, insofar as they establish the length scales and pattern over which all the chemical impurities and alloy components would be concentrated. This scale of chemical microsegregation is of immense practical importance in determining key engineering properties of materials that solidify dendritically. In pure materials, on the other hand, such as the three crystals illustrated in Figure 1, the patterns observed during crystal growth are best thought of as revealing the spatial "microsegregation" of the enthalpy. Such spatial enthalpy "distributions," in contrast to chemical microsegregation patterns, disappear rapidly as the dendritic crystal growth process is completed. The reason for the large disparity in the time scales for the annealing, or relaxation, of the enthalpy and solute segregation patterns in dendritic growth, is simply that chemical diffusivities in crystals are much smaller than are the heat diffusivities, by as much as 6 - 10 orders of magnitude.

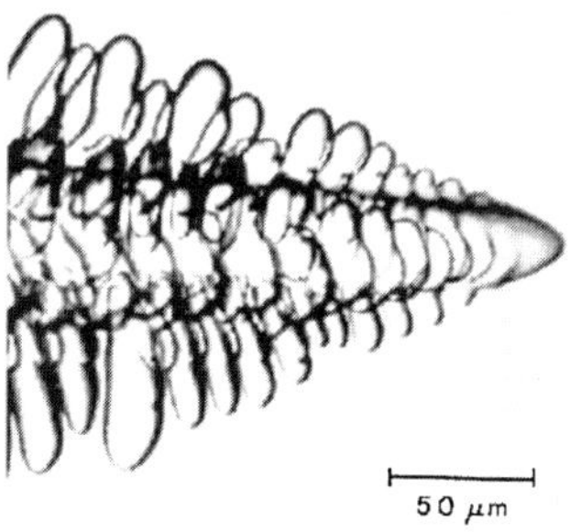

Figure 2. Overall view of a pure SCN thermal dendrite, showing the complex time-dependent development of the side branches aft of the steady-state tip. Side branches arise as instabilities behind the marginally stable tip. In a pure material the side branches fill in during the later stages of solidification, and eventually disappear. In alloys, by contrast, the side branches remain as virtually permanent "relicts" of the crystallized structure, and define the spatial scale of chemical microsegregation. Only long heat treatment times can remove dendritic microsegregation. For this reason, dendrites are usually avoided in single crystal growth.

1.4. Physico-chemical basis for dendritic growth

Dendritic crystal growth, as mentioned above, is generally acknowledged to be controlled by a diffusion-limited process. For example, in pure materials, the growth rate of a dendrite is controlled by the diffusion of latent heat away from the advancing crystal-melt interface. In addition, one must recognize that molecular or atomic transfer across the crystal-melt interface, as well as the spontaneous creation of the interface itself, requires expenditure of the free energy available for the dendritic transformation. These energetic processes encompass the basic thermophysical phenomena that lead to the formation of dendritic patterns. Unlike dendrites in pure materials, alloy dendrites propagate as the crystalline solid grows and rejects its excess solute, which flows away from the interface by chemical diffusion through the surrounding melt or solution. In addition to the solute rejection process occurring at the crystal-melt interface and the related chemical diffusion in the melt, the latent heat of fusion is also released from the creation of an orderly crystalline phase from its more random molten phase. Latent heat also must flow away from the dendritic interface by transport processes such as thermal conduction, convection, or radiation. Chemical diffusion, which is slower than thermal transport, is the rate-controlling process in alloy dendrites. In this case, the solute transport equations can be scaled to a form equivalent to that for enthalpy diffusion in pure-material dendrites. However, the presence of two coupled transport fields in alloy melts, as well as the temperature dependence of the equilibrium phase concentrations, complicates the analysis of the process somewhat. Therefore, in reviewing the salient features and theoretical approaches toward modeling of dendritic growth, we focus in this notes on the solidification of pure materials from their supercooled melts.

1.5. Thermodynamics and kinetics of dendritic crystal growth

The thermodynamic driving forces and the kinetic resistances encountered in dendritic growth were first clearly described about thirty years ago by Temkin [3] and by Bolling and Tiller [4]. These investigators considered dendritic crystal growth of a pure substance and identified three coupled *kinetic* effects: 1) transport (conduction) of latent heat; 2) molecular attachment at the crystal-melt interface; and 3) creation of interfacial area. Temkin, Bolling, and Tiller associated each of these kinetic effects with the dissipation or consumption of a fraction of the total free energy available for dendritic crystal growth. The total free energy available for crystal growth can be expressed as a quantity that is nearly proportional to the supercooling, $\Delta T = T_m - T_\infty$, where T_m is the bulk melting point, i.e., the equilibrium temperature at a stationary, planar crystal-melt interface, and T_∞ is the temperature of the supercooled melt far from the interface. The relationship between the available free energy and the supercooling is accurately linear in ΔT provided that $\Delta T/T_m \ll 1$, a condition that is well satisfied in practice. The total free energy may be subdivided among the coupled physico-chemical processes identified above, as is also indicated schematically in Figure 3 (Left). In declining order of magnitude, the undercoolings used to drive these processes are: 1) ΔT_{trans}, the temperature drop associated with thermal transport; 2) δT, the supercooling required to "drive" the net atomic or molecular transfer across the crystal-melt interface from the melt phase to the solid and, thereby, effect attachment to the crystal lattice; and 3) δT_R, the geometrically-induced small temperature depression, often called the Gibbs-Thomson capillary temperature shift, which yields the interfacial equilibrium temperature, $T_e(R)$, [5], [6], [7]. This depression of the equilibrium interface temperature from the presence of excess interfacial free energy at a curved interface may be expressed as

$$\delta T_R = T_m - T_e(R) = \frac{2\gamma\Omega}{\Delta S_f} \cdot H \tag{1}$$

In Eq.(1) we define the total interfacial curvature as equal to twice the mean curvature, $2H = \kappa_1 + \kappa_2$, where κ_1 and κ_2 are the principal curvatures at a point on the crystal-melt interface; γ , Ω, and ΔS_f are, respectively, the molar specific excess interfacial free energy, the molar volume of the crystalline phase, and the molar entropy change for melting. The total, or applied, supercooling, ΔT, distributes into several terms, each proportional to

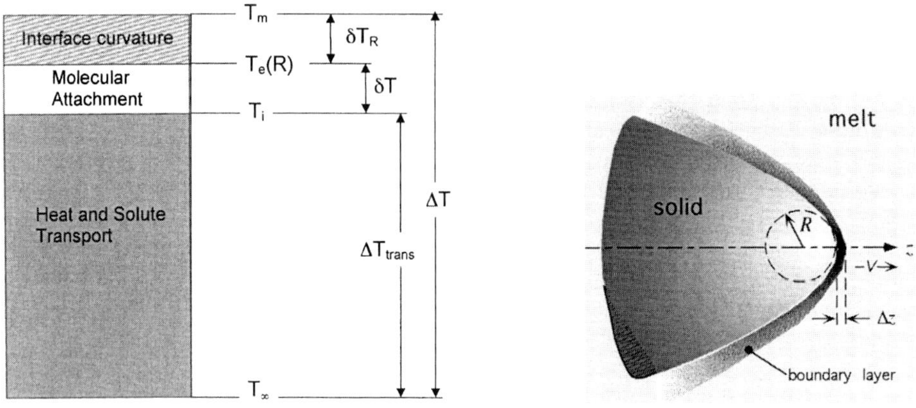

Figure 3. **Left:** Distribution of supercooling (free energy) during steady-state dendritic crystal growth. The total (applied) supercooling is ΔT, which is the difference between the bulk melting temperature, T_m and the supercooling temperature, T_∞, of the melt. The greatest portion of ΔT, needed for thermal transport, is ΔT_{trans}. The temperature drop caused by kinetic molecular attachment at the crystal-melt interface is δT. The shift in the equilibrium melting point caused by curvature of the crystal-melt interface is δT_R. Thus, the equilibrium temperature of a curved interface, such as a dendrite tip, is $T_e(R)$. The overall crystal-melt interface temperature is T_i. **Right:** Sketch of the smooth, branchless tip region of a steady-state dendrite as modeled by Ivantsov. The paraboloid is characterized by the tip radius, R, and the tip speed, V along the growth axis, z. A "boundary layer" of thickness Δz near the tip defines the region over which the temperature falls from the melting point,T_m, to the supercooling temperature, $T_\infty = T_m - \Delta T$. The thermal boundary layer is suggested by the dark gray envelope surrounding the solid-liquid interface.

the free energy dissipated or stored by the process. As Figure 3 (Left) suggests, most of the available free energy is dissipated by latent heat diffusing away from the moving dendritic interface, whereas only a relatively small amount of free energy is normally needed to activate the interfacial molecular events. Finally, we note that the Gibbs-Thomson effect corrects for the fact that a moving curved interface stores free energy, and as a result, exhibits a small depression, δT_R, that is proportional to its local mean curvature, in its thermodynamic equilibrium temperature, $T_e(R)$, compared to that of a planar interface.

1.6. Anisotropy

Real crystal-melt interfaces always exhibit some anisotropy because of the intrinsic anisotropy of the crystalline molecular fields. The surface tension anisotropy around the primary growth axis for cubic systems, such as SCN and PVA, may be expressed as

$$\gamma(\theta) = \gamma_0 + \epsilon \cos 4\theta \qquad (2)$$

Equation(2) is comprised of an energy modulus, γ_0, and the angularly dependent term that has an amplitude ϵ. In cubic systems there is the thermodynamic requirement for interfacial stability that the chemical potential, μ, must be a convex function, so $\mu \geq 0$. Thus,

$$\mu = \gamma(\theta) + \frac{\partial^2}{\partial\theta^2}\gamma(\theta) \geq 0 \tag{3}$$

If the chemical potential, μ, is evaluated by inserting Eq.(2) into Eq.(3) one obtains

$$\mu = \gamma_0 - 15\epsilon \cos 4\theta \tag{4}$$

By setting the thermodynamic condition for interface stability, viz., $\mu \geq 0$, one obtains from Eq.(4) the result that $\epsilon \leq \frac{\gamma_0}{15}$, for $\theta = 0, \frac{\pi}{2}, \pi, \frac{3\pi}{2}$, which angles correspond to four of the six $\langle 100 \rangle$ spatial directions where the chemical potential is minimized. These correspond to the six dendritic growth directions of a cubic crystal. Note that the result derived here shows that the maximum possible amplitude for anisotropy in cubic crystals is ≈ 0.06. Exceeding this amount would result in equilibrium faceting of the crystal-melt interface. The occurrence of facets changes the atomic-scale structure of the interface over the orientations so affected, and usually reduces the mobility. Dendrite-forming systems are therefore usually non-faceted.The choice of SCN and PVA as test systems for dendritic growth studies is based on the fact that for SCN $\epsilon \approx 0.005$, i.e., slightly anisotropic, whereas for PVA, $\epsilon \approx 0.05$, i.e., very anisotropic, but still non-faceting. Thus, the two systems are complementary regarding the nature of the their interfaces.

The description of free energy dissipation just described shows that the process of dendritic growth, even in pure crystals, is complex. For example, the interface temperature, T_i, depends on *both* geometrical effects (through the Gibbs-Thomson relation) and certain additional kinetic details, including the functional relationships among the interface supercooling, δT, velocity, V, orientation, θ, and mobility, M. These kinetic factors are usually expressed as a combined interface kinetic term of the form $\delta T = K(V, \theta, M)$. Every material has an unique interfacial kinetic relationship, but, fortunately, most fall into just a few broad categories. Jackson [8,9] has shown that metals, some ionic compounds, and a few organic materials, such as SCN and PVA shown in Figure 1, tend to form crystal-melt interfaces that are "rough" on an atomic scale. So-called rough interfaces easily accommodate atomic or molecular transfer and attachment from *all* interfacial orientations with respect to the principal crystal axes, so δT tends to be extremely small (high molecular mobility, M) and only weakly dependent on orientation, θ. Such materials virtually always crystallize as dendrites. Semiconductors and most covalently bonded materials, on the other hand, display much greater directionality in bonding, and therefore tend to exhibit "smooth," atomically faceted, interfaces. Covalent materials often have a δT that is small in their "rough" orientations and large in the "smooth" or faceted ones. Such materials tend to form faceted dendrites containing internal twinning defects. Finally, polymers and complex network-forming silicate materials have low mobilities so that δT is almost as large as the total supercooling, ΔT, so that transport of heat and species become relatively unimportant components of the overall crystallization process. As a consequence, dendrites seldom ever form in these materials. Although some thermal or constitutional supercooling is *always* required to form dendrites, polymers and complex oxide and sulfide melts can crystallize under extraordinarily large supercoolings in a nearly isothermal, non-dendritic manner.

2. STEADY-STATE DENDRITIC GROWTH

Numerous theoretical [10–12], and quantitative experimental dendritic crystal growth studies have been reported over the past 25 years [12,13,15–18,20,21]. Dendritic solidification requires the coupling of two independent growth processes: 1) the steady-state evolution of the dendrite tip, and 2) the non steady-state development of dendrite branches. The free energy of any system decreases as a crystal freezes from its supersaturated melt. For this to happen, the latent heat generated during crystallization must be transported from the crystal-melt front by thermal transport. Not surprisingly, "thermal" dendrites are simplest to describe as the crystallization of a pure, supercooled molten phase. By contrast, when alloy dendrites grow from a supersaturated melt, both thermal and solutal boundary layers are involved. Mathematically, however, the dendritic growth problems for pure and alloy melts are essentially identical, consisting of solving: 1) the diffusion equation, 2) boundary conditions of heat and mass conservation at the moving front, and 3) capillary effects introduced at the curved crystal-melt interface.

2.1. Transport theory

The classical theory of "diffusion-limited" dendritic growth was published in 1947 by the Russian mathematician G.P. Ivantsov [22]. Ivantsov's transport solution describes steady-state transport of the heat energy surrounding a branchless, needle-like, dendrite growing in an infinite, quiescent, supercooled melt. See Figure 3 (Right).

Ivantsov modeled the steady growth of a dendrite as a perfectly smooth, branchless, paraboloidal body of revolution. This theory may be applied to a wide range of supercooled or supersaturated melts, provided that the diffusivity for heat (or for alloys, solute) is known, along with the molar latent heat and heat capacity of the melt. Dendrites, of course, are in reality not smooth, but always exhibit side branches that change over time. Time dependent features, such as side branch evolution, are ignored in Ivantsov's theory. His theory predicts a mathematical relation between a dendrite's tip velocity, V, and its tip radius of curvature R, as functions of the supercooling (or supersaturation). Furthermore, the dendrite tip grows at a steady speed, V, into a pure melt with a spatially uniform initial supercooling, defined as $\Delta T = T_m - T_\infty$. The steady-state shape was chosen to be paraboloidal, and the crystal-melt interface was assumed to remain everywhere at the equilibrium melting temperature, T_m.

2.2. Ivantsov's transport solution

A clever insight that Ivantsov added to his analysis is that the paraboloidal shape of the dendrite tip allows the use of a separable coordinate system (confocal paraboloids). Ivantsov's solution has since been generalized to other "needle" crystals that are not bodies of revolution. These other shapes also grow at constant speed, and provide a complete family of shape-preserving elliptical paraboloidal interfaces [24]. For the paraboloid of revolution, a the "needle" crystal, Ivantsov obtained the dimensionless temperature solution $\Theta(z/R_{tip}, r/R_{tip})$ surrounding the growing dendrite. This solution may be used to relate the dimensionless supercooling, $\Delta\Theta$, to the growth Péclet number, Pe. The growth Péclet number is defined as the dimensionless quantity

$$Pe \equiv \frac{VR}{2\alpha_\ell} \tag{5}$$

where α_ℓ is the thermal diffusivity of the melt, and V and R are the steady-state dendritic tip speed and radius, respectively. The so-called *characteristic* equation connecting these transport parameters is

$$\Delta\Theta = Pe\; e^{Pe} E_1(Pe) \tag{6}$$

where $E_1(Pe)$, is the 1st exponential integral, a tabulated function which is defined as the definite integral

$$E_1(Pe) \equiv \int_{Pe}^{\infty} \frac{e^{-\zeta}}{\zeta} d\zeta \tag{7}$$

A plot of the Ivantsov relationship, Eq.(6), on linear coordinates is shown in Figure 4 (Left) and over a wider scale on logarithmic coordinates in Figure 4 (Right). If the supercooling

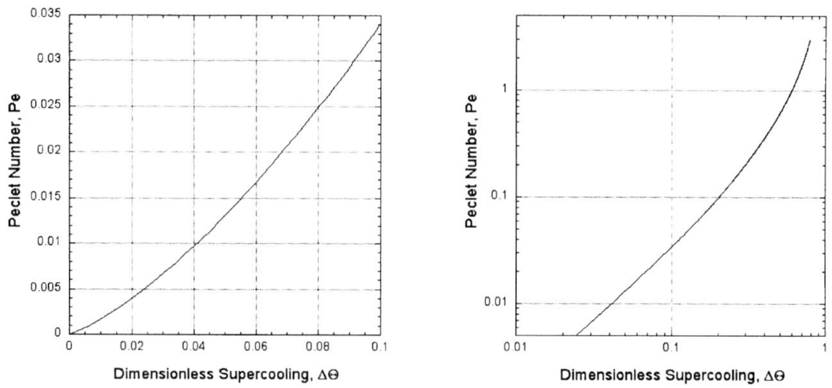

Figure 4. **Left:** Péclet number, Pe, versus dimensionless supercooling, $\Delta\Theta$, linear scales. Most practical cases of dendritic crystal growth involve small values of the dimensionless supercooling, i.e., $\Delta\Theta \ll 0.1$. At small supercoolings the Ivantsov relationship may be approximated by the transcendental form shown in Eq.(8). **Right:** Growth Péclet number, Pe, versus dimensionless supercooling, $\Delta\Theta$, log-log coordinates. The Péclet number diverges as $\Delta\Theta \to 1$.

is small, as is the case for most practical crystal growth applications, then it is also true that $Pe << 1$, and one may employ the following mathematical approximations for each of the terms appearing in Ivantsov's characteristic equation:

i. $e^{Pe} \approx 1 + Pe + \ldots$, and

ii. $E_1(Pe) \approx -\ln Pe - \gamma_E + Pe + \ldots$, where the term $\gamma_E = 0.57721\ldots$ is Euler's constant.

Substitution of these approximations into Ivantsov's transport solution, Eq.(6), yields the following transcendental approximation for the steady-state characteristic equation, valid at small growth Péclet numbers, $Pe << 1$:

$$\Delta\Theta = Pe(1 + Pe)(Pe - \ln Pe - \gamma_E) \tag{8}$$

Ivantsov's steady-state dendrite solution implies that $\Delta\Theta = \mathcal{F}(Pe)$. In solidification, however, the independent variable is the supercooling, $\Delta\Theta$, whereas the Péclet number is the dependent variable. Thus, the inverse solution to Eq.(6) would be generally more convenient. Although an analytic inverse to Eq.(6) cannot be expressed as a "closed form," that is, as a combination of a finite number of elementary functions, one can think of the formal existence of the inverse to Eq.(6) and even express it symbolically as $Pe = \mathcal{I}v^{-1}(\Delta\Theta)$. Here the symbol $\mathcal{I}v^{-1}$ represents the formal inverse of Ivantsov's expression. In any event, a specified value of the supercooling, $\Delta\Theta$, fixes the magnitude of the growth Péclet number. If the melt supercooling is specified, then the following relationship holds between the dendritic growth speed and the tip radius,

$$VR_{tip} = 2\alpha_\ell \cdot \mathcal{I}v^{-1}(\Delta\Theta) \tag{9}$$

Equation (9) contains two unknowns V and R_{tip}, so it is by itself incapable of predicting the *unique* velocity and radius of the dendrite. Ivantsov's transport solution provides only a single relationship between speed, V, and dendrite tip size, R_{tip}. An additional independent equation is needed to solve the transport problem uniquely in terms of observable dendritic characteristics. Figure 5 (Left) shows the general hyperbolic relationships established between V and R_{tip} through Ivantsov's transport solution, Eq.(9). Curves are plotted for several supercoolings at a fixed value of the melt's thermal diffusivity, α_ℓ. These curves show how the velocity of a thermal dendrite must vary with its corresponding tip radius, at constant supercooling. Thick dendrites grow slowly relative to slender ones that grow quickly. The tendency for slowly growing dendrites to have large blunt tips, and rapidly growing ones to have small sharp tips, is referred to as the "point effect" of diffusion, which is a well-known phenomenon in diffusion and electrostatics.

The Ivantsov solution cannot, however, predict which *unique* combination, (V, R_{tip}), is actually exhibited when one conducts a solidification experiment. Thus, the dendrite's "operating state," namely, $(\Delta\Theta, V_{op}, R_{op})$, remains indeterminate within the predictive power of Ivantsov's analysis. This is, of course, not a surprising outcome, because the energy transport solution, Eq.(9), is a single equation in two unknowns. Additional physics that adds an independent equation is needed to solve the operating state of a steady-state dendrite.

2.3. Interfacial physics

The growth Péclet number predictions from transport theory can be decomposed into unique speed and tip radius predictions if one introduces an additional equation that provides an independent, second length scale to the problem. The additional length scale combined with the Ivantsov transport solution selects the unique dendritic operating state. Although the physical mechanisms invoked to provide this extra length scale differ in detail in each interfacial theory [23], their effect may be expressed in terms of a "scaling

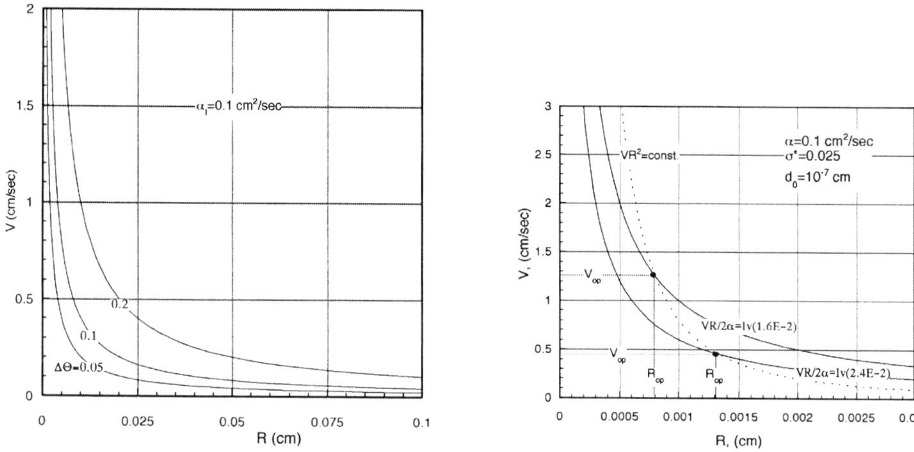

Figure 5. **Left:** Ivantsov's transport solution relating the steady-state growth speed, V, and the dendritic tip radius, R, for various dimensionless supercoolings, $\Delta\Theta$. Each curve is a hyperbola that specifies continuous pairs of V and R values that satisfy the transport of latent heat through the melt as specified by Ivantsov's characteristic equation, Eq.(6). **Right:** Operating points of a steady-state dendrite at two supercoolings. The solid hyperbolas represent Ivantsov's transport solutions at each supercooling, whereas the broken curve represents the marginal stability condition. The two intersections denote the respective operating conditions defined uniquely by V_{op} and R_{op}.

factor," $\sigma^\star$, defined as

$$\sigma^\star = \frac{2\alpha_\ell d_0}{VR^2} \tag{10}$$

The length scale d_0 appearing in Eq.(10) denotes the "capillary length", which is a microscopic quantity equal to circa 10^{-7} cm, conventionally defined as the following quantity,

$$d_0 \equiv \frac{2\gamma\Omega C_p}{\Delta H_f \Delta S_f} \tag{11}$$

where C_p is the molar specific heat of the melt, and ΔH_f and ΔS_f are the molar enthalpy and entropy changes on fusion, respectively. The term γ appearing in Eq.(11) is again the crystal-melt interfacial energy. It is now accepted that d_0 provides the required second length scale in the dendrite growth problem. This important discovery was was originally proposed by Nash in 1974, who developed the modern scaled form of the "dendrite equation" [25,26]. Remarkably, $\sigma^\star$, defined in Eq.(10), is found experimentally to fall in a narrow range for different materials over wide ranges of supercoolings. Thus, Eq.(10) may be written in the form of the dendritic "scaling law,"

$$VR^2 = \frac{2\alpha d_0}{\sigma^\star} \approx \text{const.} \tag{12}$$

Laboratory and space flight experiments, to be discussed in Section 3 of these notes, verify the unique values of V and R observed experimentally over a range of supercoolings and substances. The explicit relationships that predict V and R are derived in [1], and given as

$$V = \frac{2\alpha \Delta H_f \Delta S_f}{\gamma \Omega C_p} (Pe)^2 \sigma^\star \tag{13}$$

and

$$R = \frac{d_0}{\sigma^\star Pe} \tag{14}$$

The validity of this overall theoretical approach has a number of interesting implications for applying dendritic growth theory to practical situations, by providing microstructure rules appropriate to dendritic crystal growth processes. In fact, all the stability-based theories that have been suggested over the past 30 years , e.g., Oldfield's numerical model [27], the spherical stability model [18], Langer and Müller-Krumbhaar's paraboloidal model [10,28] eventually reduce to the identical dendritic scaling law, Eq.(12), and to the specific expressions Eqs.(13) and (14). Theoretical estimates for the interfacial scaling factor, $\sigma^\star$, may be found straightforwardly using marginal stability theory or "microscopic solvability" [29–31]. These approaches all yield similar predictions to the original marginal stability theory [10], but also take into account the anisotropy of the crystal-melt interfacial energy, and therefore include some information about the atomic bonds in the crystalline solid and their symmetry. The theoretical predictions combining transport theory and interfacial physics may be summarized for materials with nearly isotropic γ-values as

$$\sigma^\star = \frac{1}{4\pi^2} \approx 0.025 \tag{15}$$

In summary, Eqs.(13), (14), and (15) collectively permit quantitative estimation of dendritic growth kinetics in a wide variety of nearly isotropic materials. It is especially interesting to point out that this estimate for $\sigma^\star$ (from linear morphological stability theory) is almost independent of the assumed geometry of the crystal-melt interface and the physical properties of the system [11]. The surprising robustness of dendritic growth theory permits its application to a wide range of materials: from low-conductivity organics, such as SCN and PVA, to a variety of low melting point FCC metals, such as Pb, Al, Ag, Au, and Cu, and to transition metals like Ni, Fe, and Mo, semimetals and semiconductors such as Bi, Ge, and Si, and even to ceramics and ionic salt systems.

Figure 5 (Right) illustrates how the marginal stability scaling law, Eq.(12), and Ivantsov's transport solution, Eq.(6), intersect graphically, and thereby define at each supercooling the unique "operating point," of a steady-state dendritic crystal growing at the specified supercoolings. The predicted operating points of a steady-state dendrite, indicated as V_{op}, R_{op} in Figure 5 (Right), can be expressed as a pair of numerical approximations that are convenient for checking theoretical predictions against experimental observations:

$$V_{op} \cong 0.018 \, \frac{\alpha_\ell T_m \Delta S_f^2}{C_p \gamma_{s\ell} \Omega} \cdot \Delta\Theta^{2.5} \; [\mathrm{cm \cdot s^{-1}}] \tag{16}$$

and

$$R_{op} \cong 55 \, \frac{C_p \gamma_{s\ell} \Omega}{T_m \Delta S_f^2} \cdot \Delta\Theta^{-1.25} \text{ [cm]} \tag{17}$$

The expressions given as Eqs.(16) and (17) remain valid for moderate values of $\Delta\Theta$, which for typical experiments fall in the approximate range of dimensionless supercooling, $0.01 < \Delta\Theta < 0.3$. At smaller values of $\Delta\Theta$, the theory predicts that $V_{op} \approx \Delta\Theta^2$, and $R_{op} \approx \Delta\Theta^{-1}$. As a practical matter, however, at extremely small supercoolings, the thermal boundary layer surrounding a slowly growing dendrite may become sufficiently broad and diffuse that the crystal no longer can be considered as freely growing and "isolated" from the environment. At large values of $\Delta\Theta$, i.e., values of $\Delta\Theta$ approaching unity, the assumption of local equilibrium, inherent in formulating this entire transport and stability analysis, breaks down, and the theory, as presented here, becomes inaccurate.

Theoretically predicted Péclet numbers can be verified experimentally by measuring V and R simultaneously at a sequence of known supercoolings. The normal presence of gravitationally-induced convective heat transfer alters the diffusion-limited conditions under which Ivantsov's prediction of the growth Péclet number holds true. The experimental approach taken by the author and his colleagues was to provide a diffusion-controlled environment to measure dendritic growth, which will be discussed later in these notes.

3. EXPERIMENTAL VERIFICATION

3.1. Model test systems

Quantitative descriptions of dendritic microstructures, suitable for critically testing theories and hypotheses have been reported over the past 30 years. Most of the data reported involve crystal growth experiments carried out on metals and semiconductors, which are opaque, and because of their high melting points are also reactive, and generally difficult to maintain in a high state of purity. Thus, the most convincing data and observations reported derive almost exclusively from just a few critical quantitative experiments carried out on well-characterized model transparent systems [13,18], viz. SCN and PVA. These substances can be vacuum distilled and then multiply zone refined to achieve exceptional states of purity (>6-9's). The choice of these organic systems is also based on their low melting points, optical transparency, high states of purity, and lack of any facets during crystallization. In addition, these systems have extremely well-characterized melting points (±0.001K), carefully measured thermal diffusivities in the molten state (±1%), known entropies of fusion, accurately measured crystal-melt interfacial energies (±10%), also known as functions of the crystallographic orientation of the interface around the dendritic growth axis. Under terrestrial conditions, however, dendrites invariably interact with buoyancy-induced hydrodynamic flows in the melt, (see Figure 6). The buoyant melt motions are induced by the thermal conduction field surrounding a dendrite. As shown, the presence of convective flows in the melt can dramatically alter the heat and solute transport occurring at the crystal-melt interface, and modify the crystal morphology. Moreover, the interaction between melt flow and thermal transport is especially severe for melts such as SCN and PVA that have high Prandtl numbers, which is the ratio between the kinematic viscosity and the thermal diffusivity. Consequently, the basic theory of dendritic growth, i.e., the combination of Eqs.(13), (14), and (15), are of necessity best

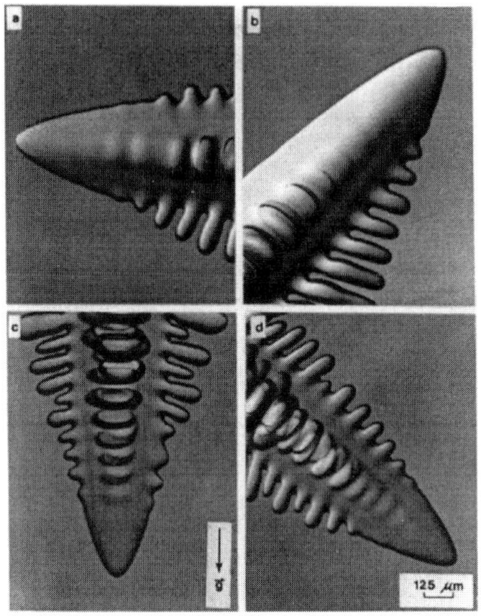

Figure 6. SCN dendrites growing under terrestrial conditions at identical supercoolings but at different orientations with respect to the gravity vector, $\vec{g_0}$. a) [100] orthogonal to $\vec{g}$; b) [100] is rotated $\approx 135°$ to $\vec{g_0}$; c) [100] parallel to $\vec{g_0}$; d) [100] is rotated $\approx 45°$ to $\vec{g_0}$. Note the influence of the orientation on the micromorphology of each dendritic crystal.

tested under strictly diffusion-controlled conditions, where gravitational acceleration that leads to convective melt motion is reduced to almost zero.

3.2. Microgravity experiments

Long-duration microgravity ($g \approx 10^{-6} g_0$, where $g_0 = 9.8\text{ms}^{-2}$ is the local gravitational acceleration at sea level) is achieved by a continuously orbiting space craft operating at sufficiently high orbits ($\geq$ 125 nautical miles). The microgravity acceleration results from the steady molecular drag imparted by the residual atmosphere on the rapidly moving space craft. An orbiting platform, such as a space shuttle traveling at orbital speeds, is actually undergoing "free fall," a phenomenon first clearly explained by Sir Isaac Newton. See Figure 7 (Left) for the details of Newton's explanation. The position within a space craft also has an effect on the local acceleration. Specifically, the lowest quasi-static (< 0.01 Hz.) acceleration levels are achieved along the flight path, defined within the space craft as the location of its centroid, or center-of-mass. This remains true irrespective of the space craft's orientation. The space craft's orientation, however, will influence the precise magnitude of the residual molecular drag experienced within, but practical operational issues, such as thermal loading from sun light, and navigational requirements, such as star sightings, usually determine the actual allowed flight orientation. An excellent primer on microgravity environments achieved in space craft is available in the book by Feuerbacher et al. [19]. For the United States Microgravity Payload Mis-

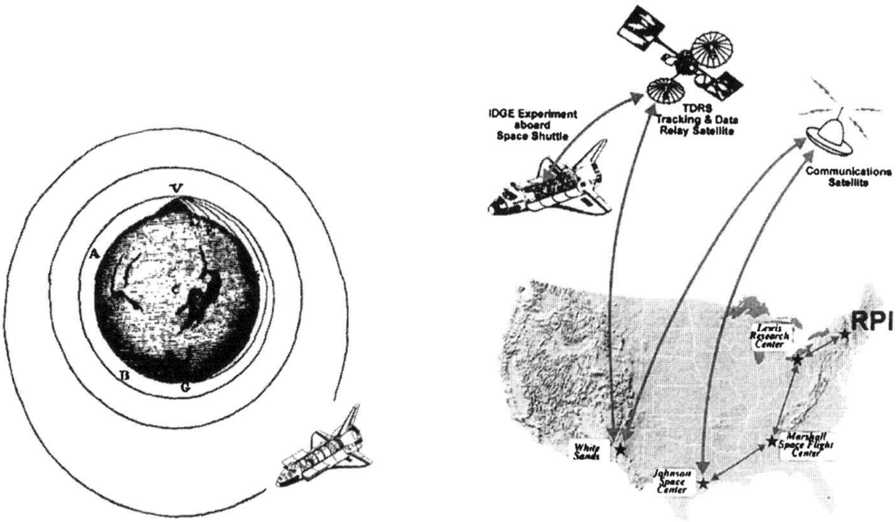

Figure 7. **Left:** Free fall was originally explained by Isaac Newton: A canon firing a ball in a horizontal direction from the mountain top, $\mathcal{V}$, falls along a ballistic path, landing some distance away, at, say, $\mathcal{F}$. As the canon ball's initial velocity increases, its landing distance grows, to, say, $\mathcal{G}$. Eventually, a critical velocity will be reached where the canon ball just makes a full rotation, or orbit, around the Earth. This is the so-called "escape" velocity, $\approx 17,500$mph. The space shuttle is given enough energy to reach the escape velocity, plus a bit more to free fall through a nearly circular orbit that is about 150 nautical miles above the Earth's surface. Thus, orbits are equivalent to ballistic free falls. **Right:** External data path for command and control of the IDGE space flight experiment aboard space shuttle/orbiter *Columbia.* The data path between the author's laboratory at Rensselaer Polytechnic Institute (RPI) in Troy, New York, USA, and the space shuttle consisted of three, wide bandwidth T-1 land lines, connected through an S-band radio channel via a commercial communication satellite and by NASA's Tracking and Data Relay Satellite (TDRS) system. The TDRS allows communication to the space shuttle. Video data downlinks required use of the high-frequency K-band antenna aboard the shuttle.

sions (USMP), typical orbits chosen were nearly circular, with apogees of approximately 150 nautical miles, with flight in the tail-to-Earth orientation. Quasi-static microgravity accelerations on the USMP missions of circa $10^{-6}g_0$ were achieved in practice and maintained for periods up to about two weeks. Vibrational accelerations, or so-called "g-jitter" (oscillatory accelerations with frequencies greater than 10^{-2}Hz.) are also encountered from crew movements, jet thrusts for attitude control, waste water dumping, rotating machinery, structural vibrations, etc. The g-jitter during flight proved to be unimportant for the purposes of achieving convection-free dendritic growth and remotely teleoperating operating the IDGE.

Flights were carried out in LEO in 1994, 1996, and 1997, at orbital apogees ranging from about 140-170 nautical miles. Thus, in LEO, under carefully arranged free-fall conditions, the quasi-steady acceleration due to gravity can be reduced by a factor of less than one-millionth. Reducing gravity to such small levels also requires that the crystal growth experiment itself must be located as close as possible (within $\approx$ 1m) from the centroid coodinate. The position of scientific hardware, along with most of the "up-mass" placed within the shuttle is decided by engineers years before the launch. This decision is critical to many scientific experiments, including some crystal growth experiments. Fortunately, the IDGE was relatively insensitive to both g-jitter and the vector orientation of the quasi-static gravitational acceleration.

3.3. IDGE

The IDGE (*Isothermal Dendritic Growth Experiment*) is a microgravity materials science space flight experiment that was designed to provide terrestrial and microgravity measurements on the kinetics, morphology, and dynamics of dendritic solidification under pure diffusion control [15,18,20,32]. To communicate and control the *Isothermal Dendritic Growth Experiment* (IDGE), a data path between Earth and orbit was required. Commands for temperature control, camera exposure, nucleation of crystals, etc. were sent to the space shuttle via the scheme shown in Figure 7 (Right). Data streams generated by the IDGE hardware were telemetered back to Earth via the shuttle's S-band antenna, excepting video data that were sent via the K-band, high-frequency antenna. The core capability for research on dendritic growth was built into the IDGE thermostat and optical data acquisition system. A simplified sketch of this system is shown above in Figure 8. The space hardware had the capability to control the temperature of a small volume (ca. 100cm^3) of ultra-pure molten SCN or PVA to better than $\pm$0.002K, both temporally and spatially. This control accuracy applied within a restricted temperature range ($\pm$2K) about each material's melting point, T_m.

The melting points themselves were known to better than $\pm$0.001K,an unusually high level of precision, because their purities exceeded 6-9's. In this manner the supercooling that controlled the kinetic of crystal growth, $\Delta T = T_m - T_\infty$, set prior to nucleation was known to an accuracy better than $\pm$0.003K. Nucleation of the crystalline phase was initiated by activating a thermoelectric element that rapidly chilled a small isolated volume of the melt just outside the growth chamber. Once nucleated, the dendritic crystals grew along a capillary "stinger" that led the crystallization front to the center of the growth chamber. Once the dendrites emerged, as shown schematically in Figure 8, the optical system detected their presence, and commenced data acquisition by flash photography and videography. See also Figure 9 for additional details of the optical system, including the arrangement for stereoscopic data recording on 35mm film and video tape.

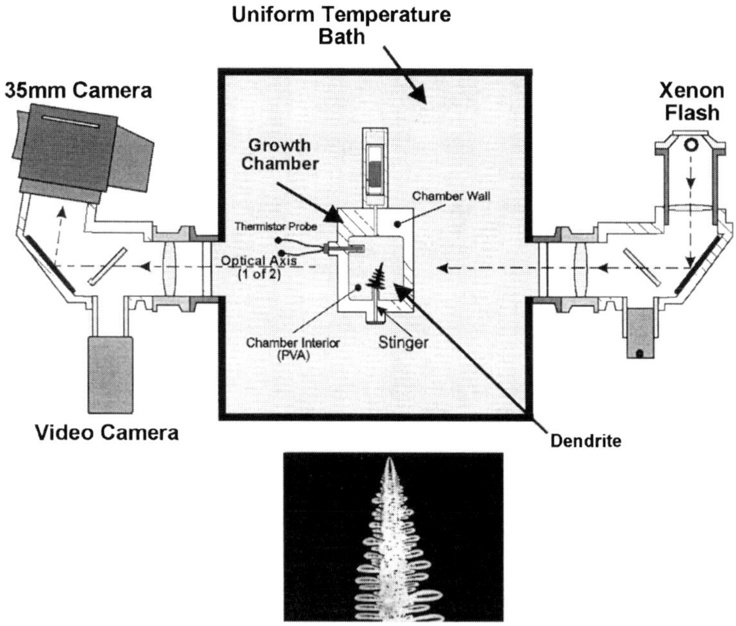

Figure 8. IDGE space flight crystal growth chamber and optical data acquisition system. A second, identical optical axis normal to the figure allowed recording of stereographic pairs of micrographs. Two light sources, the xenon flash lamp and a continuous light emitting diode (LED) permitted taking up to 250 35mm film-pair images, and unlimited gray scale video at 30 frames per second. Inset 35mm film frame shows a typical microphotograph, reclaimed and developed post flight, of a PVA dendrite grown in microgravity. When enlarged photographically, these film frames allowed measurement of the tip radii to an accuracy of $\pm5\%$. The thermostatic temperature bath had its two liter volume filled with a well-stirred heat transfer fluid that was index-matched to the molten phase within the growth chamber. The system was capable of controlling the temperature to ±0.002K within its control range around the melting points of SCN ($\approx 58.1°$)C and PVA ($\approx 36.0°$)C.

Before the advent of IDGE it was not possible to test separately and quantitatively the accuracy of the Ivantsov transport solution and the validity of the interface scaling factor, $\sigma^\star \approx 0.02$. The IDGE instrument was designed for operation in low Earth orbit (LEO) and flown on three separate missions aboard the space-shuttle orbiter *Columbia*, as part of NASAs periodic shuttle-based United States Microgravity Payload Missions: USMP-2, -3, and -4.

Collectively, IDGE space-flight crystal growth data provided the first solid evidence that Ivantsov's mathematical solution describes heat transport during dendritic growth. The two test materials, SCN and PVA, differ markedly in both the modulus and anisotropy of their crystal-melt energies. Specifically, SCN is nearly isotropic, with an interfacial

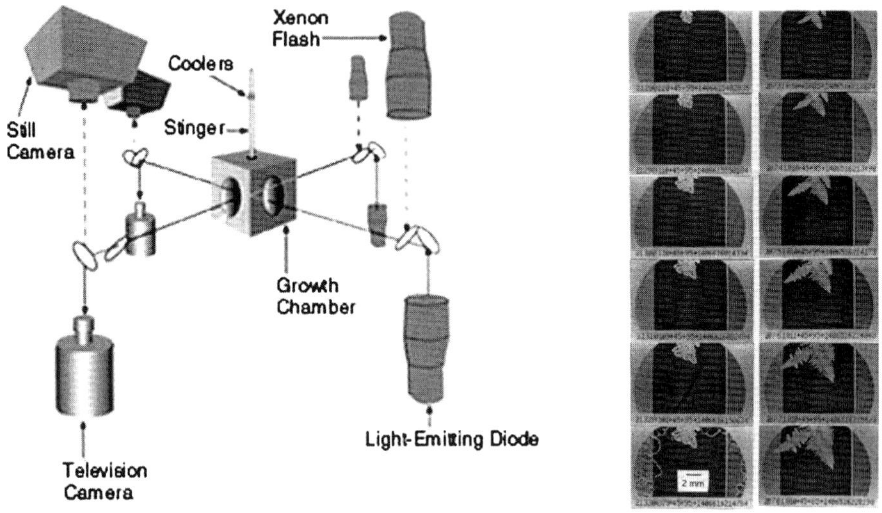

Figure 9. **Left:** Schematic of IDGE optical system. Shown are two optical axis crossing the crystal growth chamber, shown here configured for PVA crystal growth. Each axis contains a xenon flash lamp capable of exposing 35mm high-resolution film in the cameras, plus a red light-emitting diode that continually illuminated the interior of the growth chamber, allowing video recording. The development of dendrites on the "stinger" tube is suggested in the lower inset. **Right:** Stereoscopic pairs of 35mm microphotographs, showing the time evolution of the emerging PVA dendrites in microgravity. These photographs were taken aboard the space shuttle *Columbia* during its flight STS-87 in November-December 1997. The film was developed by NASA post flight, and then analyzed in the author's laboratory to obtain dendritic growth speeds and tip radii data as functions of the supercooling. Encodings beneath each film frame provide important information regarding initial supercooling, time, and RMS gravitational acceleration. On the left and right of each film frame are lens-shaped neutral filters that permit gray-scale calibration needed for precision optical image analysis.

energy $\gamma = 8.9\text{mJm}^{-2}$ and a 4-fold anisotropy about its growth axis of only $\pm 0.5\%$, whereas PVA has a smaller energy modulus, $\gamma = 1.8\text{mJm}^{-2}$, but exhibits almost ten times as much anisotropy ($\pm 5\%$) as does SCN. This difference in crystal-melt anisotropy accounts for their significantly different dendritic morphologies, as shown in Figure 1. The IDGE flight instruments provided electronic CCD images (as in-flight data), 35mm films (as post-flight data), and then for the first time on USMP-4, near-real-time, full gray-scale video data were streamed to Earth at 30 frames per second.

3.4. Verification of transport theory

On the space flights USMP-2 and USMP-3 data were gathered on dendritic growth speed and tip radius as functions of the supercooling [32], [34]. Over 200 experiments were conducted on ultra-pure SCN. These microgravity data were also combined as growth

Péclet numbers, $Pe = VR/2\alpha_\ell$, along with comparable speed and radii data measured under ordinary terrestrial conditions (i.e., at unit gravity, $g_0 = 9.8$ m/s^2).

The measurements of the steady-state tip speeds and radii taken both under microgravity and terrestrial conditions are shown plotted against the applied supercooling in Figure 10, left and right, respectively. The observed growth speed in microgravity can be much slower than the speed parallel to gravity under terrestrial conditions, particularly at small supercooling. The steady-state tip radii in microgravity are, however, always larger than those for dendrites growing parallel to gravity under terrestrial conditions. The differences of the steady-state tip speed and radius in microgravity and at $1g_0$ are due to the influence of the hydrodynamic flows induced when gravity is present. The flow around a dendrite growing under terrestrial conditions is natural convection stimulated by the buoyancy forces of the warm melt near the solid-liquid interface immersed in the denser, cooler supercooled melt. The tip speed and radius data may be combined as the

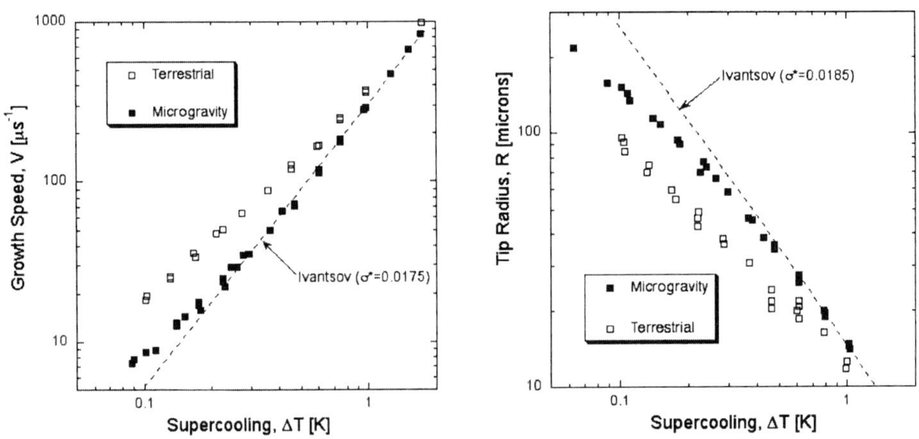

Figure 10. **Left:** Dendritic tip speed versus supercooling, ΔT, log-log coordinates. The values for the growth speeds under terrestrial conditions are above those observed in microgravity. The broken line represents a one-parameter theoretical fit to these data setting the value of $\sigma^{\star}$=0.0175. **Right:** Dendritic tip radius versus supercooling, ΔT, log-log coordinates.The values for the tip radii under terrestrial conditions are below those observed in microgravity. The broken line represents a one-parameter theoretical fit to these data setting the value of $\sigma^{\star}$=0.0185.

growth Péclet number, as shown in Figure 11 (Left), the observations may be compared to theory without any adjustable parameters. It is clear that under microgravity conditions the observed Péclet numbers are in good agreement with theory. It is important to note that there are not any adjustable parameters used in making this comparison. The terrestrial data, taken for dendrites growing parallel to gravity, result in Péclet numbers well above those measured in microgravity, where convective effects in the melt are effectively eliminated. Similar data have been calculated for pure PVA dendrites growing under microgravity conditions, and plotted along with the microgravity data for SCN in Figure 11 (Right). These data sets comprise the most exacting tests of thermal transport during dendritic growth carried out to date. The theoretical prediction for the Péclet numbers for SCN dendrites are based on Ivantsov's original paraboloidal model, whereas the theory curve shown for PVA is based on a hyperbolic tip shape. The steady-state tip shapes of dendrites (cf., Figure 1 middle, for SCN, with Figure 1 right, for PVA) clearly influence the growth Péclet number. The microgravity experiments for both SCN and PVA show that transport theory accurately characterizes the latent heat flows near the tip. Some

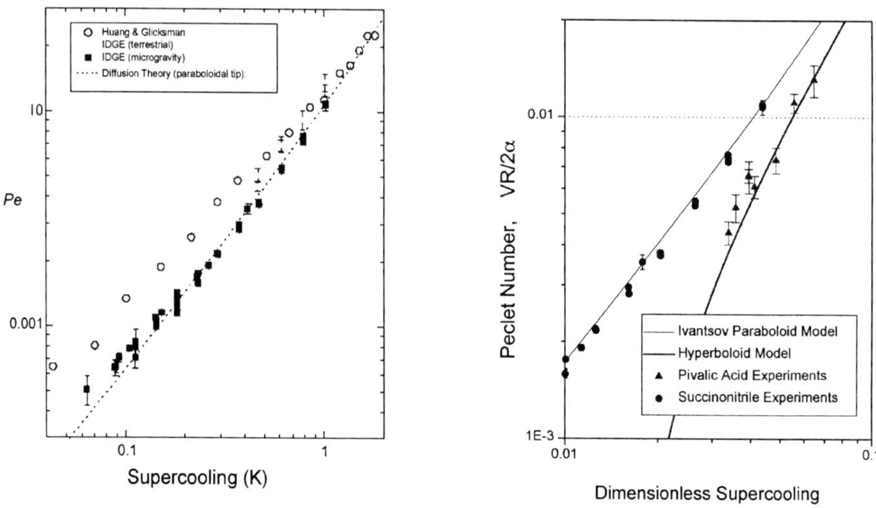

Figure 11. **Left:** Péclet number, $Pe = \frac{VR}{2\alpha}$, versus supercooling (log-log coordinates) for pure SCN. Solid symbols are data obtained from measurements in microgravity; open symbols represent terrestrial observations. The broken line is a plot of Ivantsov's transport theory, Eq.(11), so no adjustable constants are employed in this comparison. It is evident that under microgravity conditions, Ivantsov's solution provides reasonably accurate predictions of the conduction field surrounding SCN dendrites. **Right:** Péclet number, $Pe = \frac{VR}{2\alpha}$, versus supercooling (log-log coordinates) comparing pure SCN and PVA. The solid lines are predicted from transport theory using a paraboloidal tip-shape for SCN and a hyperboloidal model for PVA.

appreciation of the influence of gravity on the morphological details of dendritic crystals may be gained by studying the micrographs shown in Figure 6 of four SCN dendrites growing at the same supercooling. Here, however, the orientation of the dendritic [100] growth axis of SCN is varied with respect to the gravity vector. The symmetry of SCN dendrites would be four-fold around the growth axis, however, hydrodynamic flows, not visible in these micrographs, tend to break the symmetry dictated purely by crystallography, and alter the tip radii and growth speeds. In Figure 6a the downward pointing side branches are clearly more developed than the upward pointing branches. In Figure 6b the hydrodynamic flow suppresses side branches growing on the upper edge, whereas the side branches on the lower edge continue to grow rapidly. Figure 6c shows that when gravity and the [100] grow axis are aligned the dendrite exhibits its expected four-fold symmetry, because the melt flow itself is distributed about the growth axis with four-fold symmetry. Figure 6d shows that rotations with respect to gravity of up to about 45° have only slight effects on the dendritic morphology.

3.5. Verification of interfacial physics

Data sets similar to those presented in Figure 10 may now be used to calculate the scaling factor, $\sigma^\star = 2\alpha_\ell d_0/VR^2$. As explained in Section 2.2, all contemporary theories of dendritic growth [10,29,11] show that $\sigma^\star$ should be invariant with the supercooling. All the applicable dendritic growth data—including speed and radii measured under terrestrial and microgravity growth conditions—are shown combined as the dendritic scaling factor, $\sigma^\star$, and plotted in Figure 12. These data prove conclusively that the dendritic scaling law, $VR^2 = const.$ is valid *both* under microgravity and terrestrial crystal growth conditions. It is interesting that the hydrodynamic state of the melt has little influence on the dendritic scaling law. This behavior stands in sharp contrast with the Péclet number behavior explained in Section 2.1. Except for a slight downward drift with increased supercooling, the values of $\sigma^\star$ remain virtually constant, and independent of both the supercooling and the convection state of the melt. These experiments serve to verify the robustness of the dendritic scaling law, allowing its application to a wider variety of materials and crystal growth processes.

3.6. Scaling constants for dendritic growth

Stability analyses have since shown that $\sigma^\star$ is not strongly dependent on the details of the stability model. Table 1 lists values of $\sigma^\star$ obtained from several analyses published since the earliest numerical study by Oldfield [27], who was the first to recognize that interface stability, *per se*, is an important factor in dendritic growth. Careful experiments on dendritic growth kinetics have shown that $\sigma^\star$ may, in fact, not be strictly constant over a wide range of supercooling. It remains true, nonetheless, that the scaling law, $V_{op}R_{op}^2$ = const., is well approximated in a real systems, and the value of $\sigma^\star \cong 0.02$.

Table 1
Scaling factors, $\sigma^\star$

Stability Model	$\sigma^\star$
Planar Front (analytical)	0.0253
Parabolic Eigenstates (numerical)	0.025 ±20%
Spherical Harmonic (analytical)	0.0192
Oldfield's "force balance" (numerical)	0.02

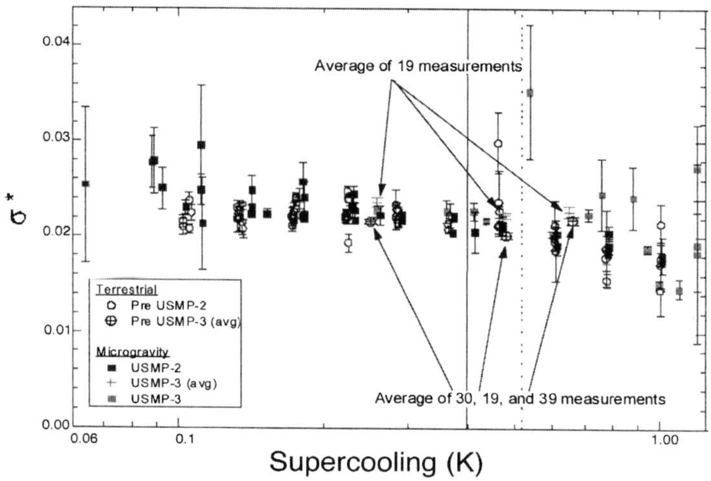

Figure 12. Scaling factor, $\sigma^{\star}$, versus supercooling. Semi-logarithmic coordinates. The scaling factor shows little if any dependence on the supercooling, verifying the theoretical scaling law that VR^2 = const. This scaling law is robust, inasmuch as dendritic growth in microgravity and under terrestrial conditions show equivalent behavior for the product VR^2. Thus, the hydrodynamic state of the melt, i.e., the influence of gravity, has a negligible effect on the special quantity VR^2.

4. APPLICATIONS OF MICROGRAVITY DATA

Microgravity experiments are difficult to perform and extremely expensive. Justification for performing these experiments must devolve on their scientific impact and utility. A few of the scientific implications of the IDGE results were already discussed in these notes, and, as explained, they functioned to test critically the two fundamental theoretical tenets of steady-state dendritic growth [32]: 1) Ivantsov's transport theory, and 2) interfacial stability physics. In addition to these primary scientific applications, the accumulation of high-quality dendritic growth data stimulated additional investigations on fundamentals of crystallization. For example, by using IDGE-generated "benchmark" quality data Dantzig et al. [35] were able to compare steady-state 3-d tip-shapes for PVA measured in microgravity using the IDGE with calculations of the tip shape accomplished with phase-field computations. A beautiful rendering of a PVA dendrite computed independently by Karma and Plapp is shown in Figure 13.

When the correct anisotropy for PVA [See again Eq.(2).] is inserted into the phase-field calculations, quantitative agreement with experiment is achieved between the computer simulations and the IDGE results. Another example where the IDGE fundamental dendritic growth experiments stimulated additional investigations concerns the observations of the influence of gravity on dendritic growth. This work stimulated Tonhardt and Am-

Figure 13. Phase field rendering of a PVA dendrite. Use of adaptive grids and other numerical techniques to reduce computation time allow extraordinary calculations of dendritic crystal forms. Good agreement was found between the 3-d steady-state tip shape of PVA dendrites measured in microgravity and phase field calculations. Three-dimensional phase-field rendering of this dendrite is presented through the courtesy of A. Karma and M. Plapp [36].

berg to calculate the interactions of a melt flowing past a downward growing SCN dendrite, as shown earlier in Figure 6c. These investigators used phase field computational methods to solve the thermal and fluid velocity fields surrounding a downward growing SCN dendrite. The fields calculated by Tonhardt and Amberg are illustrated in Figure 14. A readable and complete recent reference covering phase field and level set techniques and their applications to solidification and crystal growth is available in the doctoral thesis of I. Loginova [33].

5. SUMMARY AND CONCLUSIONS

1. The theory of dendritic growth consists of two fundamental components: a) Ivantsov's transport theory that describes how heat and (if present) solute are redistributed around a growing dendrite. Transport theory provides the mathematical relationship connecting the growth Péclet number, Pe, with the melt's dimensionless supercooling, $\Delta\Theta$ for a given steady-state tip shape. b) Interfacial stability theory that leads to the surprising relationship $\sigma^\star \propto 1/VR^2 \approx$ const. Stability theory suggests that a dendrite grows with the largest tip radius not exceeding that for which the tip would split. The direction of growth generally occurs in the $\langle 100 \rangle$ for cubic substances, which coincides with the direction where the chemical potential at the crystal-melt interface is a minimum.

2. When combined, the two theoretic components labelled (a) and (b), above, yield quantitative predictions of the steady-state dendritic speed, V, and tip radius, R, as functions of the applied supercooling, ΔT. Estimates based on this approach are in good agreement with carefully designed quantitative experiments conducted under microgravity conditions, where convection in the melt phase is absent. Under terrestrial conditions (unit gravity, g_0), by contrast, where buoyancy-induced convection

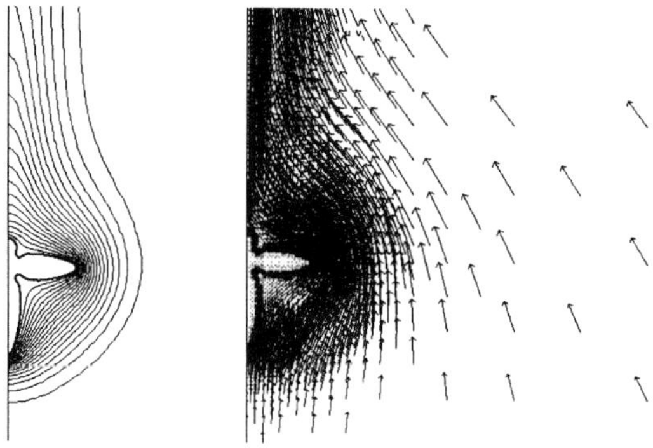

Figure 14. **Left:** Thermal field surrounding a free-growing SCN dendrite. Image is left-right mirror symmetric about a vertical line, with isotherms indicated as gray contours. The vertical pair shown grow parallel and antiparallel to gravity, with the downward member growing faster; the horizontal pair of primaries (only the right-hand primary crystal is shown) grow at equal rates orthogonally to gravity. Isotherms wrap around the dendrite tip and are drawn upward by the buoyant melt flow. **Right:** 2-d fluid velocity field surrounding a downward growing SCN dendrite. Image is left-right mirror symmetric about a vertical line. The hydrodynamic interactions between the melt and the dendrite suggested by the velocity field show that cool melt is brought near the tip, increasing its speed, whereas warm melt rises and slows the upward growing primary crystal. Data computed via phase field techniques provided through the courtesy of G. Amberg [37].

is always present, the steady-state growth speed of a dendrite is a function of the orientation angle of the growth axis with respect to $\vec{g_0}$. As shown by detailed modeling, the sensitivity to gravity results from the hydrodynamic alteration of the thermal and/or solutal fields surrounding the dendrite tip. The strength of the interaction between the dendritic growth process and the hydrodynamics depends sensitively on the melt's Prandtl number (ratio of its thermal diffusivity to its kinematic viscosity). The test substance SCN, for example, has a relatively large Prandtl number, $Pr \approx 23$, and its tip growth speeds diminish by a factor of almost 20 for crystal growth parallel and anti-parallel to $\vec{g_0}$. Metals and semiconductors, with their much lower Prandtl numbers, tend to show less sensitivity of their dendritic growth rates to their orientation angle to gravity. The careful comparisons made between kinetic and morphologic data observed terrestrial and under microgravity conditions has stimulated modeling of the hydrodynamic interactions between dendrites and their melts. Recent studies show that the hydrodynamic interactions take place over a range of length scales from the tip radius (micrometers) to the primary stem length (millimeters).

3. Interfacial stability theories predict an important "scaling law" that relates the dendritic tip speed and the tip radius: viz., $VR^2 \approx$ const. Microgravity experiments

confirm this scaling law quantitatively, insofar as the value of $\sigma^\star \approx 0.02$ is nearly independent of the supercooling, gravitational level, and even the material. Robustness of this crucial scaling law allows reliable estimates to be made of the dendritic growth responses, including tip speed, tip radius and side-branch spacing during crystal growth from a supercooled melt in a wide variety of materials. Inclusion of this basic dendritic scaling law into crystal growth and alloy solidification models allows prediction of the microstructure in cast materials. Such predictions remain valid down to length scales as small as the radius of curvature of the tip. Solidification models lacking such scaling laws seldom are capable of predicting crystallization events on scales smaller than several millimeters.

4. Microgravity experiments directly verify Ivantsov's thermal transport solution for paraboloidal dendrites. Péclet number data calculated for SCN dendrites are in agreement with predictions based on Ivantsov's theory for paraboloidal tip shapes. This provides a compelling test of theory insofar as no adjustable parameters may be used in making the comparison, once the tip shape is chosen. Similar experiments carried out in microgravity on PVA, which has 10 times the anisotropy of its interfacial energy than does SCN, shows that the growth Péclet number can be similarly calculated for a hyperboloidal tip shape that is appropriate for dendrites of cubic phases exhibiting a distinctive cruciform shape. Good agreement is found between experiment and theory when this adjustment for the shape is made.

5. The detailed steady-state tip shapes of PVA dendrites, and their micro-morphologies are markedly different from those for SCN dendrites, due primarily, as mentioned above, to their very different anisotropies of the interfacial energy. Actually, both test materials differ from the ideal, smooth, paraboloids of revolution assumed in Ivantsov's transport theory, the differences being primarily in their tip shapes. Although tip shape effects do indeed impose quantitative influences on the transport behavior they do not change the basic agreement with transport theory. The presence of side-branches on real dendrites is also neglected in the steady-state theories. It is believed that side-branches do affect the speed and tip shape of dendrites, but a comprehensive, quantitative study of these effects is lacking at present.

6. Dendritic scaling laws, which derive from the fundamental theory discussed in these notes, are starting to be incorporated into engineering applications in the field of casting and welding as well as for further scientific enquiry. Ideally, one would like to be able to predict significant attributes of a cast microstructure, including the primary and secondary branch spacings and the length scales for microsegregation in alloys. As theoretical modeling efforts continue, using advanced numerical techniques such as level sets and phase-field, further progress will unfold concerning our understanding and control of this ubiquitous form of crystallization.

Acknowledgment

The author thanks his colleague, Dr. Afina Lupulescu, Materials Science & Engineering Department, Rensselaer Polytechnic Institute, Troy, NY, for her able assistance in preparing these study notes.

REFERENCES

1. M.E. Glicksman and S.P. Marsh, Handbook of Crystal Growth D.T.J. Hurle (ed.) Elsevier Science Publishers, Amsterdam, pp. 1,075-1,122, 1993.
2. T. Fujioka and R.F. Sekerka, J. Crystal Growth, 24-25 (1974) 84.
3. D.E. Temkin, Dokl. Akad. Nauk., SSSR, 132 (1960) 1307.
4. G.F. Bolling and W.A. Tiller, J. Appl. Phys. 32 (1961) 2587.
5. J.W. Cahn and D.W. Hoffman, Acta Metall., 22 (1974) 1205.
6. D.W. Hoffman and J.W. Cahn, Surf. Sci., 31 (1972) 368.
7. M.E. Glicksman, J. Cryst. Growth, 42 (1977) 347.
8. K.A. Jackson and B. Chalmers, Can. J. Phys., 34 (1956) 473.
9. K.A. Jackson, in Growth and Perfection of Crystals, R.H. Doremus (ed.) p. 319, John Wiley, New York, 1958.
10. J.S. Langer and H. Muller-Krumbhaar, Acta Metall., 26 (1978) 1,681-1,687.
11. J.S. Langer, Princeton Series in Physics. Critical Problems in Physics: Chapter 2, Princeton University Press, 1996.
12. W.W. Mullins and R.F. Sekerka, J. Appl. Phys., 34 (1963) 323-329.
13. M.E. Glicksman, R.J. Schaefer and J.D. Ayers, Metall. Trans. A, 7A (1976) 1,747-1,759.
14. M.E. Glicksman, Crystal Growth of Electronic Materials, Chapter 5, E. Kaldis (ed.) Elsevier Science Publishers, Amsterdam, pp. 57-69, 1985.
15. M.E. Glicksman, M.B. Koss, V.E. Fradkov, M. Rettenmayr and S.S. Mani, Journal of Crystal Growth, 137 (1994) 1-11.
16. M.E. Glicksman, M.B. Koss, L.T. Bushnell, J.C. Lacombe and E.A. Winsa, ISIJ International, 35, no. 6 (1995) 604-610.
17. M.E. Glicksman, M.B. Koss, L.T. Bushnell, J.C. Lacombe, Winsa E.A., *Materials Science Forum*, 215-216 (1996) 179-190.
18. S.C. Huang, M.E. Glicksman, Acta Metall., 29 (1981) 701-715.
19. "Materials Sciences in Space," Part I, B. Feuerbacher, H. Hamacher, and R.J. Naumann (eds.) Springer-Verlag, Berlin, 1986.
20. J.C. LaCombe, M.B. Koss, V.E. Fradkov, M.E. Glicksman, Physical Review, E, 52 Nr. 3, (1995) 2,778-2,786.
21. R. Trivedi and J.T. Mason, Metall. Trans. A, 22A (1991) 235-249.
22. G.P. Ivantsov, Dokl. Akad. Nauk, USSR, 58 (1947) 56.
23. K. Kassner, "Pattern Formation in Diffusion-Limited Crystal Growth," Ch. 4, World Scientific Publishing, Singapore, 1996.
24. G. Horvay and J.C. Cahn, Acta Metall., 9 (1961) 695.
25. G.E. Nash and M.E. Glicksman, Acta Metall., 22 (1974) 1283.
26. G.E. Nash and M.E. Glicksman, *Acta Metall.*, 22 (1974) 1291.
27. W. Oldfield, Mater. Sci. Eng., 11(1973) 211.
28. R.D. Doherty, B. Cantor and S. Fairs, Metall. Trans. A., 9 (1978) 621.
29. D. Kessler and H. Levine, Phys. Rev. B, 33 (1986) 7867.
30. D. Kessler and H. Levine, Phys. Rev. Lett., 57 (1986) 3069.
31. D. Kessler and H. Levine, Acta Metall., 36 (1987) 2693.
32. M.B. Koss, L.A. Tennenhouse, J.C. LaCombe, M.E. Glicksman and E.A. Winsa,

Metall. and Mater. Trans. A, 30A (1999) 3177-3190.

33. I. Loginova, "Phase-field modeling of diffusion controlled phase transformations," Ph.D. Thesis, Royal Institute of Technology, KTH, ISSN 0348-467X, Stockholm, Sweden, 2003.
34. M.E. Glicksman and A. Lupulescu, "Dendritic Growth," in Proceedings of the International Conference on The Science of Casting and Solidification, D.M. Stephanescu, R. Ruxanda, M. Tierean, and C. Serban (eds.), pp. 15-21, Editura Lux Libris, Brasov, Romania, 2001.
35. J.A. Dantzig, N. Provatas, N. Goldenfeld, J.C. LaCombe, A. Lupulescu, M.B. Koss, and M.E. Glicksman, "A Comparison of Phase-Field Computations with Experimental Microgravity Measurements for Dendritic Growth in Pure Materials," in Modeling of Casting, Welding and Advanced Solidification Processes IX, Peter R. Sahm, Preben N. Hansen, and James G. Conley (eds.) pp. 453-460, Shaker Verlag, Aachen Germany, 2000.
36. A. Karma, Northeastern University, Private Communication, 2002.
37. G. Amberg and R. Tonhardt, KTH Stockholm, Sweden. Private Communication, 2000.

Crystal Growth – from Fundamentals to Technology
G. Müller, J.-J. Métois and P. Rudolph (Editors)

Modeling of Crystal Growth Processes

Jeffrey J. Derby[a] and Andrew Yeckel[a]

[a]Department of Chemical Engineering & Materials Science and Minnesota Supercomputing Institute, University of Minnesota, Minneapolis, MN 55455, U.S.A.

Successful crystal growth process modeling relies on an artful blend of rigor and practicality, balancing what is wanted with what can be done. We provide an overview of the current practice of modeling melt and solution crystal growth processes. The continuum transport equations describing these systems are reviewed, numerical solution strategies are outlined, and the special modeling challenges posed by crystal growth are explored. Sample results representing the state of the art of melt growth modeling are shown.

1. INTRODUCTION

The growth of bulk crystalline materials remains one of the most challenging and astonishing of technical endeavors, with objectives ranging from the one-time creation of milligrams of single-crystal protein pharmaceuticals to the annual production of metric tons of electronic-grade silicon. Due to these broad applications, the great variety of crystals needed, and the exacting quality typically required of single-crystal materials, their successful growth ranks among the most difficult challenges of modern materials processing. Toward addressing these challenges, numerical modeling has proven to be a powerful tool to analyze and optimize crystal growth processes.

The physics of bulk crystal growth encompasses a wide variety of phenomena that occur over a vast spectrum of length and time scales, literally ranging from atoms to cubic centimeters of material. One cannot pose and solve a model for crystal growth completely from first-principles due to these disparate scales. Instead, many different approaches and tools are applied to model different aspects of crystal growth, as depicted in Figure 1. Also indicated in this figure are sizes of various important features for two prototypical bulk crystal growth systems, involving the melt growth of silicon (Si) and the solution growth of potassium dihydrogen phosphate (KDP).

At a microscopic scale of tens of nanometers or less, *ab initio* molecular dynamics (MD) methods can be employed to study atomic behavior. For example, such techniques have been employed to study the liquid state of cadmium telluride to better understand some of this material's unusual crystallization behavior [1]. Unfortunately, while quite rigorous in their approach, these methods are too computationally expensive to be applied to systems of more than a few hundred atoms or for describing times scales of greater than tens of picoseconds. Molecular dynamics methods based on classical potentials can compute for much larger ensembles and longer time scales. For example, these methods have been

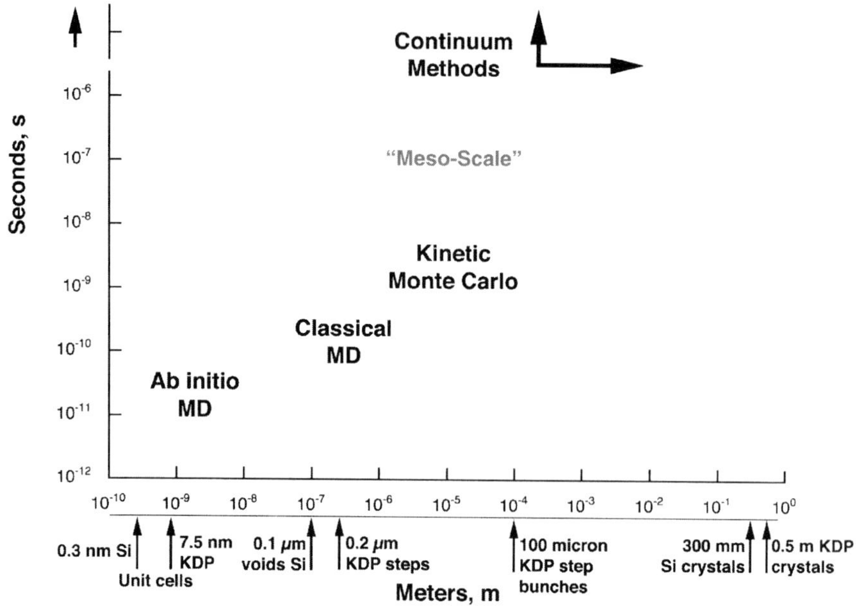

Figure 1. Different modeling approaches are appropriate for various length and time scales. Typical sizes of features from bulk melt and solution growth are indicated.

used with great success to understand the dynamics of microdefects and void formation in electronic-grade silicon [2]. To simulate even larger length scales and longer time scales, the kinetic Monte Carlo (KMC) method has been applied to many systems; see, e.g., [3].

On a macroscopic scale, typically tens of microns and larger, continuum methods are gainfully applied to describe physical phenomena. There are many important phenomena associated with crystal growth that occur on a "meso-scale" comprising hundreds of nanometer to tens of microns and occurring over long times scales (e.g., microseconds or longer). These phenomena are difficult to model by atomistic methods (MD and KMC), due to the long time scales involved, and challenging for continuum methods, due to the very small length scales. Such phenomena require innovative, "multi-scale" models; examples of such for crystal growth are the step growth models presented in [4, 5] for describing the growth of vicinal facets from liquid solution.

This challenge of scales motivates the need to formulate crystal growth models that include enough physics to make realistic, usable predictions, yet that are simple enough to remain tractable with today's computational capabilities. In the case of bulk crystal growth, the greatest purpose that modeling can serve is to directly connect processing conditions to final outcomes. Since bulk crystals usually are incorporated into electronic, optical, or optoelectronic devices, final outcome is most often measured in terms of the crystal properties relevant to these devices. These crystal properties are in turn determined by the chemical composition and structure of the crystal, particularly the types and distributions of various crystalline defects. For the crystal growth modeler, the ultimate task is to connect processing conditions to the morphology, chemical composition, and defect structure of the crystal.

2. HISTORICAL OVERVIEW

Prior to the 1980's nearly all crystal growth modeling consisted of analytical and semi-analytical models. These early models provided great insight but were limited in scope and utility, and no attempt at a review is made here. The 1980's witnessed a flowering of computer-aided analysis of bulk crystal growth, aided by rapid advances in computing hardware. Much of this work is summarized in a still-relevant review article by Brown [6].

By the end of the 1980's, continuum transport modeling had reached a state where it was routine to solve stationary problems in two space dimensions that included coupled fluid dynamics, heat and mass transport in moderately complicated geometries, for multiphase problems in which the location of the crystal-melt interface position was determined in a self-consistent manner. Developments proceeded apace and within a few years it became feasible to integrate a time-dependent transport model for growth of an entire crystal. Around this time the first significant calculations of three-dimensional transport phenomena also began to appear.

The mid 1990's was a period of great optimism that *accurate* three-dimensional calculations, including time-dependent phenomena, would soon also become routine. Much of this optimism was based on the ever-advancing capabilities of computers. Gordon Moore, co-founder of Intel, remarked in an interview in 1965 that the number of transistors on a single integrated device roughly doubles every 18 months. This observation, now called "Moore's law," implies that processor capability, i.e., the speed at which operations can be performed, should advance at the same pace. Indeed, this has largely been the case, and each year brings more computer power to apply to solving scientific problems, such as the modeling of crystal growth processes. This is pointed out in Figure 2, where the number of grid points employed in the largest models of welding and casting processes (problems which share challenges similar to crystal growth modeling) are plotted for years advancing from 1980, i.e., Y = year − 1980 [7]. Also included on this graph are computations from several of our prior bulk crystal growth models [8–10], plotted as solid circles. These data are quite well represented by the scaling of Moore's law. These advances in computer power should continue to bear fruit for modeling. However, there is an important caveat, namely that we are able to continue to take full advantage of these faster computers. With the advent of parallel high-performance computing architectures, this caveat is especially significant.

Indeed, in the past few years there appears to have been a slowing of developments in the modeling of bulk crystal growth. We attribute this slowdown to three factors: greater difficulty and overhead in solving three-dimensional problems than expected, disappointing developments in computing hardware, and a lack of progress in algorithm development for solving large sets of algebraic equations. At every stage, ranging from grid generation to visualization, the modeling of three-dimensional phenomena has proven to be subtantially

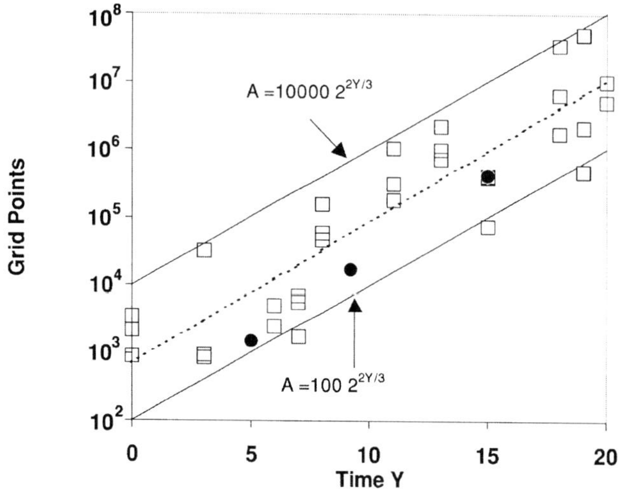

Figure 2. Moore's law describes grid size as a function of years since 1980 for modeling metal casting processes; adapted from [7]. Solid circles indicate selected studies of the modeling of crystal growth [8–10].

more difficult than for two-dimensional phenomena. This difficulty is also tied in part to slow improvement in usability of the basic tools for development of parallel computing applications. Several paradigms for parallel computing development have come and gone, as well as several specialized machine architectures, making it difficult for all but the most determined of developers to keep up with changes. Although the Message Passing Library (MPI) interface has matured into something resembling a portable standard for parallel code development, a large burden remains on the user to carefully and laboriously design algorithms that are suitable for parallel platforms. Common numerical methods used in computational transport phenomena do not parallelize easily, and the ideal of a compiler that can automatically and effectively parallelize a typical serial code simply has not emerged. Aside from these issues of usability of parallel computing platforms, even the advance of sheer computing power has been slower than anticipated. Parallel computing machines have proven an order of magnitude more expensive to build and maintain than was believed a decade ago. This reality, coupled with significant economic problems experienced by supercomputer makers in the post-Cold War era, has resulted in limited availability of clock cycles for many users of these machines, further hampering progress.

Although the issues just described have slowed progress, none has been as troublesome as the lack of a much needed breakthrough in practical methods for solving very large sets of algebraic equations of the kind encountered in continuum transport. Historically, direct methods based on variants of Gaussian elimination have proven practical, robust, and efficient for two-dimensional calculations on serial machines, to the point where the vast majority of two-dimensional problems can be solved with ease using cheap personal computers. Direct methods do not scale well to the sizes required for three-dimensional problems [11], however, making it necessary to use iterative methods [12]. Iterative solvers have two major advantages over direct solvers for solving three-dimensional problems: they are much easier to parallelize and generally require far less memory to run [13]. Unlike direct methods, however, iterative methods have failed the requirement of robustness: for transport modeling in crystal growth in particular, significant nonlinearity due to convective transport greatly slows, and sometimes completely halts, convergence to a solution. The key issue is preconditioning of the matrix that describes the set of equations [12]. To date the robustness of preconditioners has been correlated to complexity, to the point where the most robust preconditioners often resemble direct solvers [14].

3. MODELING APPROACHES

Computer modeling of crystal growth processes has often been described as an art[1] as well as a science, to acknowledge the seemingly endless difficulties that arise in the application of numerical methods to this subject. Two core competencies are required for effective use of modeling in crystal growth. One is a thorough grasp of the fundamentals of continuum transport phenomena. The other is a general understanding of the numerical methods that are used to discretize and solve the governing equations of transport phenomena. For the truly serious modeler the need to engage in code development requires additional competencies, including knowledge of modern computer languages and parallel computing architectures, as well as a solid grounding in at least one numerical method for discretization of partial differential equations. In the following discussion, we will present a synopsis of the major challenges to be faced in the modeling of crystal growth processes.

3.1. Governing equations for continuum transport

Continuum transport modeling is the most highly developed aspect of crystal growth modeling and much has already been written on the subject. The seminal treatise on the general subject of continuum transport phenomena is the textbook of Bird, Stewart, and Lightfoot [16]. At least one book devoted to modeling of transport phenomena specific to crystal growth has been published [17], featuring chapters by different authors writing on various topics. A number of chapters on the subject can be found within other works, including a general introduction to the subject by Derby [18], an extensive review of convection in melt growth by Müller and Ostrogorsky [19], and some more recent accounts of specific topics in crystal growth modeling [20–22]. Although fifteen years old now, the review article by Brown [6] remains remarkably current in many respects and is essential reading for anyone contemplating modeling. Of particular note are the proceedings of the series of International Workshops on Modeling in Crystal Growth [23, 24]. These workshops have been immensely successful at bringing together leading crystal growth modelers from around the world, and their proceedings provide an indispensable resource for tracking the state of developments in modeling of bulk crystal growth.

Since so much has already been written on transport modeling in crystal growth we

[1]Maroudas has recently suggested that modelers of materials processing problems be referred to as artisans, rather than artists, with the connotation of highly skilled craftsmanship rather than pure creativity [15].

attempt here to only briefly summarize the governing equations commonly used for this purpose. These equations are based on a number of unstated assumptions and by no means are they universally applicable to bulk crystal growth, but in the majority of circumstances they provide an accurate and rigorous account of transport. The symbols used throughout this chapter are defined in the Nomenclature section.

For a fluid that is Newtonian and incompressible, the consideration of buoyant effects (employing the Boussinesq approximation) leads to the following equations:

$$\rho_o \left(\partial \mathbf{v}/\partial t + \mathbf{v}\cdot\nabla\mathbf{v}\right) - \nabla\cdot\mathsf{T} = \rho_o \mathbf{g}\left[1 - \beta(T - T_o) + \beta_s(c - c_o)\right] + \mathbf{F}(\mathbf{v},\mathbf{x},t) \quad (1)$$

$$\nabla\cdot\mathbf{v} = 0 \quad (2)$$

where

$$\mathsf{T} \equiv -p\mathsf{I} + \mu(\nabla\mathbf{v} + (\nabla\mathbf{v})^{\mathrm{T}}) \quad (3)$$

is the total stress tensor for the fluid and the term $\mathbf{F}(\mathbf{v},\mathbf{x},t)$ represents a general body force which may be acting on the fluid.

The transport of energy and species is described by the following convection-diffusion equations:

$$\rho_o \mathrm{C_p}\left(\partial T/\partial t + \mathbf{v}\cdot\nabla T\right) = \nabla\cdot(\mathrm{k}\nabla T) \quad (4)$$

$$\partial c/\partial t + \mathbf{v}\cdot\nabla c = \nabla\cdot(\mathrm{D}\nabla c) \quad (5)$$

These equations assume that energy and species transport is dominated by forced or natural convection effects, rather than diffusion-induced flow. These assumptions are usually valid for bulk crystal growth from the melt. The situation is not so clear in solution crystal growth, where the crystallizing species is present in high concentrations. In this case diffusion-induced convection can be important, particularly at the growth interface [25–28]. Also, Fick's first law of diffusion, used to derive Eq. (5), might not be accurate in ionic solutions due to charged transport effects.

The equations above are written for a single phase. Bulk crystal growth systems always have at least two phases (crystal and liquid) but often more, in which case the equations are applied to each phase using physical parameters appropriate to the material of that phase. Thus the physical properties and field variables have an implicit index denoting the material of each phase. Also, if there are more than two chemical species of interest, Eq. (5) has an implicit index denoting each species.[2] For a more detailed presentation of the equations for the general multiphase, multispecies situation, see [30].

The momentum balance in Eq. (1) includes several contributions to the body force. Forces due to thermal and solutal buoyancy are written explicitly in terms of the Boussinesq approximation, which is standard for slightly compressible liquids, leading to the first terms of the right-hand-side of the equation. To this we add a general body force term $\mathbf{F}(\mathbf{v},\mathbf{x},t)$ which may have contributions from effects such as the application of a magnetic

[2]To apply Eq. (5) to multiple species implies pseudo-binary transport, which, strictly speaking, is valid only for dilute species. Non-dilute multispecies transport can be modelled using either the Stefan-Maxwell equations or Fick's law of multicomponent diffusion [16], but this is rarely done because little information is available on multicomponent diffusion coefficients. For an example of modeling multicomponent species transport in melt crystal growth, see [29].

field to a conducting liquid, discussed in Sec. 3.5. It should be noted that the Boussinesq approximation does not include inertial effects that can arise in certain systems, for example high rotation rates at low gravity [31, 32]. Lee and Pearlstein [32] generalize the Boussinesq approximation to include centrifugal effects for a set of equations written in a rotating reference frame, but the same result can be obtained to first order by applying a variable density throughout Eq. (1) rather than simply in the body force terms.

The equations above are written in *dimensional* form. The equations are easier to interpret and to solve when the variables are scaled to order one, so it is customary to *non-dimensionalize* the equations by scaling each of the variables with a characteristic quantity. Many different dimensionless forms of the equations are possible in models of bulk crystal growth (Brown [6] identifies thirteen dimensionless parameters relevant to crystal growth). For example, in buoyancy-dominated flows there is no obvious characteristic velocity and a few different choices are in common use. One example is $v_c = \mathrm{k}/\rho \mathrm{C_p L}$, which leads to the following dimensionless form of momentum equation:

$$\left(\partial \mathbf{v}/\partial t + Pr^{-1}\mathbf{v}\cdot\nabla\mathbf{v}\right) - \nabla\cdot\mathsf{T} = Ra(T-1)\mathbf{g} \tag{6}$$

where $Pr \equiv \mu \mathrm{C_p/k}$ and $Ra \equiv \rho^2 \mathrm{C_p g L^3} \beta T_o/\mu \mathrm{k}$. Another velocity scaling, $v_c = \sqrt{\beta \mathrm{gLT_o}}$, is suggested by the solution to the problem of a vertical heated plate in an infinite bath. This choice yields the dimensionless form:

$$\left(\partial \mathbf{v}/\partial t + \sqrt{Gr}\mathbf{v}\cdot\nabla\mathbf{v}\right) - \nabla\cdot\mathsf{T} = \sqrt{Gr}(T-1)\mathbf{g} \tag{7}$$

where $Gr = Ra/Pr$. Other forms are possible, for example setting the characteristic velocity using the rotation rate of the crystal or crucible in systems where rotational effects are dominant. Here we avoid discussing dimensionless forms in detail, only making the point that there is usually no unique best scaling. It is necessary to apply thought and experience to determine the best characteristic scalings for a given situation.

3.2. Boundary conditions

There are a plethora of boundary conditions which can be employed for modeling bulk crystal growth. Certain requirements are dictated by mathematical well-posedness; however, the most important issue to remember is that *all boundary conditions come from nature.* A judicious choice of boundary conditions is needed to realistically represent any particular system: their proper choice and implementation are indeed some of the greatest challenges for modeling.

Some typical boundary conditions for bulk crystal growth are briefly presented below (some obvious conditions have been omitted from the table, such as, for example, matching and symmetry conditions). The equations are written in a general three-dimensional form with surface directions indicated using only the unit normal vector $\mathbf{n}$ and identity tensor I, to avoid defining tangent directions. The unit normal points outward by convention.

Boundary conditions for the momentum balance are typically expressed in terms of velocity specifications, such as:

$$\mathbf{v} - (\mathbf{n}\cdot\mathbf{v})\mathbf{n} = \mathbf{V}_\mathrm{b} - (\mathbf{n}\cdot\mathbf{V}_\mathrm{b})\mathbf{n} \tag{8}$$

$$\mathbf{n}\cdot\mathbf{v} = \mathbf{n}\cdot\mathbf{V}_\mathrm{b} \tag{9}$$

Equations (8) and (9) represent no-slip and no-penetration conditions, respectively, for the fluid velocity at a solid surface. All variables are written with respect to the computational reference frame; the velocity $\mathbf{V}_\mathrm{b}$ is a rigid boundary motion that can include both normal and tangential components. The computational frame may translate at an arbitrary velocity with respect to the laboratory frame, so care must be taken to note that $\mathbf{V}_\mathrm{b}$ may have a contribution due to translation of the reference frame in addition to contributions from motion of the boundary within the reference frame.

In many cases simplifications to these boundary conditions will be applicable. For example, the right hand side of the no-slip and no-penetration conditions, Eqs. (8) and (9), will be zero wherever rigid boundaries are stationary with respect to the reference frame. The two conditions taken together are often referred to singly as the no-slip condition, written in vector form $\mathbf{v} = \mathbf{V}_\mathrm{b}$, but we separately identify the vector components because there are situations in which it is appropriate to specify only one component. An example is during solidification where there is a change of density at the phase interface: here the no-slip condition is applied, but the no-penetration condition is replaced by

$$\mathbf{n} \cdot \mathbf{v} = \mathbf{n} \cdot \left[\mathbf{V}_\mathrm{b} + \mathbf{V}_\mathrm{sl}\left(1 - \rho_s/\rho_l\right)\right] \tag{10}$$

which we call the penetration condition representing the crystallization velocity at the interface ($\mathbf{V}_\mathrm{sl}$ is measured in the reference frame of the stationary crystal). Note that Eq. (10) reverts to the no-penetration condition if the solid and liquid densities are equal.[3]

When a free boundary between gas and liquid is present, for example in meniscus-defined growth, it is necessary to account for the effects of interfacial tension. This is accomplished by invoking the following boundary condition for the momentum balance:

$$\begin{aligned} \mathbf{n} \cdot \mathsf{T}|_l &= \mathbf{n} \cdot \mathsf{T}|_g + \left[\gamma_o + (d\gamma/dT)(T - T_o) + (d\gamma/dc)(c - c_o)\right] 2\mathcal{H}\mathbf{n} \\ &\quad + (\mathsf{I} - \mathbf{nn}) \cdot \left[(d\gamma/dT)\nabla T + (d\gamma/dc)\nabla c\right] \end{aligned} \tag{11}$$

This equation represents the force balance at the interface: the normal component accounts for capillary pressure and the tangential component accounts for stress caused by the surface tension gradient (known as the Marangoni effect). The condition is often simplified by assuming that the stress induced by the gas phase is negligible. In this case the momentum balance in the gas phase is discarded and the stress term $\mathbf{n} \cdot \mathsf{T}|_g$ is set equal to an ambient pressure (denoted $\mathrm{p_a}$, often simply zero). A tangential stress $\mathbf{t} \cdot \mathsf{T}|_g$ can also be induced by an RF field for electrically conducting melts [33]. When solving a free boundary problem it is also necessary to apply Eq. (9), which acts as a constraint that determines the location of the free boundary in a self-consistent manner (in this context it is usually referred to as the kinematic condition). When Eq. (9) is used in this way, $\mathbf{V}_\mathrm{b}$ is unknown and must be calculated as part of the solution. It is worth noting that confined

[3]It is occasionally misunderstood that the proper boundary condition for the equal density case is $\mathbf{n} \cdot \mathbf{v} = 0$ when solving the problem in a reference frame fixed with the crystal. The notion that movement of the interface causes flow seems to stem from confusion with the oft-solved problem of an infinitely long system in a reference frame that moves with the interface. But in the equal density case the liquid simply freezes in place, and therefore no flow is induced in the reference frame fixed with the crystal. When the densities are not equal the situation is more complicated, because there is a net gain or loss of volume proportional to $\rho_s/\rho_l - 1$. Should the solid be denser than the liquid, for example, the melt will translate towards the crystal as a whole to accommodate the loss of volume.

melt growth systems, e.g., Bridgman systems, often have a free boundary between the melt surface and a gas-filled head space, so to be rigorous the capillary and kinematic conditions should also be applied here. But it is usually reasonable to assume that gravity keeps this free boundary nearly flat, in which case the normal momentum balance is automatically satisfied and only the tangential component of Eq. (11) is applied.

Finally, appropriate boundary conditions must be applied at the crystal growth interface to conserve energy and species across the two phases:

$$\mathbf{n}\cdot\left(-\mathrm{k_l}\nabla T|_l + \mathrm{k_s}\,\nabla T|_s\right) = \rho_s \mathrm{H_f}\,\mathbf{n}\cdot\mathbf{V}_{\mathrm{sl}} \tag{12}$$

$$c|_s = \mathrm{K}\,c|_l \tag{13}$$

$$\mathbf{n}\cdot\left(-\mathrm{D_l}\nabla c|_l + \mathrm{D_s}\nabla c|_s\right) = -\left(\mathrm{K} - \rho_s/\rho_l\right) c|_l\,\mathbf{n}\cdot\mathbf{V}_{\mathrm{sl}} \tag{14}$$

Equation (12) accounts for heat fluxes and latent heat effects, and equations (13) and (14) are needed to account for the partitioning and flux balances of species across the interface, respectively. These last conditions give rise to the *segregation* of species.

Finally, heat transfer to the ampoule must be represented by some sort of boundary condition. It is often convenient to specify a condition such as:

$$\mathbf{n}\cdot\left(-\mathrm{k_a}\nabla T|_a\right) = \mathrm{h}(T - T_f(\mathbf{x},t)) + \sigma_R\epsilon(T^4 - T_f^4(\mathbf{x},t)) \tag{15}$$

where T_f is furnace temperature profile specified external to the modeling domain. Referring to our admonition at the beginning of this section that all boundary conditions come from nature, recent efforts have been directed at replacing this condition by a self-consistent coupling of furnace-scale heat transfer, such as accomplished by the codes used in [34, 35], with the "within-the-ampoule" governing equations presented here [36, 37].[4]

3.3. Interface growth

The manner in which the growth interface is represented is a central feature of bulk crystal growth models from both a physical and a numerical point of view. At its simplest this can mean using an assumed shape, which might for example be based on the known growth habit of a given crystal. But a self-consistent growth model requires that the interface geometry be computed as part of the solution to the transport problem.

For melt growth, the normal velocity of the growth interface can be represented in terms of a thermodynamic driving force, an undercooling ΔT:

$$\mathbf{n}\cdot\mathbf{V}_{\mathrm{sl}} = \beta_{\mathrm{kin}}\Delta T, \tag{16}$$

where β_{kin} denotes a kinetic coefficient, and ΔT is usually is written in terms of some variant of the Gibbs-Thomson relation at the interface, for example [38]:

$$\Delta T = T_i - T_m\left(1 - \frac{\gamma}{\rho_s \mathrm{H_f}}\mathcal{H}\right) \tag{17}$$

where T_i is the interface temperature, T_m is the melting temperature of a planar interface, γ is a capillary coefficient, and $\mathcal{H}$ is the local mean curvature of the interface. The

[4]Indeed, Eq. (15) is an example of a boundary condition chosen more for the sake of convenience than fidelity. In high-temperature crystal growth systems, heat transfer beyond the walls of the ampoule, i.e., through the furnace, is critically important.

relation in Eq. (16) is quite general, but is written in a form that conceals considerable complexity. The linear form of the kinetic term is misleading because the coefficient $\beta_{\rm kin}$ often depends strongly on interface velocity. Also, both $\beta_{\rm kin}$ and the surface tension γ depend on interface morphology and are often anisotropic. The kinetic and anisotropic nature of these processes can lead to facetting in melt growth.

For an atomically rough interface the kinetic coefficient becomes large enough that the undercooling ΔT goes to zero and $T_i = T_m(1 - \gamma\mathcal{H}/\rho_s \mathrm{H_f})$. Here, the rate of interface movement is controlled by the flow of latent heat away from the interface. Capillary effects are only important when the interface curvature is large compared to the reciprocal of the capillary length, for example in dendritic growth. It is preferable to avoid this situation in bulk crystal growth (although perhaps not always possible), so capillary effects are often neglected in bulk melt growth models. Under these conditions Eq. (16) reduces to its simplest and most widely used form,

$$T = T_m, \tag{18}$$

commonly referred to as the melting-point isotherm condition.

In certain melt growth systems, particularly oxides, growth kinetics are important and facetting is observed, but only recently have models been developed that incorporate growth kinetics-driven facetting into a continuum transport model of bulk crystal growth. Brandon et al. [39–41] use an approach in which $\beta_{\rm kin}$ varies sharply but continuously near to singular orientations of the interface. Lan and Tu [42] use a different approach that is more geometric in nature, in which the locations of facet planes of fixed orientation are iteratively updated until the heat balance is satisfied. These treatments represent a step forward in the use of transport models to predict interface morphology, but their use requires the correct form of $\beta_{\rm kin}$, which in turn requires careful experimental measurement.

Modeling the growth velocity of a crystalline surface in bulk solution growth is much more problematic than in melt growth systems, since interfacial kinetics are much more important. The simplest representation of interface velocity and growth is

$$\mathbf{n} \cdot \mathbf{V}_{\rm sl} = \beta_\sigma \sigma \tag{19}$$

where β_σ denotes a kinetic coefficient and σ is the supersaturation. The supersaturation is defined as $\sigma \equiv \Delta\mu_g / k_B T_a = \ln(c/c_{eq})$, where $\Delta\mu_g$ is the change in the chemical potential between the crystal and liquid, k_B is the Boltzmann constant, and c and c_{eq} are the actual and equilibrium concentrations. The kinetic coefficient in this expression varies strongly as a function of the detailed nature of the surface, posing great challenges for realistic modeling. Indeed, predicting the shape of crystals growing from the solution phase is still a formidable undertaking [43, 44]. Recent efforts [4, 5] to integrate a mesoscale model of step flow with a macroscale model of transport have shown promise.

3.4. Radiation heat transfer

The equations presented here treat radiation heat transfer only in the most superficial way: as a boundary condition (see Eq. 15) that represents flux of heat in terms of an externally specified temperature profile presumed to be available as a model input. Although this simple approach is often useful, particularly if good experimental measurements of temperature within the furnace are available, it is limiting nonetheless. If it is desired to simulate the coupling of transport phenomena within the growth vessel to furnace heat transfer, then it is necessary to consider radiation heat transfer at a significant level of detail, for example by calculation of view factors, by ray-tracing techniques, or by some other rigorous approach. Calculation of radiation heat transfer is particularly important in simulation of Czochralski systems, where small changes in crystal and melt shape have a highly nonlinear effect on radiation [45]. For some materials, particularly oxides, internal radiation transfer can also be important [46, 47].

Several models of melt crystal have been developed that include detailed calculations of furnace radiation. For example, Brown and coworkers [45, 48, 49] developed sophisticated models for global heat transfer in Czochralski systems. Subsequent, more general efforts by the groups led by Dupret and Müller have led to commercial codes, FEMAG [34] and CrysVUN [35, 50], respectively. An evaluation of these models can be found in [51]. These codes are currently restricted to solving two-dimensional, axisymmetric problems.

3.5. Magnetic fields

Use of magnetic fields to influence convection is of ongoing interest and has attracted considerable attention from modelers in recent years. A general review of magnetohydrodynamics in materials processing is given by Davidson [52]. Some examples include use of magnetic fields to suppress convection in microgravity Bridgman experiments [53, 54], to control instability in Czochralski growth [22], and to promote mixing in melt growth systems [55, 56]. A typical model consists of the governing equations discussed previously with the Lorentz body force included in the momentum balance, Eq. (1):

$$\mathbf{F} = \sigma_c(-\nabla\mathbf{\Phi} + \mathbf{v} \times \mathbf{B}) \times \mathbf{B} \tag{20}$$

where $\mathbf{B}$ is the magnetic field and σ_c is the electrical conductivity, and a scalar equation added to determine the electric potential $\mathbf{\Phi}$:

$$\nabla^2\mathbf{\Phi} = \nabla \cdot (\mathbf{v} \times \mathbf{B}) \tag{21}$$

where it has been assumed that the magnetic field is unaffected by the flow. The equations are straightforward to solve, but the magnetic fields cause thin boundary layers that can be challenging to resolve numerically. The behavior can also be non-intuitive due to the complicated interaction of flow kinematics with the flow of electric current.

3.6. Turbulence

The form of the Navier-Stokes equation given in Eqns (1)–(3) is suitable for conditions of laminar flow. Many bulk crystal growth systems exhibit laminar flow, including small-scale float-zone and most Bridgman systems, but turbulence is common in Czochralski systems and in some larger solution growth systems as well. The crudest approach to incorporating turbulent effects is to use enhanced transport properties (e.g., viscosity, thermal conductivity, mass diffusivity) in a laminar flow model. A more accurate picture of transport can be obtained using a turbulence model. The Reynolds-averaged Navier-Stokes (RANS) approach has been used most often. Robey [57, 58] used a standard $k - \epsilon$ model to model turbulent transport in solution growth of KDP. A variety of $k - \epsilon$ models have also been used to model turbulent melt convection in Czochralski growth of silicon. Lipchin and Brown compared three $k - \epsilon$ models for turbulent viscosity and concluded that a low Reynolds number form worked best [59]. Low Reynolds number $k-\epsilon$ models have also been used in [60–62]. An alternative to the RANS approach is large eddy simulation (LES) [63, 64]. Recently Ivanov et al. [65] have proposed a hybridized method in which RANS-derived equations are applied in wall regions and LES-derived equations are applied in the core region. Each of these attempts to model turbulent transport has been a mixed success, and no clear choice of turbulence model has emerged. Given the importance of turbulent transport to the industrial production of silicon, turbulence modeling likely will long remain an active research area.

4. NUMERICAL METHODS

The model equations described in the previous sections do not yield to analytical mathematical methods and must therefore be studied using computer-aided numerical techniques. There are many methods in use and to describe them all is not feasible, so in the following sections we focus on those issues in numerical analysis that we deem of special importance to crystal growth modeling. We illustrate some of our points with examples drawn from our experience using finite element methods, but in most cases an analogous situation or procedure applies to other popular discretization methods such as finite volume or finite difference.

Many commercial codes exist that are devoted to solving problems in fluid mechanics and transport phenomena, mostly based on finite volume or finite element methods. Two codes specialized for crystal growth modeling are FEMAG [34] and CrysVUN [35]. Examples of crystal growth modeling performed using general purpose codes can be found in [66] (CAPE), [57] (CFX4), [67] (FIDAP), [68] (FLUENT), and [69] (CFD-ACE). It is not necessary for users to completely understand the numerical methods used in these codes, but a general understanding will greatly facilitate their correct use. Even when using commercial codes, however, it is always critical to understand the model itself, as embodied in the equations described in Sec. 3.

4.1. Discretization of field equations

Several algorithmic requirements are of paramount importance for crystal growth modeling. Complicated geometries must be easily accommodated. The algorithm must also be amenable to the solution of a variety of governing equations, especially those of incompressible fluid flow, and the approach must allow for implementation of physically relevant boundary conditions. Above all, the methods must be able to reliably compute solutions to extremely nonlinear problems and must be numerically convergent to the solutions of the underlying governing equations.

From a pragmatic viewpoint, any consistent discretization technique, carefully applied, may be up to these challenges. Finite difference and spectral methods have been used in crystal growth modeling but are sometimes difficult to apply to complicated domain shapes. Finite volume techniques ameliorate these problems somewhat, since they can be readily implemented on unstructured meshes. We have relied primarily upon the Galerkin finite element method [70, 71] and the Galerkin/Least-squares method [72] to solve the governing equations of these systems. These algorithms can be applied to complicated geometries with relative ease. In addition, employing the weak form of these formulations allows for the convenient specification of realistic boundary conditions, since physically relevant quantities, such as stresses or heat fluxes, naturally appear in surface terms. Newton–Raphson techniques are employed to solve problems arising from steady-state formulations and implicit temporal integration approaches.

The first step in the numerical solution of the governing equations is to convert the partial differential equations into a set of differential-algebraic equations (DAE's) by discretizing spatial derivatives. These DAE's can then be integrated in time using any one of several standard methods. We have typically employed the second-order trapezoid rule [73] or the backward Euler method to integrate the resulting DAE's. Regardless of the methods used, the result is a set of algebraic equations which must be solved at each time step. The equations are often nonlinear and must be solved iteratively. Our approach is to use Newton's method, which reduces the equations to a linear form at each iteration. Solving the linearized equations at each Newton iteration is by far the costliest step in computational terms. It is usually preferable to solve these equations using a direct method based on Gaussian elimination when solving two-dimensional problems, but it is almost mandatory that an iterative method be used when solving three-dimensional problems. Our choice has been to use preconditioned GMRES [12, 74], a Krylov subspace projection method. Details on our methods and software can be found in [13, 30].

4.2. Interface representation

Many methods have been devised to represent phase interfaces in discretized transport models. We cannot describe them all here, but we try to make some useful generalizations about their nature. Methods generally are of two basic types: fixed grid methods and deforming grid methods. Each type has its advantages and disadvantages, and each introduces its own complexities to constructing the discretized model. In fixed grid methods all fields (e.g., temperature, flow, etc.) are computed on a fixed grid of computational cells (e.g., finite elements, finite volumes). In general the interface will pass through the interior of cells, which introduces two complications. Cells at the interface are located partially in each phase, which requires an interpolation of material properties within the cells. Also interface boundary conditions need to be applied along a curve that does not coincide with the underlying grid used to discretize the field variables, which introduces another interpolation of some sort. These complications are absent from deforming grid methods, the essence of which are to allow discretizing the interface conditions on the same grid used to represent the field variables. In these methods the grid is deformed so that the interface lies along the edges of computational cells everywhere. Each cell lies entirely inside of one phase and remains in that phase (unless re-gridding with topology change is needed), allowing material properties to be represented locally without interpolation.

A fixed grid discretization cannot directly represent the interface, which therefore must be represented with a supplemental discretization. This can take many forms: the phase-field [38], enthalpy [75], and level set [76] methods all use an implicit representation of the interface, but each has a distinct approach to the problem.

The phase-field method is a true diffuse interface model in which the interface is implicitly represented by the order parameter Φ. Models of Φ are phenomenological in origin, but much emphasis has been placed in recent years on formulation of phase-field models that are thermodynamically consistent and that asymptotically match the sharp interface limit, putting these models on firm theoretical ground [38]. Interfacial flux conditions arise naturally as volumetric conditions within the interfacial zone defined by Φ, and physical properties can be represented in terms of simple functions of Φ. One interpretation is that the diffuse interface of the phase-field model corresponds to a real interfacial transition zone. In practice it is not possible to resolve such a thin region, however, and the interface thickness parameter is chosen based on numerical considerations.

The enthalpy method resembles the phase-field method in some superficial respects, but is better described as a sharp interface method discretized in a diffuse manner rather than as a true diffuse interface model. The sharp interface model results in a discontinuous enthalpy field at the phase boundary, but this discontinuity cannot be resolved by the underlying fixed grid discretization, so in practice the enthalpy varies rapidly but continuously across an interfacial zone of non-zero thickness (sometimes this region takes on physical meaning as a mushy zone for computation of metallurgical casting problems). Also, it is not possible to directly implement sharp interface conditions using the underlying fixed grid representation, so it is customary to implement interfacial flux conditions as pseudo-volumetric conditions within the interfacial zone.

The level set is a true sharp interface method, but as with the enthalpy method it is not possible to directly implement sharp interface conditions using the underlying fixed grid representation. Hence it holds in common with the enthalpy and phase-field methods an artificial smearing of interfacial flux conditions, blurring the practical distinction between the terms sharp and diffuse.

Whereas the aforementioned fixed grid techniques all rely on an implicit characterization of the interface, a host of fixed grid methods also exist in which the interface is tracked explicitly, for example front-tracking [77], volume-of-fluid [78], and sharp interface [79] methods. Since the location of the interface is arbitrary with respect to the fixed grid it is necessary to use a discretization scheme to track the interface that is independent of the underlying fixed grid discretization. Then it is necessary to interpolate material properties and interfacial flux conditions between these discretization schemes. Thus these methods share in common with their implicit fixed grid cousins some degree of artificial smearing of interfacial flux conditions [80]. Udaykumar et al. [79] describe a fixed grid method that uses special discretization procedures to obtain second order accurate interpolation of interfacial conditions to avoid this problem, but even so the method represents the interface position with only first order accuracy, a common characteristic of fixed grid methods. Methods limited to first order accuracy demand a high degree of grid refinement to resolve the interfacial region.

Deforming grid methods make it is possible to achieve high accuracy with much less grid refinement than required by fixed grid methods, provided that the grid can represent the interface without excessive distortion of the computational cells. Interfacial conditions and physical properties are accurately represented, eliminating key sources of error; typically the accuracy of the interface position is comparable to the accuracy of the underlying field variables such as temperature. Figure 3 shows a comparison of the enthalpy-based fixed grid method of CrysVUN to the deforming grid method employed in Cats2D [30] for a model of gallium arsenide growth by the vertical gradient freeze method. A global furnace model is included in the CrysVUN model. Temperature data at the ampoule wall from the CrysVUN calculations are used to construct thermal boundary conditions for the Cats2D

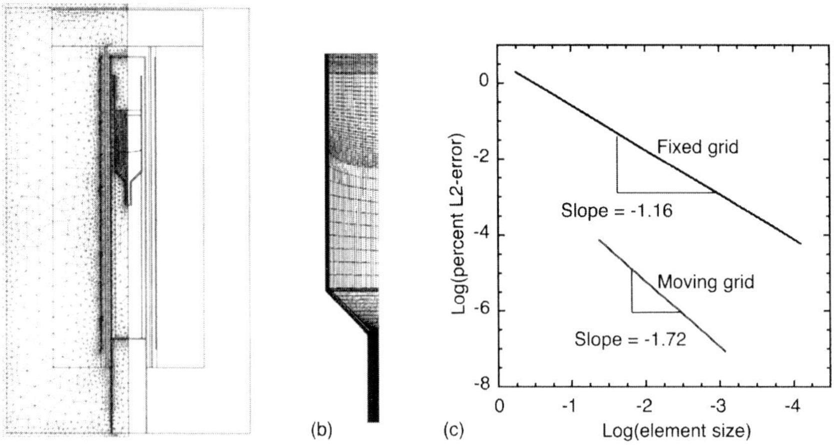

Figure 3. Comparison of fixed grid and deforming grid methods for modeling the VGF growth of GaAs. (a) Grid for CrysVUN global heat transfer calculation, with interface represented by enthalpy method (fixed grid). (b) Grid for local analysis using Cats2D, with interface represented by deforming grid (outer ampoule temperatures are provided by CrysVUN). (c) Convergence of interface position versus grid refinement.

calculations. There is good agreement in the predicted interface shape provided the grids are sufficiently refined, but as shown in Figure 3(c), the fixed grid calculations achieve only first order convergence in interface position, whereas the deforming grid calculations achieve nearly second order convergence with a much lower error in interface position. Deforming grid methods are subject to failure in many situations, however, when the interface shape becomes too complicated. In such cases it becomes necessary to employ fixed grid approaches such as the phase-field method.

4.3. Deforming grids and ALE methods

When using a deforming grid method, some means of numerical grid generation is required to reposition nodes in a way that avoids excessive deformation of the computational cells. Grid generation methods fall into two general categories, algebraic grid generation and partial differential equation-based grid generation [81]. Use of algebraic grid generation, for example the method of spines [82], has declined in favor of PDE-based methods such as pseudo-solid domain mapping [83] and elliptic grid generation [81]. In these methods the nodal positions (or computational cell boundaries, depending on the point of view) are characterized by a continuous vector field $\mathbf{x} = (x, y, z)$, which defines a mapping $\boldsymbol{\xi} = (\xi, \eta, \zeta)$ between the physical domain and a reference, or *parent*, domain (the elemental coordinate systems defined by standard finite element methods is a particularly convenient choice for this parent domain). The vector field $\mathbf{x} = \mathbf{x}(\boldsymbol{\xi})$ is obtained by solving a system of partial differential equations, many varieties of which have been contrived to achieve various desired effects in grid generation. These equations can be solved in the same parent domain $\boldsymbol{\xi}$ as field variables such as temperature and flow, a convenient arrangement that simplifies algorithm development.

In the pseudo-solid method, $\mathbf{x}$ is not computed directly, but is updated by solving a set of PDE's for displacement $\boldsymbol{\delta}$ of a solid body undergoing deformation (usually treated as linear elastic). Given an initial grid $\mathbf{x}_o$ a new grid is computed from $\mathbf{x} = \mathbf{x}_o + \boldsymbol{\delta}$. We emphasize that despite the physical basis of these equations, they are used in a purely artificial manner here, as a convenient way to deform the grid in a controlled manner. The method is straightforward to implement and has the major advantage that it makes no assumptions about the structure of the grid. A disadvantage is the strong tendency of the pseudo-solid PDE's to conserve the volume of the computational cells, a property that can cause severe distortion of the cells if there is a large evolution in interface shape.

When using deforming grid methods in time-dependent problems, it is important to recognize that the time derivatives in the governing equations are taken with respect to the computational reference frame, whereas time derivatives of the discretized equations are computed with respect to the reference frame of the parent domain, which moves locally at the velocity of the nodes. In order to evaluate the time derivative of a scalar field f we must convert it to the frame of the parent domain [84]:

$$\frac{\partial f}{\partial t} = \dot{f} - \dot{\mathbf{x}} \cdot \nabla f \tag{22}$$

where the overdot indicates time derivatives with respect to the reference frame of the parent domain.

Then the energy equation is discretized in the form (compare to Eq. 4):

$$\rho_o \mathrm{C_p} \left(\dot{T} + (\mathbf{v} - \dot{\mathbf{x}}) \cdot \nabla T \right) = \nabla \cdot (\mathrm{k} \nabla T) \tag{23}$$

where the effect of the moving grid appears as a correction to the convective velocity. Analogous changes appear in momentum and species conservation equations (Eqs. 1 and 5). Because the reference frame of the parent domain is neither Eulerian nor Lagrangian, the approach described here is often referred to as an arbitrary Lagrangian-Eulerian (ALE)[5] method [85]. Several excellent references [86–88] describe the use of ALE methods with elliptic grid generation for solving time-dependent moving boundary problems.

4.4. Quasi-steady-state models

Interface motion in bulk crystal growth systems is generally very slow compared to the time scale of transport phenomena [89]. In many systems energy and momentum transport undergo large transients at the start of growth, but after a relatively short time these transients decay. Subsequently the transport equations remain nearly at steady-state, and the only significant effect of interface motion relates to balances of heat or material across the interface. This observation naturally leads to consideration of a quasi-steady-state (QSS) model, in which the geometric position of the interface is fixed, and the steady-state equations for transport are solved to obtain a snapshot in time. For melt

[5]Caution should be exercised to not confuse ALE methods with *mixed Eulerian-Lagrangian* methods, which are a type of fixed grid method with explicit interface tracking. In these methods the field equations are solved in an Eulerian frame and the interface is tracked in a Lagrangian frame, hence the term *mixed*.

growth an assumed value of interface velocity $\mathbf{V}_{sl}$ is applied to account for the release of latent heat in Eq. (12). After sufficiently long time, $\mathbf{V}_{sl}$ reaches a constant equal to the translation rate of the furnace relative to the growth vessel, under which condition the QSS model is valid. From a computational point of view, this approach is expedient, since one need only solve a set of nonlinear algebraic equations for the model rather than a set of time-dependent DAE's.

After the initial transients of the transport equations have decayed, there is an intermediate stage of growth during which the interface velocity is far from its final value and the conventional QSS model is not valid. Brandon and Vizorub [90] have developed a method intermediate to the QSS model and a fully time-dependent model that extends the QSS model to this intermediate stage of growth. The method consists of an outer iteration that starts with an assumed interface velocity. The QSS model is solved twice within each iteration, at two slightly different positions relative to the furnace. The interface positions provided by these two solutions are used to compute the interface velocity by finite differences. The iteration is repeated using the updated interface velocity until the procedure converges. Brandon and Vizorub outline conditions under which the method is accurate and show that the method can be used to extend the QSS model to considerably shorter growth times in many systems. For two-dimensional problems, it is by now quite fast to solve the fully time-dependent model using almost any computer, which reduces the attraction of the method. A much greater benefit is expected when solving three-dimensional problems, however.

The situation with species mass conservation in melt growth is somewhat different than that for energy and momentum conservation. During melt growth processes, energy and momentum is exchanged between the system and the surroundings, but from the standpoint of mass these are closed systems. The effect of partitioning (Eq. 13) is to cause either a progressive enrichment or depletion of a species from the melt during growth under most conditions. Mass transfer is inherently time-dependent in these systems. To formulate a solvable QSS model of segregation, it is necessary to add material to the melt to compensate for depletion of a partitioning species (or remove it to compensate for accumulation). The usual practice is to imagine that the melt within the computational domain communicates with a far-field condition at uniform concentration [91]:

$$\mathbf{n} \cdot \mathrm{D_l} \nabla c|_l = -\mathbf{V}_{sl}(c - c_o) \tag{24}$$

where $\mathbf{V}_{sl}$ is the QSS interface velocity (usually the furnace translation rate) and c_o is the far-field concentration. This boundary condition applies to long growth vessels in which a well-mixed region at the far-field boundary is isolated from the influence of the interface region. Xiao et al. [92] discuss issues with the validity of this boundary condition when these assumptions do not hold. The QSS segregation model can predict radial segregation, but only under conditions of no axial segregation, a significant restriction.

5. SAMPLE MODELING RESULTS

We'd like to finish with two examples of what modern process modeling can do. Both examples consider the vertical Bridgman growth of cadmium zinc telluride (CZT), a semi-

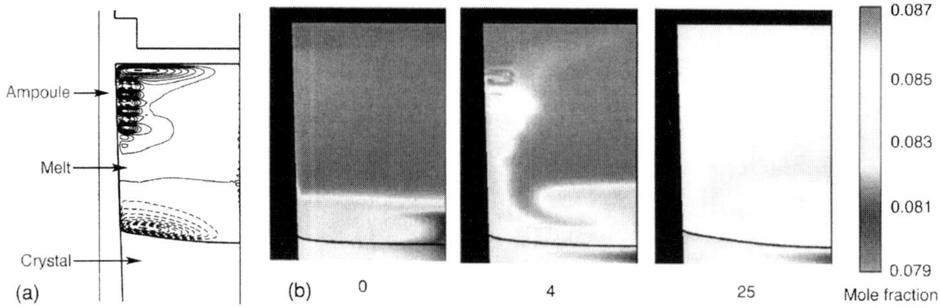

Figure 4. Accelerated crucible rotation technique applied to vertical Bridgman growth of cadmium zinc telluride. (a) Streamlines in meriodonal plane at peak deceleration. (b) Zinc distribution in melt versus number of ACRT cycles completed. Adapted from [95].

conductor crystal used in fabrication of infrared and radiation detectors. CZT is commercially grown by the vertical Bridgman method, but it has not yet been possible to reliably produce CZT in quality and quantity sufficient for mass production of detectors. Modeling studies present an opportunity for better understanding and process improvement.

5.1. Axisymmetric analysis: Effects of ACRT

Reducing compositional gradients via mixing processes can delay the onset of constitutional supercooling, thereby improving the quality of grown material and allowing higher growth rates. One of the methods used to control mixing in Bridgman growth systems is to rotate the ampoule about its axis during growth. Both steady [93, 94] and accelerated rotation [95–97] have been studied via computer modelling. Steady rotation tends to suppress mixing by damping convective flows in the meriodonal plane; however, the accelerated crucible rotation technique (ACRT) [98] *induces* flow in the meriodonal plane.

Although the flow is three-dimensional, it depends only on r, z space coordinates and therefore can be discretized on a two-dimensional mesh. The model equations are solved using the finite element code Cats2D [30]. The Galerkin finite element method is employed, with biquadratic basis functions on nine-node quadrilateral elements used to discretize all field variables except pressure, which is discretized using linear discontinuous pressure basis functions. The grid is controlled by elliptic grid generation, using an ALE deforming grid technique with sharp interface representation. Grids with up to 40 734 degrees of freedom were used. The growth of 10 cm diameter crystals is considered [95], and experimental data was employed to construct a furnace temperature profile for Eq. (15).

Figure 4(a) shows streamlines of the meriodonal flow induced by ACRT at peak deceleration. A Taylor-Görtler type flow instability is triggered by rotational deceleration near the ampoule wall, as shown by the stack of toroidal vortices along the upper ampoule wall. Close spacing of the streamlines shows that flow in these vortices is quite strong. These flow effects bring about excellent mixing of zinc in the system, as shown in Fig. 4(b) where zinc concentration in the melt is shown as a function of the number of ACRT cy-

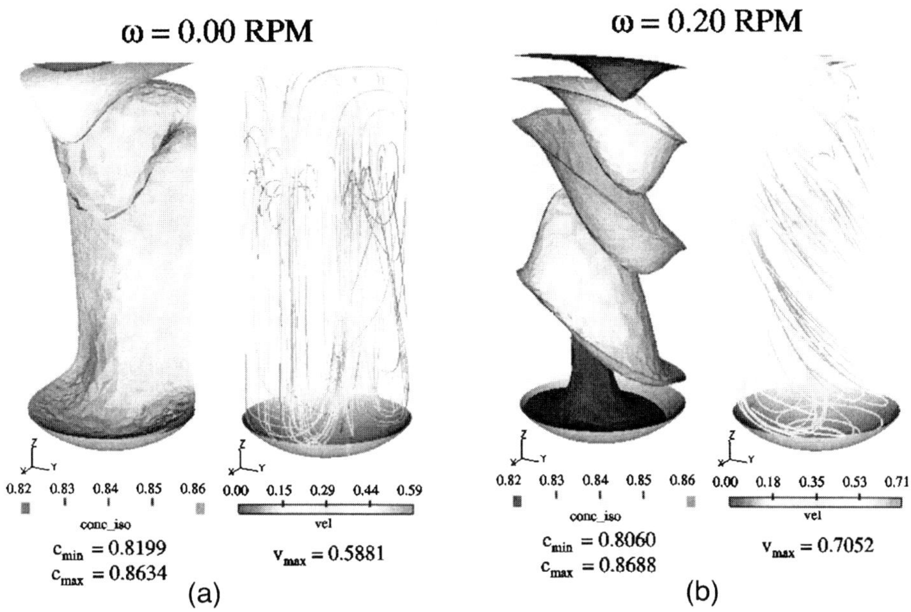

Figure 5. Bridgman system tilted 5° from vertical. In each, right panel shows iso-concentration surfaces in melt and left panel shows pathlines for flow in melt. (a) Without rotation; (b) with slow rotation at 0.2 rpm. Adapted from [99].

cles performed. Axial motion of the separation streamline, which we refer to as Ekman pumping, allows the zinc-poor region adjacent to the interface to exchange with zinc-rich liquid from the upper zone of the melt. Zinc-poor liquid that is carried into the upper region by Ekman pumping is mixed by the intense flow of the Taylor-Görtler instability. After 25 rotation cycles (2 hours) the diffusion layer at the separation streamline has been largely eliminated and zinc concentration in the melt is nearly homogenized.

5.2. Three-dimensional analysis: Effects of ampoule tilt and slow rotation

Vertical Bridgman systems are usually designed to favor axisymmetric conditions in the ampoule, but many factors can result in three-dimensional behaviors. These can arise from intrinsic three-dimensional fluid mechanical instabilities due to nonlinear effects or from extrinsic system features that violate cylindrical symmetry. We have studied several such extrinsic features, which we refer to as imperfections. These include tilting of the ampoule with respect to gravity, misalignment of the ampoule within the furnace, deviation of ampoule shape from cylindrical, and localized furnace heating [99]. Some of these effects have also been studied by Lan and coworkers [100, 101].

We consider a fully three-dimensional model for quasi-steady-state operation of a small-scale vertical Bridgman system growing CZT. Heat transfer boundary conditions are supplied from a furnace model solved using CrysVUN. The model equations are discretized using the Galerkin-Least Squares method for the Navier-Stokes equations [72], and the Streamline-Upwind Petrov-Galerkin method for the energy and species equations [102]. Four-noded tetrahedral elements with linear basis functions are used to represent all field variables. The discretized equations are solved using a MPI-based parallel implementation of preconditioned GMRES [12, 74]; more details are available in [13, 103]. A mesh containing 512 302 elements with a total of 632 107 degrees of freedom is used.

We consider the effects of tilting the axis of the ampoule away from the direction of gravity, creating a three-dimensional flow and segregation. To counter this effect, a slow ampoule rotation is applied. Sample results of iso-concentration surfaces and pathlines for a system tilted 5° from vertical are shown in Fig. 5. In a case without ampoule rotation, shown in Fig. 5(a), the flow is dominated by a large cell that circulates from top to bottom of the ampoule and the departure from axisymmetry is dramatic. The long vertical extent of iso-concentration surfaces indicates that considerable non-axisymmetric radial segregation is driven by the buoyant convection cell in the melt, with higher concentration at the interface occurring where zinc-rich material flows downward, and lower concentration occurring where zinc-depleted material flows upward. A significant degree of axisymmetry can be restored by rotating the system, however. When rotation is applied to the ampoule (in the counterclockwise direction, when viewed from above), the iso-concentration surfaces are significantly flattened (Fig. 5b). This effect is largely due to azimuthal averaging of mass transport effects due to linear superposition of a solid-body rotation on the flow in the system. Note that the rotation rates studied here are too weak to significantly modify the secondary flow via nonlinear Coriolis effects. These simulations suggest that beneficial effects may be realized even by modest crucible rotation rates. A far more extensive discussion of this result and others are presented in [99].

6. SUMMARY AND OUTLOOK

We have presented an overview of many topics relevant to the modeling of crystal growth processes, and we hope that the interested reader will further his or her knowledge via the citations presented here and in the general literature. Continuing advances in computers and algorithms are enabling increasingly powerful models for crystal growth processes. It is incumbent for any serious crystal growth practitioner to understand and utilize such tools, especially since software packages are now available for nearly all ranges of expertise and computers.

Future challenges for modeling must lead to more realistic representation of the multi-scale interactions important in crystal growth systems. Models must be capable of describing detailed system geometry and design (e.g., furnace heat transfer for melt growth systems), three-dimensional and transient continuum transport (flows, heat and mass transfer), phase-change phenomena (thermodynamics and kinetics), and atomistic events. Progress is being made on all of these fronts, but many challenges remain. The understanding gained from more realistic, multi-scale models for crystal growth will ultimately lead to the ability to link crystal structure and properties with growth conditions and the macroscopic factors which influence them.

ACKNOWLEDGMENTS

The authors' research programs have been supported in part by Alexander von Humboldt Foundation, Johnson Matthey Electronics, Lawrence Livermore National Laboratory, National Aeronautics and Space Administration, National Science Foundation, Sandia National Laboratories, U.S. Civilian Research and Development Foundation, Minnesota Supercomputer Institute, and Army High Performance Computing Research Center. We thank G. Compère, A. Pandy, and R.T. Goodwin for significant contributions.

NOMENCLATURE

Symbol	*Description*	*Units*
$\mathbf{B}$	magnetic field	mass/(charge × time)
c	species concentration	mass/length^3
c_{eq}	equilibrium solubility	mass/length^3
c_o	species concentration reference value	mass/length^3
$\mathrm{C_p}$	heat capacity	energy/(mass × degree)
D	diffusion coefficient	length^2/time
$\mathrm{D_s}$	diffusion coefficient in crystal	length^2/time
$\mathrm{D_l}$	diffusion coefficient in melt	length^2/time
f	scalar field	arbitrary
$\mathbf{F}$	body force per unit volume	mass/(length^2 × time^2)
g	gravitational acceleration	length/time^2
$\mathbf{g}$	gravitational acceleration vector	length/time^2
h	heat transfer coefficient	energy/(length^2 × time × degree)
$\mathcal{H}$	mean curvature	length^{-1}
$\mathrm{H_f}$	enthalpy of fusion	energy/mass
I	identity tensor	—
k	thermal conductivity	energy/(length × time × degree)
$\mathrm{k_l}$	thermal conductivity of melt	energy/(length × time × degree)
$\mathrm{k_s}$	thermal conductivity of crystal	energy/(length × time × degree)
$\mathrm{k_a}$	thermal conductivity of ampoule/crucible	energy/(length × time × degree)
k_B	Boltzmann constant	energy/degree
K	partition coefficient	—
L	characteristic length	length
$\mathbf{n}$	outward normal unit vector	—
p	pressure	mass/(length × time^2)
$\mathrm{p_a}$	ambient pressure	mass/(length × time^2)
$\mathbf{t}$	unit tangent vector	—
t	time	time
T	stress tensor	mass/(length × time^2)
T	temperature	degree
T_a	absolute temperature	degree
T_f	temperature profile of furnace	degree
T_o	reference temperature	degree
$\mathbf{v}$	velocity	length/time
$\mathbf{V}_\mathrm{b}$	velocity of boundary	length/time
$\mathbf{V}_\mathrm{sl}$	velocity of melt-crystal interface	length/time
$\mathbf{x}$	position vector	length
$\mathbf{x}_o$	initial position	length
x, y, z	coordinates	—
β	thermal compressibility of melt	degree^{-1}

β_s	solution expansivity of melt	length^3/mass
β_{kin}	kinetic coefficient for melt growth	length/degree × time
β_σ	kinetic coefficient for solution growth	length/time
γ_o	surface tension at reference temperature	force/length
$d\gamma/dT$	thermal variation of surface tension	force/(length × degree)
$d\gamma/dc$	compositional variation of surface tension	force × length^2/mass
$\boldsymbol{\delta}$	displacement	length
ϵ	emissivity of ampoule/crucible	—
μ	dynamic viscosity	mass/(length × time)
μ_g	chemical potential	energy
$\boldsymbol{\xi}$	reference domain position vector	—
ξ, η, ζ	coordinates in reference domain	—
ρ_l	liquid density	mass/length^3
ρ_o	melt density at reference temperature	mass/length^3
ρ_s	crystal density	mass/length^3
σ	Supersaturation	—
σ_c	Electrical conductivity	time/(energy × length)
σ_R	Stefan-Boltzmann constant	energy/(length^2 × time × degree^4)
Φ	phase-field	—
$\boldsymbol{\Phi}$	electric potential	force/charge
Gr	Grashof number	—
Pr	Prandtl number	—
Ra	Rayleigh number	—

REFERENCES

1. V. Godlevsky, J. Derby, J. Chelikowsky, Phys. Rev. Lett. 81 (1998) 4959.
2. T. Sinno, E. Dornberger, W. von Ammon, R. Brown, F. Dupret, Material Science and Engineering Reports 28 (2002) 149.
3. G. H. Gilmer, K. A. Jackson, in: E. Kaldis, H. J. Scheel (Eds.), Crystal Growth and Materials, North-Holland, Amsterdam, 1977, p. 80.
4. H. Lin, R. Vekilov, F. Rosenberger, J. Crystal Growth 158 (1996) 552.
5. Y.-I. Kwon, J. Derby, J. Crystal Growth 230 (2001) 328.
6. R. Brown, AIChE J. 34 (1988) 881.
7. V. Voller, F. Porté-Agel, J. Comput. Phys. 179 (2002) 698.
8. J. Derby, R. Brown, F. Geyling, A. Jordan, G. Nikolakopoulou, J. Electrochem. Soc. 132 (1985) 470.
9. P. Sackinger, R. Brown, J. Derby, Intern. J. Numer. Methods Fluids 9 (1989) 453.
10. Q. Xiao, J. Derby, J. Crystal Growth 152 (1995) 169.
11. A. Yeckel, J. Smith, J. Derby, Intern. J. Numer. Methods Fluids 24 (1997) 1449.
12. Y. Saad, Iterative Methods for Sparse Linear Systems, PWS Publishing Company, Boston, 1996.
13. A. Salinger, Q. Xiao, Y. Zhou, J. Derby, Comput. Methods Appl. Mech. Engrg. 119 (1994) 139.
14. M. Benzi, J. Comput. Phys. 182 (2002) 418.
15. D. Maroudas, *AIChE Annual Meeting*, San Francisco, CA, November 16–21 (2003).
16. R. Bird, W. Stewart, E. Lightfoot, Transport Phenomena, John Wiley, New York, 1960.
17. J. Szmyd, K. Suzuki (Eds.), Modelling of Transport Phenomena in Crystal Growth,

Vol. 6 of Developments in Heat Transfer, WIT Press, Southampton, England, 2000.
18. J. Derby, in: J. van der Eerden, O. Bruinsma (Eds.), Science and Technology of Crystal Growth, Kluwer Academic Publishers, Dordecht, The Netherlands, 1995, p. p. 97.
19. G. Müller, A. Ostrogorsky, in: D. Hurle (Ed.), Bulk Crystal Growth, Growth Mechanisms and Dynamics, Handbook of Crystal Growth Vol. 2b, North-Holland, Amsterdam, 1994, p. 708.
20. A. Yeckel, J. Derby, in: H. Scheel, T. Fukuda (Eds.), Crystal Growth Technology, John Wiley & Sons, West Sussex, UK, 2003, Ch. 6.
21. V. I. Polezhaev, in: H. Scheel, T. Fukuda (Eds.), Crystal Growth Technology, John Wiley & Sons, West Sussex, UK, 2003, Ch. 8.
22. K. Kakimoto, in: H. Scheel, T. Fukuda (Eds.), Crystal Growth Technology, John Wiley & Sons, West Sussex, UK, 2003, Ch. 7.
23. F. Dupret, J. Derby, K. Kakimoto, G. Müller, N. V. D. Bogaert, A. Wheeler (Eds.), Proceedings of the Second International Workshop on Modelling in Crystal Growth, Vol. 180 of J. Crystal Growth, Elsevier, 1997.
24. A. Yeckel, V. Prasad, J. Derby (Eds.), Proceedings of the Third International Workshop on Modelling in Crystal Growth, Vol. 230 of J. Crystal Growth, Elsevier, 2000.
25. W. R. Wilcox, J. Crystal Growth 12 (1972) 93.
26. G. H. Westphal, F. Rosenberger, J. Crystal Growth 43 (1978) 687.
27. W. R. Wilcox, J. Crystal Growth 65 (1983) 133.
28. W. R. Wilcox, Prog. Crystal Growth and Charact. 26 (1993) 153.
29. J. Derby, N. Ponde, V. de Almeida, A. Yeckel, in: A. Roósz, M. Rettenmayr, D. Watring (Eds.), Solidification and Gravity 2000, Vol. 329 of Material Science Forum, Trans Tech Publications Ltd, Zurich, 2000, p. 93.
30. A. Yeckel, R. T. Goodwin, unpublished (available at http://www.msi.umn.edu/~yeckel/cats2d.html) (2003).
31. C. Lan, J. Crystal Growth 229 (2001) 595.
32. H. Lee, A. Pearlstein, J. Crystal Growth 240 (2002) 581.
33. A. Sneyd, J. Fluid Mech. 92 (1979) 35.
34. F. Dupret, P. Nicodème, Y. Ryckmans, P. Wouters, M. Crochet, Int. J. Heat Mass Transfer 33 (1990) 1849.
35. M. Kurz, A. Pusztai, G. Müller, J. Crystal Growth 198 (1999) 101.
36. A. Yeckel, A. Pandy, J. Derby, in: G. D. V. Davis, E. Leonardi (Eds.), Advances in Computational Heat Transfer II, New York Academy of Sciences, 2001, p. 1193.
37. J. Derby, P. Daoutidis, Y.-I. Kwon, A. Pandy, P. Sonda, B. Vartak, A. Yeckel, M. Hainke, G. Müller, in: M. Breuer, F. Durst, C. Zenger (Eds.), High Performance Scientific and Engineering Computing, Lecture Notes in Computational Science and Engineering, Springer Verlag, Berlin, 2003.
38. H. Emmerich, The Diffuse Interface Approach in Materials Science: Thermodynamic Concepts and Applications of Phase-Field Models, Springer-Verlag, Heidelberg, 2003.
39. Y. Liu, A. Vizorub, S. Brandon, J. Crystal Growth 205 (1999) 333.
40. A. Vizorub, S. Brandon, Modelling and Simulation in Materials Science and Engineering 10 (2002) 57.
41. S. Brandon, A. Vizorub, Y. Liu, in: H. Scheel, T. Fukuda (Eds.), Crystal Growth

Technology, John Wiley & Sons, West Sussex, UK, 2003, Ch. 3.
42. C. Lan, C. Tu, J. Crystal Growth 233 (2001) 523.
43. D. Winn, M. Doherty, AIChE J. 46 (2000) 1348.
44. E. van Veenendaal, A. J. Nijdam, J. van Suchtelen, J. Crystal Growth 235 (2002) 603.
45. L. Atherton, J. Derby, R. Brown, J. Crystal Growth 84 (1987) 57.
46. S. Brandon, J. Derby, Int. J. Num. Meth. Heat Fluid Flow 2 (1992) 299.
47. S. Kuppurao, J. Derby, Numer. Heat Transfer, Part B: Fundamentals 24 (1993) 431.
48. R. Brown, T. Kinney, P. Sackinger, D. Bornside, J. Crystal Growth 97 (1989) 99.
49. D. Bornside, T. Kinney, R. Brown, G. Kim, J. Stenzenberger, B. Weinert, Int. J. Numer. Methods Eng. 30 (1990) 133.
50. R. Backofen, M. Kurz, G. Müller, J. Crystal Growth 199 (2000) 210.
51. E. Dornberger, E. Tomzig, A. Seidl, S. Schmitt, H.-J. Leister, C. Schmitt, G. Müller, J. Crystal Growth 180 (1997) 461.
52. P. Davidson, Annu. Rev. Fluid Mech. 31 (1999) 273.
53. N. Ma, J. S. Walker, Physics of Fluids 8 (1996) 944.
54. C. Lan, I. Lee, B. Yeh, J. Crystal Growth 254 (2003) 503.
55. R. Barz, G. Gerbeth, U. Wunderwald, E. Buhrig, Y. Gelfgat, J. Crystal Growth 180 (1997) 410.
56. N. Ma, D. Bliss, G. Iseler, J. Crystal Growth 259 (2003) 26.
57. H. Robey, D. Maynes, J. Crystal Growth 222 (2001) 263.
58. H. Robey, J. Crystal Growth 259 (2003) 388.
59. A. Lipchin, R. Brown, J. Crystal Growth 205 (1999) 71.
60. J. Virbulis, T. Wetzel, A. Muiznieks, B. Hanna, E. Dornberger, E. Tomzig, A. Mühlbauer, W. Ammon, J. Crystal Growth 230 (2001) 92.
61. V. Kalaev, I. Y. Evstratov, Y. N. Makarov, J. Crystal Growth 249 (2003) 87.
62. A. Krauze, A. Muiznieks, A. Mühlbauer, T. Wetzel, W. von Ammon, J. Crystal Growth 262 (2004) 157.
63. I. Y. Evstratov, V. Kalaev, A. Zhmakin, Y. N. Makarov, A. Abramov, N. Ivanov, E. Smirnov, E. Dornberger, J. Virbulis, E. Tomzig, W. Ammon, J. Crystal Growth 230 (2001) 22.
64. V. Kalaev, D. Lukanin, V. Zabelin, Y. Makarov, J. Virbulis, E. Dornberger, W. von Ammon, J. Crystal Growth 250 (2003) 203.
65. N. Ivanov, A. Korsakov, E. Smirnov, K. Khodosevitch, V. Kalaev, Y. N. Makarov, E. Dornberger, J. Virbulis, W. von Ammon, J. Crystal Growth 250 (2003) 183.
66. L. Vujisic, S. Motakef, J. Crystal Growth 174 (1997) 153.
67. J. Santailler, T. Duffar, F. Théodore, P. Boiton, C. Barat, B. Angelier, N. Giacometti, P. Dusserre, J. Nabot, J. Crystal Growth 180 (1997) 698.
68. G. Ratnieks, A. Muižnieks, A. Mühlbauer, J. Crystal Growth 230 (2001) 48.
69. T. Wetzel, A. Muiznieks, A. Mühlbauer, Y. Gelfgat, L. Gobunov, J. Virbulis, E. Tomzig, W. Ammon, J. Crystal Growth 230 (2001) 81.
70. T. Hughes, The Finite Element Method, Prentice Hall, Englewood Cliffs, NJ, 1987.
71. P. Gresho, R. Sani, Incompressible Flow and the Finite Element Method, John Wiley and Sons, West Sussex, England, 1998.
72. T. J. R. Hughes, L. P. Franca, G. M. Hulbert, Comput. Methods Appl. Mech. Engrg. 73 (1989) 173.

73. P. Gresho, R. Lee, R. Sani, in: C. Taylor, K. Morgan (Eds.), Recent Advances in Numerical Methods in Fluids, Vol. 1, Pineridge Press, Ltd., Swansea, U.K., 1980, p. p. 27.
74. Y. Saad, M. H. Schultz, SIAM J. Sci. Stat. Comp. 7 (1986) 856–869.
75. V. Voller, A. Brent, C. Prakash, Int. J. Heat Mass Transfer 32 (1989) 1719.
76. J. Sethian, P. Smereka, Annu. Rev. Fluid Mech. 35 (2003) 341.
77. J. Glimm, J. Grove, X. Li, K. Shyue, Y. Zeng, Q. Zhang, SIAM J. Sci. Comput. 19 (1998) 703.
78. M. Bogdanov, S. Demina, S. Y. Karpov, A. Kulik, M. Ramm, Y. N. Makarov, J. Comput. Phys. 39 (1981) 201.
79. H. Udaykumar, R. Mittal, W. Shyy, J. Comput. Phys. 153 (1999) 535.
80. H. Ji, D. Chopp, J. Dolbow, Int. J. Numer. Methods Eng. 54 (2002) 1209.
81. J. Thompson, Z. Warsi, C. Mastin, Numerical Grid Generation, Elsevier, New York, 1985.
82. S. F. Kistler, L. E. Scriven, Int. J. Num. Meth. Fluids 4 (1984) 207.
83. P. Sackinger, P. Schunk, R. Rao, J. Comput. Phys. 125 (1996) 83.
84. D. Lynch, K. O'Neill, Int. J. Num. Meth. Engg. 17 (1981) 81.
85. T. Hughes, W. Liu, T. Zimmerman, Comput. Meth. Appl. Mech. Eng. 29 (1981) 329.
86. K. Christodoulou, S. Kistler, P. Schunk, in: S. Kistler, P. Schweizer (Eds.), Liquid Film Coating, Chapman and Hall, London, 1997, p. p. 297.
87. J. de Santos, Ph.D. thesis, University of Minnesota, available through University Microfilms, Inc. (www.umi.com). (1991).
88. K. Christodoulou, L. Scriven, J. Comput. Phys. 99 (1992) 39.
89. S. Kuppurao, S. Brandon, J. Derby, J. Crystal Growth 155 (1995) 93.
90. A. Vizorub, S. Brandon, J. Crystal Growth 254 (2003) 267.
91. C. Chang, R. Brown, J. Crystal Growth 63 (1983) 343.
92. Q. Xiao, S. Kuppurao, A. Yeckel, J. Derby, J. Crystal Growth 167 (1996) 292.
93. A. Yeckel, F. Doty, J. Derby, J. Crystal Growth 203 (1999) 87.
94. C. Lan, J. Crystal Growth 197 (1999) 983.
95. A. Yeckel, J. Derby, J. Crystal Growth 209 (2000) 734.
96. A. Yeckel, J. Derby, J. Crystal Growth 233 (2001) 599.
97. C. Lan, J. Chian, J. Crystal Growth 203 (1999) 286.
98. E. Schulz-Dubois, J. Crystal Growth 12 (1972) 81.
99. A. Yeckel, G. Compère, A. Pandy, J. Derby, J. Crystal Growth 263.
100.M. Liang, C. Lan, J. Crystal Growth 167 (1996) 320.
101.C. Lan, M. Liang, J. Chian, J. Crystal Growth 212 (2000) 340.
102.A. Brooks, T. Hughes, Comput. Methods Appl. Mech. Engrg. 32 (1982) 199.
103.A. Yeckel, J. Derby, Parallel Computing 23 (1997) 1379.

Crystal Growth – from Fundamentals to Technology
G. Müller, J.-J. Métois and P. Rudolph (Editors)

Modeling of fluid dynamics in the Czochralski Growth of Semiconductor Crystals

Koichi KAKIMOTO

Research Institute for Applied Mechanics,
Kyushu University
6-1, Kasuga-Koen, Kasuga 816-8580, JAPAN

This course is aimed at showing how to understand and solve problems of melt flow during crystal growth from the melt. The course is divided into five parts as follows. The first part of the course focuses on an analytical approach for determining the effects of external forces based on gravitational acceleration and of rotations of a crystal and a crucible on convection. Analysis of the effects of electric, magnetic and electromagnetic forces on the melt convection will be also introduced. Then we will treat an actual system in which a combination of the forces should be taken into account. The second part of this course shows how to express actual phenomena by mathematical equations based on continuum equations: mathematical modeling. Then the course focuses on how to descritize continuum equations such as momentum and energy equations to differential equations. For discretization of continuum equations, only the finite differential method and the finite volume method are treated. The focus of the third part of the course is on how to solve descritization equations on a computer. A large computation system is needed since an actual system of crystal growth is based on a three-dimensional configuration and the phenomenon is time-dependent. Therefore, high-performance computing is needed for large-scale calculations. This part of the course deals with parallel computation by using Open-MP on a system with dual CPUs and MPI (Message Passing Interface) using a system with a PC cluster for massive computation. The fourth part of this course shows how to present the results of calculation in three-dimensional space. The results of calculation contain three-dimensional data, and the three-dimensional data must be converted into two-dimensional images since images can only be recognized in two-dimensional space.

Some movies using data calculated in a three-dimensional and time-dependent condition will be demonstrated in the last part of this course.

1. INTRODUCTION

Over the past fifty years, single crystals of semiconductors such as silicon (Si), gallium arsenide (GaAs) and indium phosphide (InP) have become increasingly important materials in the fields of computer and information technology. Attempts to produce pure silicon (i.e., a defect-free single crystal of silicon) were motivated by the desire to obtain ultra large-scale integrated circuits (ULSIs) in which micro-voids of about 10 nm diameter [1] are formed during crystal growth. Research over the past decade on crystal growth of silicon has focused on analysis of the formation of such micro-voids during crystal growth using mass transfer and reaction equations and on temperature field in the crystals, obtained from global modeling. Control of solid-liquid interface shapes of GaAs and InP has been extensively studied to find a way to prevent formation of dislocations and poligonization during crystal growth [2-5].

Micro-voids are formed by agglomeration of vacancies that are introduced at a solid-liquid interface of silicon. In most past studies, it has been difficult to reduce the total number of such micro-voids in a whole crystal because the vacancy flux in silicon crystals must be controlled to reduce the probability of agglomeration. One of the key points for controlling the vacancy flux in crystals, especially that near a solid-liquid interface, is control of the convection of melt, by which the shape of the solid-liquid interface can be controlled. From the above point of view, efforts have been made to control the periodic and/or turbulent flow of melt inside a crucible of large diameter. Crystal growth industries have mainly focused on quantitative prediction of a solid-liquid interface, point defect distribution, oxygen concentration, and dislocation free growth. Steady (DC) and/or dynamic (AC) [6-45] magnetic fields including electromagnetic fields are opening up new fields to meet an increasing demand for large-diameter crystals.

In this paper, the effects of rotations of a crystal and a crucible during crystal growth of CZ silicon are described. The effects of magnetic fields such as vertical and transverse magnetic fields applied to the CZ method on convection of the melt are also discussed. A means for solving the problem of convection mathematically by using computers is also shown in this paper.

2. EFFECTS OF INTERNAL AND EXTERNAL FORCES

2.1. Effects of temperature and of crystal and crucible rotations

To express convection in a two-dimensional configuration as shown in Fig. 1, equations for mass conservation (1), momentum (2-4) and energy (5) for the thermal flow of melt in the crucible must be solved. Boussinesq approximation was used to solve the problem of natural convection in the present model.

$$\frac{1}{r}\frac{\partial}{\partial r}(ru_r)+\frac{1}{r}\frac{\partial u_\phi}{\partial \phi}+\frac{\partial u_z}{\partial z}=0 \tag{1}$$

$$\frac{du_r}{dt}-\frac{u_\phi^2}{r}=-\frac{1}{\rho}\frac{\partial p}{\partial r}+\nu\left(\nabla^2 u_r-\frac{u_r}{r^2}-\frac{2}{r^2}\frac{\partial u_\phi}{\partial \phi}\right) \tag{2}$$

$$\frac{du_\phi}{dt}+\frac{u_r u_\phi}{r}=-\frac{1}{\rho}\frac{1}{r}\frac{\partial p}{\partial \phi}+\nu\left(\nabla^2 u_\phi+\frac{2}{r^2}\frac{\partial u_r}{\partial \phi}-\frac{u_\phi}{r^2}\right) \tag{3}$$

$$\frac{du_z}{dt}=-\frac{1}{\rho}\frac{\partial p}{\partial r}-g\beta(T-T_m)+\nu\nabla^2 u_z \tag{4}$$

$$\frac{dT}{dt}=\frac{\lambda}{\rho C_p}\nabla^2 T \tag{5}$$

$$\frac{d}{dt} = \frac{\partial}{\partial t} + u_r \frac{\partial}{\partial r} + \frac{(u_\phi - u_G)}{r} \frac{\partial}{\partial \phi} + u_z \frac{\partial}{\partial z}$$

$$\nabla^2 = \frac{\partial^2}{\partial r^2} + \frac{1}{r}\frac{\partial}{\partial r} + \frac{1}{r^2}\frac{\partial^2}{\partial \phi^2} + \frac{\partial^2}{\partial z^2} \quad ,$$

where r, ϕ and z are common coordinates in a cylindrical system, u_r, u_ϕ and u_z are components of melt velocity in r, ϕ and z directions, respectively, p and T are pressure and temperature of the melt, respectively, β is thermal expansion coefficient, ν is kinematic viscosity, ρ is melt density, C_p is specific heat, λ is thermal conductivity, g is acceleration due to gravity, and u_G is velocity of the grid in the ϕ direction, which is equal to the crucible rotation rate.

Fig. 1 shows velocity profiles of Czochralski growth of silicon without the effect of temperature. Fig. 2 shows calculated temperature and velocity profiles of Czochralski growth of silicon with temperature effect. This configuration contains melt and a crystal of silicon. Operating conditions of crystal (ω_S) and crucible (ω_C) rotation rates in Figs. 1 and 2 are listed in Table 1.

Table 1 Operating conditions of crystal (ω_S) and crucible (ω_C) rotation rates.

Fig. 1	(a)	(b)	(c)	(d)
ω_S	0	-3	0	-3
ω_C	0	0	10	10
Fig. 2	(a)	(b)	(c)	(d)
ω_S	0	-3	0	-3
ω_C	0	0	10	10

Fig. 1 (a) has no flow since there are no external and internal forces in the melt. Fig. 1 (b) shows the velocity profile only with crystal rotation, in which velocity is small. This is due to the small viscosity of the melt of silicon, which cannot diffuse momentum effectively from a crystal to the melt. Figs. 1 (c) and (d) show similar profiles, though the crucible rotates. This is because of the large effect of crucible rotation on convection of the melt.

Fig. 2 shows temperature profiles in the conditions listed in Table 1. Figs. 2 (a) and (b) show almost the same profiles of temperature and velocity, similar to the relationship between Figs. 1 (a) and (b). This is because of the small viscosity of the melt. Figs. 2 (c) and (d) show small velocity and a profile of temperature that is similar to that of almost conduction dominant case. The small velocity is attributed to a conservation rule of angular momentum in the rotating melt [46]. The momentum equations Navier-Stokes equation of rotating melt contains terms of the Coriolis and the centrifugal forces as shown in equation (6) in the rotating coordinate system,

$$\frac{\partial u}{dt} = -u\nabla u - 2(\Omega k)\times u + (\Omega k)\times(\Omega k)\times r + \frac{1}{\rho}\nabla p + \mu/\rho\Delta u + g\beta(T - T_0) \tag{6}$$

where u and r are the vectors of relative velocity on a rotational basis and position, respectively, Ω denotes the crucible rotation rate, p and μ represent pressure and viscosity of the melt, and g, β, and T_0 are the vector of gravitational acceleration, the volume expansion coefficient and the reference temperature corresponding to specific mass, respectively. The second and third terms of the right-hand side of equation (6) express the Coriolis force and centrifugal accelerations, respectively. The k in eq. (6) is a unit vector in the z-direction.

The centrifugal acceleration vector (a_{cen}) can be expressed as equation (7).

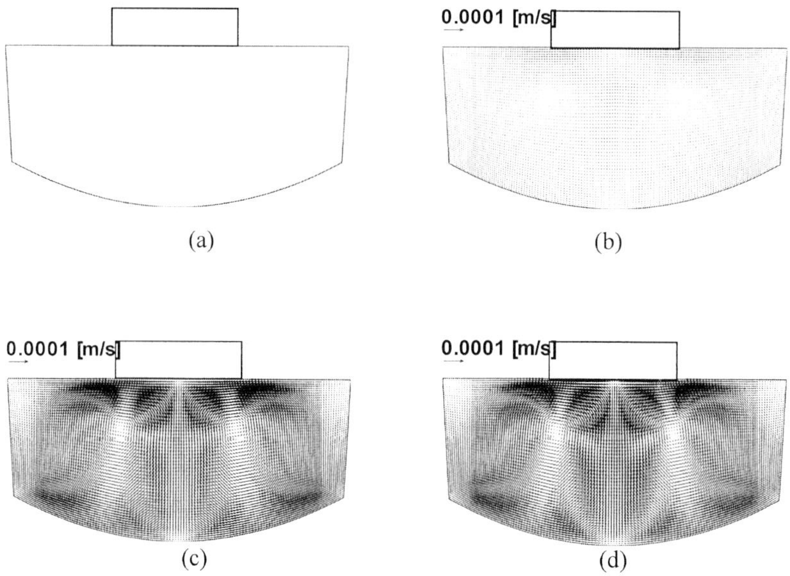

Figure 1. Velocity profiles of Czochralski growth of silicon without temperature effect. Velocity profiles for various rotation rates of the crystal (ω_s) and the crucible (ω_c); ω_s and ω_c are given in the Table 1.

$$a_{cen} = \Omega^2 r = L^2 / r^3 \tag{7}$$

where L is the angular momentum of the melt.

When a small volume element moves instantaneously from position r to r' (=r+Δr), excess force shown in eq. (8) is caused by the conservation of angular momentum.

$$\Delta a_{cen} = L^2(1/r^3 - 1/r'^3) \tag{8}$$

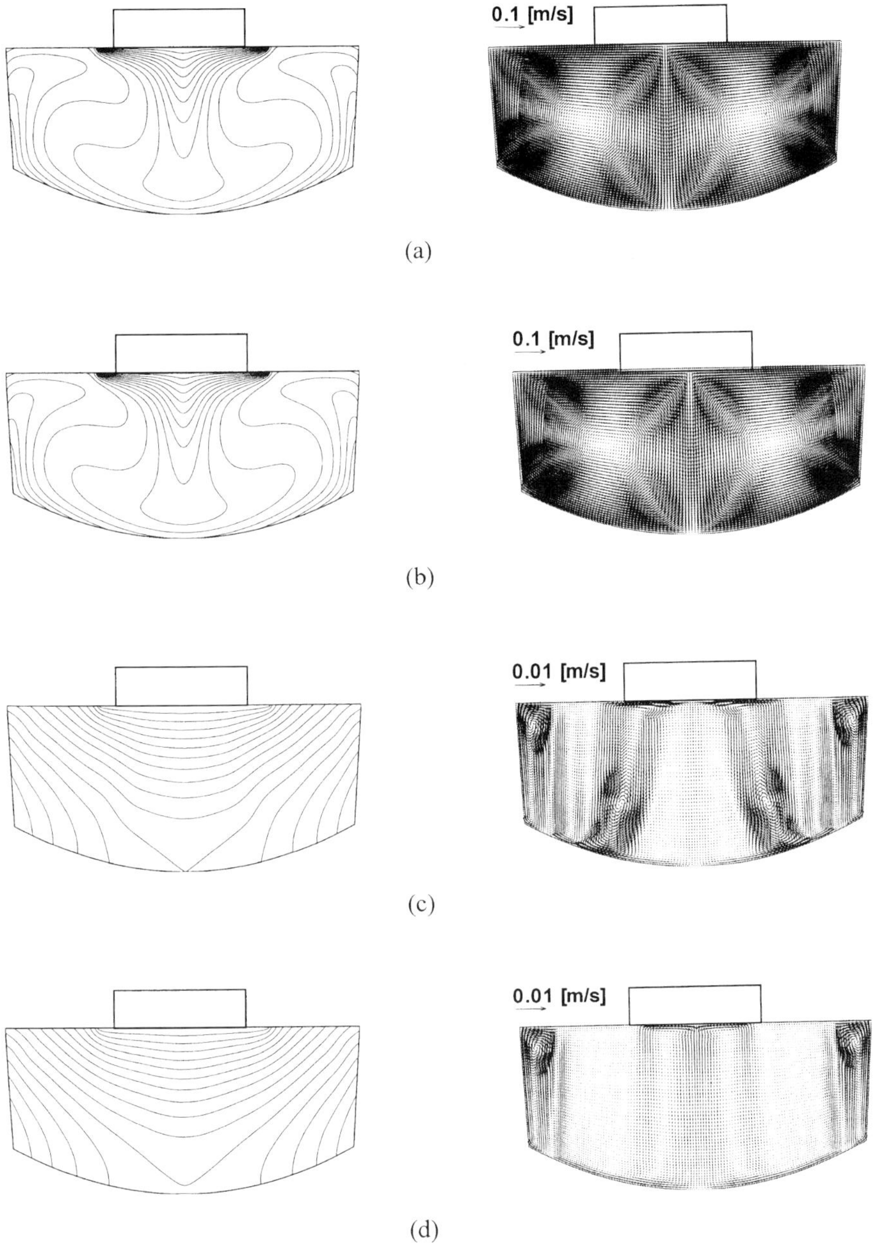

Figure 2. Temperature (left) and velocity profiles (right) of CZ growth with temperature effect. Temperature difference between the contours is ΔT= 5 K.

Consequently, the centrifugal force always acts in the opposite direction. This means that the melt motion in the radial direction is suppressed by the crucible rotation.

2.2. Effects of steady electromagnetic forces

Research on electromagnetic hydrodynamics has a long history in the fields of steel and metal manufacturing processes. Since molten silicon, like molten steel or metal, has many free electrons, electromagnetic hydrodynamics can be used to control convection in metallically conducting melt of molten silicon when magnetic and or electric fields are applied to the metallically conducting melt.

Electric current (J) in the melt and Lorentz force (F), which is induced by the current, in the case of a steady electromagnetic field are shown in eqs. (9) and (10), respectively, where σ_e, E, B and v are electric conductivity of the melt, electric field, magnetic flux density and velocity of the melt, respectively.

$$J = \sigma_e \left(\mathrm{E} + \mathrm{v} \times \mathrm{B} \right) \tag{9}$$

$$F = J \times B \tag{10}$$

Due to the continuity condition of an electric current in the melt, eq. (11), which is a Poisson-type equation, should be satisfied, since there is no source of charge in this case.

$$\nabla \cdot J = 0 \tag{11}$$

2.3. Effects of dynamic electromagnetic forces

When dynamic magnetic fields are used, electric fields in the melt can be expressed by eqs. (12) and (13). Faraday's equation, eq. (13), can be expressed by a vector potential in eq. (14).

$$E = -\nabla \Phi - \frac{\partial A}{\partial t} \tag{12}$$

$$\nabla \times E = -\frac{\partial B}{\partial t} \tag{13}$$

$$\nabla \times A = B \tag{14}$$

By combining eqs. (9) and (12), the following equation for electric current in melt is obtained.

$$J = \sigma_e \left(-\nabla \Phi - \frac{\partial A}{\partial t} + v \times B \right). \tag{15}$$

Then the following equation for Lorentz force can be obtained.

$$F = \sigma_e \left(-\nabla\Phi - \frac{\partial A}{\partial t} + v \times B \right) \times B. \tag{16}$$

The equation for Lorentz force includes the terms such as electric fields, velocity fields and time dependence of vector potential, which corresponds to the second term on the right-hand side of eq. (16). The time dependence of vector potential plays an important role in convection of melt through Lorentz force due to the time dependence of magnetic field density by rotating magnetic fields. The following equation can be derived from eqs. (11) and (15):

$$\nabla \cdot \left(-\nabla\Phi - \frac{\partial A}{\partial t} + v \times B \right) = 0. \tag{17}$$

Therefore, the Poisson equation, eq. (17), can express electric potential in the melt.

$$\nabla^2\Phi = -\frac{\partial}{\partial t}(\nabla \cdot A) + \nabla \cdot v \times B. \tag{18}$$

Let us consider the mechanisms by which static and dynamic magnetic fields suppress and enhance the motion of metallically conducting melts such as silicon and gallium arsenide (GaAs) semiconductors.

Fig. 4 shows moving metallically conducting melts inside parallel walls that are electrically insulated. Static magnetic fields are applied in the z-direction. Due to the coupling of magnetic field density and motion of the melt, the melt, which has positive charge, is subjected to Lorentz force in the x-direction, and the electrons therefore move in the opposite direction. However, since the walls are electrically insulated, electrons accumulate near the wall. Consequently, electric potential is formed as is schematically shown in Fig. 4.

This phenomenon resembles the Hall effect in semiconductor solids, although the melt has smaller viscosity than the solid. If we can measure the electric potential difference, we have a possibility of estimating flow velocity from the electric potential difference; however, it is very difficult to measure the electric potential difference in an actual system of crystal growth of semiconductors because semiconductor melts are chemically reactive and their temperatures are high.

When the wall is electrically conductive, as shown in Fig. 5, electric current of the melt flows through the wall. Consequently, Lorentz force works effectively in the opposite direction to that of the melt motion. The force thus reduces the velocity of the melt. The difference in electric potential becomes almost zero due to a large electric current that flows in the electrically conductive walls.

2.4. Vertical magnetic fields

The vertical magnetic field-applied Czochralski (VMCZ) method was one of the methods used in early works on magnetic field-applied crystal growth. Fig. 6 shows a schematic diagram of a VMCZ system with a solenoid coil.

Magnetic fields are applied in the z-direction; therefore, motion of the melt in the radial and/or azimuthal directions reacts with the magnetic field, while melt motion in the vertical direction does not react with the field. Fig. 7 schematically shows how electric currents flow and Lorentz forces in the melt work. Electric current is induced in the azimuthal direction by the radial motion of the melt under a vertical magnetic field. Therefore, Lorenz force works in the opposite direction. This force suppresses velocity of the melt in the radial direction. Radial current, which is induced by a coupling with azimuthal velocity and magnetic field, cannot flow through a crucible wall due to the electric insulation of the wall. Thus, the Lorentz force cannot work, and the melt therefore flows freely in the azimuthal direction.

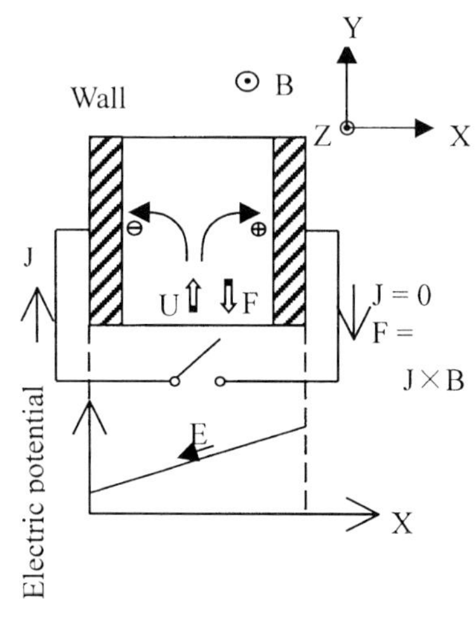

Figure 4. Electric current and electric potential in metallically conducting melts with electrically insulated walls under the condition of a magnetic field.

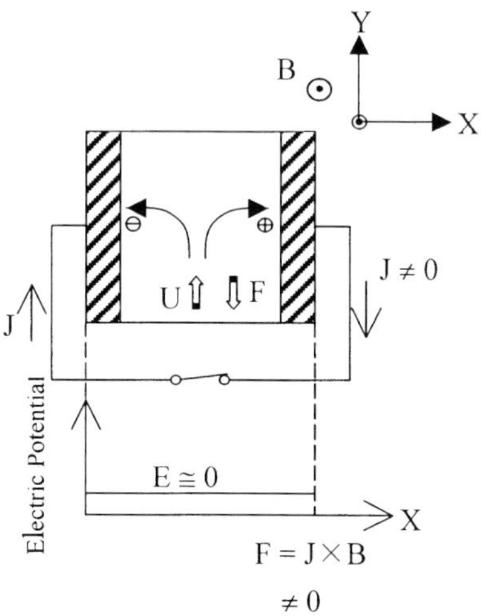

Figure 5. Electric current and electric potential in metallically conducting melts with electrically conductive walls under the condition of a magnetic field.

Fig. 8 shows experimental results for melt velocity in a meridional plane obtained by a visualization technique using an X-ray radiography method [47]. The dots show experimental data, and the lines show results of numerical calculation using a three-dimensional configuration of the melt. This figure clearly shows that there is a reduction in melt motion in a meridional plane. It was clarified from visualization that the motion in the azimuthal direction was not suppressed.

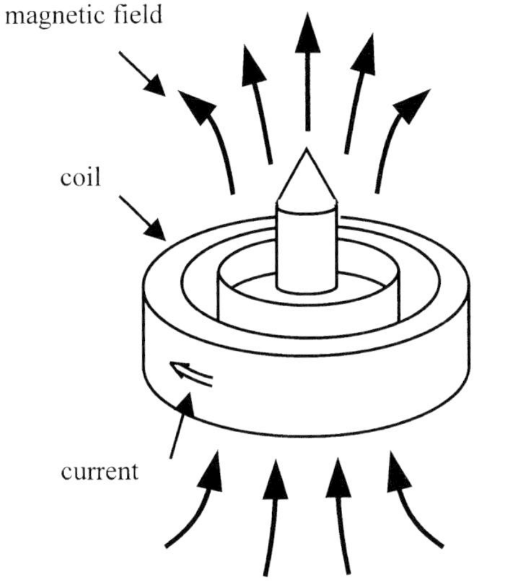

Figure 6. Schematic diagram of a VMCZ system. Static current is applied to the cylindrical coil.

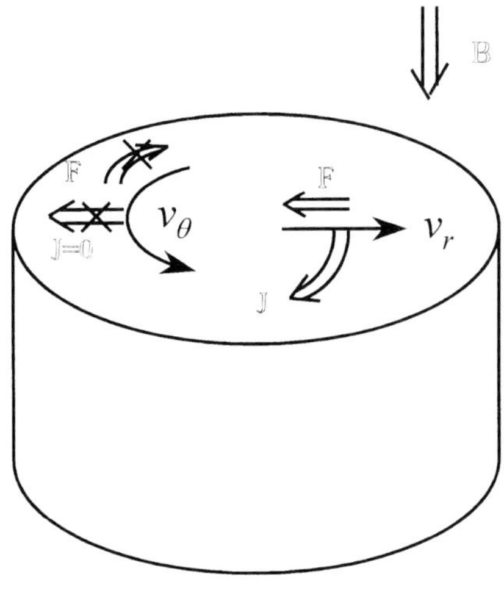

Figure 7. Schematic diagram of electric currents and Lorentz forces under vertical magnetic fields.

2.5. Transverse magnetic fields

The transverse magnetic field-applied CZ (TMCZ) method is only one case of MCZ to produce commercially available crystals in the present stage. Fig. 9 shows a schematic diagram of a TMCZ system.

The TMCZ system has a non-axisymmetric configuration, and temperature and velocity fields therefore have two-folded symmetry. Although this system has such asymmetry, it has been used for actual production of silicon for charge-coupled devices (CCDs), since the system enables crystals with a low oxygen concentration to be produced. The CCDs should have homogeneous and low oxygen concentration for reduction of inhomogeneity of image cells in the devices.

Since the system has asymmetry as shown in Fig. 9, the flow and temperature fields in the melt become three-dimensional. Fig. 10 shows a schematic diagram of electric current and Lorentz force at the initial stage of application of transverse magnetic fields to the melt. An electric current induced by the magnetic field can flow in a direction parallel to the crucible wall at positions A and B in the figure. Therefore, the melt motion is effectively suppressed by Lorentz force.

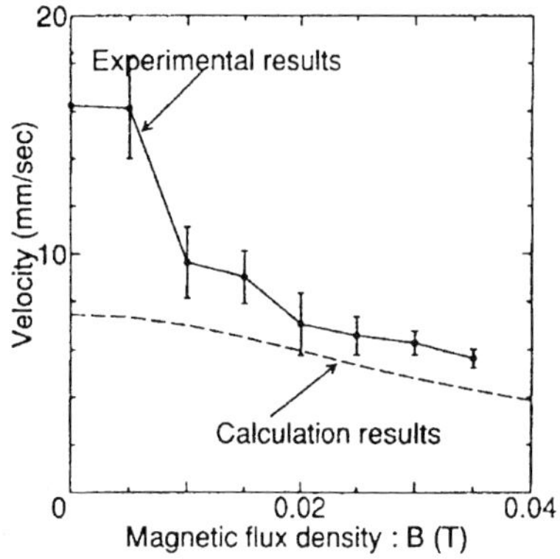

Figure 8. Relationship between magnetic fields and flow velocity in a meridional plane under vertical magnetic fields.

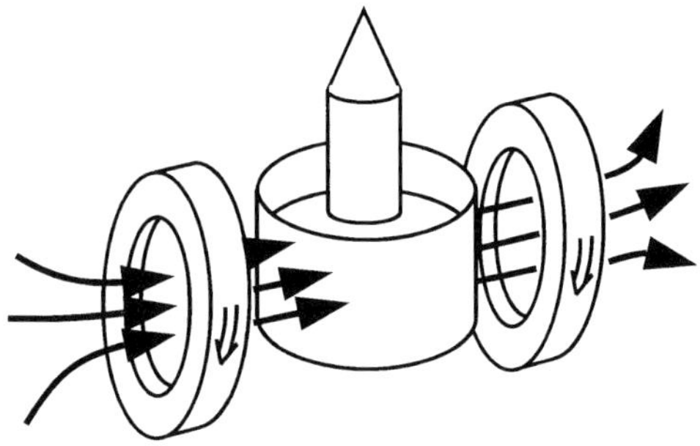

Figure 9. Schematic diagram of a TMCZ system.

The situation regarding current flow and Lorentz force at positions C and D is the opposite to that of the above case. The electric current near the wall at positions C and D cannot flow into or from the electrically insulated wall. Therefore, the Lorentz force vanishes at these positions. Consequently, the melt motion cannot effectively be suppressed as shown in Fig. 11(a). Only a downward flow remains in the y-z plane, while natural convection still exists in the x-z plane as shown Fig. 11(b). Consequently, two roll cells remain in the melt in the case of a TMCZ system. Due to the above electric boundary conditions, two main rolls aligned in parallel in the x direction remain. This means that the heat transfer in the melt in the x direction is larger than that in the y direction; therefore, an asymmetric temperature profile is formed in the melt.

Time-dependent calculation of such temperature and velocity distributions in the melt shows that these asymmetric profiles are fixed in a laboratory frame except for a layer close to the crucible wall as shown in Fig. 12. Fig. 12 (a) shows temperature distribution at the top of the melt, and Figs. 12 (b), (c) and (d) show velocity distributions in the vertical plane along the magnetic field, at the top of the melt and in the vertical plane perpendicular to the magnetic field, respectively. The interface between the crystal and melt is deformed, since the calculation contains deformation of the interface.

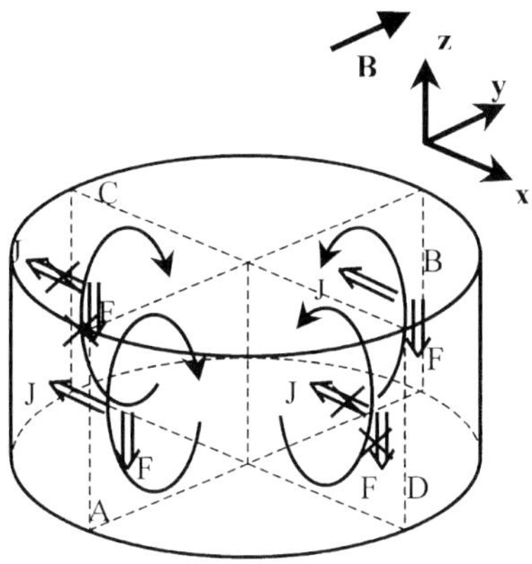

Figure 10. Schematic diagram of electric current and Lorentz force at the initial stage of application of transverse magnetic fields to the melt.

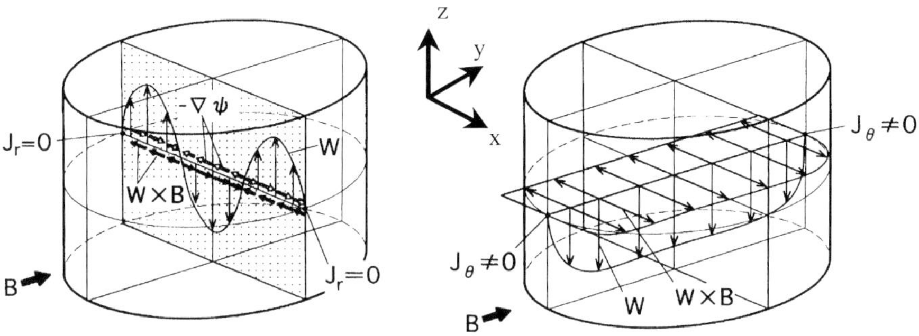

(a) (b)

Figure 11. Schematic diagram of electric current and Lorenz force in the melt under transverse magnetic fields. Cross sections in planes perpendicular and parallel to the magnetic field (a) and (b),respectively. J_r and J_θ, are electric current in radial and azimuthal directions. W and ψ are velocity of the melt in a vertical direction and electric potential, respectively.

An elliptic temperature distribution due to inhomogeneous heat transfer in the melt can be seen in Fig. 12 (a). As schematically shown in Figs. 11 (a) and (b), only down-flow is formed in a plane parallel to the magnetic field, while two roll cells are formed in a plane perpendicular to the field. Consequently, thin boundary layers of velocity, temperature and oxygen near the crucible wall are formed.

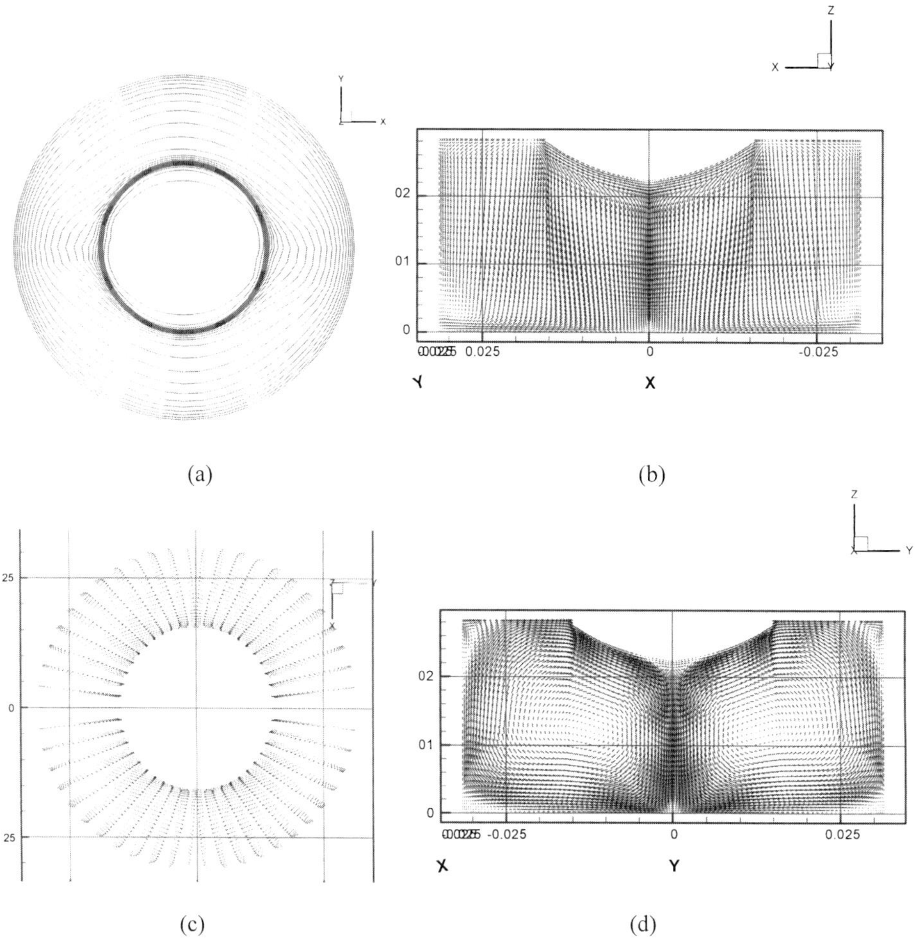

Figure 12. Temperature and velocity distribution in silicon melt with a transverse magnetic field that is applied in X-direction. Distributions of temperature (a) and velocity (c) at the top of the melt. Velocity distributions in planes perpendicular (b) and parallel (d) to the magnetic field.

This phenomenon is a characteristic of a transverse magnetic field, which is static and non-axisymmetric. If an axisymmetric magnetic field such as a vertical or a cusp-shaped field is used, the melt rotates with the same angular velocity as that of the crucible.

3. PARALLEL COMPUTING

When performing large-scale computation, parallel computers are sometimes used to enhance the speed of computation. Richardson proposed the concept of such a computing system 1922 [48]. A group of 64000 persons in a large-circle theater who act as a part of computer may predict weather in the world by sharing the calculation of heat and mass transfer on the earth. Using a large number of high-performance computers and a communication system can now make the same predictions. Therefore, we can perform large-scale computation in the fields of heat and mass transfer on a macro scale and/or atomic scale.

There are two types of parallel computer systems. One is a common memory system, and the other is a distributed memory system. Since all central processing units (CPUs) in the common memory system can use memory as a common use, it is easy to build a code with high efficiency of parallelization for a common memory system. We usually use an Open-MP [49] for data communication. However, the cost of the common memory system is much higher than that of the distributed memory system due to expensive hardware.

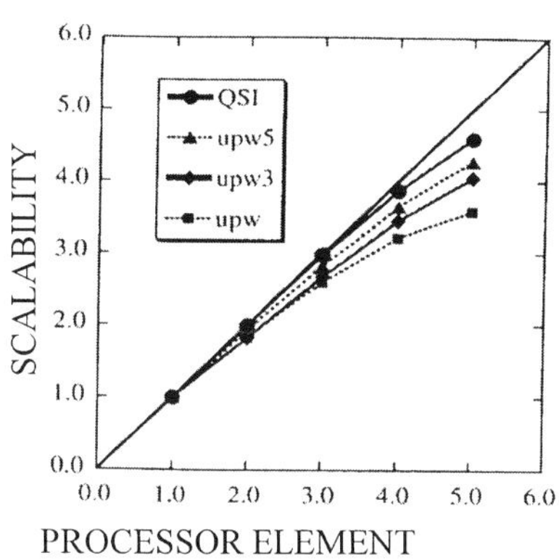

Figure 13. An example of scalability as a function of PEs for the finite difference method. QSI, upw5, upw3 and upw represents calculation of diffusion term with 4th, 5th, 3rd and 2nd order accuracy.

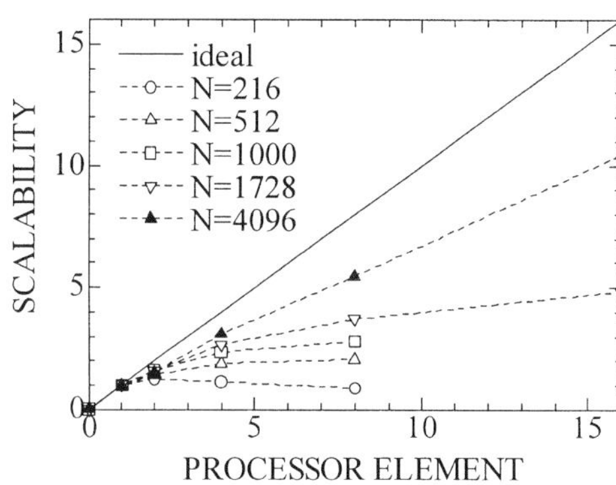

Figure 14. An example of scalability as a function of PEs for molecular dynamics. N shows total number of atoms in the calculation system.

The distributed memory system has memories in all personal computers (PCs), and data transfer between PCs must therefore be managed by using a message passing interface (MPI) [50]. Although the program code becomes complex due to the MPI, the system has no limitation of total number of PCs. Consequently, large-scale computation is possible by using this system.

Figs. 13 and 14 show examples of scalability as a function of numbers of processor elements for codes of the finite difference method and molecular dynamics simulation, respectively. When a large number of PEs is used, the time required for data communication is long. Therefore, a data communication system with high speed is required for fast computing.

4. VISUALIZATION METHOD

The calculated result should be checked after the calculation of flow. A visualization technique, especially for three-dimensional calculation, is needed since the visualization system in our eyes is only a two-dimensional system.

Figs. 15 to 17 show results obtained by three-dimensional calculation of silicon with transverse magnetic fields. This system has a large asymmetry due to transverse magnetic fields. Fig.15 shows calculated particle paths in the melt. Calculated temperature and velocity profiles are shown in Fig. 12: however, it is sometimes difficult to recognize three-dimensional structures of profiles, especially two roll cells, from the figures. We can recognize the two roll cells in Fig. 16, which are in parallel to the x-axis in which transverse magnetic fields were applied. We can also recognize the iso-surface, which is deformed as shown in Fig. 16. Such deformation is due to the two roll cells, which are shown in Fig. 15.

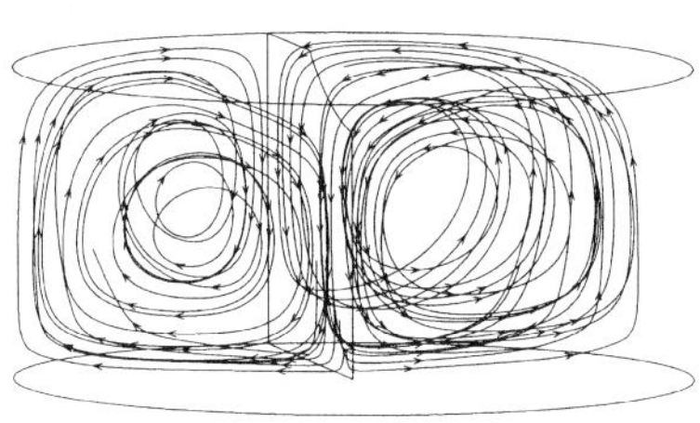

Figure 15. Calculated particle paths in silicon melt with transverse magnetic fields.

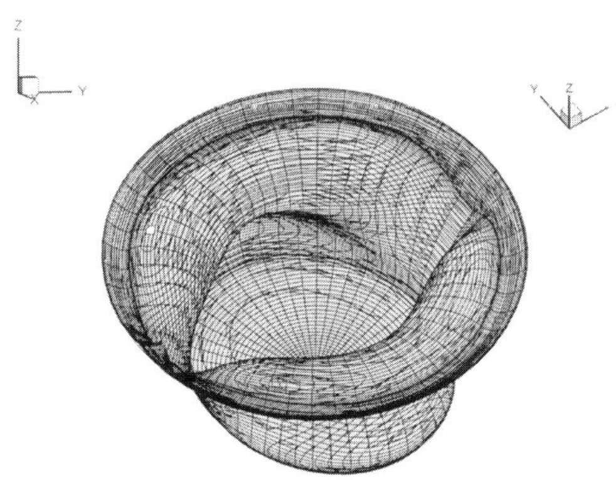

Figure 16. Iso-surface of temperature in silion melt with transverse magnetic fields.

Although Figs. 15 and 16 shows profiles only in the melt, these data were obtained from a global calculation, which includes conductive, convective and radiative heat transfer in a Czochralski furnace. Since heat transfer in a furnace is affected by all parts such as the crystal, crucible, melt, heater and thermal shields, heat transfer based on conduction, convection and radiation should be solved self consistently.

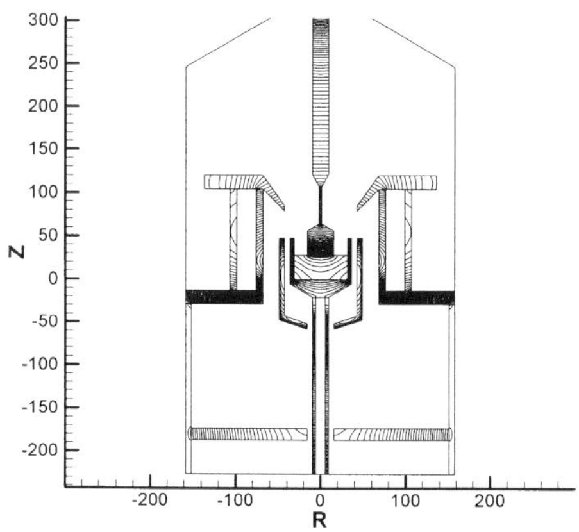

Figure 17. Temperature distribution in a global model with a two-dimensional configuration.

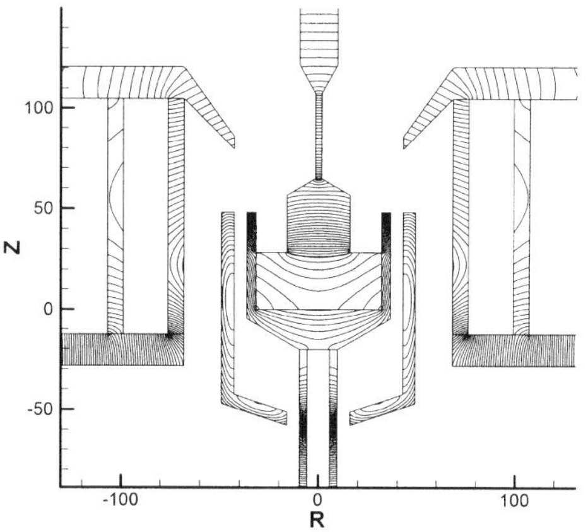

Figure 18. A zoomed temperature distribution in a global model with a two-dimensional configuration.

Figs. 17 and 18 [51] show two-dimensional temperature distributions in a furnace in Kyushu University. We can recognize temperature distributions in each part of the crystal, crucible, melt, heater and thermal shields. If the system is completely axisymmetric, a two-dimensional configuration can be imposed due to the almost axisymmetric configuration of a Czochralski furnace. However, many studies have shown that flow in the melt has a three-dimensional structure. Therefore, we have to change the configuration from two-dimensional to three-dimensional as shown in Fig. 19 [52]. Fig. 19 shows a three-dimensional temperature profile in a crystal, the melt and a crucible with transverse magnetic field. One half of a crucible wall has been removed so as to be able to see the temperature profile in the melt. We can easily recognize the three-dimensional structure of temperature distribution in the whole system. Such three-dimensional visualization is necessary to enhance understanding of three-dimensional structures, especially the structure in a case with asymmetric configuration with transverse magnetic fields.

5. SUMMARY

Modeling of melt flow is important for determining appropriate conditions for obtaining crystals with required quality. The use of numerical simulation also enables an appropriate magnetic field to be selected from various types of magnetic fields in order to control melt flow.

However, we need to understand the phenomena occurring in the melt under the various conditions such as various magnetic fields and various rotation conditions of a crystal and a crucible. The most appropriate magnetic field can be determined by taking into account how the melt flow couples not only the magnetic fields but also temperature distribution in the furnace. If we select one type of magnetic field, computer simulation helps us to predict temperature, velocity and impurity concentration more quantitatively. Although current computer power is still not sufficient to enable re production of almost all of the phenomena occurring in a furnace in the case of a magnetic field-applied system, more quantitative prediction of the phenomena should soon become possible by the use of new algorithms and hardware.

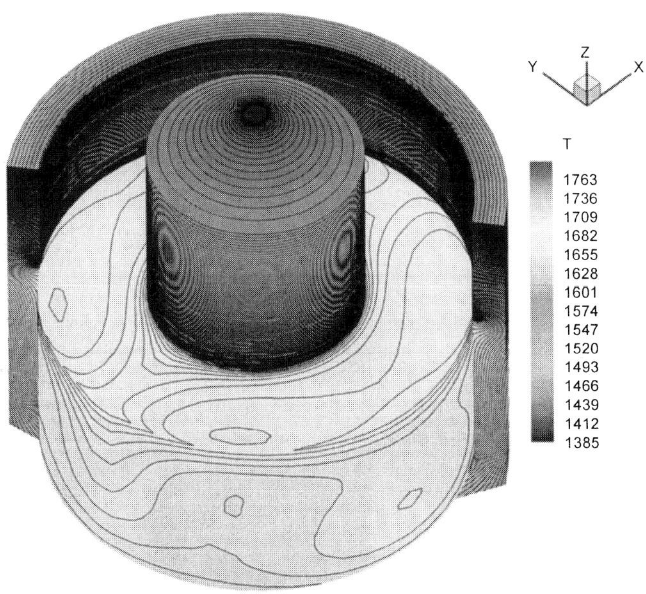

Figure 19. Three-dimensional temperature profile in a crystal, the melt and a crucible with transverse magnetic fields. Half of the crucible wall has been removed.

REFERENCES

1. M. Itsumi, H. Akiya, and T. Ueki, J. Appl. Phys. 78 (1995) 5984.
2. G. Müller, Convection and Inhomogenieties in Crystal Growth from the Melt, crystals, Vol. 12, Springer, Berlin, 1988.
3. T. A. Kinney, D. E. Bornside, R. A. Brown and K. M. Kim, J. Crystal Growth 126 (1993) 413.
4. K. Koai, A. Seidel, H.-J. Leister, G. Müller and A. Koehler, J. Crystal Growth 137 (1994) 41.
5. H.-J. Leister and M. Peric, J. Crystal Growth 123 (1992) 567.
6. H. Yamagishi, M. Kuramoto, and Y. Shiraishi, Solid State Phenom. 57-8 (1997) 37.
7. Y. C. Won, K. Kakimoto, H. Ozoe, Numerical Heat Transfer A36 (1999) 551.
8. Kyung-Woo Yi, Masahito Watanabe, Minoru Eguchi, Koichi Kakimoto and Taketoshi Hibiya, Jpn. J. Appl. Phys. 33 (1994) L487.
9. M. G. Williams, J. S. Walker and W. E. Langlois, J. Crystal Growth 100 (1990) 233.
10. A. E. Organ and N. Riley, J. Crystal Growth 82 (1987) 465.
11. J. S. Walker and M. G. Williams, J. Crystal Growth 137 (1994) 32.
12. J. Baumgaartl, M. Gewald, R. Rupp, J. Stierlen and G. Müller, Proceedings of the VIIth European Symposium on Materials and Fluid Sciences in Microgravity, Oxford, UK, pp.10 (1989).
13. L. N. Hjellming and J. S. Walke, J. Crystal Growth 87 (1988) 18.
14. S. Kobayashi, J. Crystal Growth 85 (1987) 69.
15. M. Akamatsu, K. Kakimoto, H. Ozoe, Transport phenomena in thermal science and process engineering 3 (1997) 637.
16. K-W Yi, K. Kakimoto, M. Eguchi, M. Watanabe, T. Shyo and T. Hibiya, J. Crystal Growth 144 (1994) 20.
17. K. Kakimoto and H. Ozoe, J Crystal Growth 212 (2000) 429.
18. R. A. Brown, T. A. Kinney, P.A. Sackinger and D. E. Bornside, J. Crystal Growth 97 (1989) 99.
19. H. Hirata and N. Inoue, Jpn. J. Appl. Phys. 23 (1984) L527.
20. H. Hirata and K. Hoshikawa, J. Crystal Growth 96 (1989) 747.
21. H. Hirata and K. Hoshikawa, J. Crystal Growth 98 (1989) 777.
22. H. Hirata and K. Hoshikawa, J. Crystal Growth 113 (1991) 164.
23. K. Hoshi, T. Suzuki, Y. Okubo and N. Isawa, Extended Abstracts Electrochem. Soc. Spring Meeting (The Electrochem. Soc., Pennington, 1980) vol.80-1, p.811.
24. K. Hoshikawa, Jpn. J. Appl. Phys. 21 (1982) L545.
25. K. Hoshikawa, H. Kohda and H. Hirata, Jpn. J. Appl. Phys. 23 (1984) L37.
26. K. Kakimoto and L. J. Liu, Crystal Research and Technology 38 (2003) 716.
27. Kakimoto K., Proceedings of 2nd Workshop on High Magnetic Fields, (1995) Florida, ed. by H. Schneider-Muntau, World Scientific, USA (1997) 21.
28. K. Kakimoto, Prog. Crystal Growth and Charact. 30 (1995) 191.
29. K. Kakimoto, Y. W. Yi and M. Eguchi, J. Crystal Growth 163 (1996) 238.
30. K. Kakimoto, and M. Eguchi, J. Crystal Growth 116 (1996) 1257.
31. K. M. Kim and W. E. Langlois, J. Electrochem. Soc. 133 (1986) 2586.
32. A. E. Organ and Riley, J. Crystal Growth 82 (1987) 465.

33. Z. A. Salnick, J. Crystal Growth 121 (1992) 775.
34. T. Suzuki, N. Isawa, Y. Okubo and K. Hoshi, Semiconductor Silicon 1981, eds. H. R. Huff, R. J. Kriegler and Y. Takeishi (The Electrochem. Soc., Pennington, 1981) p.90.
35. R. N. Thomas, H. M. Hobgood, P. S. Ravishankar and T. T. Braggins, Solid State Technol (April 1990) 163.
36. M. Watanabe, M. Eguchi, K. Kakimoto and T. Hibiya, J. Crystal Growth 128 (1993) 288.
37. M. Watanabe, M. Eguchi, K. Kakimoto and T. Hibiya, J. Crystal Growth 131 (1995) 285.
38. A. F. Witt, C. J. Herman and H.C.Gatos, J. Mater. Sci. 5 (1970) 882.
39. K.-W.Yi, M. Watanabe, M. Eguchi, K. Kakimoto and T. Hibiya, Jpn. J. Appl. Phys. 33 (1994) L487.
40. Y. Gelfgat, J. Krumins, B. Q. Li, J. Crystal Growth 210 (2000) 788.
41. Y. Gelfgat, Jpriede, Magneto- hydrodynamics 31 (1995) 102.
42. R. U. Barz, G. Gerbeth, Y. Gelfgat, J. Crystal Growth 180 (1997) 410.
43. T. Kaiser and K. W. Benz, Phys. Fluids 10 (1998) 1104.
44. F.-U. Brucker and K. Schwerdtfeger, J. Crystal Growth 139 (1994) 351.
45. J. Virbulis, Th. Wetzel, A. Muiznieks, B. Hanna, E. Dornberger, E. Tomzig, A. Muhlbauer, W. v. Ammon, Proceedings of the Third International Workshop on Modeling in Crystal Growth (2000) 31.
46. L. J. Liu, T. Kitashima, K. Kakimoto, Proceedings of the International Symposium on Processing Technology and Market Development of 300mm Si Materials (ISPM-300mm Si), Sept. 8-9, 2003, Beijing, China (10), (1922) 2551.
47. K. Kakimoto, H. Watanabe, M. Eguchi and T. Hibiya, J. Crystal Growth 88 (1988) 365.
48. L. F. Richardson, Weather Prediction by Numerical Process (Cambridge University Press, London, 1922).
49. http://www/openmp.org
50. http://www/mpi-forum.org
51. K. Kakimoto and R. Liu, Cryst. Growth Res. Technol. 38 (7-8), (2003), 716.
52. R. Liu and K. Kakimoto, submitting to J. Crystal Growth (2003).

Crystal Growth – from Fundamentals to Technology
G. Müller, J.-J. Métois and P. Rudolph (Editors)

Molecular simulations of crystal growth processes

Jan P.J.M. van der Eerden
Condensed Matter and Interfaces, Debye Institute, Utrecht University
P.O. Box 80.000, 3508 TA Utrecht, The Netherlands.

This lecture is an extension of an earlier lecture [1] that was presented in Mysore in 2003. It will be discussed what are the possibilities and limitations of molecular simulation of crystal growth. It will be shown that molecular simulation takes an intermediate position between theory and experiment, bridging and enriching both. An overview of several simulation methods will be given. Using examples of applications the students will get a feeling of how and when to choose a suitable simulation method. Molecular simulation will be put in the proper perspective as compared to other theoretical and experimental approaches.

Different approaches are presented for numerical simulations of crystal growth. The common bottleneck for such simulations is the limited efficiency of even the fastest computers. In order to cope with this difficulty a hierarchy of models and modelling approaches is being developed. At the present state of the art one aims at qualitative, generic understanding of the important issues, structures and processes for a given crystal growth system. Therefore one has to reflect first on an adequate level of description, specifically on the length and time scales at which information and understanding is requested. This guides the selection of the appropriate simulation method.

This lecture summarizes the state of the art and sketches some promising developments for crystal growth simulation. Though in reality a crystal cannot exist without attractive interactions, these interactions are the main cause of simulations to die in jammed states. In such states no further mesoscale dynamics like crystal growth is possible. Therefore, as a main point of research one attempts to develop simulation techniques that keep the essential physics while softening the repulsive forces and replacing the attractive forces by overall forces like pressure. Techniques, measurable quantities and results are discussed.

Examples are given that demonstrate the width of this research field. It is hoped that this work helps the interested reader to distinguish promising and hopeless simulation approaches for his or her own research.

1. INTRODUCTION

Numerical simulation is a powerful tool to increase our understanding and mastering of crystal growth processes. There exist a variety of different techniques, models and methods. Though *ab initio* models and methods can be formulated to describe the time evolution of materials at the atom and electron levels, such approaches are not often of much use. The reason is the limited size and speed of any imaginable computer. Indeed all simulation methods have in common that the required size of the computer increases proportional to the simulated system size or faster, and the required computational time increases proportional to the system size and to the simulated physical time or faster. This puts a serious limit to the crystal growth phenomena that can be studied with a given method. Generally, the more detailed the models are, the smaller are the systems and the shorter are the time intervals systems that can be simulated.

At present *ab initio* quantum chemical calculations can be carried out for systems containing up to a few hundreds of electrons, and dynamical processes of up to nanoseconds. Important crystal growth length and time scales range from these atomic scales up to meters and days. In practice the choice of simulation models and methods has to be a sensible compromise between computational accuracy and speed. Therefore the simulation approach of choice strongly depends on the type of question.

It is the purpose of this lecture to give some insight in the methods that are available, and in the kind of results which can be obtained. Also the kind of reasoning which one could follow to find a profitable simulation approach will be illustrated.

2. COMPUTER SIMULATION *VS* COMPUTER EXPERIMENT

Once you have realized that a quantitatively accurate atomic scale simulation of crystal growth is far out of reach, you have to ask yourself what is the aim of your simulation [2-4]. Do you want to improve your specific crystal growth system, e.g. to optimise the crystal quality or the yield of a crystallization reactor, or do you aim at better understanding of some aspect of crystal growth at a fundamental level. In the first case your approach will be to develop an accurate reproduction of your crystal growth system, transferring as much as possible of the physical properties of real system to your simulation model. This approach may be referred to as *computer simulation.* In contrast, if you feel that you have to develop a better theory and more fundamental understanding, then you first develop a *generic* model, the simplest (and computationally most efficient) model which keeps the essentials of your real system; next you treat your model calculations as measurements on a well known material, trying to derive a theory which accurately reproduces the outcomes of these *computer experiments.*

An example of computer simulation in this restricted sense is the solution of the Navier-Stokes equations to calculate the heat and mass transport fluxes in large industrial crystallization reactors. An example of computer experiments of liquids is the simulation of hard sphere models. Comparing computer and real experiments one was led to understanding general properties of diffusion and of local structure. Moreover one concluded, after much debate, that repulsive interactions alone are sufficient to produce a first order liquid-solid phase transition.

Examples of *computer simulations* in crystal growth are the search for effective crystallization inhibiting species in industrial crystallization, and the large-scale heat and mass transport calculations to locate convection instabilities during single crystal growth from the melt. Examples of *computer experiments* are the finding and understanding of surface roughening during growth from vapour or solution, and the subtle role of surface tension anisotropy in dendrite formation during growth from the melt.

In the context of this lecture I cannot discuss the whole field, rather I will focus on simulation models at molecular up to colloidal size scales, leaving out the subatomic *ab initio* modelling, as well as the macroscopic heat and mass transport modelling.

3. GENERIC CRYSTAL GROWTH MODELS: KOSSEL AND LENNARD-JONES

3.1. The Kossel model, for growth from vapour and from solution

Every crystal grower knows that crystals often but not always are faceted, that the dependence of the growth on supersaturation usually is non-linear, and that the crystal shape may be quite compact, but also often is very anisotropic (needles or platelets). In order to understand such observations the Kossel [5] model was put forward by Stranski and Kaischew [6]. The development of the Kossel model for crystal growth illustrates the idea of finding a generic model for a complex system, to treat the relevant questions. The model had to explain the experimental observation that some crystals do have natural faces, whereas others don't. The orientations of these faces were clearly related to the atomic scale (X-ray) structure. Hence the model was built on a simple cubic crystal lattice. The structure of the mother phase (melt, vapour, solution etc.) seemed not to be very relevant. Therefore it was sufficient to allow only two states n_j for each lattice site j, solid ($n_j = 1$) to represent the crystalline atoms in the system, and fluid ($n_j = 0$) to take into account the mother phase, in an average sense. In general flat faces were likely to appear for high heats of crystallization and small supersaturation. Nearest neighbour interaction was an easy way to bring the heat of crystallization into play. Supersaturation was represented by different probabilities for isolated solid and fluid lattice sites. The solid on solid (SOS) condition (solid lattice points only above other solid points) led to a sharp and unambiguous definition of the surface position and some mathematical simplification.

In this way a very simple model was constructed, containing only two dimensionless parameters (and the Boltzmann constant k_B): $\Delta H/k_BT$, to represent the inverse temperature and $\Delta\mu/k_BT$, for the supersaturation. These parameters were directly linked to experimentally accessible quantities: the crystallization energy ΔH of the crystal, the temperature T and the free energy difference $\Delta\mu$ between growth units in the mother phase and in the crystal. Studies of this model paved the way for enormous progress in our understanding of the factors that influence crystal morphology. In addition this model turned out to be a rich source of inspiration for developing general insight in surface structure and surface phase transitions.

The model was proposed long before the advent of modern computers. In 1951 Burton, Cabrera and Frank [7] used the model to introduce the concept of flat *vs* rough faces. However, only when in the period 1965-1985 several groups all over the world carried out extensive computer simulations, the concept of the surface roughening was developed. Indeed, Kossel crystal simulations have led to understanding the roughening transition. It became clear that systems for which the surfaces have energetic preference for discrete positions along their normal axis have a thermodynamically well-defined surface phase transition [8]. Below the transition temperature the surface is atomically smooth, the crystals develop facets, they grow with step motion [9] and dislocation free crystals grow significantly only beyond a certain finite supersaturation (the so-called surface nucleation dip). Above that temperature the surface is rough on an atomic scale, the crystal face becomes macroscopically curved, and it grows linearly with supersaturation. It could also be understood that below the roughening point the most pregnant impurity effect is the formation of macrosteps *via* step bunching. Above the roughening point morphological instability usually is most important.

3.2. The Lennard-Jones model for growth from a melt

A short reflection is sufficient to see that, at the atomic level growth from a melt cannot be accurately described by lattice models. Indeed, in this case the rearrangement of atoms and molecules towards their crystalline lattice positions cannot be described in a lattice model. Probably the best-known and best-understood off-lattice model for crystal growth and nucleation is the Lennard-Jones model. This model possesses a stable (hexagonal close packed, HCP) crystal phase, a liquid phase and a gas phase. It is the generic model for crystal growth from the melt and it can be used for growth from the vapour as well. The potential energy depends on the particle positions $\mathbf{r}^N$. It is built up from isotropic pair potentials, which are strongly repulsive at short distance, attractive at intermediate distance and negligible at large distance:

$$E_{pot}(\mathbf{r}^N) = \sum_{i=1}^{N-1} \sum_{j=i+1}^{N} V\left(\left|\mathbf{r}_i - \mathbf{r}_j\right|\right) \tag{1}$$

$$V(r) = 4\varepsilon\left(\left(\frac{\sigma}{r}\right)^{12} - \left(\frac{\sigma}{r}\right)^{6}\right) f^{CO}\left(\frac{r}{r^{CO}}\right) \tag{2}$$

Here a smooth cut-off function ($f^{CO} = 0$ for $r > r^{CO} \approx 2.5\ \sigma$) is introduced, $\varepsilon > 0$ is the bond energy and σ the atom diameter.

Like the Kossel model, the Lennard-Jones model contains two dimensionless parameters: $\varepsilon/k_B T$, to represent the inverse temperature and either $P\sigma^3/k_B T$ for isobaric systems or $N\sigma^3/V$ for isochoric systems, to fix the supersaturation. These parameters were directly linked to experimentally accessible quantities: the crystallization energy ΔH of the crystal, the temperature T and the free energy difference $\Delta\mu$ between growth units in the mother phase and in the crystal.

In the Lennard-Jones model crystal growth and nucleation from the melt have been simulated. It should be mentioned, however that below the melting point the HCP phase is only slightly more stable than the face centred cubic (FCC) phase, whence in simulation experiments often randomly stacked close packed layers are found (rcp phase) rather than a real stable phase. Unfortunately, using straightforward MC or MD, these processes can be studied only under conditions that are, from an experimental point of view exotic, to say the least. Indeed growth rates of the order of 1 m/s and nucleation rates of the order of $10^{25}\ s^{-1}mm^{-3}$ were accessible [10].

Lennard-Jones simulations have been used and will be used further to check and to improve the classical crystal growth theories. The growth rate R turned out to depend linearly on the supercooling $\Delta T = T_m - T$ over a large range of supercooling. This result was in line with the theory developed for the lattice models. Indeed the dimensionless heat of melting of melting $\Delta H_m/(k_B T_m) \sim 1.8$ is so small that all crystal-melt surfaces were predicted to be above the roughening temperature. As expected the crystal-vapour and crystal-melt surface tensions were found to be nearly isotropic [11-16]. The surprisingly strong growth rate anisotropy turned out to be related to the atomic scale structural difference in the interfacial region, as we will see below.

Crystal nucleation from a Lennard-Jones melt has been studied as well. With MD and MC simulation one has studied nucleation at an supercooling of about 20K [17], [18]. This however, was possible only with the help of artificial enhancement of the simulations. Umbrella sampling, transition path sampling and ultra-fast temperature and pressure sweeps [19] have been applied. The shape, and shape fluctuations of crystal nuclei in the supercooled melt have been examined. These observations show that the actual nuclei and their temporal evolution are quite distant from the structures assumed in classical nucleation theories. Nevertheless it turned out that classical nucleation theory, if properly parameterised, describes the relations between experimentally accessible quantities quite well.

4. BASIC STATISTICAL THERMODYNAMICS

In most cases the relevant size ranges for crystal growth phenomena vary from a few tenth of nm to a few μm. Hence usually quantum mechanical effects are negligible. Both the interactions and the dynamics are described accurately enough by classical mechanics.

In particle-based models, like the Lennard-Jones model, system states φ correspond to a combination of single particle states for each particle i with $i = 1,\ldots, N$ in the system. If the particles are structureless ("atoms" in the proper sense of the word) then a single particle state φ_i is fully specified by the position and velocity of particle i. This leads to a number of different, yet equivalent, ways to denote a system state:

$$\varphi \equiv \left(\mathbf{r}^N, \mathbf{p}^N\right) \equiv \left(\mathbf{r}_1, \mathbf{p}_1, \ldots\ldots, \mathbf{r}_N, \mathbf{p}_N\right) \equiv \left(\varphi_1, \ldots\ldots\ldots, \varphi_N\right) \equiv \varphi^N \tag{3}$$

If a particle has internal structure (e.g. polarisability, interaction centres, vibrations) then a single particle state φ_i includes the internal degrees of freedom as well.

In lattice models like the Kossel model the state of a system is given by a list of occupation numbers n_j (e.g. $n_j = 0$ for a fluid site and $n_j = 1$ for a solid site) for each lattice position j with $j = 1,\ldots, M$.

$$\varphi \equiv \left(n_1, \ldots., n_M\right) \equiv n^M \tag{4}$$

Macroscopic system properties are ensemble averages of state properties. E.g. in the so-called *canonical* ensemble, which describes a unary system at constant N (number of particles), V (volume) and T (temperature), the thermodynamic energy U of a system is defined as

$$U \equiv \langle E \rangle = \sum_{\varphi} E(\varphi) P(\varphi) = \frac{1}{h^{3N} N!} \int_{-\infty^{3N}}^{\infty^{3N}} \int_{V^N} E\left(\mathbf{p}^N, \mathbf{r}^N\right) P\left(\mathbf{p}^N, \mathbf{r}^N\right) d\mathbf{r}^N d\mathbf{p}^N \tag{5}$$

Here $E(\varphi)$ is the energy of state φ, and $P(\varphi)$ its probability. Averaging is over the whole *phase space*, i.e. over all degrees of freedom of the system. For lattice and quantum mechanical models this leads to summation, for classical systems to integration and a combinatorial factor $1/(h^{3N}N!)$ to compensate for overcounting states of indistinguishable identical particles, h is the Planck constant. It should be noted that for some ensembles one may have to integrate over macroscopic system variables as well, e.g. over all possible V for isobaric systems and over all possible N for open systems.

The probabilities are known from general statistical mechanics. In the canonical ensemble the states should appear with the Boltzmann probability:

$$P(\varphi) = \exp\left(-E(\varphi)/k_B T\right)/Q \tag{6}$$

where k_B is the Boltzmann constant and Q the canonical partition function, defined as:

$$Q(N,V,T) \equiv \sum_{\varphi} \exp\left(-E(\varphi)/k_B T\right) \tag{7}$$

where the same summation or integration is meant as in the definition of U. Statistical thermodynamics and classical thermodynamics are linked by the relation

$$A(N,V,T) \equiv -k_B T \ln(Q(N,V,T)) \tag{8}$$

between the Helmholtz free energy A and the canonical partition function Q. Starting from this relation microscopic expression for other macroscopic variables can be derived. E.g. it can be shown that the expression used above for U is consistent with this expression for A.

5. MOLECULAR DYNAMICS AND MONTE CARLO SIMULATION

Molecular simulation methods are meant to replace the exact calculation of ensemble averages, which usually is impossible analytically. Simulation often is more efficient, and moreover it bears some similarity to the natural approach of an experimentalist. A simulation produces a large set of system states φ that can be treated as *snapshots* of the system under consideration. The idea is to estimate system properties by "measuring and averaging" them in these snapshots, in the same way as an experimentalist would do. The simulations have the advantage that *exact* snapshots of the simulated system are available. A simulation is physically relevant, if the set of generated states is representative for states that would occur in a real experiment. E.g. for the study of equilibrium properties of a canonical (*NVT*) system the states should appear with the Boltzmann probability.

The basic simulation methods are Molecular Dynamics (MD) and Monte Carlo (MC). MD generates a continuous path in phase space that closely resembles the actual time evolution of the system. MC hops discretely along a path in such a way that each state is visited with the appropriate probability, the path may or may not approximately follow a temporal evolution path.

5.1.Measuring macroscopic quantities

A sound computer simulation produces a set of representative states, representative meaning that the relative probability $P(\varphi)/P(\psi)$ for two states φ and ψ to be present in the set and in the real system are equal. Consequently the average over the n generated states φ_i in the set is a good approximation of the ensemble average:

$$\overline{B} \equiv \frac{1}{n}\sum_{i=1}^{n} B(\varphi_i) \approx \langle B \rangle \equiv \sum_{\varphi} P(\varphi) B(\varphi) \tag{9}$$

As in experiments a sufficient number of statistically independent states has to be generated for a given average to have statistical significance. It is in the first place the skill and expertise of the researcher to judge whether reasonable sampling has been achieved.

The averaging procedure can be directly employed for those physical quantities that are meaningful for an individual state. Examples are basic thermodynamic variables, like the energy, the pressure, the volume and the system density $\overline{\rho} = N/V$.

A slightly more complicated situation occurs when different states have two be compared. This happens e.g., for heat capacities and for elastic constants, which involve fluctuations of the energy and of the pressure respectively. For dynamic quantities like diffusivity and viscosity the relaxation of displacements or velocities with time has to be obtained. Usually for these multi-state properties many more states have to be generated before sufficiently accurate averages can be obtained.

5.2.Molecular Dynamics Simulation

The MD simulation method is deterministic. It solves Newton's equation of motion for the particle positions $\mathbf{r}_i(t)$:

$$m_i d^2\mathbf{r}_i / dt^2 = -\partial E_{pot}(\mathbf{r}^N) / \partial \mathbf{r}_i \tag{10}$$

Here the right hand side is the force on particle i, given by the derivative of the potential energy with respect to $\mathbf{r}_i$. The equation of motion is solved numerically. In order to have a detailed description of the motion of each particle, the time step dt should be such that the atomic displacements between t and $t + dt$ are 1-10% of the atomic size or less (~ .005 nm in condensed systems). At a typical thermal velocity of particles at room temperature (~ 500 m/s) a reasonable choice is dt ~ .01 ps. For a simple pair-wise interaction potential the interaction range typically comprises some 100 atoms, which implies some $100N$ force calculations per integration step. This practically limits the simulation of liquid and solid systems to a few thousand atoms during a few ns. Within these limits usually thermodynamic parameters, the equation of state and the phase diagram can be studied, as well as dynamic properties like diffusivity and viscosity.

5.3.Monte Carlo Simulation

The MC simulation method is not deterministic but probabilistic. The system jumps discretely from one state φ to another state ψ. The transition probabilities $p(\phi \rightarrow \psi)$ are chosen in such a way that on the long run the Boltzmann probability $P(\varphi)$, see Eq. (2) for each configuration φ will be obtained. A sufficient condition on the transition probabilities is that they satisfy the condition of *detailed balance*:

$$P(\varphi)p(\varphi \rightarrow \psi) = P(\psi)p(\psi \rightarrow \varphi) \tag{11}$$

A popular choice to satisfy detailed balance is the Metropolis scheme, where one starts in state φ to select a possible new stable ψ with a symmetric generation probability $p_{gen}(\varphi, \psi) = p_{gen}(\psi, \varphi)$. In the simplest form of MC of an atomic system one displaces one atom in state φ to generate state ψ. This state ψ is accepted as the new state with an acceptation probability $p_{acc}(\varphi \rightarrow \psi)$ which depends on the energy difference of the two states:

$$p(\varphi \rightarrow \psi) = p_{gen}(\varphi, \psi)\ p_{acc}(\varphi \rightarrow \psi) \tag{12}$$

$$p_{acc}(\varphi \rightarrow \psi) = \min(1, P(\psi)/P(\varphi)) = \min(1, \exp(-(E(\psi) - E(\varphi))/k_B T)) \tag{13}$$

In some cases it is advisable to change to another generation rule for a new state, then a corresponding change in the acceptation rule has to be made as well. An important case is Configuration Bias Monte Carlo simulation, which we shall discuss below.

5.4.Comparison of Molecular Dynamics and Monte Carlo

With the state generating procedure mentioned above MC and MD simulation paths will be quite similar [20] and qualitative conclusions on the system dynamics can be deduced from both types of simulation. In practice the computational times usually are comparable. A computational advantage of MD is that it is easier to apply parallel programming. With MC on the other hand it is less critical to have a very small time step *dt*. The most important advantage of MC is its greater flexibility. With MC one is not restricted to continuum models but also lattice models for crystal growth can be (and are frequently) applied. Also *non-physical state changes* like creation, annihilation and exchange of particles and instantaneous chemical reactions, can be implemented without loosing physical relevance, i.e. leaving the relative probabilities of states unchanged.

6. GENERIC CRYSTAL MORPHOLOGY THEORIES

Our fundamental understanding of nucleation and growth is based on studies of the structure and dynamics of crystal surfaces. One of the first objects of research has been the shape of the crystal. We are now in the position to compare the morphology predictions of some classical theories with generic simulation results on a simple model.

6.1.Classical morphology rules

With the advent of X-ray crystallography, the relation was proven between the possible *orientations* of natural crystal faces and the atomic structure of the crystal unit cell. Following this, crystallographers have put forward practical rules for the relative *size* of different faces. For simple crystal structures Bravais, Friedel, Donnay and Harker [21] hypothesized (BFDH-rule) that the crystal growth rate *R(hkl)* and the face area size in the direction [*hkl*] are inversely proportional to the interplanar distance *d(hkl)*:

$$\text{BFDH - rule:} \quad R(hkl) \div d(hkl)^{-1} \tag{14}$$

For more complicated structures the Hartman-Perdok theory [22] was developed. This theory led Hartman and Bennema [23] to their HB-rule that *R(hkl)* is proportional to the attachment energy, E^{att}, the energy change when a new growth unit is attached to a flat *(hkl)*-surface:

$$\text{HB - rule:} \quad R(hkl) \div E^{att}(hkl) \tag{15}$$

This rule relates the atomic scale bond structure directly to the macroscopic growth shape of crystals.

6.2.Lattice models

In the field of lattice models the central result is the understanding and application of the roughening concepts. Combining the theory of the roughening transition with the Hartman-Perdok theory it was understood that the BFDH and HB growth laws should only be applied for faces below the roughening temperature. Faces grow much faster above the roughening point, but MC simulations did not support suggestions that the growth rate would be approximately proportional to T - TR. The growth morphology can be obtained on the basis of MC simulation results. Tests for simple cases showed that the HB-rule is qualitatively correct. Commercially available software packages are based on BFDH- or HB-rules.

For more complicated lattices the Hartman Perdok analysis is difficult to do by hand. Therefore algorithms have been developed to study arbitrary lattices. The results, still being amended by various authors [24], form the basis of modern crystal morphology predictions.

Parallel to the gradual increase in understanding of the roughening transition one has tried to test the predictions of the Kossel model in experimental systems. To this end one had to replace the simple cubic lattice by general lattices. Algorithms, invoking graph theory, have been developed and implemented in computer software. Also the SOS condition was modified, and different types of solid particles and interactions were introduced. The next step is that nowadays programs like MONTY [25, 26] are developed. With this program arbitrary surface orientations on arbitrary lattices with arbitrary (short range) interaction parameters can be simulated. In this way one can predict and understand growth forms.

Using this program, several different crystal growth experiments could be interpreted qualitatively, with quite satisfactory success. As an example the spherulitic growth of aspartame crystals [27] could be explained [28].

In retrospect, the discussion above illustrates nicely how generic computer experiments evolved towards directly applicable computer simulations. Initially qualitative experimental observations led to the development of a generic model, the Kossel model. Next computer experiments with this model generated fundamental understanding (of the roughening transition). This in turn allowed generalizations that were implemented in computer simulations with the MONTY simulation package. These simulations now pave the way to quantitative predictions of crystal growth shapes.

6.3. Lennard-Jones morphology

Broughton and Gilmer [14-16] were the first to study two-phase systems. One of their results was that the solid-melt surface free energy was almost isotropic. Hence the equilibrium form of a Lennard-Jones crystal in its melt at $T = T_m$ is almost spherical. We found [29] a strong orientation dependence of the growth rate *R(hkl)*:

$$\text{Monte Carlo Simulation}: R(111) \div R(100) \div R(110) = 1 \div 1.99 \div 1.30 \tag{16}$$

This came as a surprise. Moreover this orientation dependence was different from both the BFDH and the HB predictions. The BFDH-rule predicts

$$\text{BFDH-rule}: \ d(111)^{-1} \div d(100)^{-1} \div d(110)^{-1} = 1 \div 1.150 \div 1.633 \tag{17}$$

The HB prediction is, when only nearest neighbour interactions are taken into account

$$\text{HB-rule}: E^{att}(111) \div E^{att}(100) \div E^{att}(110) \approx 1 \div 1.333 \div 1.667 \tag{18}$$

When next nearest neighbour bonds are used that are half as strong as the nearest neighbour bonds then the HB-rule gives

$$\text{HB-rule}: E^{att}(111) \div E^{att}(100) \div E^{att}(110) \approx 1 \div 1 \div 1.333 \tag{19}$$

Apparently the growth rate of the cubic (100)-face in the Lennard-Jones model is much larger than expected from lattice model approaches.

To appreciate this difference, note that the predicted growth form of the Lennard-Jones model is a regular octahedron bounded by the pyramidal (111)-faces only, whereas the growth form that is predicted by BFDH, and also by HB is a cubo-octahedron with cube faces of about the same size as the pyramidal faces. Although a full explanation of this result is still not given, detailed studies of the interface structure during growth [2, 20] have shown that structuring in the melt near the interface is responsible.

7. SMART CHOICE OF MODELS AND EXPERIMENTS

In retrospect it is clear that for realistic studies of the crystal growth process itself the achievable physical times are too short and system sizes are too small. For a good direct description of crystal growth and nucleation under practical circumstances, both the simulated time and system sizes have to be increased by several orders of magnitude. This is not within the reach of present computers for straightforward MC or MD simulations. Several techniques and approximations have been developed to improve this situation. All of these have their specific advantages and disadvantages. I will now shortly describe and comment some promising developments.

One way to approach the problem is to choose the most effective model and (simulation) experiment. If this is not sufficient, both the model and the dynamics can sometimes be modified in a controlled way. The general strategy is to first identify the structures and processes that are most relevant for the crystallization problem under consideration. Next, look for the simplest model and simulation method that can handle these degrees of freedom. It is favourable, but not always possible to use models and methods whose parameters can be directly related to physically accessible parameters of the real system. Then study and simulate the model chosen until its general properties are well understood. Finally translate the gained insight back to the original system.

7.1. Choosing a smart model: striped phases in biomembranes

The first possibility to find a sensible effective model is to give up some of the atomic scale resolution, e.g. by replacing complicated molecules consisting of many atoms by simple structures with a few interaction sites only. This means effectively moving from atomic to molecular or even mesoscopic length scales. This approach is especially useful when the relevant structures and the rate determining processes take place at these larger scales. It can often be applied when dealing with biomolecules, colloids or quantum dots (clusters of a few nm in size).

The study of membranes is an important issue in biochemistry. Real bio-membranes consist of bilayers built up from many different lipids. In and on these bilayers different proteins organise the biological functions. To obtain fundamental insight, model membranes are prepared, consisting of only one or a few different lipids. Combining these with well-defined and pure peptides or small proteins progress is being made.

Usually biochemists concentrate on the chemical aspects of such biomolecules to explain the macroscopic properties. An example of another approach occurred when we found special structures in Atomic Force Microscope (AFM) images of model biomembranes [30]. Striped patterns were obtained of a well-defined spacing of about 9 nm, see Figure 1A. Complementary experiments showed that these stripes were alternating lines of peptides and lipids. Explanations in terms of biochemical properties failed. This was in line with the observation that the same striped domains, with the same stripe spacing, were found over a whole range of biochemically quite different peptides and lipids.

At this point we had to look for generic properties. We developed a simulation model [31] with two different types of elongated molecules. One type represented lipid dimers, having relatively small Lennard-Jones interaction centres spaced in such a way that in equilibrium an ordered layer of tilted molecules was formed (as for the real lipids). The other type represented peptides, having relatively large and overlapping Lennard-Jones centres, which led to an equilibrium ordered layer of vertical molecules (as for the real peptides). The centres at the end, representing hydrophilic regions, had stronger interactions than those in the middle, which represented the hydrophobic part of the molecules. The mutual interactions between the model peptides and the model lipids favoured mixing (in the real peptides tryptophane residues induce strong lipid-peptide binding). It is topologically impossible to randomly mix vertical and tilted sticks in a horizontal layer. As a compromise the striped structure of Figure 1B is formed. This generic explanation, to which we refer as *frustrated mixing*, forms the basis of further research, aiming e.g. at understanding the value of the spacing of the stripes.

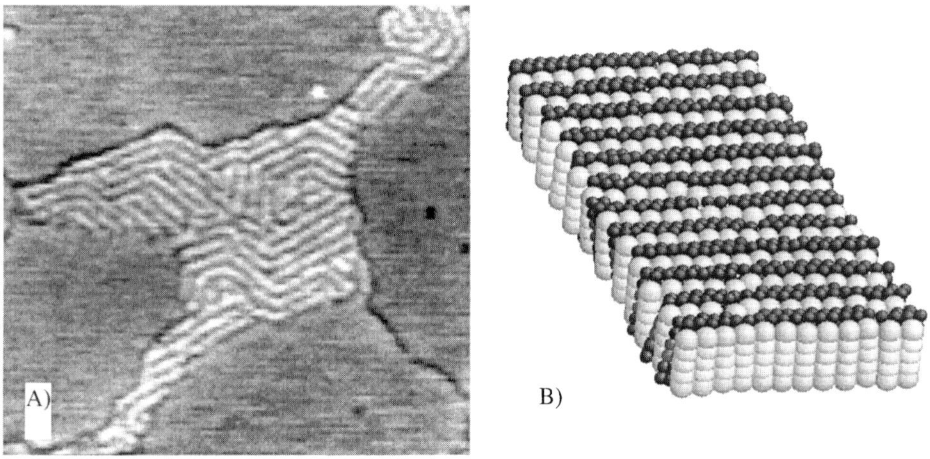

Figure 1. A) AFM image (400nm×400 nm) of a model biomembrane of DPPC lipid bilayers with α-helical transmembrane WALP23 peptide showing striped domains and channels separating DPPC tilt domains. B) A generic computer experiment: tilted "lipids" and vertical "peptides" spontaneously form a striped pattern.

One of the differences between the real system and the model is the relative size of the lipids and peptides. To check that frustrated mixing still occurs when molecules of significantly different size are mixed, we studied mixtures of long thin and shorter thick spherocylinders. Again, depending on the mixing energy phase separation or stripe formation was found. From the real experiments there are also indications that the striped phase disappears close to the temperature where the pure lipid bilayer melts (in the biochemical sense where liquid refers to disordered tails). Therefore we also develop a model with flexible tails, and study this with the DPD-technique (see below) [32, 33].

7.2.Choosing a smart experiment: double-pulse nucleation

The liquid phase can be made meta-stable as compared to the crystalline state by applying supercooling or overpressure. If one stays near the melting line a slow nucleation process is rate determining. Whenever a nucleus has been formed it is likely to grow to a large size (at least in the absence of impurities). If one is too far into the metastable region, nucleation is so fast that polycrystalline, amorphous or glassy material will be formed.

So there seems to be a fundamental problem for nucleation studies. Either nucleation is so slow that it cannot be studied within accessible computational times, or it is so fast that no recognizable nuclei are formed and can be studied. The boundary of the metastable zone in the (*PT*) - plane corresponds to the border-line values of the supercooling or overpressure.

A similar problem occurred, 25 years ago, in homogeneous [34] and heterogeneous [35] electrocrystallization of silver. Then invoking the double-pulse technique solved the problem. In electrochemistry changing electric potential differences between electrodes changes supersaturation very rapidly and accurately. In the experiments first a relative large overvoltage is applied during a short time. During this pulse a number of nuclei is formed, but they are too small to be detected by (optical) microscopy. Next during a much longer time interval a small overvoltage is applied. In this period no new crystals are formed, but the already existing ones grow to visible size. Using this technique, nuclei as small as a few atoms could be quantitatively studied.

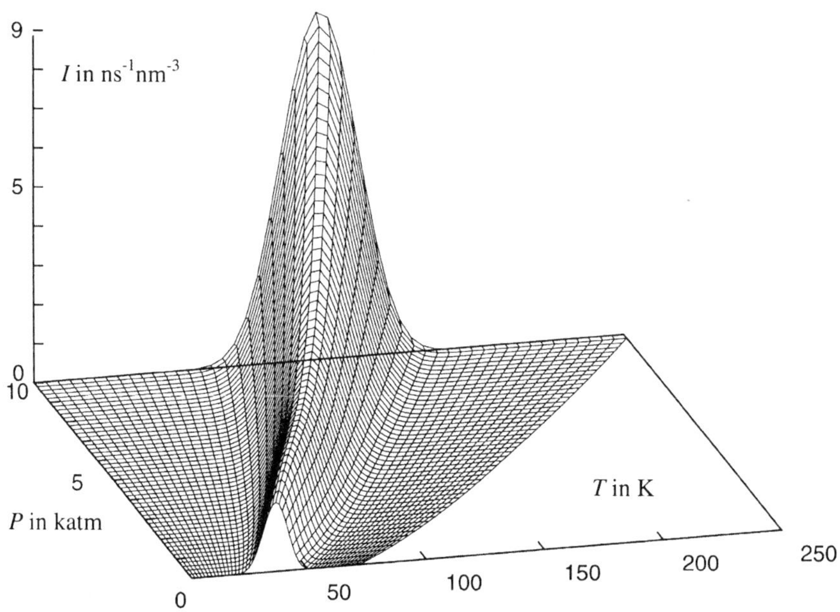

Figure 2. Nucleation rate *I(P,T)* in the metastable zone of the Lennard-Jones model. The melting line is the end of the drawn raster. The curved surface is a least squares fit of the data of supersaturation sweep experiments to a classical nucleation rate expression.

In real experiments this technique is restricted to electrochemistry as in other systems supersaturation cannot be switched on and off rapidly and accurately. In computer simulation however, we can apply the same idea to nucleation in the melt. Using rapid sweeps of temperature or pressure deep into the meta-stable zone, a few nuclei are formed at the largest supersaturation, and in the next period of relatively modest supersaturation these nuclei start to grow out.

With this technique we succeeded to determine nucleation rates for a large part of the metastable zone [19]. Independently we determined $\Delta\mu\,(P,T)$ for this zone and used this in a classical expression for the nucleation rate. Using the surface tension as fit parameters Figure 2 shows that the results were reasonably well described by classical nucleation theory.

8. SMART APPROXIMATIONS FOR MODELS AND DYNAMICS

An alternative to choosing the optimal model, is to modify the actual physical situation or dynamics to better manageable circumstances and to correct the artefacts afterwards. This works especially well if clearly identifiable thermodynamic or kinetic barriers are dominating the molecular scale motion.

8.1.Coarsening the temporal resolution: DPD simulation

Often we are not interested in the detailed motion of the individual atoms and molecules. We would be satisfied to see the gross motion of the molecules along relatively smooth paths. This gross motion corresponds to a sensible time-average of the exact motion. This is the idea of Dissipative Particle Dynamics (DPD) simulation [36-39].

DPD is a technique that may be employed for crystal growth from a liquid phase. With the present computational resources, conventional molecular dynamics and Monte Carlo techniques are fast enough to simulate growth of atomic crystals from its melt, at least if the calculation of particle interactions is not too complicated. E.g. Lennard-Jones (for noble gases or nearly spherical organic molecules) or effective medium (for metals) interaction models are accessible. It should be realised that for spherical frictionless particles, crystal growth from the melt is rate-limited by translations over interparticle distances. The oscillation period of an atom in the "cage" formed by its neighbours, sets the time scale for this translational motion.

In the case of larger growth units, like proteins, colloids or quantum dots, friction and steric hindrance strongly couple the rotational motion of neighbouring particles. The time scale for this concerted rotational motion may be orders of magnitude smaller than the time scale for translation of frictionless spherical particles. Moreover, to really contribute to crystal growth, a growth unit hitting a surface position still has to rotate towards its proper crystal orientation. So straightforward modeling of molecular crystal growth suffers both from very small maximal time steps *dt* and small "docking" probabilities of growth units hitting the surface.

In the DPD method smooth conservative forces and additional friction and random forces replace the actual steep interaction forces. This is done in such a way that hydrodynamic interactions are not effected. In this way the maximum time step *dt* for an atomic liquid in DPD could be increased by a factor of 50 as compared to MD. Interfacial properties and demixing have been studied. For molecular liquids and solids even much larger reductions are being obtained [33]. With DPD, molecular crystallization has come within reach.

8.2.Coarsening the spatial resolution: continuum dynamics

In other cases one is only interested in the velocities, positions and orientations of some of the molecules. E.g. when growing colloidal crystals or assembling nanocrystals from a solution, the focus will be on the large solute particles, as the effect of the smaller solvent particles mainly will be to generate Brownian motion of the colloidal particles. This idea has led to *Brownian dynamics simulation* [40], [41]. Here the large solute particles, which may have steep interaction potentials (e.g. hard sphere), are simulated explicitly and the solvent manifests itself by random forces and friction with the solvent bath. In this way the size and time scales change from those of the solvent molecules of 0.1 nm and ps to colloidal scales of μm and ms. A more precise description of the motion of colloidal particles in a solvent can in principle be obtained using *Stokesian dynamics simulation* [42], [11, 43]. There the complete hydrodynamic equations of the solvent are taken into account to estimate the forces and torques on the solute particles. The method can be useful to describe the influence of flow on in reactors.

A further approximation can be to refrain totally from molecular level descriptions and to move on to macroscopic *transport equations* [44]. Here Navier-Stokes equations for the hydrodynamics of the liquid are combined with heat transfer and reaction-diffusion equations. This leads to coupled partial differential equations for which sophisticated algorithms have been developed. Simulations on this basis implement the effects of reactor shapes, furnaces, coolers and stirring methods. They are useful e.g. for the design and optimisation of melt growth reactors for large semiconductor single crystals.

8.3.Modifying the interaction potential: Umbrella Sampling

The discussions in the previous sections pertain to efficient simulation techniques to take solvent effects on the dynamics and interactions of solute particles (crystal growth units in general) into account. However, also interactions of the growth units often pose enormous difficulties for a reliable simulation. To cope with these difficulties various model modifications have been proposed whose influence can be corrected afterwards [4].

The first possibility is to carry out simulations on a modified model and to correct the results afterwards. Umbrella sampling is one of the techniques here. It is based on a simple relation between statistical averages in different models. In a canonical system the average value of a quantity B is derived from the Boltzmann probability distribution, Eq. (6):

$$< B > \equiv \sum_{\varphi} B(\varphi)\exp\left(-\beta E(\varphi)\right) / \sum_{\varphi} \exp\left(-\beta E(\varphi)\right) \tag{20}$$

If one modifies a model, replacing $E(\varphi)$ by $E(\varphi) + \delta E(\varphi)$, then in the modified model the average value of B is given by:

$$< B >_{E+\delta E} \equiv \sum_{\varphi} B(\varphi)\exp\left(-\beta\{E(\varphi)+\delta E(\varphi)\}\right) / \sum_{\varphi} \exp\left(-\beta\{E(\varphi)+\delta E(\varphi)\}\right) \tag{21}$$

and simple manipulations reveal

$$< B >_E = \frac{< B\exp(\beta\delta E) >_{E+\delta E}}{< \exp(\beta\delta E) >_{E+\delta E}} \tag{22}$$

Therefore, from a simulation of the modified system we can, at least in principle, obtain the thermodynamic parameters of the original system. This method is effective if the averages can be calculated, which means in practice that δE should be a small correction to E.

A very useful application of umbrella sampling is to decrease the energy of the transient states in activated processes. This idea has been used to study nucleation in the Lennard-Jones model [17, 18]. Here a rough idea of the structure of the critical nucleus was obtained from previous simulations at very large supercooling. This information was used to deduce a form of δE that favoured transient states along the nucleation path. In this way structural and thermodynamic information was derived for the whole transition path from the homogeneous melt to the crystalline state. Especially the free energy of nucleation was determined, and turned out to be comparable to the value predicted by classical nucleation theory.

8.4. Modifying the state generation method: Configuration Bias Monte Carlo

For large molecules it is not easy to find a change of state $\varphi \rightarrow \psi$ for which the energy increase $E(\psi) - E(\varphi)$ is so small that ψ has a reasonable probability to be accepted. Consider a long polymer in a dense phase, melt or crystal (other macromolecules have the same problem). Rotation over a small angle already leads to a large displacement of some of the atoms of the polymer and to overlap with atoms of other polymers, and hence to a large energy increase. Also the displacement of one of the atoms or of a group of atoms of the polymer will lead to high stretching and/or bending energies. In both cases the acceptation probability will become prohibitively small.

A solution is Configuration Bias Monte Carlo (CBMC). Here one Monte Carlo step corresponds to building a new polymer configuration ψ segment by segment. For each segment first a number of conformations is selected in accordance with a part of the total interaction energy (e.g. the intra-molecular interactions). The "segment weight" is the average value of the Boltzmann factor for the remaining part of the interaction energy. One of the conformations is selected with the corresponding Boltzmann probability. In this way a new polymer configuration ψ is generated, and a "Rosenbluth weight" $W(\psi)$, equal to the product of the segment weights is ascribed to it. The same procedure is repeated with the original polymer configuration φ with the difference that one of the segment conformations is the real segment conformation in φ. Finally the acceptation rule is adjusted as compared to the usual Metropolis rule:

$$p_{acc}(\varphi \rightarrow \psi) = \min\left(1, W(\psi)/W(\varphi)\right) \tag{23}$$

In practice several modifications [45] are proposed to enhance the performance of CBMC. As a result reliable free energy calculations can be carried out for melts of polymers consisting of up to 10 segments.

8.5. Using only successes: Transition Path Sampling

In crystal growth we are dealing with a first order phase transition. This means that we have a metastable mother phase and a stable crystal phase (unless we are outside the metastable zone of the mother phase). Usually these phases are reasonably well understood separately. The problem is how to transfer from one phase to the other. Indeed, in a first order phase transition the phases are necessarily separated by a thermodynamic (free energy) barrier, otherwise one of the phases would be thermodynamically unstable. Therefore, in a straightforward simulation as appropriate for the individual phase, the probability of spontaneous phase transformation is very small. Of course in principle one could invoke a MC-step that transforms at once the whole system from a mother phase state φ to a crystalline state ψ. This would not change the overall thermodynamics but it bears no relation to the physical transformation process. We have seen before that umbrella sampling may be effective to find a reasonable transition path.

What one would like to do is to modify this path in such a way that more probable transition paths are found. This is precisely what Transition Path Sampling [46] aims to do. This technique effectively samples the *transition path ensemble*, the collection of all possible sequences of MC-steps that start from a given initial state φ and end in a given final state ψ. This sampling is done in such a way that the relative probabilities for given paths are the same as in the underlying straightforward MC or MD simulation. The method allows one to calculate the thermodynamic barrier and the hopping rate between the initial and final states. Though this approach still is computationally demanding, some studies pointing at the possibility to study crystal nucleation already have been carried out [47].

9. CHARACTERIZING ATOMIC SCALE STRUCTURE

When we discuss crystal growth processes at an atomic level, we refer to relatively small rearrangements of atoms and molecules. Therefore it is reasonable to expect that the key to understanding crystal growth at this level lies in an accurate description of the neighbourhood of each atom. This information is comprised in the *local density*

$$\rho(\mathbf{r}) = \left\langle \sum_{i=1}^{N} \delta(\mathbf{r} - \mathbf{r}_i) \right\rangle \tag{24}$$

where δ denotes the Dirac-distribution. The local density can be viewed as an atomic resolution microscopic image of the system. The translation invariant *bond vector density* is

$$\rho_2(\mathbf{r}) \equiv \frac{1}{N} \left\langle \sum_{i=1}^{N} \sum_{j \neq i=1}^{N} \delta(\mathbf{r} - (\mathbf{r}_i - \mathbf{r}_j)) \right\rangle \tag{25}$$

It is the basis of several important quantities. The *pair correlation function* $g(r)$ is the rotational average of $\rho_2(\mathbf{r})$, scaled such that $g(r \to \infty) = 1$

$$g(r) \equiv \frac{1}{\bar{\rho}} \frac{1}{4\pi} \int_0^{2\pi} \int_0^{\pi} \rho_2(r, \theta, \varphi) \, r^2 \sin(\theta) \, d\theta \, d\varphi \tag{26}$$

The structure factor $S(k)$, which is the Fourier transform of $g(r)$ serves to detect long range translation order in the system. For crystals g has several peaks, for liquids usually there is one recognizable peak, followed by an (almost) monotonous approach to $g(r \to \infty) = 1$

9.1. Definition and characterization of the neighbourhood of a particle

One definition of the neighbourhood of a particle is the set of particles in the spherical shell around particle i with a cut-off radius r_{co}. The number of these particles is called the coordination number $Z(i)$. Often r_{co} is chosen equal to the first minimum of g, but also a value may be chosen that minimizes the fluctuations of $Z(i)$. Note that, even for an optimal choice of r_{co}, this method does not guarantee a constant number $Z(i)$ of neighbour particles.

An alternative definition of the neighbourhood of particle i is "the M nearest neighbours of i" (in this case $Z(i) = M$) ,e.g. $M = 12$. This definition eliminates fluctuations in the number of neighbours. The neighbourhood volume $V(i)$ however, is not constant in general.

A disadvantage of these methods, especially for multi-phase systems, is that it is not always obvious which value should be taken for the parameters r_{co} or M. The optimal choice should depend on the phase and temperature.

These ambiguities are absent with the parameter free *Voronoi method*. The Voronoi polyhedron of particle i is the region of space that is closer to i than to any other particle. This region is the neighbourhood of i. Each face of the polyhedron is related to a neighbour j. This method can handle the local anisotropy at interfaces in a natural way. Indeed, to understand growth and nucleation mechanisms at an atomic level, the Voronoi method is to be preferred.

These definitions are simple to implement, but near the melting point they suffer from large fluctuations. The reason is that there are always particles close to the neighbourhood boundary. They may jump in and out of the neighbourhood. With the cut-off method the co-ordination number $Z(i)$ fluctuates strongly. For FCC or HCP Lennard-Jones crystals near the melting point the standard deviation of $Z(i)$ is at least 0.43. With the M-neighbours method $Z(i) = M$ is constant but the neighbourhood volume $V(i)$ fluctuates. And with the Voronoi method both $Z(i)$ and $V(i)$ fluctuate. For all three definitions the neighbourhood does not always consist of the same particles.

To reduce fluctuation effects, we introduced a "neighbour weight" $C(i,j)$ to each possible neighbour j of the central atom i, satisfying

$$\sum_{j\neq i} C(i,j) = 1 \tag{27}$$

The $C(i,j)$ must be chosen such that they are large for particles j that are clearly inside the neighbourhood, small for the j that are close to the neighbourhood boundary and zero for the j that are far away from i. The definitions discussed so far correspond to a constant value of $C(i,j)$ for the particles inside the neighbourhood and $C(i,j) = 0$ for the remaining particles in the system.

The effect of fluctuations is reduced when a smooth decay of $C(i,j)$ is chosen. This can be done in a natural way within the Voronoi method. For each neighbour j the distance r_{ij} and the area $A(i,j)$ of the corresponding side face of the polyhedron are known. We chose $C(i,j)$ proportional to the viewing angle $A(i,j)/r_{ij}^{\,2}$. The co-ordination number is defined as the inverse of the average value of $C(i,j)$, explicitly:

$$Z(i) \equiv 1 \Big/ \sum_{j\neq i} C(i,j)^2 \tag{28}$$

Note that the co-ordination number is a continuous function of the environment of a particle.

The *local bond orientation density* describes the neighbourhood of i:

$$\rho_b(i,\hat{\mathbf{r}}) \equiv \sum_{j\neq i} \delta_{S1}\left(\hat{\mathbf{r}} - \hat{\mathbf{r}}_{ij}\right) C(i,j) \tag{29}$$

The sum gets its main contributions from particles in the neighbourhood of i, δ_{S1} denotes the Dirac distribution on the unit sphere, $\hat{\mathbf{r}}$ is an arbitrary unit vector and $\hat{\mathbf{r}}_{ij} = (\mathbf{r}_j - \mathbf{r}_i)/r_{ij}$ is a unit vector pointing from the central particle i to the neighbouring particle j. The local bond orientation density can be decomposed in (complex) spherical harmonics [48]:

$$\rho_b(i,\hat{\mathbf{r}}) \equiv \sum_{l=0}^{\infty} \sum_{m=-l}^{l} Q_m^l(i) \mathrm{Y}_m^l(\hat{\mathbf{r}})^* \tag{30}$$

where the local bond order parameters are weighted averages of the spherical harmonics

$$Q_m^l(i) \equiv \sum_{j \neq i} Y_m^l(\hat{\mathbf{r}}_{ij}) C(i,j) \tag{31}$$

The local bond order parameters, notably rotation invariant combinations of them, can be used as "finger prints" for given structures. To investigate the interfaces during growth and nucleation, one has to recognize different local structures around individual atoms, especially near the melting point. Instantaneous values of local bond order parameters for individual atoms or for groups of atoms are used to this end. Both the bipolar product

$$Q^l(i,j) \equiv \frac{4\pi}{2l+1} \sum_{m=-l}^{l} Q_m^l(i) Q_m^l(j)^* \tag{32}$$

and the tripolar product, involving Wigner $3j$ symbols,

$$W^{l_1 l_2 l_3}(i,j,k) = \sum_{m_1=-l_1}^{l_1} \sum_{m_2=-l_2}^{l_2} \sum_{m_3=-l_3}^{l_3} \begin{pmatrix} l_1 & l_2 & l_3 \\ m_1 & m_2 & m_3 \end{pmatrix} Q_{m_1}^{l_1}(i) Q_{m_2}^{l_2}(j) Q_{m_3}^{l_3}(k) \tag{33}$$

only depend on the relative orientations of the bond vectors around the particles i, j and k. Average values and distributions of $Q^l(i,i)$ and $W^{lll}(i,i,i)$ with $l = 4$, $l = 6$ and $l = 8$ have been used to recognize hexagonal close packing (HCP), face centred cubic (FCC), body centred cubic (BCC) and liquid local structures. It has been reported that the interior of a Lennard-Jones nucleus has an FCC-structure, whereas the surface tends to a BCC-structure [49]. In this study however, only FCC, BCC and liquid phases were taken into account. As BCC and HCP turn out to be quite difficult to distinguish we consider the BCC structure at the surface as questionable.

It is important, especially close to the melting point, to choose an optimal order parameter for the type of investigation at hand. We have carried out a thorough investigation, involving several hundred thousands order parameters. Our study revealed that different order parameters have to be chosen, depending on which phases may occur and have to be recognized. We first determined, on the basis of independent simulations, the distribution functions $h_\alpha(O)$ for n different "pure" phases, e.g. $n = 4$: α = FCC, HCP, BCC or liquid [50] (as BCC is unstable in the Lennard-Jones model, a BCC Einstein crystal with the bond strength that gave correct bond order parameters for FCC and HCP was used as reference for BCC). In subsequent simulations an atom i with the value O_i of O will be assigned to the phase α for which $h_\alpha(O_i)$ is the largest. We define the *selection quality* $K_{sel}(O)$ of O as the fraction of wrong assignments. Clearly $K_{sel}(O) = 0$ if the h_α do no overlap and $K_{sel}(O) = 1\text{-}1/n$ if they completely overlap. We also defined the *decomposition quality* that is a measure for the accuracy of estimating the composition of a mixed phase. Both quality parameters are found to give roughly, but not exactly the same result when ordering the bond order parameters.

When distinguishing four phases, the order parameters $Q^6(i,i)$ and $W^{666}(i,i,i)$ with $M = 12$ neighbours, which are frequently used in the literature, led to $K_{sel}(Q^6) = 0.36$ and $K_{sel}(W^{666}) = 0.35$ respectively. The best result, $K_{sel}(W^{668}) = 0.27$, was obtained with an $M = 11$ neighbour selection. Distinguishing pairs of phases, we found that FCC is easy to recognise: a best $K_{sel}(O) \approx 0.02$ is found when one of the two is FCC. On the other hand a best $K_{sel}(O) \approx 0.07$ is obtained for pairs made of HCP, BCC and liquid.

As explained above overlapping of the order parameter distributions causes the rather poor performance. The overlapping parts of the distributions however are different for different order parameters. Therefore we may expect that choosing intelligent combinations of two or more order parameters would be useful. We found that the best joint order parameter with Voronoi neighbourhood definitions for distinguishing four phases is $K_{sel}(W^{468}, W^{668}) = 0.09$.

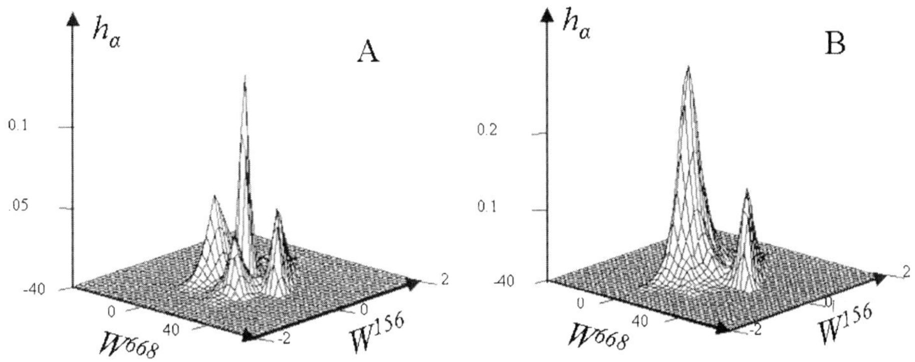

Figure 3. Order parameter histograms $h_\alpha(\mathbf{O})$ for the 2D bond order parameter $\mathbf{O} = (W^{l56}, W^{668})$. A) Reference histograms of the pure phases: the peak at $\mathbf{O} = (-0.1, 15)$ is from α = HCP, the peak at $\mathbf{O} = (-0.3, 32)$ from α = FCC, the peak at $\mathbf{O} = (-0.3, 20)$ from α = BCC and the peak at $\mathbf{O} = (0, 0)$ from α = liquid. Note that 1D histograms for $O = W^{l56}$ or $O = W^{668}$ would strongly overlap. B) Histogram of an FCC / melt interface near T_m: some enhancement is observed where the FCC and liquid peaks overlap, but no BCC peak is recognized.

In Figure 3A plots of the $h_\alpha(\mathbf{O})$ are shown that illustrate this idea for the joint order parameter $\mathbf{O} = (W^{156}, W^{668})$. The most frequent errors are assignments of HCP in a BCC crystal (16%) and BCC in an HCP crystal (9%). The remaining errors are below 3%. We used this joint order parameter for the characterization of a 3022-atom simulation of an FCC(111) / melt interface at T_m. Initially the crystal consisted of 8 layers of 128 atoms, the remaining 1998 atoms were liquid. After equilibration we see in Figure 3B that the FCC and liquid structures are dominant. Only 180 atoms are ascribed to HCP and 240 to BCC, and nowhere in the system the BCC structure was the most prominent. This result could also be expected if the interface were intermediate between FCC and liquid. Therefore such results do not give a clear indication for BCC structures near the interface.

A further application of bond order parameters is to test whether two particles have a similar environment, by checking whether the *connectivity factor*

$$\gamma^l(i,j) \equiv Q^l(i,j) / \sqrt{Q^l(i,i)\, Q^l(j,j)} \tag{34}$$

for the two atoms exceeds a predetermined threshold. This allows one to define clusters and to study the size and the shape of a nucleus.

9.2. Structure assessment by Ensemble of Force Networks

The structure of condensed matter is to a large extent determined by short-range intermolecular repulsive forces. This was realized already in the early days of liquid phase statistical mechanics. One of the main conclusions was that many of the properties of liquids could be understood using a hard sphere model with an appropriately chosen hard sphere radius in a *NPT* or *NVT* ensemble.

For crystals, in view of the large variety of crystal structures, one might expect that details of the attractive forces are most relevant. But already the hard sphere model possesses a first order liquid-solid phase transition at a well-defined density. Also, 25 years ago, I asked professor Hartman, the founder of crystal morphology theory, which interactions should be taken into account in a Hartman-Perdok morphology analysis. He gave me the puzzling answer that only the short range repulsive interactions should be considered.

In line with this philosophy one may consider the possibility to describe the essential structural and dynamical properties in terms of short-range repulsive forces. This is precisely what the Ensemble of Force Network (EFN) method [51] does. In my view this method is very promising for the further development of our understanding of the liquid-solid transformation, i.e. crystal growth from the melt.

The EFN method has been developed recently to describe the stability of granular material. The elementary particles of granular matter are small, but macroscopic, grains which dominantly interact with a very short-range repulsive interaction (almost hard spheres or other hard bodies). As the interaction changes from zero to a strong repulsion over very short differences in distance, very small variations in the particle positions may lead to a large change in the pattern of forces between the grains. Thus for one position configuration many different force configurations are conceivable (or at least might occur if there would be a small dispersion of the individual grain properties). The set of all repulsive force configurations that leave a given configuration of particles in rest (i.e. zero total force on each particle) is called the ensemble of force networks. Note that the repulsive forces in each member of the ensemble may be, and in general are, quite different from the actual interparticle forces.

The key observation is that the force probability distributions of ensembles obtained from "jammed" (immobile, frozen) configurations are characteristically different from those obtained from "unjammed" (floating, mobile, liquid) configurations. In unjammed systems the distribution is monotonically decreasing, whereas a peak near the average force appears for jammed structures. A similar difference is also observed for the real force distribution. An explanation of this similarity could be that in a real solid or liquid system that is close to thermodynamic equilibrium, the individual forces between pairs of particles tend to compensate and hence are usually much larger than the net force on a single particle. The advantage of using the EFN, rather than the actual force distribution is that the EFN gives a reliable force distribution already for a single configuration, whereas one would need a large number of configurations to get this distribution for the real system. Thus the EFN is much more efficient. Moreover further studies of the physics underlying the success of the EFN method may lead to better understanding of the underlying physics of growth from the melt.

Currently modifications of the EFN method are being developed. Non-central forces are considered to study shape and friction effects. These are absent in atomic crystals, but may be quite strong for molecular crystals and for crystals growing from finite size particles like colloids, proteins or quantum dots.

10. ESTIMATING FREE ENERGIES AND SUPERSATURATION

One of the central parameters in crystal growth, the supersaturation σ is a dimensionless free energy difference, namely the chemical potential difference of growth units in the mother phase and in the crystalline phase (note that traditionally its sign is chosen such that $\sigma > 0$ for growth):

$$\sigma \equiv \left(\mu^{mother\ phase} - \mu^{crystal}\right)/\left(k_B T\right) \tag{35}$$

From thermodynamics we know how μ is related to the Gibbs energy G and to the Helmholtz energy A. Specifying to a one-component system we have

$$\mu = \left(\frac{\partial G}{\partial N}\right)_{P,T} = \left(\frac{\partial A}{\partial N}\right)_{V,T} = \frac{G}{N} = \frac{A+PV}{N} \tag{36}$$

Like the internal energy U, the free energies A and G have no natural zero, and they have to be specified as the difference with respect to a suitably chosen reference state. For real systems the third law assures that the entropy vanishes for $T \to 0\text{K}$, hence an absolute entropy can be defined by taking $T = 0$ as the reference. The third law is the macroscopic consequence of the quantum mechanical result that the ground state of any real system is non-degenerate. The third law is not valid in classical systems. Indeed in these systems the specific heat $C_P(T)$ remains finite, even when $T \to 0\text{K}$ (at least there is the contribution $3Nk_B/2$ of the kinetic energy). The general relation $(\partial S/\partial T)_P = C_P(T)/T$ then implies that $S \to -\infty$ for $T \to 0\text{K}$.

In thermodynamics the ideal gas phase of the same molecules is the standard reference point for gases. The free energy and chemical potential of an ideal gas can be obtained analytically [52]. It is sensible to use this reference for all phases. To this end we define the intra-molecular energy E_{intra} as the single particle contribution to the system energy, i.e. E_{intra} comprises the kinetic energy and the intra-molecular interaction energy. The intermolecular interaction energy E_{inter} is defined as the remaining part of E:

$$E(\varphi) = \sum_{i=1}^{N} E_{intra}(\varphi_i) + E_{inter}(\varphi) \tag{37}$$

The ideal gas reference system is obtained by neglecting the intermolecular interaction energy, $E_{inter} = 0$. Using Eq. (8) it seems that the free energy of a real system in reference to that of an ideal gas, i.e. the difference $A - A^{idg}$, can be obtained from an ensemble average:

$$\exp\left(\beta\left(A - A^{idg}\right)\right) = \frac{\sum_{\varphi} \exp\left(-\beta E_{intra}(\varphi)\right)}{\sum_{\varphi} \exp\left(-\beta E(\varphi)\right)} \overset{??}{=} \left\langle \exp\left(\beta E_{inter}\right)\right\rangle \tag{38}$$

The problem with the right hand side expression is that states with a large positive interaction energy, e.g. states with overlapping atoms, have a very small probability to occur, but if they occur they give a very large contribution to the average. In fact, it can be shown that the standard deviation is infinite, hence the expression is useless. The problem of large fluctuations is typical for all quantities that involve entropies.

There are two ways out. The first possibility is to use an expression for the chemical potential, which involves the interaction energy of one particle with the remaining particles in the system. This approach is the basis of the *virtual particle exchange* and the *overlapping distribution* methods. These methods are powerful for gases, liquids and solid-vapour surfaces. The second method is to choose a reference system for which the free energy is exactly known, and transform this gradually to the real system, while keeping track of the free energy changes during the transformation. This leads to *thermodynamic integration* methods, which often work well for crystalline phases.

10.1. Virtual particle insertion and removal

The *virtual particle insertion* method starts from the differential form of the definition of a chemical potential:

$$\mu = \left(\frac{\partial A}{\partial N}\right)_{V,T} = -k_B T \ln\left(\frac{Q(N+1,V,T)}{Q(N,V,T)}\right) \tag{39}$$

Straightforward application of the definitions leads to an expression for the chemical potential of a real system in reference to the ideal gas:

$$\exp\left(-\beta\left(\mu - \mu^{idg}\right)\right) = \left\langle \exp\left(-\beta\varepsilon_{ins}\right)\right\rangle_{N,V,T} \tag{40}$$

The right hand side is the ensemble average of the Boltzmann factor for virtually inserting a particle at an arbitrary position in an N-particle system, ε_{ins} is the energy increase for such an insertion. The virtual insertion method is effective for gases, and for liquids up to moderate densities. The reason is that in such systems enough positions can be found, where a new particle could be inserted without being strongly repelled by the particles that are present already. These positions contribute significantly to the ensemble average.

An analogous approach could be imagined by virtually removing an arbitrary particle, but the *virtual particle removal* method turns out to suffer from the same unbounded fluctuations in the inverse Boltzmann factor $\exp(\beta\varepsilon_{rem})$ (ε_{rem} is the energy increase when arbitrarily one of the particles is removed from the system) as we discussed in the previous section for the total system free energy.

A more refined and more powerful method built on the same ideas is the *overlapping distribution* method [53, 54].This method is based on measuring the probability distributions for the energies of virtual insertion and removal:

$$f_{N\to N+1}(\varepsilon) \equiv \left\langle \delta\left(\varepsilon - \varepsilon_{ins}\right)\right\rangle_{N,V,T} \text{ and } f_{N\to N-1}(\varepsilon) \equiv \left\langle \delta\left(\varepsilon - \varepsilon_{rem}\right)\right\rangle_{N,V,T} \tag{41}$$

The following remarkable expression can be derived for the chemical potential:

$$\ln\left(f_{N\to N-1}(\varepsilon)\right) - \ln\left(f_{N-1\to N}(\varepsilon)\right) + \beta\varepsilon = \beta\left(\mu - \mu^{idg}\right) \tag{42}$$

This expression is valid for all values of the replacement energy ε. Often a finite energy interval can be found where both the removal and the insertion energy *distributions* can be accurately obtained. Averaging over this interval a reliable estimate for the chemical potential can be obtained. In practice it is usually assumed that the system is sufficiently large for the difference between the insertion distributions $f_{N-1\to N}$ and $f_{N\to N+1}$ to be negligible. Then both distributions at the left hand side can be obtained from the same simulation. The "overlapping distribution method" works well for quite dense liquids. For solids usually the overlap interval usually is quite small, and moreover the assumption $f_{N-1\to N} \approx f_{N\to N+1}$ often is problematic.

10.2. Thermodynamic integration methods

The basic idea for thermodynamic integration method is borrowed from experimental methods of determining free energies. There too, direct measurement of entropies and free energies is impossible, but derivatives with respect to V, T and P often can be measured. One prepares the system in a reference state where the free energy or entropy is known. Next one brings the system slowly, along an appropriate path, from the reference state to the actual state. Integrating the measured derivative leads to the desired result. If the path is crossing a phase boundary then additional corrections have to be taken into account.

The Gibbs-Helmholtz relation $\partial(G/T)/\partial T = -H/T^2$ is used to get the temperature variation of the Gibbs energy G:

$$\frac{G(T,P,N)}{T} = \frac{G(T_{ref},P,N)}{T_{ref}} - \int_{T_{ref}}^{T} \frac{H(T',P,N)}{T'^2} dT' \tag{43}$$

Analogous expressions can be found and applied for the volume and pressure dependence of G. This approach can be used if the actual state can be reached from the reference state without passing a phase transition. In real experiments latent heats can be taken into account at phase transitions, but in simulations these quantities usually are not accessible. Thus it is not immediately clear how the free energy of a crystal phase can be related to the ideal gas reference.

Fortunately there is a way out. Indeed in simulations we can not only vary temperature and other physical parameters, but also non-physical parameters can be used to travel from a known reference phase to the desired point in phase space. A reference phase that is exactly known and can often be used is the Einstein crystal model. In this model the actual interactions between the particles are replaced by a constant energy $E_{E,0}$, a simple harmonic potential to bind each particle to a fixed position in space and, for non-spherical particles, an orienting potential [55] (here we only treat non-linear molecules, the modification for linear molecules is found in [52]):

$$E_{Einstein}(\varphi) = E_{intra}(\varphi) + E_{E,0} + \sum_{i=1}^{N} C_i \left|\mathbf{r}_i - \mathbf{r}_{i,E}\right|^2 - \sum_{i=1}^{N} D_i \,\mathrm{Tr}\left(\mathbf{R}_{i,E}\mathbf{R}_i^{-1}\right) \tag{44}$$

Here $\mathbf{r}_i$ and $\mathbf{r}_{i,E}$ are the actual and the reference position in the Einstein crystal of the centre of mass of particle i respectively. Similarly for non-spherical particles, $\mathbf{R}_i$ and $\mathbf{R}_{i,E}$ are the actual and the reference orientation matrix for particle i and Tr(**M**) is the trace of the matrix **M** (note that $\mathrm{Tr}(\mathbf{R}_{i,E}\mathbf{R}_i^{-1}) = \cos(\beta)+(1+\cos(\beta))\cos(\alpha+\gamma)$, where α, β and γ are the Euler angles over which particle i has to be rotated to reached its reference orientation). Values for the parameters $E_{E,0}$, C_i and D_i can be chosen afterwards to optimise the thermodynamic integration accuracy.

As the free energy A_{idg} of the ideal gas is can be calculated [52], the free energy of the Einstein crystal is known as well:

$$A_{Einstein}(N,V,T) = A_{idg}(N,V,T) - k_B T \ln(N!) + \delta A(N,V,T) \tag{45}$$

The second term at the right hand side is necessary because the particles in the Einstein crystal are distinguishable, whereas they are not in the ideal gas. The last term describes the influence of the Einstein reference potential field

$$\delta A(N,V,T) = E_{E,0} - k_B T \sum_{i=1}^{N} \left\{ \frac{3}{2} \ln\left(\frac{\pi k_B T}{C_i V^{2/3}} \right) + D_i \ln\left[\mathrm{I}_0\left(\frac{2D_i}{k_B T} \right) - \mathrm{I}_1\left(\frac{2D_i}{k_B T} \right) \right] \right\} \tag{46}$$

where I_0 and I_1 are Bessel functions.

The path from the Einstein reference crystal to the actual crystal is given by a parameter λ which is such that for $\lambda = 0$ we have the Einstein crystal and for $\lambda = 1$ we are at the actual system:

$$E(\lambda,\varphi) = (1-\lambda)E_{Einstein}(\varphi) + \lambda E(\varphi) \tag{47}$$

From the definition of the Helmholtz energy it is easy to see that

$$\left(\frac{\partial A}{\partial \lambda} \right)_{N,V,T} = \langle E - E_{Einstein} \rangle_{\lambda,N,V,T} \tag{48}$$

and therefore

$$A(N,V,T) = A_{Einstein}(N,V,T) + \int_0^1 \langle E - E_{Einstein} \rangle_{\lambda,N,V,T} \, d\lambda \tag{49}$$

The method of Einstein integration has been applied to many different systems, both for atomic and for molecular crystal models. It may be worthwhile to note that in the literature often slightly different reference crystal potentials are found, usually of the Einstein type [56] but sometimes also a Debye crystal was taken as reference [57]. This should not change the final result, but is a matter of convenience. It is in practice also important to choose the parameters of the Einstein potential such that the integrand along the integration path is small and varies smoothly.

10.3. Example: ice and water phase diagram for rigid H_2O models

Although H_2O is one of the most important materials for our live, it is difficult to get a reliable simple simulation model which can be used to study the crystallization of ice. As a matter of fact the so-called TIP4P model is widely used to describe water in protein and other biomolecular crystals. This model also has been used to simulate the ice crystallization. Indeed in the literature references are found which describe surface melting [58], crystal growth [59] and even nucleation [60] at temperatures of 230 - 250K. The problem however, is that we have shown [61] that the normal melting point of TIP4P is around 210K. Careful analysis of the literature learned that unphysical boundary conditions (leading to a very large negative pressure) were the cause of observing ice growth above 210K [60].

The recently proposed TIP5P [60, 62] model has an acceptable melting point. But, the equilibrium ice structure of both the TIP4P and the TIP5P model turns out to be an anti-ferroelectrically ordered proton distribution, whereas real ice is proton disordered. Therefore we developed the TIP6P H_2O model. As shown in Figure 4 this model does not have these unphysical properties.

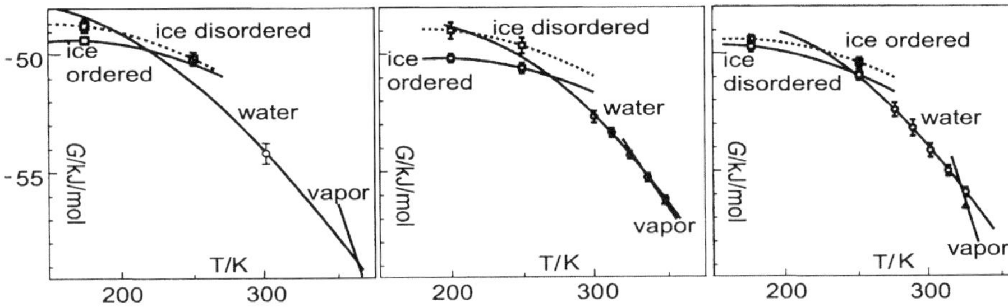

Figure 4. Gibbs energies for three different rigid models for H_2O. Left TIP4P model: T_m too low, proton-ordered phase stable. Middle TIP5P model: T_m correct but proton-ordered phase stable. Right TIP6P model: T_m correct, proton-disordered ice stable

11. CONCLUSION

In this lecture an overview has been given over different promising simulation methods. It has been shown that different approaches are necessary depending on the question one is looking at. It has become clear that in all cases computational capacity is a bottleneck for straightforward implementation of simulation research in crystal growth. Different groups have worked on these problems and several new and efficient methods have been developed.

At present we have reached the point where an outlook emerges over the possibilities, which will come within reach in the coming years. Growth and nucleation of atomic crystals is reasonably well understood at the molecular scale level. It can be expected that a similar molecular scale understanding of the generic structural and dynamical aspects of growth and nucleation phenomena of molecular crystals will be developed.

However, the moment that computer simulation is able to predict what kind and what quality of crystals will emerge from a given crystalliser is still far away.

ACKNOWLEDGEMENT

This paper is based on the support and the dedicated work of my co-workers Dragomir Ganchev, Edzer Huitema, Jeroen Makkinje, Hiroki Nada, Hilde Rinia, Margot Snel, Margot Vlot and Thijs Vlugt.

REFERENCES

1. J.P.J.M. van der Eerden, Possibilities And Limitations For Numerical Simulation Of Crystal Growth, in: K. Byrappa, T. Ohachi, M. Klapper, and R. Fornari (eds.), Crystal Growth of Technologically Important Electronic Materials (Allied Publishers PVT. Limited, New Delhi, 2003) 3.
2. W.J. Briels and H.L. Tepper, Phys. Rev. Lett. 79 (1997) 1496.

3. D. Frenkel and B Smit, Understanding Molecular Simulations. From Algorithms to Applications, (Academic Press, New York, 1996).
4. D. Frenkel and B Smit, Understanding Molecular Simulations. From Algorithms to Applications, (Academic Press, New York, 2002).
5. W Kossel, Nachr. Ges. Wiss. Göttingen, (1927) 135.
6. I. Stranski and R. Kaischew, Z. Phys. Chem. B26 (1934) 100.
7. W.K. Burton, N. Cabrera and F.C. Frank, Trans. Roy. Soc. A243 (1951) 299.
8. J.M. Kosterlitz and D.J. Thouless, J. Phys. Chem. C6 (1973) 1181.
9. J.P. Van der Eerden, in: D.T.J. Hurle (ed.), Handbook of Crystal Growth 1a (North-Holland, Amsterdam, 1993) 307.
10. E. Burke and J.Q. Broughton, J. Chem. Phys. 89 (1988) 1030.
11. J.Q. Broughton and G.H. Gilmer, J. Chem. Phys. 79 (1983) 5095.
12. J.Q. Broughton and G.H. Gilmer, J. Chem. Phys. 79 (1983) 5105.
13. J.Q. Broughton and G.H. Gilmer, J. Chem. Phys. 79 (1983) 5119.
14. J.Q. Broughton and G.H. Gilmer, J. Chem. Phys. 84 (1986) 5741.
15. J.Q. Broughton and G.H. Gilmer, J. Chem. Phys. 84 (1986) 5749.
16. J.Q. Broughton and G.H. Gilmer, J. Chem. Phys. 84 (1986) 5759.
17. J.S. van Duijneveldt and D. Frenkel, J. Chem. Phys. 99 (1992) 4655.
18. P.R. ten Wolde, M.J. Ruiz-Montero and D. Frenkel, J. Chem. Phys. 104 (1996) 9932.
19. H.E.A. Huitema, J.P.J.M. van der Eerden, J.J.M. Janssen and H. Human, Physical Review B. 62 (22) (2000) 14690.
20. H.E.A. Huitema and J.P.J.M. van der Eerden, J. Chem. Phys. 110 (1999) 3267.
21. A. Bravais, J. Ecole Polytechn. Paris. 19 (1850) 1.
22. P. Hartman, in: I. Sunagawa (ed.), Morphology of Crystals (Terra, Tokyo, 1987) 269.
23. P. Hartman and P. Bennema, J. Crystal Growth. 49 (1980) 145.
24. R.F.P. Grimbergen, H. Meekes, P. Bennema, C.S. Strom and L.J.P. Vogels, Acta Cryst. A54 (1998) 491.
25. S.X.M. Boerrigter, Thesis: Modeling of Crystal Morphology: Growth Simulation on Facets in Arbitrary Orientations, (Dept of solid state chemistry, Nijmegen, 2002).
26. H. Meekes, S.X.M. Boerrigter, G.P.H. Josten, J. van de Streek, F.F.A. Hollander, J. Los, H.M. Cuppen and P. Bennema, J. Phys. Chem., (2004) submitted.
27. Tetsushi Mori, Noriaki Kubota, Sou Abe, Shin'ichi Kishimoto, Satoshi Kumon and Masayoshi Naruse, J. Crystal Growth. 133 (1-2) (1993) 80.
28. R van Eerd: Explanation of growth morphology of aspartame, in: Proceedings ICCG14, Grenoble, 2004 (Elsevier).
29. H. E. A. Huitema, M. J. Vlot and J. P. J.M. van der Eerden, Journal of Chemical Physics. 111 (10) (1999) 4714.
30. H. A. Rinia, R. Kik, R. A. Demel, M. M. E. Snel, J. A. Killian, J. J. P. M. van der Eerden and B. de Kruijff, Biophysical Journal. 78 (2000) 323a.
31. J. P. J. M. van der Eerden, M. M. E. Snel, J. Makkinje, A. D. J. van Dijk and H. A. Rinia, J. Crystal Growth. 237 (2002) 111.
32. F. Yarrow: Phase transitions in mixed bilayers, in: Proceedings ICCG14, Grenoble, 2004 (Elsevier).
33. M. Kranenburg, Thesis: Phase transitions of lipid bilayers, a mesoscopic approach, (Faculty of science, mathematics and informatics, Amsterdam, 2004).
34. V. Bostanov, W. Obretenov, G. Staikov and E. Budevski, J. Electroanal. Chem. 146 (1983) 303.

35. A. Milchev and E. Vassileva, J. Electroanal. Chem. 107 (1980) 337.
36. R. D. Groot and P. B. Warren, J. Chem. Phys. 107 (11) (1997) 4423.
37. I. Pagonabarraga and D. Frenkel, Molecular Simulation. 25 (3-4) (2000) 167.
38. I. Pagonabarraga and D. Frenkel, J. Chem. Phys. 115 (11) (2001) 5015.
39. I. Pagonabarraga, M. H. J. Hagen and D. Frenkel, Europhysics Letters. 42 (1998) 377.
40. D.L. Ermack and J.A. McCammon, J. Chem. Phys. 69 (1978) 1352.
41. D.M. Heyes and J.R. Melrose, J. Non-Newtonian Fluid Mech. 46 (1993) 1.
42. J.F. Brady and G. Bossis, Annu. Rev. Fluid Mech. 20 (1988) 111.
43. B. Cichocki, R.B. Jones, R. Kutteh and E. Wajnryb, J. Chem. Phys. 112 (2000) 2548.
44. G. Mueller, J. Crystal Growth. 237-239 (2002) 1628.
45. S. Consta, T.J.H. Vlugt, J. Wichers-Hoeth, B. Smit and D. Frenkel, Mol. Phys. 97 (12) (1999) 1243.
46. C. Dellago, P.G. Bolhuis, F.S. Csajka and D. Chandler, J. Chem. Phys. 108 (2000) 1964.
47. P.G. Bolhuis, C. Dellago and D. Chandler, Faraday Discuss. 110 (1998) 421.
48. D.A. Varshalovich, A.N. Moskalev and V.K. Khersonskii, Quantum Theory of Angular Momentum, (World Scientific, Singapore).
49. P.R. ten Wolde, M.J. Ruiz-Montero and D. Frenkel, Phys. Rev. Lett. 75 (1995) 2714.
50. J. Makkinje and J.P.J.M. van der Eerden, J. Chem. Phys., (2004) in progress.
51. J. H. Snoeijer, T. J. H. Vlugt, M. van Hecke and W. van Saarloos, Physical Review Letters. 92 (2004) 054302.
52. D.A. McQuarrie and J.D. Simon, Physical Chemistry, A molecular approach, (University Science Books, Sausalito, California, 1997).
53. S. Shing and K.E. Gubbins, Molec. Phys. 46 (1982) 1109.
54. S. Shing and K.E. Gubbins, Molec. Phys. 49 (1983) 1121.
55. H. Nada and Y. Furukawa, Surf. Sci. 446 (2000) 1.
56. M. J. Vlot, J. Huinink and J. P. van der Eerden, J. Chem. Phys. 110 (1999) 55.
57. G. T. Gao, X. C. Zeng and H. Tanaka, J. Chem. Phys. 112 (2000) 8534.
58. G.-J. Kroes, Surf. Sci. 275 (1992) 365.
59. H. Nada and Y. Furukawa, J. Phys. Chem. B 101 (1997) 6163.
60. M. Matsumoto, S. Saito and I. Ohmine, Nature. 416 (2002) 409.
61. H. Nada and J.P.J.M. van der Eerden, J. Chem. Phys. 118 (16) (2002) 7401.
62. H. Tanaka, Nature. 380 (1996) 328.

Crystal Growth – from Fundamentals to Technology
G. Müller, J.-J. Métois and P. Rudolph (Editors)

Dislocation Patterns in Crystalline Solids - Phenomenology and Modelling

Michael Zaiser[a]

[a]The University of Edinburgh,
Center for Materials Science and Engineering,
The King's Buildings, Sanderson Building,
Edinburgh EH93JL, United Kingdom

The paper gives an overview of dislocation patterning phenomena in metals and semiconductors and their modelling using dislocation dynamics approaches. After a brief introduction of some fundamental concepts, experimental observations are reviewed and some fundamental scaling relations are discussed which characterize dislocation patterns under widely varying conditions. The emergence of dislocation patterns is related to the non-equilibrium dynamics of the system of interacting dislocation lines. Modelling approaches include discrete dislocation dynamics simulations as well as models based on ideas from non-equilibrium thermodynamics and statistical mechanics.

1. INTRODUCTION

Plastic deformation of crystalline solids is intimately related to the motion and proliferation of linear lattice defects, i.e. dislocations. Preservation of the lattice structure during plastic deformation implies that plastic displacements can occur only by shifting adjacent lattice planes against each other by a lattice vector ('slip vector') $\vec{b}$. If the slipped area is bounded, the boundary line corresponds to a lattice dislocation and the slip vector is the Burgers vector of this dislocation. For energetic reasons, in a given lattice structure slip occurs preferentially on the closest-packed lattice planes and the slip vector is usually a vector connecting nearest neighbors (i.e. $b = |\vec{b}|$ is the interatomic spacing). A set of parallel lattice planes on which slip occurs ('slip planes') is characterized by their normal vector $\vec{n}$; a slip plane normal plus slip vector define a slip system. In the following we will formally distinguish different slip systems by an index β; each value of β corresponds to a particular pair $(\vec{b}, \vec{n})$.

As crystalline solids deform plastically, the expansion of slipped areas leads to an increase in the dislocation line length within the crystal ('dislocation multiplication'). This proliferation of dislocations goes along with their spatial re-distribution and, more often than not, the formation of heterogeneous dislocation patterns (Figure 1). The characteristics of such patterns depend on material properties and deformation conditions. Two extreme cases are illustrated in Figure 1 (a) and (b). Although in both cases the material (monocrystalline Cu), stress state (uniaxial tension/compression) and crystal orientation

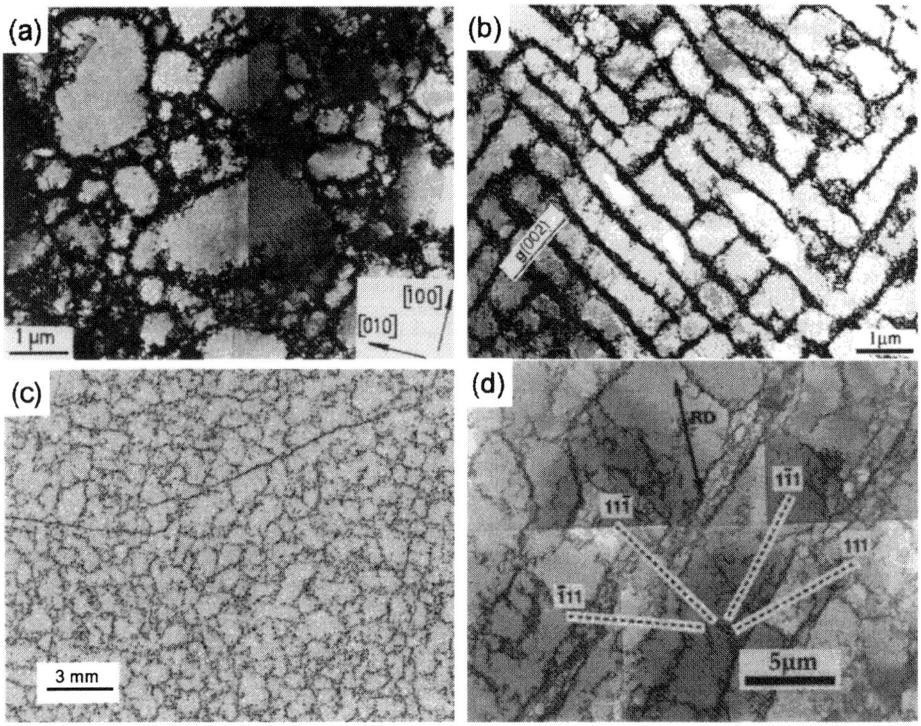

Figure 1. Dislocation patterns formed in different materials and under different deformation conditions. (a) Fractal dislocation cell structure formed in a Cu single crystal deformed in tension along the [100] axis; deformation was carried out at room temperature to a stress (resolved shear stress in each of the 8 active slip systems) of 75.6 MPs; transmission electron micrograph (TEM micrograph) after Mughrabi et. al. [1]. (b) TEM micrograph of the 'labyrinth' structure formed in a [100] oriented Cu single crystal after cyclic deformation (tension/compresssion) at a strain amplitude $\hat{\gamma} = 1.8 \times 10^{-3}$ to a saturation stress amplitude of 40 MPa, after Gong et. al. [2]; (c) Dislocation cell structure in a GaAs single crystals; the pattern has formed after plastic relaxation of thermal stresses during crystal growth; dislocation positions are mapped by etch pits, courtesy of P. Rudolph. (d) Dislocation cell structure in an Al polycrystal after rolling, TEM micrograph after Liu et. al. [3].

(stress axis along the [100] lattice direction) are the same, two different deformation modes (unidirectional versus cyclic straining) yield completely different morphologies of the dislocation pattern: while cyclic deformation leads to highly regular 'labyrinth' patterns of intersecting dislocation-dense walls with a well defined spacing, unidirectional deformation yields a very irregular pattern which is characterized by fractal scale-invariance rather than any characteristic length scale [4]. Such scale-invariance is, however, an exception rather than the rule - the majority of dislocation patterns in unidirectionally deformed crystals pertains to the cellular type illustrated in Figure 1 (c) and (d); these patterns, while irregular, exhibit a well-defined characteristic length scale (mean cell size λ).

2. DISLOCATION DYNAMICS: FUNDAMENTALS

2.1. Forces and interactions in dislocation systems

Dislocations move under the action of stresses. Specifically, the so-called Peach-Koehler force acting on a dislocation segment of unit length, Burgers vector $\vec{b}$ and tangent vector $\vec{t}$ is given by

$$\vec{f}_{\rm PK} = (\vec{b}\boldsymbol{\sigma}) \times \vec{t} \tag{1}$$

where $\boldsymbol{\sigma}$ is the stress tensor at the location of the dislocation segment. From a dynamical point of view it is important to distinguish glide motion of a dislocation line within its slip plane (which corresponds to the plane spanned by $\vec{b}$ and $\vec{t}$, with normal vector $\vec{n}$) from out-of-plane 'climb' motions. The latter cannot take place without inserting or removing material, and therefore dislocation climb is possible only at temperatures where long-range self-diffusion may occur. The glide component of the Peach-Koehler force can be written as

$$\vec{f}_{\rm G} = \tau_{\vec{b}\vec{n}} b(\vec{n} \times \vec{t}) \quad . \tag{2}$$

$\tau_{\vec{b}\vec{n}} = \vec{s}\boldsymbol{\sigma}\vec{n}$ is called the *resolved shear stress* in the slip system $(\vec{b}, \vec{n})$; $\vec{s} = \vec{b}/b$ is the unit slip vector.

The key problem of dislocation dynamics is to evaluate the stresses at the positions of all dislocation line segments. The local stress at a given point in the crystal lattice is a sum of externally applied and internal stresses. As 'external' we classify any stresses acting from outside on the dislocation system - due to tractions applied to the surface of the dislocated body, temperature gradients giving rise to thermal stresses, etc. Such stresses act as external driving forces which may induce dislocation motion and plastic flow (see below).

Dislocation interactions, on the other hand, are embodied within the internal stress field created by the dislocation system itself. Dislocations are sources of internal stresses. The stress field of a straight dislocation with Burgers vector $\vec{b}$ and tangent vector $\vec{t}$ is given by

$$\boldsymbol{\sigma}_{\vec{b},\vec{t}}(\vec{r}) = \frac{Gb\boldsymbol{f}_{\vec{b},\vec{t}}(\phi)}{r} \quad , \tag{3}$$

where $\vec{r}$ is a vector of modulus $|\vec{r}| = r$ in the plane normal to $\vec{t}$, and G is the shear modulus of the material. The dislocation stress field decays radially like b/r and may be strongly anisotropic; its dependence on the azimuthal angle ϕ is governed by the angle θ between the Burgers vector $\vec{b}$ and line direction $\vec{t}$, which defines the *type* of the dislocation. For instance, dislocations with $\vec{b}$ parallel to $\vec{t}$ are referred to a screw dislocations, dislocations with $\vec{b}$ perpendicular to $\vec{t}$ edge dislocations, and dislocations with $\theta = \pm\pi/3$ or $\pm 2\pi/3$ are called 60° dislocations. Explicit expressions for the stress fields of edge, screw and general ('mixed') dislocations in isotropic media can be found in standard works (e.g. [5]). Here we only note that the stress field of a straight dislocation is antisymmetric in both $\vec{b}$ and $\vec{r}$: $\boldsymbol{\sigma}_{\vec{b},\vec{t}}(\vec{r}) = -\boldsymbol{\sigma}_{-\vec{b},\vec{t}}(\vec{r}) = -\boldsymbol{\sigma}_{\vec{b},\vec{t}}(-\vec{r})$.

For a quasi-two-dimensional system of straight parallel dislocations with common line direction $\vec{t}$, the internal stress at the point $\vec{r}$ in the plane perpendicular to the dislocation lines is simply obtained by summation over all dislocations,

$$\boldsymbol{\sigma}^{\rm int}(\vec{r}) = \sum_j \boldsymbol{\sigma}_{\vec{b}_j,\vec{t}}(\vec{r} - \vec{r}_j) \quad . \tag{4}$$

For a three-dimensional system of curved dislocation lines, the situation is slightly more complicated. A dislocation segment of tangent vector $\vec{t}$, Burgers vector $\vec{b}$, and unit length situated at the origin creates at the point $\vec{r} = (r, \phi, \psi)$ a stress

$$\boldsymbol{\sigma}(\vec{b}, \vec{t}, \vec{r}) = \frac{bG\boldsymbol{f}([\vec{b}, \vec{t}], \phi, \psi)}{r^2} \quad . \tag{5}$$

Explicit expressions for the stress fields of short straight segments of general orientation can be found in [6]. The internal stress field at $\vec{r}$ is obtained by summation over all segments,

$$\boldsymbol{\sigma}^{\rm int}(\vec{r}) = \sum_j \int_{\mathcal{C}_j} \boldsymbol{\sigma}(\vec{b}_j, \vec{t}(s), \vec{r} - \vec{r}(s)) {\rm d}s \quad . \tag{6}$$

Here, $\vec{r}(s)$ is the parametrization of a three-dimensionally curved dislocation $\mathcal{C}_i$ with Burgers vector $\vec{b}_i$, and $\vec{t}(s)$ is the corresponding bundle of tangent vectors. Again, the summation runs over all dislocation lines [1].

2.2. Dislocation motion and plastic flow

Dislocations move under the action of Peach-Koehler forces. Moving dislocations are subject to electron and phonon drag leading to a friction force per unit length $\vec{f}_{\rm F} = \chi\vec{v}$ which is proportional to the dislocation velocity. Since the effective mass of a dislocation is small, dislocation glide often takes place in an over-damped manner and can therefore be described by 'Aristotelian' dynamics,

$$\vec{v} = \vec{f}_{\rm G}(\vec{r})/\chi \quad . \tag{7}$$

This description ceases to be correct when dislocation motion is controlled by thermal activation processes. Thermal activation of dislocations may be important for the overcoming of localized obstacles such as solute atoms, radiation- or deformation-induced

[1] For topological reasons, dislocations cannot terminate inside the crystal and the Burgers vector is constant along the dislocation line. Any dislocation must either form a closed loop or connect two triple nodes where $\sum_i \vec{b}_i = 0$. In a finite crystal, one or both of these nodes may be replaced by a point at the surface, and in an infinite crystal by the point at infinity.

point defects and their agglomerates, or small precipitates. Thermal activation is also crucial for the motion of dislocations in crystals with high Peierls barriers: If a dislocation moves along its slip plane, its core configuration and hence its energy changes in a lattice-periodic manner. In certain materials the resulting energy barriers to dislocation motion ('Peierls barriers') are significant and at sufficiently low temperatures, dislocation motion is governed by their overcoming via a thermally activated double-kink mechanism. In cases where dislocation motion is controlled by thermal activation, the velocity (defined on a scale which is large in comparison with the spacing of the relevant obstacles) is given by an Arrhenius-type expression,

$$v = \nu b \exp\left[-\frac{H(\tau)}{kT}\right] \quad , \tag{8}$$

where ν is a characteristic frequency of the order of magnitude of the Debye frequency. The stress dependence of the activation energy $H(\tau)$ reflects the way how the energy profile of the relevant obstacles is tilted by the resolved shear stress acting on the dislocation.

Whatever the stress dependence of its velocity, a gliding dislocation (dislocation segment) of tangent vector $\vec{t}$ moves in the direction $\vec{n} \times \vec{t}$ of the glide component of the Peach-Koehler force. For dynamic purposes it is often convenient to distinguish dislocations according to their direction of motion into different 'populations'. For a 2D dislocation system of straight parallel dislocations, $\vec{t}$ can take only two values which define two possible 'signs' $s := \mathrm{sign}(\vec{b}.\vec{t})$ of a dislocation. For each slip system β and sign s we may formally introduce a discrete dislocation density $\rho_{\mathrm{D}}^{\beta,s}$ by

$$\rho_{\mathrm{D}}^{\beta,s}(\vec{r}) = \sum_j \delta_{ss_j}\delta_{\beta\beta_j}\delta(\vec{r} - \vec{r}_j) \quad , \tag{9}$$

where the sum runs over all dislocations. The local shear strain rates on the different slip systems are related to the dislocation densities and dislocation velocities by Orowan's relation,

$$\dot{\gamma}^{\beta}(\vec{r}) = \sum_s s\rho^{\beta,s}(\vec{r}) b(\vec{n} \times \vec{t})\vec{v}(\vec{r}) \quad . \tag{10}$$

Depending on the level of description (discrete vs. continuum), $\rho^{\beta,s}(\vec{r})$ in this equation is understood as the discrete density given by Eq. (9), or an appropriate coarse-grained version thereof. Analogous expressions can be formulated in 3D.

The plastic strain rate tensor is obtained from the shear strain rates on the different slip systems via

$$\dot{\varepsilon}(\vec{r}) = \sum_\beta \boldsymbol{M}^{\beta}\dot{\gamma}^{\beta}(\vec{r}) \tag{11}$$

where the projection tensors $\boldsymbol{M}^{\beta} = [\vec{s}^{\beta} \otimes \vec{n}^{\beta}]_{\mathrm{sym}}$ are the symmetrized tensor products of the respective unit slip vectors and slip plane normals.

2.3. Scaling relations for dislocation patterns

Irrespective of their morphology, dislocation patterns in plastically deformed crystals obey some fairly universal scaling relations. We discuss these scaling relations first for the simplified case of a system of straight parallel dislocations which can be envisaged as point particles in a plane normal to their line direction. We assume that temperatures are sufficiently low such that dislocations move only by glide, and that Peierls barriers are low such that glide may take place at arbitrarily small stresses. Under these circumstances a necessary and sufficient condition for a system of N dislocations to be in static equilibrium at an applied stress $\boldsymbol{\sigma}_{\text{ext}}$ is that the resolved shear stresses acting on all dislocations must be zero, i.e.

$$\vec{b}_i \left[\boldsymbol{\sigma}_{\text{ext}} + \sum_{j \neq i} \boldsymbol{\sigma}(\vec{r}_i - \vec{r}_j) \right] \vec{n}_i = 0 \quad , \quad i \in [1 \dots N] \ . \tag{12}$$

Now assume that we increase the applied stress by a factor η and simultaneously decrease all dislocation spacings by the same factor. From Eqs. (3) and (12) it follows immediately that the new dislocation arrangement is also in static equilibrium. This simple scaling invariance also holds in the limit $N \to \infty$ in an infinite crystal. In this case the re-scaling by a factor η simply increases the dislocation density (number of dislocations per unit area) by a factor of η^2.

From this scaling property the following conclusions derive:

- If a given dislocation arrangement of average density ρ is stable under a certain applied stress $\boldsymbol{\sigma}_{\text{ext}}$, then there exists a one-parameter manifold of dislocation arrangements of density ρ' with the same morphology that are stable under stresses $(\rho'/\rho)^{1/2}\boldsymbol{\sigma}_{\text{ext}}$.

- If a dislocation arrangement exhibits inhomogeneous patterns with a characteristic length λ (for examples see (b)-(d) in Figure 1) and is stable under the applied stress $\boldsymbol{\sigma}_{\text{ext}}$, then morphologically equivalent arrangements with characteristic length λ' are stable under stresses $(\lambda/\lambda')\boldsymbol{\sigma}_{\text{ext}}$.

- Assume that the applied stress is slowly increased until sustained dislocation motion occurs on a slip system $(\vec{b}, \vec{n})$. The corresponding critical resolved shear stress $\tau^{\text{f}}_{\vec{b},\vec{n}}$ is also called the flow stress of this slip system, and the critical resolved shear stress of the slip system(s) where long-range dislocation motion first occurs is simply called the flow stress $\tau^{\text{f}} = \min[\tau^{\text{f}}_{\vec{b},\vec{n}}]$. From scaling it follows that, for dislocation patterns of similar morphology but different total density, flow stresses are proportional to the square root of dislocation density. This is often expressed in terms of Taylor's relation

$$\tau^{\text{f}} = \alpha G b \sqrt{\rho} \ , \tag{13}$$

where G is the shear modulus of the material, and the constant α depends weakly on the morphology of the dislocation arrangement (typically $0.2 < \alpha < 0.4$). In general, because of dislocation multiplication (cf. below) plastic deformation goes along with an increase in dislocation density, and this in turn implies an increasing flow stress (work hardening).

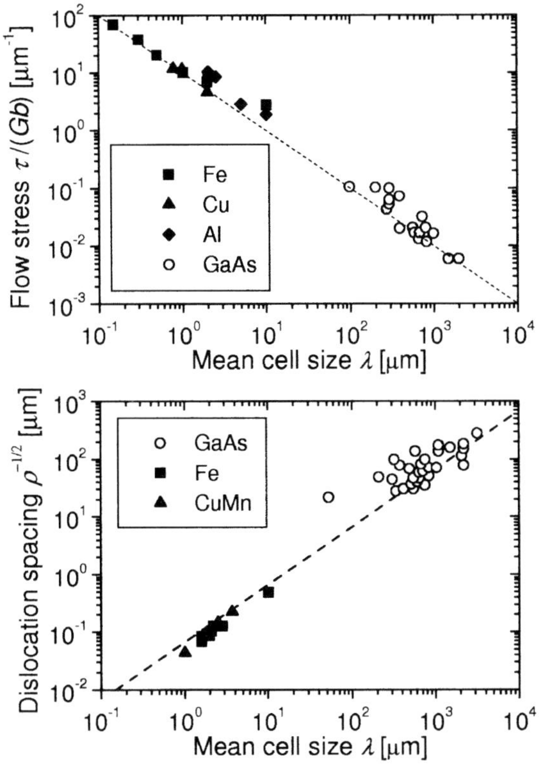

Figure 2. Scaling of dislocation cell patterns in metals and semiconductors. Top figure: cell size vs. flow stress (resolved shear stress in the active slip system(s)). Bottom figure: cell size vs. mean dislocation spacing. Data for pure metals after Holt [7], data for CuMn after Neuhaus and Schwink [8], data for GaAs after Rudolph et. al. [9].

The above arguments can be generalized to three-dimensional systems of curved dislocation lines. For a 3D dislocation arrangement in static equilibrium, the internal stress field (6) must balance the externally applied stress everywhere along the dislocation lines. Again, by increasing the applied stress by a factor η, and shrinking all distances including the length ds of a line element (cf. Eq. (6)) by the same factor, we obtain another static dislocation arrangement. This procedure increases the dislocation density (now defined as line length per unit volume) by a factor of η^2, and we recover the same scaling relations as above.

The experimental data in Figure 2 illustrate the relations between dislocation densities, characteristic lengths of dislocation cell patterns, and the flow stresses at which these patterns have formed. It is seen that the general scaling relations $\tau^{\mathrm{f}} \propto \sqrt{\rho} \propto 1/\lambda$ are fulfilled over a wide range of materials and deformation conditions, with proportionality factors that are remarkably insensitive to all peculiarities of the deformation process. Often, in the course of deformation the flow stress and dislocation density increase and the characteristic length of the dislocation pattern decreases in accordance with these scaling relations, while the morphology of the dislocation arrangement remains more or less unchanged. Such self-similar behavior of the dislocation pattern during hardening has been termed the 'law of similitude' [10].

At this point we have to address an apparent paradox: The scaling arguments we have formulated pertain to *static* dislocation arrangements while the formation of heterogeneous dislocation patterns necessarily involves the *motion* of dislocations. By looking at Eqs. (7) or (8) we see, however, that the dynamic behavior of dislocations is *not* invariant under the re-scaling procedure discussed above.

For situations where the motion of dislocations is governed by drag forces or, more generally, the dislocations are highly mobile (e.g. at high temperatures), this paradox is readily resolved. In this case, numerical estimates show that at each given moment plastic deformation is carried by only a few dislocations (dislocation segments). These 'active' segments move rapidly in regions where the dislocation-induced internal stresses assist the externally applied stress. At the same time, the overwhelming majority of the dislocations is practically at rest. It has been estimated that, in a typical fcc crystal deforming at a characteristic experimental strain rate of 10^{-3} s^{-1}, only a fraction of 10^{-6} of all dislocations at each given moment appreciably contribute to the deformation [11,12]. Under these circumstances, scaling relations for static dislocation arrangements retain their validity even during plastic flow.

The situation is more complicated in crystals where the motion of dislocations is limited by localized barriers that are *not* dislocation-related, such as Peierls barriers or localized obstacles. At low temperatures the overcoming of such obstacles may require substantial stresses (cf. Eq. (7); for appreciable dislocation motion to occur on typical experimental timescales, stresses must be high enough to reduce the activation enthalpy $H(\tau)$ to about $25kT$). This may lead to a temperature-dependent flow stress contribution that is independent on dislocation density, and in such situations the scaling relationships of the 'law of similitude' do not apply.

3. DISCRETE DISLOCATION DYNAMICS (DDD) SIMULATIONS

The most straightforward way of dealing with the dynamics of a many-dislocation system is to directly solve the equations of motion of the dislocation lines, keeping track of all the dislocation interactions. We first briefly discuss fully three-dimensional dislocation dynamics simulations which attempt to give a true representation of the evolution of the system of interacting dislocation lines, and then the less realistic but computationally much less intensive alternative of studying idealized quasi-two-dimensional systems consisting of parallel dislocations.

3.1. DDD simulation of 3-dimensional dislocation systems

Several different schemes have been proposed for simulating the 3D dynamics of a system of interacting dislocations. These schemes mainly differ in the way how the dislocation

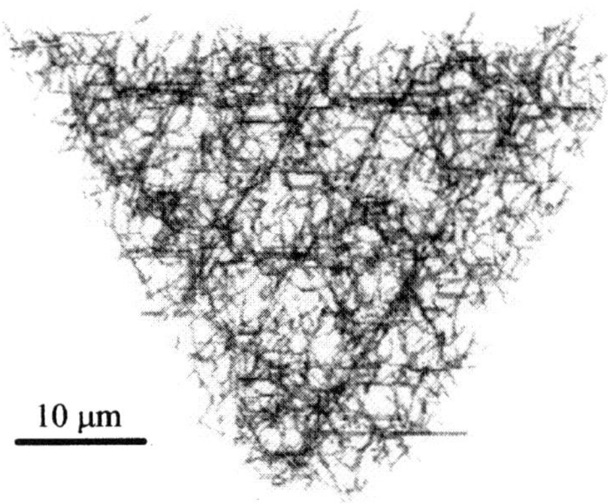

Figure 3. Dislocation pattern observed in a three-dimensional DDD simulation (simulation of an fcc crystal deformed along a [100] axis to a total strain of 0.003; the image shows the dislocations contained within a 'TEM foil' of 5 μm thickness and [111] orientation, after Madec et. al. [16].

lines are represented. Representations in terms of parametric splines have been adapted, using the spline parameters as generalized line coordinates [13]. Other simulations represent a dislocation line by a sequence of straight segments of variable orientation [14], using the nodes between segments as line coordinates. In the following we discuss in more detail the approach by Kubin, Devincre and co-workers [6,15–17] who use a representation of dislocation lines in terms of sequences of linear segments for which only a limited number of discrete orientations are allowed.

Elastic interactions between dislocation segments are mediated by the corresponding segment stress fields which according to Eq. (5) decay in space like $1/r^2$. In this sense the computational problem is very similar to the molecular dynamics simulation of a system of particles with long-range (e.g. gravitational or electrostatic) interactions[2]. Once the stress acting on a dislocation segment is known, the Peach-Koehler force and the segment velocity can be calculated. In moving the segments, however, it is important to maintain connectivity of the dislocation line which for curved dislocations requires continuous adjustment of the segment length and/or the insertion and removal of segments.

Additional problems arise if two dislocation segments get very close. Two dislocation segments of the same slip system but opposite sign may annihilate spontaneously if they

[2]Since piecewise linear representations of the dislocation lines give rise to artificial stress singularities at the 'corners', it may be necessary to apply corrections in order to correctly evaluate the interaction of adjacent segments.

approach each other below a critical distance (typically a few interatomic spacings)[3]. Another process may take place if two dislocation lines gliding on different slip planes intersect. If the Burgers vectors $\vec{b}_1$ and $\vec{b}_2$ and line configuration of the dislocations are such that the two lines attract each other, then upon contact they will form a new 'junction' segment with Burgers vector $\vec{b}_1 + \vec{b}_2$, thereby reducing the total elastic energy and dislocation line length. As a consequence, the intersection takes an H- rather than an X-shape and two triple nodes emerge[4]. The junction segment initially runs along the direction of intersection of the slip planes of the two dislocations. If only discrete segment orientations are allowed, it is important that these include the orientation of potential junction segments. For instance, in fcc crystals where the Burgers vectors are directed along [110] lattice directions and the slip planes are of [111] type, junction segments can have screw or 60° orientations and these orientations should be included in the discretization [17].

There are two reasons why junction formation is of crucial importance for the dynamical behavior of a dislocation system. On the one hand, junctions act as dislocation obstacles. A dislocation moving on its slip plane will continuously form junctions with 'forest' dislocations threading through this plane, and dislocation motion can only proceed if the stress acting on the dislocation is high enough to destroy these junctions. In metals deforming in multiple slip, this mechanism yields a major contribution to the flow stress. On the other hand, dislocation segments pinned at two anchoring points may bulge out and nucleate dislocation loops that expand during deformation. Expansion of dislocation loops increases the total dislocation line length and is thereby responsible for the increasing dislocation density in a deforming crystal. Through this mechanism junction formation (or, more general, the 'knitting' of a 3D dislocation network) is also crucial for dislocation multiplication and work hardening.

Three-dimensional dislocation dynamics simulations correctly represent the processes of dislocation multiplication and junction formation. They account without additional assumptions for the multitude of conceivable bowing and intersection processes which may occur in the dislocation network. However, the neccessity to continuously re-evaluate the stresses acting on all dislocation segments tends to make such simulations computationally extremely intensive. Because of their high computational cost, present-day simulations of dislocation dynamics in bulk metals are restricted to comparatively small system sizes (typically about $(10\mu m)^3$, though periodic boundary conditions may be imposed to mimic a bulk crystal), small strains (typically far less than 1% total deformation), and low dislocation densities ($\rho L^2 < 5 \times 10^2$ where L is the characteristic linear dimension of the simulated volume). Under these circumstances, only the very first stages of dislocation cell patterning can be studied (Figure 3). Since dislocation densities are low, the cell size of the incipient dislocation cell patterns is of the same order as the size of the simulation cell (e.g., Figure 3 shows a section of the simulation cell plus its first periodic images), and therefore not much information about the spatial morphology of the emergent patterns can be obtained.

---

[3]In case of two edge segments, annihilation may leave an agglomerate of point defects in the crystal.

[4]Only triple nodes are stable in a dislocation system; intersections which do not lead to junction formation imply that the two dislocations repel each other and, hence, no quadruple nodes are formed.

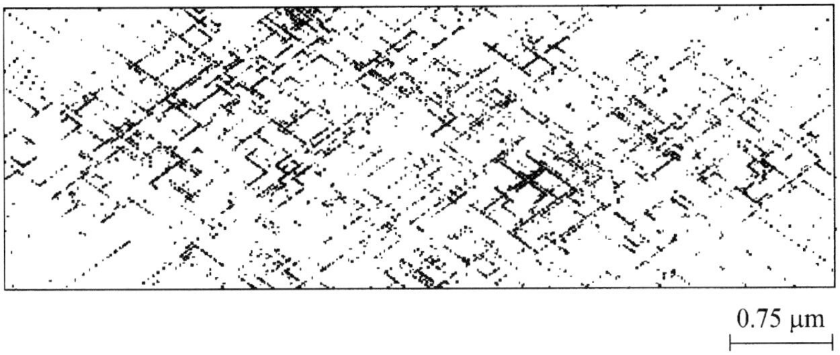

Figure 4. Dislocation pattern observed in a two-dimensional DDD simulation of deformation in symmetrical double slip; shear strain in each of the two slip systems $\gamma = 0.073$, resolved shear stress $\tau = 2.6 \times 10^{-3} G$, total dislocation density $2 \times 10^{14} \mathrm{m}^{-2}$; after Benzerga et. al. [18].

3.2. DDD simulation of 2-dimensional dislocation systems

Because of the tremendous reduction in degrees of freedom, two-dimensional dislocation systems can be simulated to much larger strains and dislocation densities. Since the basic scaling relations for dislocation patterns are similar in two and three dimensions, one may ask whether such simulations are a way of modelling the later stages of dislocation pattern formation and to the evolution of these patterns during sustained deformation.

A 2D dislocation system is essentially a system of interacting point particles moving in a plane. For 2D dislocation dynamics to be representative of the behavior of a real, three-dimensional crystal, two conceptual problems must be addressed: (i) In a system of straight dislocations there is no mechanism for increasing the dislocation density (dislocation multiplication) which in 3D occurs by bulging of segments in the dislocation network and expansion of the ensuing dislocation loops. Since straight dislocations can neither bulge nor expand, one has instead to devise some phenomenological rule for throwing new dislocations into the system. (ii) Since dislocations cannot intersect in 2D, the formation/destruction of junctions and the corresponding work hardening have to be accounted for in a phenomenological manner.

The problem is complicated by the fact that dislocation loop generation and destruction of junctions in a 3D dislocation system may be 'many-body' processes which in general simultaneously involve several dislocation lines. The approach adopted in the most recent 2D simulations is to model junction formation as a dislocation pair reaction: as soon as two dislocations of different slip systems approach beyond a certain collision radius, they form a sessile junction. This junction can be broken if a critical stress is exceeded which is related to the dislocation configuration in its surrounding. Source formation and operation are modelled in a similar spirit. This procedure makes it necessary to isolate and parameterize a limited set of dislocation mechanisms. It introduces into the models a number of phenomenological parameters which have to be determined either by matching to 3D simulations or to experimental data [18].

The advantage of 2D simulations lies in the possibility to achieve strains and dislocation densities sufficiently high such that at least the early stages of dislocation pattern formation and evolution can be observed (Figure 4) and statistically significant information on the morphology of the dislocation patterns can be deduced. However, even if the problems relating to mapping 3D dislocation processes onto a 2D geometry can be solved, the approach is restricted to certain glide geometries since it is difficult to represent simultaneous dislocation glide on more than two intersecting slip planes in a 2D model.

4. CONTINUUM DISLOCATION DYNAMICS APPROACHES

In view of the computational complexity of discrete dislocation dynamics simulations, repeated attempts have been made to arrive at some kind of coarse-grained description of dislocation dynamics. There are two fundamental problems: a) how to define coarse-grained dislocation densities in a manner which retains the relevant information about the kinetic properties of the discrete dislocation lines, and b) how to obtain the corresponding equations of motion. Two main approaches may be followed - an 'energetic' approach which starts out from the elastic energy associated with the dislocation system, and a 'dynamic' approach which determines equations of motion for dislocation densities in a phenomenological manner.

4.1. Linear irreversible thermodynamics and energy minimization

Dislocations move under the action of forces, which implies that they move 'downhill' in an energy functional. Any static dislocation arrangement necessarily corresponds to a local minimum of the elastic energy[5]. It is therefore tempting to address dislocation patterning in terms of energy minimization arguments.

An early attempt in this direction was made by Holt [7] who studied the stability of a spatially homogeneous dislocation system with conserved total dislocation density. In the following we present a slightly generalized version of this model which works analogous to the Cahn-Hilliard theory of spinodal decomposition [19].

The elastic energy of the internal stress field is given by

$$E_{\mathrm{el}} = \int_V \boldsymbol{\sigma}(\vec{r})\boldsymbol{C}^{-1}\boldsymbol{\sigma}(\vec{r})\mathrm{d}^3r \tag{14}$$

where $\boldsymbol{C}^{-1}$ is the tensor of elastic compliances and V is the crystal volume. In the following we consider a system of straight parallel dislocations. If we denote by $\sigma^\beta(\vec{r})$ the stress field of an isolated straight dislocation of slip system β, then the elastic self-energy per unit length of this dislocation is given by

$$E^\beta_{\mathrm{self}} = \int_V \boldsymbol{\sigma}^\beta(\vec{r})\boldsymbol{C}^{-1}\boldsymbol{\sigma}^\beta(\vec{r})\mathrm{d}^2r \tag{15}$$

where the integration is carried over the plane normal to the dislocation. The dislocation stress field is given by Eq. (3), and accordingly the dislocation self-energy diverges with

[5]Note that the elastic energies associated with the dislocation system are usually so large that entropy effects can safely be neglected.

increasing crystal radius R like $E^{\beta}_{\text{self}} = K^{\beta}Gb\ln(R/b)$ where K^{β} is a constant of the order of unity which depends on the type of the dislocation. The energy divergence which occurs according to Eqs. (3,15) as one approaches the dislocation core may be truncated at radius b since, on scales comparable to the atomic spacing, local linear elasticity ceases to be valid and Eq. (3) no longer applies. Far away from the dislocation, on the other hand, linear elasticity is perfectly valid and, hence, the logarithmic divergence of the self-energy is limited only by the finite crystal size.

The interaction energy of two dislocations of slip systems β and β' and signs s and s' at the respective positions $\vec{r}$ and $\vec{r}'$ is given by

$$E^{\beta\beta' ss'}_{\text{int}}(\vec{r}-\vec{r}') = ss' \int_V \boldsymbol{\sigma}^{\beta}(\vec{r}-\vec{r}'')\boldsymbol{C}^{-1}\boldsymbol{\sigma}^{\beta'}(\vec{r}'-\vec{r}'')\mathrm{d}^2r'' \quad . \tag{16}$$

This interaction energy can be written as $E^{\text{int}} = ss'K^{\beta\beta'}(\phi)Gb\ln(R/|\vec{r}-\vec{r}'|)$ where $K^{\beta\beta'}(\phi)$ is a non-dimensional function that depends on the relative orientation of the two dislocations in the plane. Note that the dislocation interaction is strongly anisotropic; its angular average vanishes, $\int K^{\beta\beta'}(\phi)\mathrm{d}\phi = 0$.

Now the energy of a discrete dislocation system characterized by the discrete densities $\rho^{\beta,s}_{\text{D}}(\vec{r})$ as defined in Eq. (9) is readily written as

$$E = \int_V \rho^{\beta,s}_{\text{D}}(\vec{r})E^{\beta}_{\text{self}}\mathrm{d}^2r + \sum_{\beta\beta' ss'} ss' \int\int_V \rho^{\beta,s}_{\text{D}}(\vec{r})\rho^{\beta',s'}_{\text{D}}(\vec{r}')E^{\beta\beta' ss'}_{\text{int}}(\vec{r}-\vec{r}')\mathrm{d}^2r\mathrm{d}^2r' \quad . \tag{17}$$

To arrive at a continuum description we average over an ensemble of statistically equivalent dislocation systems. This leads to continuous densities $\rho^{\beta,s}(\vec{r}) = \langle\rho^{\beta,s}_{\text{D}}(\vec{r})\rangle$. Averaging the product of discrete densities in the interaction term of Eq. (17) leads to pair densities $\rho^{\beta\beta',ss'}_{(2)}(\vec{r},\vec{r}') = \rho^{\beta,s}(\vec{r})\rho^{\beta',s'}(\vec{r}')[1 + d^{\beta\beta',ss'}(\vec{r},\vec{r}')]$ where $d^{\beta\beta',ss'}(\vec{r},\vec{r}')$ are pair correlation functions.

In a spatially homogeneous dislocation arrangement, because of symmetry the pair correlation functions depend on the relative position of the two dislocations only. We also note that because of the general scaling relations discussed above the 'range' of the pair correlation functions is proportional to the mean dislocation spacing, i.e., $d^{\beta\beta',ss'} = d^{\beta\beta',ss'}([\vec{r}-\vec{r}']\sqrt{\rho})$ where $\rho = \sum_{\beta,s}\rho^{\beta,s}$ is the total dislocation density. Furthermore it has been shown that pair correlations in a statistically homogeneous dislocation arrangement are short-ranged, i.e., the pair correlation functions $d^{\beta\beta',ss'}$ decay to zero faster than algebraically if $|\vec{r}-\vec{r}'|$ exceeds a few dislocation spacings [20]. Hence, if dislocation density variations occur on a scale that is large as compared to the dislocation spacing, we may adopt a procedure similar to density functional theories of many-electron systems or inhomogeneous fluids, i.e., we approximate the pair correlation functions by those of a spatially homogeneous dislocation arrangement.

Using this approximation, the elastic energy density of an inhomogeneous dislocation arrangement can be written as

$$\rho(\vec{r})\bar{E}(\vec{r}) = \rho(\vec{r})\left[\bar{E}_{\text{self}} + \int \rho(\vec{r}')\bar{E}_{\text{int}}(\vec{r}-\vec{r}')\mathrm{d}^2r'\right] \quad . \tag{18}$$

Here the average self- and interaction energies of a dislocation are given by

$$\bar{E}_{\text{self}} = \sum_{\beta,s} f^{\beta,s}E^{\beta}_{\text{self}} \qquad \bar{E}_{\text{int}}(\vec{r}) = \sum_{\beta\beta',ss'} ss' f^{\beta,s}f^{\beta',s'}E^{\beta\beta',ss'}_{\text{int}}(\vec{r})d^{\beta\beta',ss'}(\vec{r}\sqrt{\rho}) \quad , \tag{19}$$

where $f^{\beta,s} = \rho^{\beta,s}/\rho$. In a random dislocation arrangement without correlations, the average interaction energies vanish and the energy density (18) diverges logarithmically. The presence of dislocation pair correlations re-normalizes this divergence [12].

To derive the temporal evolution of the dislocation pattern from the elastic energy functional we make the following assumptions: (i) only the total dislocation density changes whereas the 'composition' of the dislocation arrangement as expressed by the density fractions $f^{\beta,s}$ remains invariant, (ii) the total number of dislocations is preserved, i.e., the time evolution of ρ is given by the continuity equation

$$\partial_t \rho(\vec{r}) = -\nabla \boldsymbol{J}(\vec{r}) \tag{20}$$

(iii) the dislocation flux J is proportional to the gradient of the average dislocation energy, $\boldsymbol{J} = -\chi\rho\nabla\bar{E}(\vec{r})/\chi$ where χ is an effective friction coefficient. The problem is very similar to the linear irreversible thermodynamics treatment of a system with conserved order parameter as developed, for instance, in the Cahn-Hilliard theory of spinodal decomposition.

Now we consider a homogeneous dislocation arrangement of total density ρ_0 and study the time evolution of small space-dependent density fluctuations $\delta\rho(\vec{r})$. Retaining in the equation of evolution only terms of linear order in $\delta\rho$, we obtain

$$\chi\partial_t\delta\rho(\vec{r}) = \rho_0\Delta\int\delta\rho(\vec{r}\,')\bar{E}_{\text{int}}(\vec{r}-\vec{r}\,')\mathrm{d}^2r' \approx \rho_0\Delta[D^{(0)}\delta\rho(\vec{r}) + \boldsymbol{\nabla}\boldsymbol{D}^{(2)}\boldsymbol{\nabla}\delta\rho(\vec{r})] \tag{21}$$

where

$$D^{(0)} = \int\bar{E}_{\text{int}}(\vec{r})\mathrm{d}^2r \quad , \quad \boldsymbol{D}^{(2)} = \int\frac{\vec{r}\otimes\vec{r}}{2}\bar{E}_{\text{int}}(\vec{r})\mathrm{d}^2r \quad . \tag{22}$$

In the last step we have performed a Taylor expansion of $\rho(\vec{r}\,')$ around $\vec{r}\,' = \vec{r}$, using the fact that the dislocation pair correlation functions and therefore also the average interaction energy $\bar{E}_{\text{int}}(\vec{r}-\vec{r}\,')$ are short-ranged functions, cf. Eq. (19). Re-writing the evolution equation for $\delta\rho$ in Fourier space we get

$$\chi\partial_t\delta\rho(\vec{k}) = \rho_0\vec{k}^2[-D^{(0)} + \vec{k}\boldsymbol{D}^{(2)}\vec{k}]\delta\rho(\vec{k}) \tag{23}$$

Since dislocations arrange such as to reduce their total energy, the coefficients $D^{(0)}$ as well as the components of the matrix $\boldsymbol{D}^{(2)}$ are negative. This implies a long-wavelength instability similar to the instability encountered in spinodal decomposition. The amplification of fluctuations is in general anisotropic; the direction(s) of maximum amplification corresponds to the Eigenvector(s) of $\boldsymbol{D}^{(2)}$ with the largest Eigenvalue $D^{(2)}_{\text{max}}$. The corresponding wavelength of dislocation density fluctuations is given by $\lambda = \pi\sqrt{8D^{(2)}_{\text{max}}/D^{(0)}}$. Note that, since the range of the dislocation pair correlations is proportional to the dislocation spacing, $D^{(0)} \propto 1/\rho_0$ and $\boldsymbol{D}^{(2)} \propto 1/(\rho_0)^2$. Hence the wavelength of maximum instability which characterizes the emergent patterns is proportional to (but in general much larger than) the dislocation spacing, in agreement with the principle of similitude.

This 'energetic' approach to dislocation patterning has certain attractive features, as the treatment of dislocation patterning within the general framework of linear irreversible thermodynamics leads to a simple continuum theory which correctly accounts for the observed scaling relation $\lambda \propto \rho^{-1/2}$, while the possibility of describing the entire dislocation population in terms of a single partial differential equation significantly reduces the computational complexity. Furthermore, the approach offers a simple explanation for dislocation patterning: the elastic energy of a dislocation is reduced if the dislocation density increases, and therefore the dislocation system segregates into regions of high and low dislocation density. On the other hand, the approach rises several problems:

- The assumption that the dislocation flux follows the gradient of the elastic energy may not be warranted. The motion of dislocations is subject to dynamic constraints, e.g., at low temperatures their glide is confined to the slip planes. Even at higher temperatures, where out-of-plane motions are possible, the effective mobilities for in-plane and out-of-plane motions are rather different.

- As dislocations of different slip systems may move in different slip planes, their fluxes have different directions.

- The approach cannot account for dislocation motions driven by an external stress. Under an externally applied stress, positive and negative dislocations may move in opposite directions, and the evolution of the dislocation arrangement cannot be described in terms of a single dislocation flux even if only dislocations of a single slip system are present.

- During plastic deformation the dislocation density continually increases, and hence the assumption that the dislocation density behaves like a conserved order parameter is not justified.

- It has also been criticized that the predicted proportionality of the pattern wavelength and the dislocation spacing is actually a consequence of the assumed scaling properties of the pair correlation functions.

However, a recent study has to a certain extent justified the energy-based approach. By statistically averaging the equations of motion of a discrete dislocation system, Zaiser and co-workers arrived at evolution equations which under certain restrictive assumptions (dislocations of a single slip system, no external stress) are similar to Eq. (22). The dynamic constraints to dislocation motion (glide motion can occur on a single slip plane only) introduce additional anisotropies; for the case of a system of straight parallel edge dislocations it has been found that the instability leads to the formation of periodic walls perpendicular to the dislocation glide direction [20]. Such walls are indeed observed in cyclically deformed metals (Figure 1 (b)). In prolonged cyclic deformation the dislocation system reaches a steady state where the dislocation density remains practically constant, and one may conjecture that periodic reversal of the external stress reduces the influence of directed dislocation fluxes on the dislocation patterning. In unidirectional deformation, on the other hand, evolution of the dislocation microstructure is characterized by directed dislocation fluxes and increasing dislocation densities, and it is difficult to see how these features can be accommodated within the 'energetic' approach.

4.2. Synergetic models

A plastically deforming crystal is a driven system which by the continuous influx and dissipation of mechanical work is kept in an out-of-equilibrium state. This observation has prompted attempts to model dislocation patterning as a 'synergetic' pattern formation phenomenon, in analogy with spatial patterning phenomena observed in many driven far-from-equilibrium systems, e.g., the Belushov-Zhabotinskii reaction in chemical kinetics or the Taylor instability in hydrodynamics (for overviews of synergetic patterning, see [21,22]). In the theoretical description of these phenomena, spatial patterns emerge from symmetry-breaking instabilities in systems of nonlinear reaction-diffusion-transport equations. It has been proposed to adopt this framework to dislocation systems by distinguishing dislocation populations ρ_k of different specification and writing down reaction-transport equations of the type [23–25]

$$\partial_t \rho_k + \boldsymbol{\nabla} \boldsymbol{J}_k = f_k([\rho_l]) \quad . \tag{24}$$

Here, the non-linear reaction terms f_k model local interactions between dislocation populations and the flux terms describe long-range dislocation transport. Several phenomenological models of this type have been proposed in the literature, assuming different transport mechanisms to be operative. For instance, Kratochvil considered the flux of edge dislocation dipoles induced by their 'sweeping' by moving screw dislocations [26]. Walgraef and Aifantis [24] consider single-slip cyclic deformation and describe the net transport arising from the forward-backward motion of dislocations during a stress cycle by diffusion-like terms of the form $\boldsymbol{J}_k = \boldsymbol{\nabla D} \rho_k$. In its simplest one-dimensional form, the Walgraef - Aifantis model reads

$$\begin{aligned} \frac{\partial \rho_{\rm m}}{\partial t} &= D_{\rm m} \frac{\partial^2 \rho_{\rm m}}{\partial x^2} + f(\rho_{\rm i}, \rho_{\rm m}) \quad , \\ \frac{\partial \rho_{\rm i}}{\partial t} &= D_{\rm i} \frac{\partial^2 \rho_{\rm i}}{\partial x^2} + g(\rho_{\rm i}) - f(\rho_{\rm i}, \rho_{\rm m}) \quad . \end{aligned} \tag{25}$$

Here, $\rho_{\rm m}$ and $\rho_{\rm i}$ are the densities of mobile and immobile (dipole) dislocations, respectively, each population consisting of dislocations of both signs. The non-linear reaction term is assumed as $f(\rho_{\rm i}, \rho_{\rm m}) = C_1 \rho_{\rm i} - C_2 \rho_{\rm m} \rho_{\rm i}^2$ where C_1 and C_2 were introduced as phenomenological coefficients.

The Walgraef-Aifantis model allows to obtain periodic solutions corresponding to the ladder-like dislocation arrangement in persistent slip bands (PSB, see Figure 5, right). An extension of the model to more than one dimension has been also used to investigate the competition between matrix and PSB patterns as observed experimentally (Figure 5, left) [24]. In either case, the model leads to estimates of the pattern wavelength in terms of the diffusionlike coefficients $D_{\rm m}$ and $D_{\rm i}$ entering the reaction-diffusion scheme (25).

A general problem with the phenomenological synergetic approach is that, in the absence of a well-defined procedure for coarse-graining dislocation dynamics, the definition of the dislocation populations in Eq. (24), the determination of the functional form of the reaction terms, and the mathematical specification of the dislocation fluxes are all left to educated guess. With some mathematical training it is possible to devise nonlinear equations which produce a particular type of spatial patterns but, since the results are in a sense pre-determined by the phenomenological 'input', it is arguable whether this really

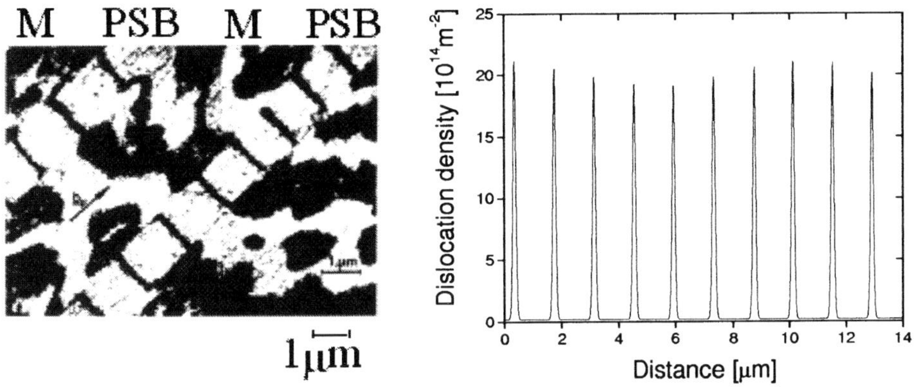

Figure 5. Left: dislocation structure in single-slip cyclic deformation; the ladder-like structure corresponds to a persistent slip band (PSB); the surrounding patchy structure is referred to as 'matrix' (M). Right: One-dimensional dislocation density pattern in a persistent slip band obtained from the Walgraef-Aifantis model; Courtesy of E.C. Aifantis.

solves the problem of dislocation patterning. Nevertheless the synergetic approach has been important from a conceptual point of view: If the dislocation system is envisaged as a driven out-of-equilibrium system, one cannot simply take for granted the applicability of energy minimization concepts or of linear irreversible thermodynamics, and the driven dynamics of the dislocation ensemble is put into the focus of interest.

5. STOCHASTIC APPROACHES

5.1. Discrete stochastic dislocation dynamics

The discrete stochastic dislocation dynamics approach proposed by Groma and co-workers [27,28] considers the stress-driven motion of discrete dislocations but uses a statistical description of the force created by the internal stress field. As discussed in detail in [12], in many situations the internal stress field associated with a dislocation arrangement can be split into a long-wavelength part ('mean-field stress') which varies in space on the scale λ of dislocation density variations and can be expressed as a functional of the dislocation densities [12,27], plus a short-wavelength part which fluctuates on the scale $\rho^{-1/2}$ of the average local dislocation spacing. The probability distribution of this fluctuating part of the internal stress has been evaluated in [27], and the short-range nature of its spatial correlations has been demonstrated in [12].

The basic idea of a discrete stochastic dislocation dynamics simulation is to calculate explicitly only the mean-field stresses, while replacing the short-wavelength flucutations of the internal stress field by a statistically equivalent stochastic process. The mean-field stresses are functionals of the dislocation densities which change slowly in space and time. Since these stresses need not be updated in every time step and, moreover, can be evaluated from the dislocation densities by convolution using Fast Fourier Transformation (FFT), a very considerable reduction in the computational cost of the simulation can in principle be achieved.

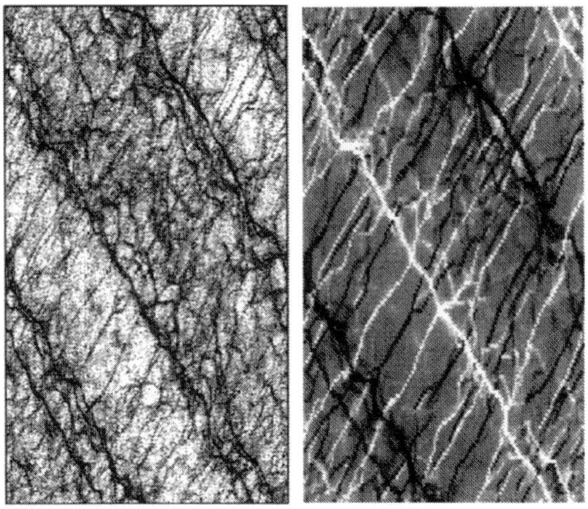

Figure 6. Simulated dislocation pattern using a stochastic dislocation dynamics code; left: dislocation density pattern, right: greyscale pattern of the excess Burgers vector density; after Groma [28]

The crucial problem with this method is to correctly represent the statistical properties of the fluctuating stresses acting on the dislocations. Groma et. al. use a strong approximation by representing the temporal statistics of the internal stresses at the dislocation positions in terms of the statistics of the internal stress fluctuations at a random point in space. This is feasible if (a) only a small fraction of the dislocations is mobile at each given moment and (b) these mobile dislocations move as ***individuals***, each of them individually scanning the internal stress 'landscape' created by the majority of the dislocation arrangement which remains at rest. While assertion (a) is presumably valid, it is rather difficult to assess the validity of assumption (b). In fact, analytical arguments [11,12,29] as well as simulations [30,31] and experiment [30,32] indicate that dislocations move in a strongly correlated, avalanche-like manner where dynamic interactions among the moving dislocations may be of crucial importance. An adequate theoretical understanding of these collective effects and an appropriate mathematical description of the corresponding fluctuation phenomena are still missing.

In spite of these problems, simulations of deformation-induced dislocation patterns using the discrete stochastic dislocation dynamics approach yield interesting results. An example is given in Figure 6 which shows the dislocation pattern forming in a 2D dislocation system deforming in symmetrical double slip. The hierarchical nature of the dislocation pattern with long straight walls carrying substantial excess Burgers vector (and, hence, substantial lattice rotations) compares well with certain experimentally observed cellular microstructures (see e.g. Figure 1 (d)).

5.2. Continuum stochastic dislocation dynamics

The continuum stochastic dislocation dynamics method formulated by Hähner [29] and elaborated by Hähner and Zaiser [4,11,12,33] approaches the problem of dislocation patterning on a much more phenomenological level. This approach starts out from phenomenological differential equations which describe dislocation density evolution and work hardening on a macroscopic scale. The complex dynamics of plastic flow on mesoscopic scales is then taken into account by considering the shear strain rates on the different slip systems [i.e. the dislocation fluxes, cf. Eq. (10)] as fluctuating functions of space and time. Since dislocation accumulation and dislocation reactions are driven by the dislocation fluxes, this implies that dislocation density evolution is described in terms of a set of stochastic differential equations.

For illustration, we discuss the simplest model of this type [34]. We assume deformation to occur in symmetrical double slip. Dislocations are stored in the crystal after travelling a mean glide path $L = B/\sqrt{\rho}$ that is proportional to, but much larger than, the average dislocation spacing. The increase of the total dislocation density ρ is then given by

$$\partial_t \rho = \frac{B}{b}\sqrt{\rho}\dot{\gamma} = \frac{B}{b}\sqrt{\rho}[\langle\dot{\gamma}\rangle + \delta\dot{\gamma}] \tag{26}$$

where in the second step we have split the (local) strain rate in either of the slip systems into the mean strain rate and a spatio-temporally fluctuating contribution.

Fluctuations in the local strain rates are modelled by assuming that straining proceeds in the form of a shot noise where the strain rate is composed of discrete random 'events',

$$\dot{\gamma}(\vec{r}, t) = \sum_i \hat{\gamma}\delta(t - t_i) g(\vec{r} - \vec{r}_i) \tag{27}$$

The times t_i and locations $\vec{r}_i$ of the different 'slip events' are assumed as statistically independent random variables. The shape function $g(\vec{r})$ and event amplitude $\hat{\gamma}$ are chosen to reflect general scaling relations and phenomenological observations of collective dislocation motion: a) In agreement with the scaling relations underlying the 'law of similitude' we assume that the range of spatial correlations in plastic flow decreases in inverse proportion with the flow stress. This implies that g is a function of $\vec{r}/\tau^{\mathrm{f}}$. b) The event amplitude $\hat{\gamma}$ increases in proportion with the flow stress. This reflects the scaling relation $\hat{\gamma} = \rho b L$ [cf. Orowan's relation, Eq. (10)] where L can be understood as the mean path travelled by a dislocation during a collective slip event. Since $\rho \propto (\tau^{\mathrm{f}})^2$ and $L \propto 1/\tau^{\mathrm{f}}$ it follows that $\hat{\gamma} \propto \tau^{\mathrm{f}}$. c) Correlations in plastic flow are strongly anisotropic, their range in the direction of dislocation motion significantly exceeds their range normal to that direction.

Relevant information on dislocation patterning can be directly obtained by studying the stochastic differential equation (26). As discussed in detail in [34], the flow stress of an inhomogeneous dislocation arrangement can be written as $\tau^{\mathrm{f}} = \alpha G b \langle\sqrt{\rho}\rangle$ where $\langle\ldots\rangle$ denotes an average over the inhomogeneous microstructure. It then follows from Eq. (26) that the flow stress increases linearly with average strain, $\tau^{\mathrm{f}} = (\alpha G B/2)\langle\dot{\gamma}\rangle$. Equation (26) is conveniently re-written by introducing the transformation of variables

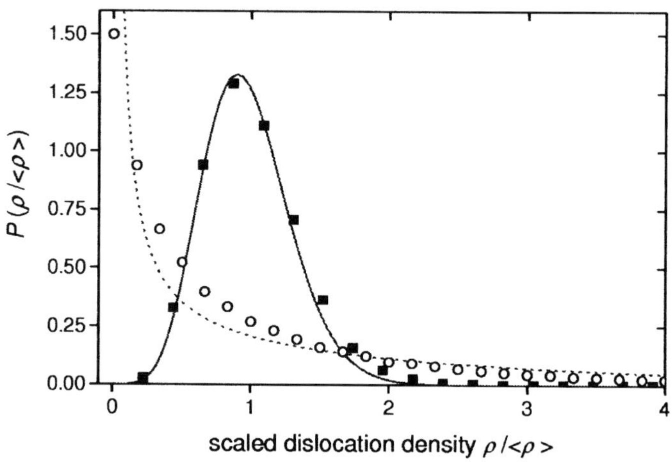

Figure 7. Probability distributions of dislocation densities calculated from Eq. (29) for low ($Q^2 = 0.1$, solid line) and high ($Q^2 = 2$, dotted line) noise amplitudes. Data points: Results obtained from simulations of deformation in discrete random events with event sizes corresponding to the same effective noise amplitudes.

$\rho = [\tau^{\mathrm{f}}/(\alpha G b)]^2\tilde{\rho}, t = [\tau^{\mathrm{f}}/(\alpha G B\langle\dot{\gamma}\rangle)]\tilde{t}$. [6] Furthermore we idealize the stochastic process $\delta\dot{\gamma} = \dot{\gamma} - \langle\dot{\gamma}\rangle$ by a Gaussian white noise process $\dot{w}(t)$. This leads to the stochastic differential equation

$$\partial_{\tilde{t}}\tilde{\rho} = \sqrt{\tilde{\rho}} - \tilde{\rho} - Q\tilde{\rho}\dot{w} \quad . \tag{28}$$

The 'noise amplitude' Q reflects the coarseness of plastic deformation. In scaled variables, Q is time-independent and proportional to $\hat{\gamma}/\tau$. The Fokker-Planck equation corresponding to the stochastic differential equation (28) can be solved analytically; using Ito calculus its steady-state solution is found as

$$p(\tilde{\rho}) = \frac{N}{\tilde{\rho}}\exp\left[\frac{2}{Q^2}\left(2\sqrt{\tilde{\rho}} - \tilde{\rho}\right)\right] \tag{29}$$

[6]Note that this transformation is again consistent with the scaling relations of the 'law of similitude'

where N is a normalization constant. The shape of this probability distribution depends on the value of the 'noise amplitude' Q which acts as a control parameter of the system. Below a critical noise amplitude $Q^2 = 1/2$ the probability density exhibits a maximum which, at small noise amplitudes, lies close to the average density. Above the critical noise amplitude this maximum disappears and gives way to a hyperbolic decay with an exponential cut-off. This ***noise-induced transition*** has been interpreted as a transition towards the formation of cell structures. This is illustrated in Figures 7 and 8.

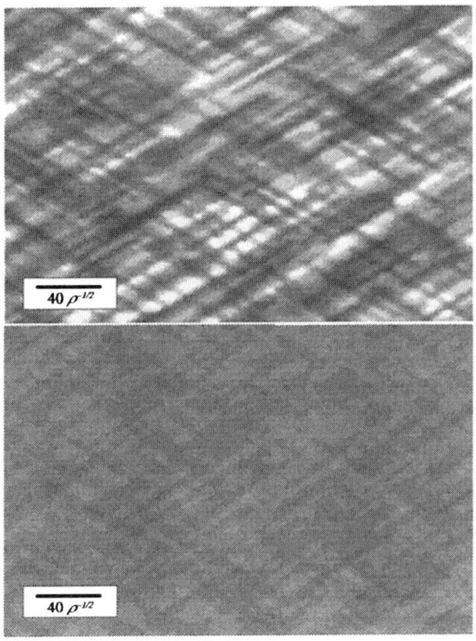

Figure 8. Dislocation density patterns obtained from simulations of the continuum stochastic dislocation dynamics model discussed in the text; top: large noise amplitude, bottom: small noise amplitude. The probability distributions of dislocation densities corresponding to the top and bottom patterns are given by the open and full symbols in Figure 7, respectively.

Figure 7 shows scaled probability distributions (all densities have been normalized by their strain-dependent average values). The full line has been calculated from Eq. (29) for $Q^2 = 0.1$, and the dotted line for $Q^2 = 2$. The data points show the results of simulations assuming discrete 'deformation events' with amplitudes corresponding to the same noise strengths. The simulations have been carried out by assuming that the function $g(\vec{r})$ in Eq. (27) decays exponentially as a function of $|\vec{r} - \vec{r}_i|$ with a decay length of one average dislocation spacing in the direction perpendicular to the glide direction and a decay length of 10 average dislocation spacings in the glide direction.

Spatial patterns obtained from these simulations are shown in Figure 8. The top image obtained for the large noise amplitude shows a pattern of dislocation-dense walls enclosing dislocation-depleted cell interiors. The walls follow the direction of the slip planes and the overall morphology of the pattern is very similar to the patterns observed in two-dimensional DDD simulations (Figure 4). For the small noise amplitude, on the other hand, the dislocation pattern is more or less homogeneous and exhibits no discernable features.

Irrespective of the technical details, stochastic dislocation dynamics approaches lead to an interpretation of dislocation patterning that is very different from the energetic approaches outlined earlier. According to the stochastic interpretation, dislocation patterning is a far-from-equilibrium phenomenon driven by fluctuating dislocation fluxes. These fluxes occur only if an external driving is provided in terms of an applied stress or imposed macroscopic strain rate. Due to dislocation interactions, the dislocation fluxes on mesoscopic scales are heterogenous in space and time and this heterogeneity is ultimately responsible for the emergence of heterogeneous dislocation patterns.

6. CONCLUSIONS

In spite of the fact that the spontaneous formation of dislocation patterns is an ubiquitous phenomenon in plastic deformation of crystalline solids, to date there exists no commonly accepted approach towards computational modelling and/or theoretical understanding of dislocation patterning. This observation is particularly astonishing in view of the fact that dislocation patterns obey simple scaling relations which can be observed over at least five decades in pattern 'wavelength'. Furthermore, the same scaling relations are observed to hold in widely differing materials (metals, ionic solids, semiconductors). In spite of this remarkable degree of universality in the phenomenology of dislocation patterns, a commonly accepted theory does not (yet) exist.

This is partly due to the fact that three-dimensional dislocation dynamics simulation, which is the most straightforward approach towards computational modelling of dislocation patterns, is still limited by the substantial computational cost of simulating systems of long-range interacting, flexible and reactive lines. Two-dimensional simulations reduce the computational cost but increase the phenomenological 'input' required in the models and are restricted to a limited set of deformation geometries. Continuum dislocation dynamics approaches constitute an alternative, but suffer from the fact that there exists no systematic procedure for coarse-graining systems of flexible lines with long-range interactions. Stochastic models reduce the computational cost by using 'mean-field-plus-fluctuation' approaches to the internal stresses and/or dislocation fluxes. In this case, the fundamental problem is that despite recent progress the fluctuation properties of driven many-dislocation systems are theoretically poorly understood, and therefore the models have to rely on crude approximations or heavy phenomenological input.

Is there a way out of this impasse? Of course, the exponential growth of available computing power will ultimately beat the power-law increase of computational cost as three-dimensional simulations are carried to larger system sizes, strains and dislocation densities. However, the experience of the last decade has shown that progress along this direction has not been as rapid than initially expected. In the opinion of the present author, an alternative strategy may be to further develop our understanding of the non-equilibrium statistical mechanics of systems of interacting lines in order to obtain coarse-grained descriptions and/or fluctuation properties of dislocation systems. Ultimately, it should not be necessary to know the position of every single dislocation segment in order to understand how dislocation patterns emerge.

ACKNOWLEDGEMENTS

The author expresses particular thanks Prof. Peter Rudolph for drawing his attention towards the formation of dislocation cell patterns in semiconductor crystal growth and for inspiring discussions on this subject. The present overview could not have been written without discussions, collaborations and controversies involving numerous other groups and individuals. These scientific interactions have been greatly helped by financial support of the European Commission through two TMR/RTN networks under Contracts No ERB-FMRX-CT-960062 and HPRN-CT-2002-00198, which hereby is gratefully acknowledged.

REFERENCES

1. H. Mughrabi, T. Ungar, S. Kienle and M. Wilkens, Phil. Mag. A 53 (1986) 793.
2. B. Gong, Z. Wang and Z.G. Wang, Acta Mater. 45 (1997) 1365, 1379.
3. Q. Liu, D. Juul Jensen and N. Hansen, Acta Mater. 46 (1998) 5819.
4. M. Zaiser, K. Bay and P. Hähner, Acta Mater. 47 (1999) 2463.
5. J.P. Hirth and J. Lothe, Theory of Dislocations, Wiley, New York, 1982.
6. B. Devincre and M. Condat, Acta Metall. Mater. 40 (1992) 2629.
7. D. Holt, J. Appl. Phys. 41 (1970) 3197.
8. R. Neuhaus and Ch. Schwink, Phil. Mag. A 65 (1992) 1463.
9. P. Rudolph, Journal of Crystal Growth, in the press. (Data presented at the 15th American Conference on Crystal Growth and Epitaxy, Keystone, July 20-24, 2003.)
10. N. Hansen and D. Kuhlmann-Wilsdorf, Mater. Sci. Engng. 81 (1986) 141.
11. M. Zaiser, Mater. Sci. Engng. A 309-310 (2001) 304.
12. M. Zaiser and A. Seeger, in: Dislocations in Solids, Vol. 11, eds. F.R.N. Nabarro and M.S. Duesbery (North-Holland, Amsterdam, 2002) p. 1.
13. N.M. Ghoniem, S.H. Tong and L.Z. Sun, Phys. Rev. B 61 (2000) 913.
14. O. Politano and J.M. Salazar, Mater. Sci. Engng. 309 (2001) 261.
15. B. Devincre and L.P. Kubin, Mater. Sci. Engng. A 234-236 (1997) 8.
16. R. Madec, B. Devincre and L.P. Kubin, Scripta Mater. 47 (2002) 689.
17. R. Madec, B. Devincre and L.P. Kubin, Phys. Rev. Letters 89 (2002) 255508.
18. A.A. Benzerga, Y. Brechet, A. Needleman and E. Van der Giessen, Model. Simul. Mater, Sci. Engng. 12 (2004) 159.
19. J.W. Cahn and J.E. Hilliard, J. Chem. Phys. 28 (1958) 258.
20. M. Zaiser, M.-C. Miguel and I. Groma, Phys. Rev. B 64 (2001) 224102.

21. H. Haken, Synergetics (Springer, Berlin, 1978).
22. G. Nicolis and I. Prigogine, Self-Organisation in Non-Equilibrium Systems (Wiley, New York, 1987).
23. E.C. Aifantis, Mater. Sci. Engng. 81 (1986) 563.
24. D. Walgraef and E.C. Aifantis, Int. J. Engng. Sci. 23 (1985) 1351, 1359, 1365.
25. E.C. Aifantis, Int. J. Plasticity 3 (1987) 211.
26. J. Kratochvil, Scripta Metall. Mater. 26 (1993) 113.
27. I. Groma and B. Bako, Phys. Rev. B 58 (1998) 2969.
28. I. Groma and B. Bako, Phys. Rev. Letters 84 (2000) 1487.
29. P. Hähner, Appl. Phys. A 62 (1996) 473.
30. M.-C. Miguel, A. Vespignani, S. Zapperi, J. Weiss amd J. Grasso, Nature 410 (2001) 667.
31. M.-C. Miguel, A. Vespignani, M. Zaiser and S. Zapperi, Phys. Rev. Letters 99 (2002) art. no. 165501
32. H. Neuhäuser, in: Dislocations in Solids, Vol. 6, ed. F.R.N. Nabarro, North-Holland, Amsterdam, 1984, p. 319
33. P. Hähner, K. Bay and M. Zaiser, Phys. Rev. Letters 81 (1998) 2470.
34. M. Zaiser, Mater. Sci. Engng. A 249 (1998) 145.

Crystal Growth – from Fundamentals to Technology
G. Müller, J.-J. Métois and P. Rudolph (Editors)

Silicon crystal growth

W.v.Ammon

Siltronic AG, P.O. Box 1140, 84479 Burghausen, Germany

Silicon crystal growth technology has rapidly advanced during the past decades. One of its striking features has been the diameter race which was driven by the demand for cost reduction in the device industry. However, there are now strong indications that this almost periodic increase of the crystal diameter will considerably slow down in the future. The main reason are the dramatically rising costs for crystal growth. The latter is a consequence of the technological challenges which arise from the required large charge sizes. The recently introduced 300mm diameter is expected to stay for at least one decade before the next diameter jump to 450mm will occur. The other remarkable aspect of the silicon technology is the meanwhile rather thorough understanding of the most relevant bulk defects and their behavior in device processes which has been predominantly developed during the last 10 years. In particular, the aggregation of vacancies which results in the formation of microvoids that are increasingly harmful to device performance with shrinking design rules, can be simulated with great accuracy, today. Besides the Czochralski (CZ) method, which accounts for nearly 95% of the total monocrystalline silicon production, the floating zone (FZ) technique is commerically used for niche products, such as high ohmic material for power and high frequency devices. The diameter increase of FZ crystals has proceeded slower as compared to CZ crystals which is mainly related to the complex high frequency inductive heating of the FZ technology and the availability of larger poly silicon feed rods of appropriate quality. Recently, 200mm FZ crystals have been successfully grown.

1. GENERAL ASPECTS OF SILICON CRYSTAL GROWTH

The development of the silicon crystal growth technology has been driven by two major challenges, i.e. the demand for larger wafer diameters, on one hand, and the need to improve the bulk quality of the crystals, on the other hand. The Czochralski (CZ) method has proven to be most suitable to follow this rapid development. Today, it is estimated that about 10.000 metric tons of CZ crystals are produced per year. However, the floating zone (FZ) technique is also well established as it provides crystals with superior purity which cannot be produced by the CZ method. Nevertheless, FZ crystals have remained as a niche product with an approximate share of only 5% of the total silicon wafer sales.

Since the beginning of the integrated circuit (IC) technology, the area of silicon wafers and, as a consequence, the diameter of silicon crystals has been increased in nearly periodic steps of several years. This seemingly never ending diameter race is related to the significantly higher complexity of every new IC generation which cannot be fully compensated by shrink ing design rules. This results in larger chip sizes and, hence, less chips per wafer. However, if the number of chips per wafer drops below a certain limit, the IC manufacturing costs rapidly rise and the IC industry is forced to changeover to the next larger wafer diameter. As will be shown in section 6, the lifetime cycle of a wafer diameter has become longer during the last decade and will further be extended, as the technological challenges meanwhile dramatically increase with every new diameter generation.

With regard to the bulk quality, the device manufacturers were content with silicon crystals of a defined dislocation density in the early days of silicon based semiconductor technology. However, as the crystal diameters increased beyond one inch, it soon became apparent that crystals with a defined dislocation density were neither economically nor technologically feasible. Fortunately, the invention of the seed necking technique by Dash [1,2] allows to grow fully dislocation-free crystals which are now standard in the silicon semiconductor world. Nevertheless, this was only the beginning of a still ongoing challenging period of intensive investigations devoted to the understanding and control of the defect behavior in bulk silicon.

Besides the crystalline quality, the bulk purity is of utmost importance for the device industry. In particular, metallic impurities must be reduced to extremely low residual concentrations. This is related to their mostly very low solubility in silicon which may result in the formation of metal silicides in the near surface region of the wafer during device processing. The precipitation of these silicides can severly damage device performance and must therefore be avoided by all means. Furthermore, many metals form deep traps in the energy band gap of the electrons and, thus, shorten the minority carrier lifetime.

On the other hand, certain impurities which define the electrical resistivity and the conductivity type of silicon must be added and incorporated during crystal growth. For p-type silicon, boron is usually the preferred dopant, while phosphorus, arsenic and antimony are used for n-type. The various resisitivity ranges specified by the device manufacturers vary over 5-6 orders of magnitude! For example, power devices may require either very low (few mOhmcm) or very high (nearly 1000 Ohmcm) resisitivities. The silicon wafers typically needed for memory and logic devices are in the medium range of 1 - 20 Ohm cm. For very high voltage devices like thyristors, the specified target value must be met with extreme accuracy along with a homogeneous resistivity profile across the entire wafer. In that case, the resistivity is adjusted by neutron transmutation doping in nuclear research reactors [3,4]. This technique makes use of the fact that silicon contains about 3.1% of the isotope ^{30}Si which is transformed to ^{31}Si by capture of a neutron. Subsequently, ^{31}Si rapidly decays to phosphorus ^{31}P by emission of an electron (2.6 h). The reaction can be very well controlled by neutron flux measurements and the irradiation time. Neutron transmutation doping is not applicable for p-type material, as no suitable reaction for a p-type dopant is available. Moreover, the technique is rather expensive and irradiation capacities are limited.

For the vast majority of the devices, the specified homogeneity of the dopant distribution is somewhat more relaxed, but still challenging. The incorporation of impurities for standard doping methods is mainly determined by the melt convection and the segregation of the dopant. The latter is a consequence of the phase diagram of impurities in liquid and solid silicon. For silicon, the segregation coefficient k_0, i.e. the ratio between the impurity concentrations in the solid and liquid under equilibrium conditions, is always less than one, - with the exception of oxygen where the segregation coefficient is still under discussion. During crystal growth, the advancing solid/liquid interface forms a boundary layer in the liquid phase where the dopant piles up in front of the solid interface and, thus, the above equilibrium conditions are not valid. Therefore, an effective segregation coefficient k_{eff} has to be used which depends on the growth rate. While the macroscopic pull rate can be kept rather constant, in particular in FZ growth processes, the microscopic growth rate is very difficult to control and shows strong fluctuations. This results in pronounced resistivity striations with a spacing of a few hundred μm or less which are the larger the smaller k_{eff} is. The macroscopic resistivity variation is predominantly affected by the melt convection and can be easier improved as compared to the microscopic striations. As most of the semiconductor devices require p-type silicon, the device industry can take advantage of the fact that boron is the only impurity which has a segregation coefficient close to 1 and, thus, p-type silicon exhibits only small resistivity variations.

In the following sections, we will further elucidate the defect behavior of silicon and then describe the current status of the CZ and FZ growth technology. Sections 3 – 5 will be published in more detail in ref.5.

2. TECHNOLOGICAL RELEVANCE OF CRYSTAL DEFECTS

During crystal growth, intrinsic point defects, i.e. vacancies and Si interstitials, are incorporated in a concentration of the order of 10^{13} - 10^{14} at/cm^3. While single point defects are so far not known to cause any adverse effects on device perfomance, their aggregation to large clusters can have severe impacts or be even detrimental to device functionality. This is also true for extended defects like dislocations which, however, disappeared as an issue with the introduction of dislocation-free crystal growth. On the other hand, the generation of dislocations in silicon *wafers* during high temperature heat treatments persists as a challenge and must be carefully taken into account by the design of the thermal processes and the related hardware.

As the shrink of the design rule progressed along with the demand for higher device performance, it soon became apparent that intrinsic point defects and their aggregation during the cool down phase of the crystal growth process has an increasingly negative impact on device performance and yield. Historically, one of the first severe challenges in this regard was the aggregation of Si interstitials in floating zone grown crystals which generated socalled A-swirl [6] or L-pit defects [7]. Although, the dimension of these defects only reaches several μm , they severely affected the performance of high voltage devices [8, 9]. In the second half of the 80ties, the industry began to face increasing problems with early breakdowns of the gate oxide in memory devices [10]. After intensive gate oxide integrity (GOI) investigations, it was found that the root cause for the gate oxide degradation was vacancy aggregates, so-called voids [11,12]. After wafer polishing, these voids show up as dimples or laser light scattering (LLS) defects on the wafer surface and, there, cause a local thinning of the gate oxide [13,14]. The size of voids is considerably smaller (ca. 150 nm) as compared to A-swirl defects and, thus, their impact on device performance is only seen if the locations of a void and an active element, e.g. a transistor, coincide. In addition, most of these single defective transistors can be repaired by the built-in redundancy in memory chips. Consequently, vacancy aggregates are tolerable for many devices, in particular, as long as their density is insignificant as compared to that of device process induced defects. This is in contrast to A-swirl or L-pits, as they always result in unrepairable damage to devices owing to their large size.

Empirically, it has been found that oxide layers of 40 - 50 nm thickness are most susceptible to void defects [14]. Thinner oxides show higher GOI yields, and, at a thickness of less than 5 nm, the influence of voids on the GOI yield disappears [15,16]. However, as the design rule approaches the size of voids, additional adverse effects have been identified, such as shorts between trench capacitors, lack of device reliability, etc. [17,18]. Thus, the device manufacturers may have to switch to silicon material which is essentially free of intrinsic point defect aggregates. Consequently, it is of utmost importance, that the silicon manufacturers develop a thorough understanding of the mechanisms of defect aggregation to taylor the bulk properties to the needs of the device industry. In particular during the past decade, tremendous progress has been achieved in this field which will be discussed in the next sections.

3. THERMOPHYSICAL PROPERTIES OF INTRINSIC POINT DEFECTS

In the past, the thermophysical properties of intrinsic point defects, i.e. vacancies and Si intersititials, were poorly understood and the respective data spread over many orders of magnitude [19]. However, in recent times, the experimental as well as the theoretical methods have been substantially improved to the point that the uncertainty of the values is now reduced to less than 10%.

Most of the experimental data of thermophysical properties are inferred from metal diffusion experiments as the diffusion rates of some metallic impurities are coupled to the mobility and concentration of intrinsic point defects (kick-out and Frank-Turnbull mechanism). Valuable data are also obtained from rapid thermal annealing (RTA) experiments with fast cooling rates which reveal the influence of the free vacancy concentration on oxygen precipitation [20,21]. Moreover, the careful analysis of the defect behavior of crystals grown with smoothly or abruptly changing pull rates allowed to further narrow down the uncertainty range of many data points.

There have been numerous attempts to compute point defect thermophysical properties directly by atomistic simulations. The most important element of these calculations is the accuracy of the assumed interatomic interaction. In most cases, empirical potentials , e.g. Stillinger-Weber or Tersoff potential functions, are used which disregard explicit representations of electronic interactions. An excellent overview about the current status is found in [22]. It should be noted, that the equilibrium concentrations of vacancies and Si interstitials are rather similar with a slight excess of vacancies over Si interstitials (ca. 30% at the melting point of silicon). On the other hand, the diffusivity of vacancies is significantly slower as compared to Si interstitials in the upper temperature range. Vacancies and Si interstitials can also exist in various charged states (e.g. V^{2+}, V^{+}, V^{0}, V^{-},V^{2-}), but, up to now, there is no indication that charged states have any impact on defect aggregation and, hence, they are not considered in current defect nucleation models.

Only few experimental data are available which relate to the recombination coefficient of vacancies and Si interstitials [23,24]. Nevertheless, there is general agreement that recombination between vacancies and Si interstitials is so fast that the product of both species concentrations is always in thermal equilibrium. The exact value is therefore not relevant for our further discussion.

4. AGGREGATES OF INTRINSIC POINT DEFECTS

4.1. Experimental observations

The dominant intrinsic point defect aggregates in silicon single crystals are the beforementioned voids and L-pits/A-swirl. Figure 1 shows a TEM image of a void with a 5nm oxide layer on the inner surface. The average size for standard growth processes typically ranges from 70 to 200 nm. A strain field around the void is generally not observed. Usually, voids exhibit an octahedral morphology, but they can change to a platelet or even rodlike shape, if the crystal is doped with nitrogen (see section 3.4.1). Under standard growth conditions, twin voids consisting of two partial octahedral voids are predominantly observed [25,26]. At lower oxygen contents and higher cooling rates of the growing crystals, single octahedral voids are formed preferrably [27,28]. Historically, different notations were introduced for this type of defect depending on their delineation or detection techniques: crystal originated particle (COP) [29], laser light scattering tomography defect (LSTD)[7], flow pattern defect (FPD)[30] and D-defects [31]. Today, it is generally accepted that they all denote the same defect type, i.e. vacancy aggregates. In Figure 2, the TEM image of a L-pit is depicted.

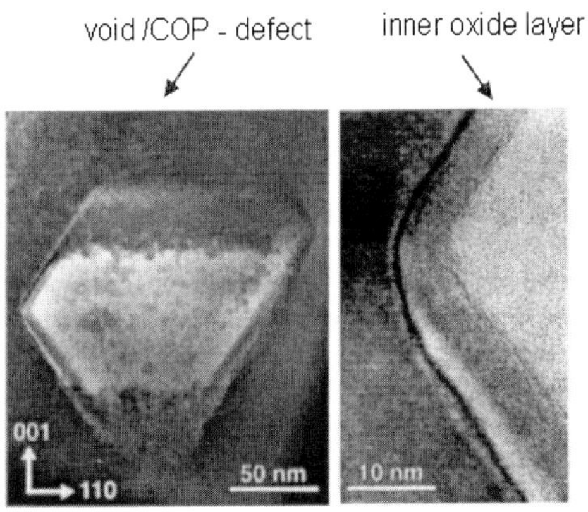

Figure 1. TEM picture of a void/COP (left). The inner surface of the void is covered by an oxide layer (right) [32].

The network of perfect dislocation loops which is characteristic for this defect type is clearly visible. The core of the defect is a self-interstitial aggregate which forms an extrinsic stacking fault [33]. If the stacking fault reaches a critical size the strain exerted on the crystal lattice is relaxed by the generation of dislocation loops. The socalled B-swirl as observed in FZ crystals [34,35,36] is believed to be related to such self-interstitial aggregates of a size below the critical limit [37,38].

As the voids are stabilized by the oxidation of their inner surface they are very difficult to annihilate. During a high temperature treatment (ca. 1200°C), it is therefore necessary to first lower the oxygen content by outdiffusion in order to dissolve the oxide layer and then the void itself. Today, annealed wafers that offer a void free subsurface region of more than

$10\mu m$ have meanwhile become a standard product. L-pits/A-swirl, on the other hand, are extremely stable and cannot be destroyed by annealing.

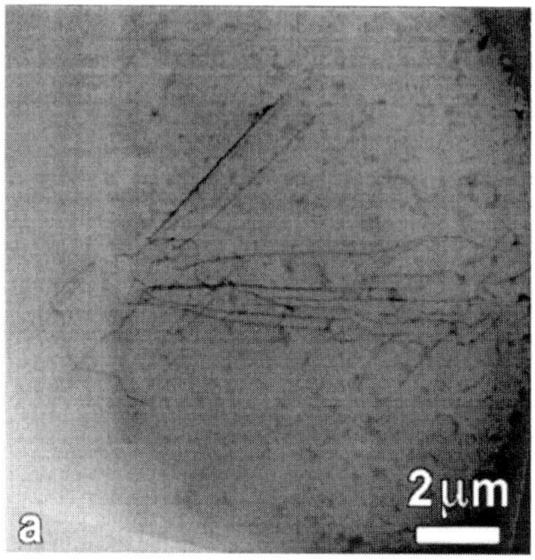

Figure 2. TEM picture of a L-pit [39].

It is well established that the pull rate V and the axial temperature gradient G at the growth interface of the growing crystal has a dominant influence on the defect types that develop in the growing crystal [40,41,42] and their spatial [43] as well as their density/size distribution [44]. At high pull rates, vacancy related defects, i.e. voids, are observed in the entire crystal volume (Figure 3). When the pull rate is reduced, oxidation induced stacking faults (OSF´s) develop in a small ring-like region near the crystal rim [45]. If the pull rate is further lowered, the ring diameter shrinks and, at the same time, L-pits are detected in the area outside the OSF ring [7]. No voids are found in this outer region. Thus, the OSF ring obviously represents the spatial boundary between vacancy and self-interstitial type defects. At a critical pull rate, the void region and the OSF ring disappear completely and L-pits are detected only. CZ and FZ crystals exhibit the same defect behavior but at different pull rate regimes. Due to their inherently low oxygen content, no OSF ring, but a defect free zone is observed as a boundary in FZ crystals [46].

The radial position of the OSF ring can approximately be described by the equation [43]

$$V/G(r) = \xi_{tr} = 1.34\text{x}10^{-3}\text{cm}^2\text{K}^{-1}\text{min}^{-1} \quad (1)$$

r being the radial position. Thus, the parameter V/G controls the type of grown-in defects: if $V/G > \xi_{tr}$, vacancy aggregates develop and, for $V/G < \xi_{tr}$, Si interstitial related defects are observed. This was first recognized by V.V.Voronkov in 1982 [40].

A further important parameter for the control of grown-in defects is the thermal history of the crystal. The data in Figure 4 clearly demonstrate that the void density increases with higher cooling rate. The defect density was derived from GOI measurements [47] which allow to detect much smaller defect sizes than the standard delineation techniques based on etching and/or laser light scattering.

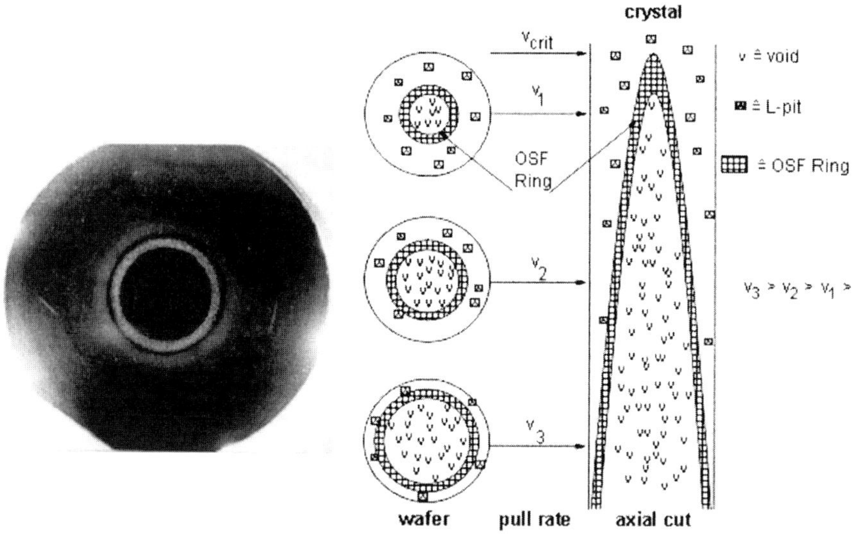

Figure 3. Variation of the radial defect behavior of a CZ crystal as a function of pull rate V ($V_1 < V_2 < V_3$). The insert on the left shows a wafer from the crystal part grown with medium pull rate. The defects were delineated by Secco etching [5].

It was further observed that the average void size decreases with higher cooling rate and vice versa (Figure 5) [44]. A striking feature is also the characteristic change in the defect density/size distribution across the crystal diameter (Figure 6) [48]. While, in the crystal center, large voids of low density prevail, the distribution gradually shifts to smaller sizes and higher density towards the boundary of the void region.

Transient growth experiments, where the growth process was abruptly halted for some time and then resumed, revealed that the nucleation temperature of voids lies between 1000 and 1100°C and varies with the growth conditions. It was also demonstrated that the nucleation and growth period of the voids occurs in a temperature interval of less than 50 K [49]. Void growth is then obviously stopped by the formation of the inner oxide layer. The oxide layer continues to grow down to relatively low temperatures. The rare data available for L-pits indicate a similar temperature range of nucleation [50]. Unfortunately, no defect density data as a function of the cooling rate are available for L-pits. The behavior of this defect type is more complex as small interstitial aggregates which do not generate dislocation loops are not detected as L-pits.

4.2. Theoretical model: Incorporation of intrinsic point defects

According to the generally accepted model of V.V.Voronkov [40], the incorporation of intrinsic point defects is a result of two competing fluxes at the growth interface: a vacancy and an interstitial flux which are both directed into the crystal (Figure 7). Each of the two fluxes

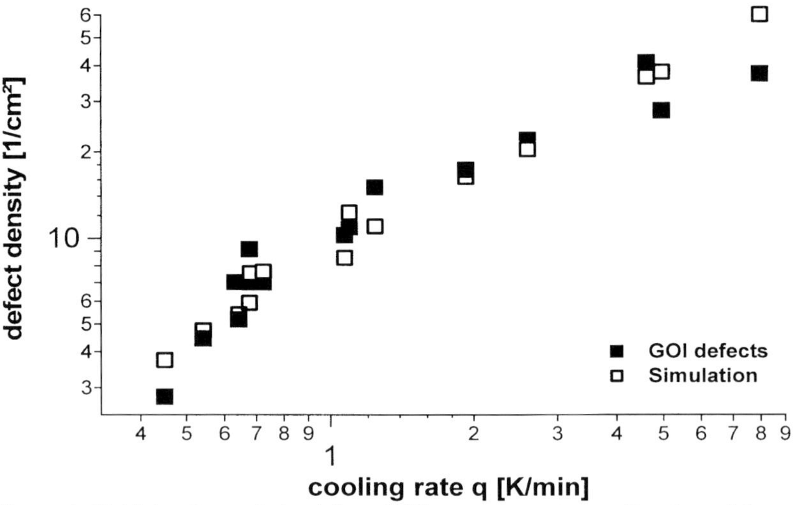

Figure 4. Void density as derived from GOI measurements as a function of the cooling rate at ca. 1100°C during growth in comparison to simulation results. The various data points relate to different growth processes. Note that the void density is shown as an area density as the GOI measurement yields area related defects.

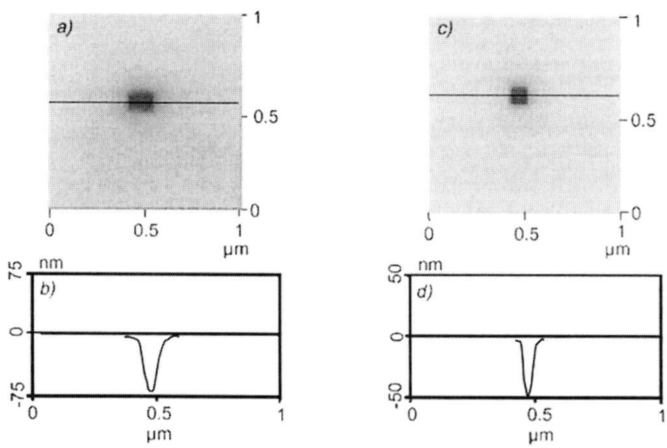

Figure 5. Atomic force microscope measurements of voids at the wafer surface. The depth of the surface void is rather close to its original size in the crystal. Samples are from a slowly cooled (left) and a fast cooled crystal (right) [51].

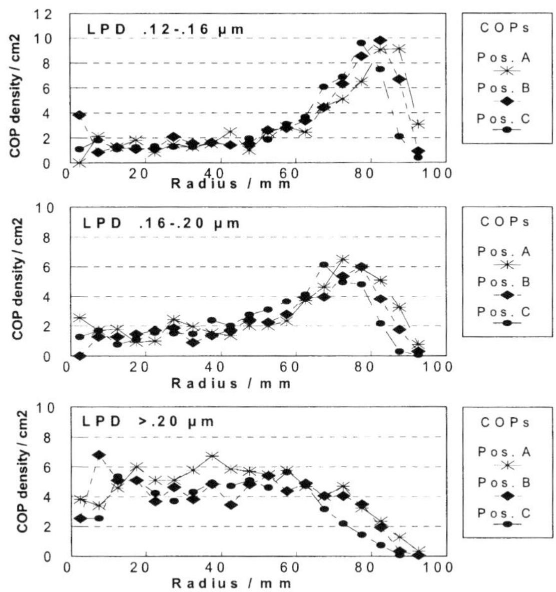

Figure 6. Radial variation of the density/size distribution of voids/COP's across a wafer measured by laser light scattering. The light point defect size (LPD) correlates well to the actual void size. The positions A,B and C relate to different axial crystal positions [5].

consists of two components: The first flux component is driven by the advancing growth interface, which generates a convective flux proportional to the pull rate V, and the second is driven by the vacancy/interstitial recombination behind the interface which creates a concentration gradient and, in turn, a Fickian diffusion flux. The latter is proportional to G. If the pull rate is low, then the Fickian diffusion dominates over the convective flux. As self-interstitials diffuse significantly faster than vacancies, the self interstitial flux wins over the vacancy flux. During crystal cooling, V-I-recombination virtually eliminates the vacancies and the surviving self interstitials are driven into supersaturation and finally aggregate. On the other hand, at fast pull rates, the convective flux dominates and, to due the larger equilibrium concentration of vacancies as compared to that of self interstitials, now more vacancies flow into the crystal resulting in vacancy aggregates. A similar change of the prevailing defect type is obtain when G is varied.

The above approach yields the previously mentioned parameter V/G which determines the type of defect that develops in a growing Si crystal. The critical transition value ξ_{tr} at which vacancy and Si interstitial flux is equal can be calculated from the expression [19]

$$\xi_{tr} = \frac{D_I(T_m)C_I^{eq}(T_m)\gamma_I - D_V(T_m)C_V^{eq}(T_m)\gamma_V}{T_m[C_V^{eq}(T_m) - C_I^{eq}(T_m)]} = 1.38 \times 10^{-3}\ \mathrm{cm^2 min^{-1} K^{-1}} \quad (2)$$

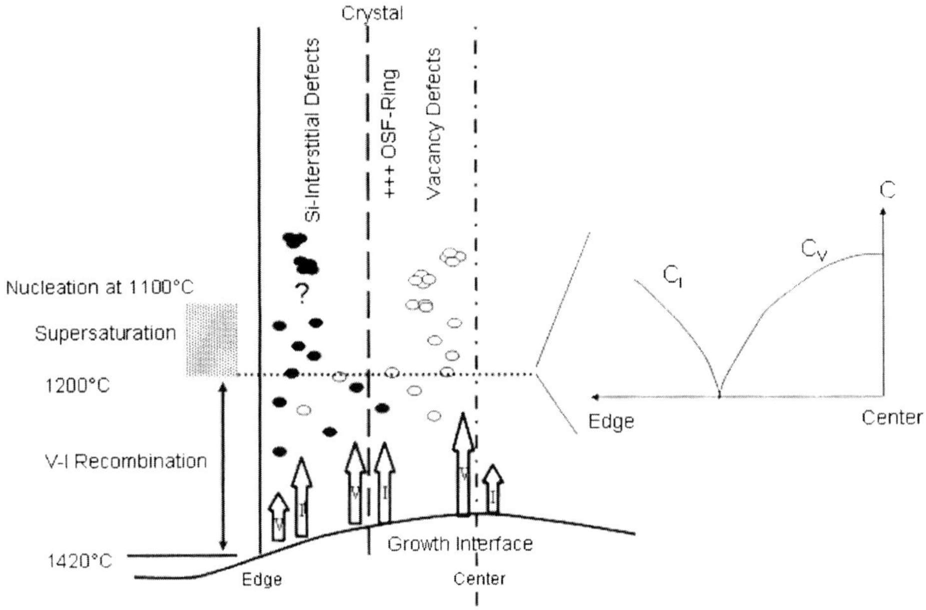

Figure 7. Schematic picture of the incorporation of vacancies (V) and Si interstitials (I), respectively, into the growing crystal. The insert on the right shows the radial variation of the remaining species after VI-recombination has ceased and supersaturation starts [5].

where γ_I and γ_V are dimensionless effective formation enthalpies and T_m denotes the melting temperature of silicon. The above theoretical value is rather close to the experimental data.

In standard growth processes, G always exhibits a notable increase from the center towards the crystal rim. This effect is a consequence of the effective heat loss by radiation just above the triphase junction at the crystal rim, while, in the crystal center, the heat must be removed by thermal conduction. At medium pull rates, this radial behavior of G results in a vacancy excess in the center, while Si interstitials dominate at the crystal rim as indicated in Figure 8, which gives a simple explanation for the experimentally found concentric defect regions and their variation as a function of pull rate. The exact value of ξ_{tr} is still under discusion because G has to be taken directly at the interface where the thermal field in the crystal changes very rapidly [52,43]. Therefore, it has not been possible up to now to accurately measure G at the growth interface [53]. Fortunately, computer simulations with sufficiently refined meshes allow a rather accurate calculation of G. Based on these calculations, a value of 1.3 x10^{-3} $cm^2min^{-1}K^{-1}$ has been determined from growth experiments with hot zones of different G´s [41] which is meanwhile widely accepted.

It should be noted that the quantitative relationship between the relevant thermophysical properties is rather peculiar for silicon as it allows the changeover in defect behavior to occur in accessible growth rate regimes.

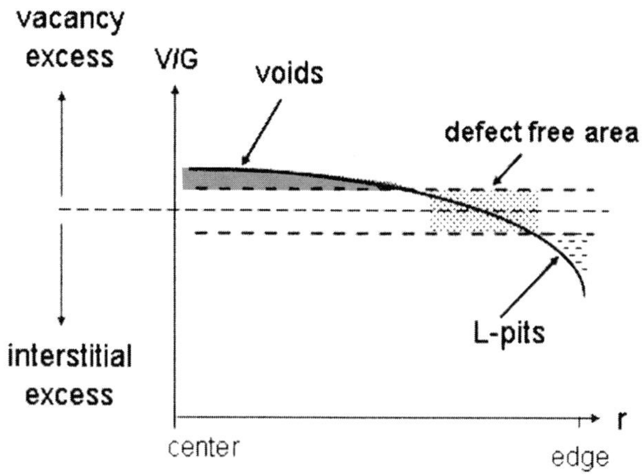

Figure 8. Schematic radial variation of defect types as a function of the radial variation of V/G.

The incorporated vacancy concentration can be calculated from a simple expression [54,55]

$$C_V(r) \approx \left(C_V^{eq}(T_m) - C_I^{eq}(T_m)\right) \cdot \left[1 - \frac{\xi_{tr} G(r)}{V}\right] \tag{3}$$

This expression is also applicable for Si interstitials (at $V/G < \xi_{tr}$). It then defines the incorporated Si interstitial concentration C_I with a minus sign.

If V/G approaches ξ_{tr}, i.e. the vacancy and Si-interstitial flux become equal, both species are annihilated by V-I recombination. Thus, the incorporated V and I concentration is negligible and no aggregates can form. This particular condition must be met for the growth of socalled perfect silicon which has recently been introduced in mass production.

4.3. Theoretical model: Aggregation of instrinsic point defects

Most aggregation models have focused on the nucleation and growth of voids as they are technologically most relevant. Since voids are not obscured by secondary defects, the experimental verification of the simulation results is also easier and more reliable. From Figure 9, it can be seen that, for a standard growth process, the V-supersaturation in the region with vacancy excess already begins to build up at around 1200°C [56] which is well above the experimentally determined nucleation temperature of voids at ca. 1100°C. Thus, the supersaturation C_V / C_V^{eq} necessary to nucleate voids is appreciable, i.e. ca. 10 [19] as estimated from Figure 9. This fact also allows to decouple the phase where V-I recombination is the major vacancy loss mechanism from the phase where nucleation dominates the vacancy loss, i.e. one can assume that the cluster formation starts with a fixed concentration which can be calculated from (3).

The first vacancy aggregation calculations were carried out with a model which was originally developed to simulate oxygen precipitation [57,44] . For the formation of small clusters (less than 20 vacancies), the kinetics are described by chemical rate equations, while, for larger clusters, the rate equations are expanded into a continuum formulation which yields a single Fokker-Planck equation. The void shape is assumed to be spherical. Input data are the computed temperature field, which can be quite accurately calculated by commercially available standard codes, and the surviving vacancy concentration after V-I recombination has cea-

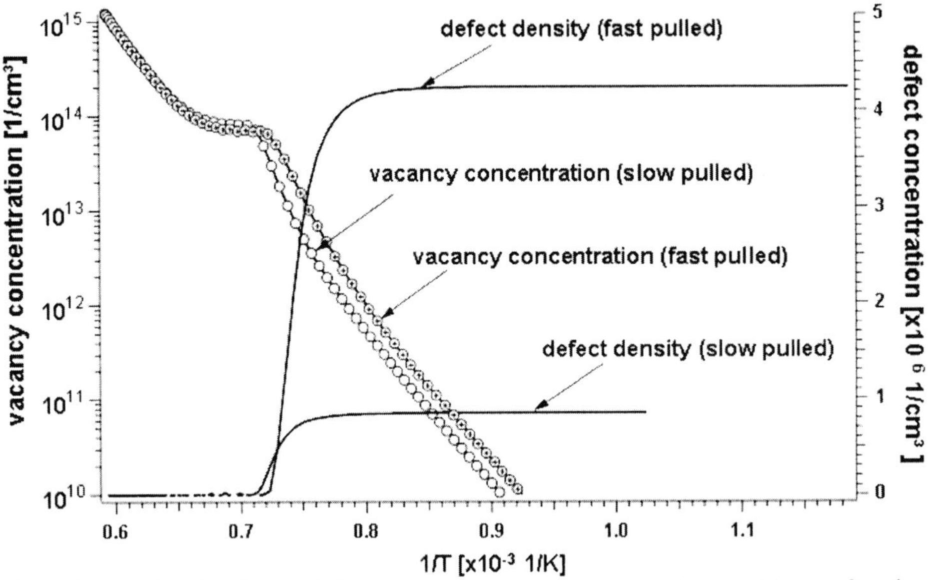

Figure 9. Simulated variation of the vacancy concentration in growing crystals as a function of temperature for a fast and a slowly cooling growth process. The density of voids nucleated at ca. 1100°C and below is shown on the right scale [5].

sed. Besides the physical properties of vacancies and Si interstials also the surface energy must be known which is estimated to be around 950 erg cm^{-2}.

As depicted in Figure 9, the vacancy concentration starts to drop rapidly at ca. 1100°C which is characteristic for the beginning of void nucleation. The correct prediction of the nucleation temperature requires a proper modeling of the initial nucleation process. A simple model which assumes the direct nucleation of voids yields a far too low aggregation temperature with the above input data as compared to the experimentally observed value [58,59]. A recent model proposes that the nucleation starts as a rather "amorphous cloud" of vacancies which quickly relaxes into an octahedron with (111) facets [60]. Atomistic calculations of this initial nucleation process indeed shift the calculated aggregation temperature into the experimentally observed range.

The results in Figure 9 also demonstrate, that a higher cooling rate obviously decreases the nucleation temperature. The origin of this effect is the higher supersaturation which builds up as there is less time for vacancy consumption. Due to the shift in the nucleation temperature, the residual vacancy concentration after void formation remains higher for fast cooling rates as compared to slow cooling. This has an important effect on the nucleation of oxygen precipitates as the latter is strongly enhanced by a higher free vacancy concentration [61,62].

The higher supersaturation at fast cooling rates also results in a larger nucleation rate as more aggregates have to be formed to effectively lower the supersaturation and the chemical potential, respectively. On the other hand, the aggregates have less time to grow and, therefore, remain small. Thus, the density of voids increases, but their size is reduced with higher cooling rates and vice versa. As shown in Figure 4, the experimental data are in excellent agreement with the simulation results.

A down shift of the nucleation temperature is also obtained when the initial vacancy concentration before the onset of nucleation is lowered [63]. Hence, more clusters of smaller size are formed again. The behavior of the size distribution as a function of the initial vacancy concentration gives a simple explanation for the experimentally found radial variation of the void size distribution in Figure 7. As the initial vacancy concentration decreases towards the crystal rim due to the decrease of V/G (Figure 8), the average size of the voids shrinks, but their density grows. This behavior is also fairly well reproduced by simulations [48].

An expression of the void density as a function of the cooling rate and the starting vacancy concentration can be found by a rather simple approach. If nucleation has stopped and the residual vacancy concentration is only further reduced by the growth of voids, the simulation results show that the vacancy supersaturation S remains fairly constant (neglecting the oxidation of voids). Thus, we can write

$$C_V(t) \approx S \cdot C_V^0 \cdot exp(\frac{-E_V}{kT}) \tag{4}$$

where E_V is the formation enthalpy for a vacancy and $T(t) = T_n - qt$ with T_n as the nucleation temperature and q the cooling rate. The vacancy loss is determined by

$$\frac{dC_V}{dt} = -4\pi \cdot D_V \cdot C_V \cdot R_{eff} \cdot N \tag{5}$$

with R_{eff} as an effective average void radius and N as the void density after nucleation is finished. If (4) is inserted into (5) we obtain

$$\frac{-E_V \cdot q}{k(T_n - q \cdot t)} = -4\pi \cdot D_V \cdot R_{eff} \cdot N \tag{6}$$

When the growth of voids has sufficiently slowed down, R_{eff} remains nearly constant and can be written as

$$R_{eff} = \left(\frac{3 \cdot C_V^i}{4\pi \cdot \rho \cdot N}\right)^{\frac{1}{3}} \tag{7}$$

where C_V^i is the initial vacancy concentration before nucleation has started and ρ is the silicon number density. If (7) is inserted into (6) we arrive with a simple expression where the void density is proportional to $q^{3/2} * C_V^{i\ -1/2}$. This result has first been derived by Voronkov et al. in a somewhat different way [64]. As shown in Figure 10, it is obviously a rather good approximation.

Most of the current aggregation models do not consider the oxidation of the inner void surface which stops void growth before the vacancies are essentially depleted. This has less impact on the void size distribution, but considerably affects the residual vacancy concentration and, hence, oxygen precipitation (see section 5).

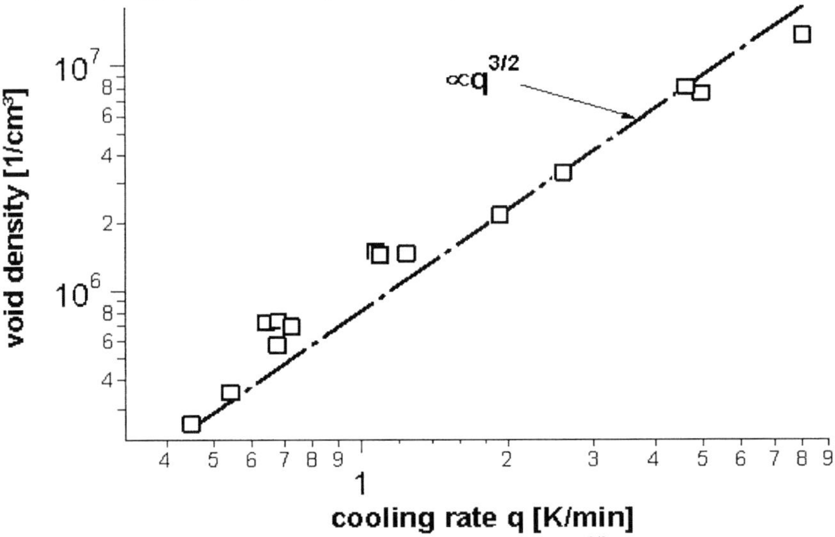

Figure 10. Simulated void density in comparison with the $q^{3/2}$ law. The simulated data points are the same as in Figure 4 but are transformed into a volume density.

The occurence of double voids is still under discussion. One attempt to explain this phenomenon is based on a partial oxidation of the (111) facets during void growth which allows the incorporation of further vacancies at the unoxidized residual facet surface [65].

4.4. Effect of impurities on intrinsic point defect aggregation

4.4.1. Nitrogen

Nitrogen doping of silicon has gained considerable technological interest during the last years. While nitrogen has long been known to simultaneously suppress interstitial and vacancy related defects, A-swirl and D-defects/voids respectively [66], in floating zone (FZ) grown crystals, it has been an undesirable dopant in Czochralski (CZ) grown silicon as it enhances the formation of oxidation induced stacking faults (OSF) [67]. In addition, the desirable suppression of intrinsic point defect aggregation by nitrogen is largely offset by the interaction of nitrogen and oxygen [68,5]. However, in more recent times, it was discovered that nitrogen still has a notable effect on the defect size, i.e. the higher the nitrogen concentration the smaller the defect size [47,69]. This effect is very favorable for defect annealing [8]. It was also observed that nitrogen not only reduces the size, but also systematically changes the void morphology from octahedron to platelet and, finally, to rods or even rod clusters with increasing concentration [70] (Figure 11). In addition, the strong enhancement of oxygen precipitation by nitrogen doping [71,72,73] can be used to improve the gettering efficiency of metallic contaminants in low thermal budget device processes [74,75].

As the necessary nitrogen concentration for the suppression of A-Swirl and D-defects is in the same range as the intrinsic point defect concentration before aggregation it was suggested that nitrogen has a direct interaction with vacancies (V) and Si interstitials (I) and, thus, pre-

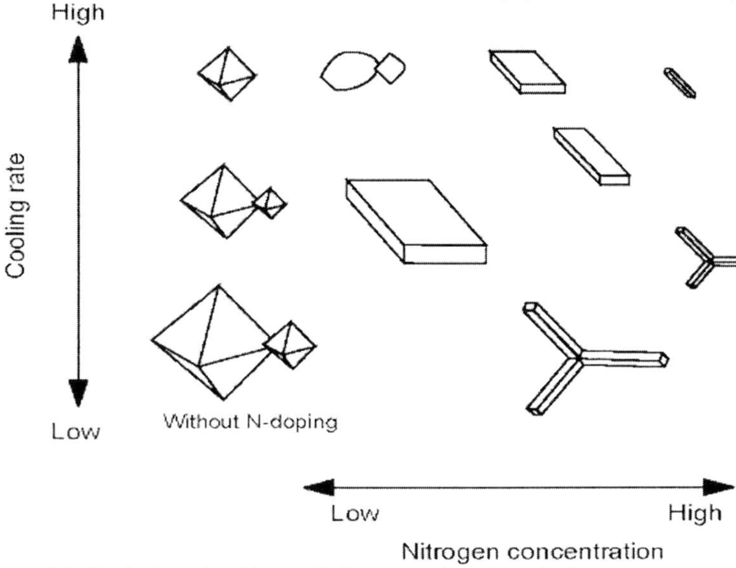

Figure 11. Variation of void morphology as a function of nitrogen concentration and cooling rate of the growing crystal [70].

prevents their aggregation [76,77]. First principle calculations of Sawada et al. and Kageshima et al. [78,79] have confirmed earlier experimental results that nitrogen exist as a dimer in silicon. They have further shown that the N-N dimer can form N_2V and N_2V_2 complexes, the latter being more stable due to a larger enthalpy of formation. Based on these results, it has been proposed that the vacancy aggregation is suppressed by the reaction

$$N_2 + V \leftrightarrow N_2V \quad \text{(I)}$$

where as the formation of Si interstitial agglomerates is prevented by the reaction

$$N_2V + I \leftrightarrow N_2. \quad \text{(II)}$$

As vacancies and Si interstitials have very similar concentrations at the melting point, reactions I) and II) mainly act as an additional V-I recombination path during crystal cooling with only marginal impact on the N_2 and N_2V level [80]. However, when recombination is completed and either vacancies or Si interstials have died out, reaction I) or II) prevents the aggregation of the surviving species. While this model works well for FZ crystals, it does not explain why the nitrogen effect on defect aggregation is strongly diminished in CZ crystals. It seems rather evident that the well-known interaction between nitrogen and oxygen, which results in the formation of N_xO_y complexes [81], probably is the origin of the vanishing nitrogen effect. However, it was early found that, in spite of oxygen levels of 5 – $6x10^{17}$at/cm^3, ample unreacted nitrogen was left in oxygen and nitrogen doped crystals to suppress defect aggregation [82]. As the experimentally confirmed N_2O complex is only stable up to about

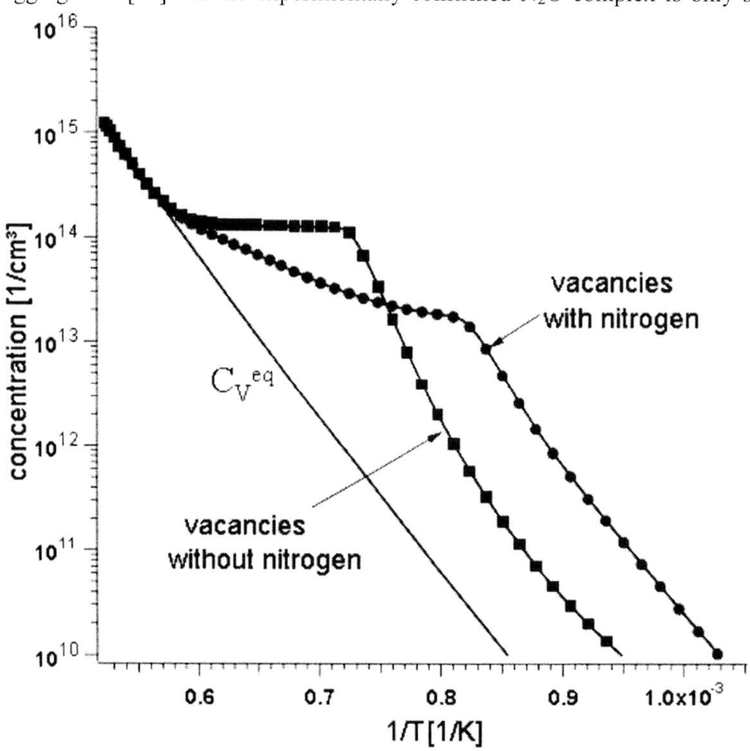

Figure 12. Simulated variation of the vacancy concentration as a function of temperature for a nitrogen doped and non doped crystal [5].

700°C [81,83], it has been suggested [80] that nitrogen reacts with oxygen at high temperatures according to

$$2NO \leftrightarrow N_2 + 2O_i \qquad \text{(III)}$$

Due to the large excess of oxygen over nitrogen, it has been assumed that the equilibrium of reaction III) is on the left hand side near the melting point and, thus, nearly no N_2 is available. At decreasing temperatures, the equilibrium is gradually shifted to the right hand side and more and more N_2 is generated. If the temperature, where N_2 starts to develop in sufficient quantity, is below the vacancy aggregation temperature, the formation of voids is not suppressed. If the N_2 formation temperature is above the vacancy aggregation temperature, it depends on how much N_2 is available in relation to the vacancy concentration at the nucleation temperature of voids. The higher the N_2 formation temperature, the more N_2 is available to reduce the number of vacancies by reaction I). According to reaction III), the formation temperature of N_2 should increase with lower oxygen and higher nitrogen concentration. The lower free vacancy concentration and, thus, the lower aggregation temperature results in voids of smaller size but higher density - in agreement with the experimental findings.

The lower free vacancy concentration before aggregation entails a lower void nucleation temperature, which, in turn, increases the residual free vacancy concentration after void formation as compared to none nitrogen doped material (Figure 12). The latter effect gives a simple explanation for the observed strong enhancement of oxygen precipitation in nitrogen doped CZ crystals (see section 5).

The remarkable variation of the void morphology is still under discussion. A similar change in morphology is also well known for oxygen precipitates. In that case, the effect can be interpreted in terms of the balance between the relaxation energy of the lattice strain and the interfacial energy. However, TEM investigations do not reveal any lattice strain around voids and, thus, another mechanism is required for the explanation of the varying void morphology.

4.4.2. Boron

Boron doping exhibits similar effects on the intrinsic point defect aggregation as nitrogen, but at much higher concentrations [84]. It was found that the COP region as well as the diameter of the OSF ring starts to shrink when the boron content exceeds a level of about 5×10^{18} at/cm^3 (Figure 13). In contrast with nitrogen, the OSF formation is not enhanced and also the width of the OSF ring remains nearly unchanged with increasing boron concentration. At ca. 10^{19} at/cm^3 , the COP's and the OSF ring disappear in the center of the crystal. Although detailed investigations with regard to the behavior of L-pits in highly boron doped silicon are still missing, it can be inferrred from the well known defect free quality of pp+ epi wafers (lightly doped epi-layer on a heavily doped substrate) that L-pits are simultaneously suppressed by the high boron level.

4.4.3. Carbon

Several effects on intrinsic point defect aggregation have also been reported for carbon doping. One observation is the shrinkage of the void region in the crystal center upon carbon doping [85,37], while the region of Si interstitial aggregates is widened, i.e. carbon does not inhibit the formation of L-pits/A-Swirl. It was further found that carbon doping reduces the size of grown-in voids [86,70]. Although the effect on size reduction is appreciable, the morphology of the voids is not changed in contrast to nitrogen [70]. Moreover, carbon is well known to enhance oxygen precipitation. However, similar to boron, the effect on defect aggregation and oxygen precipitation is only seen at significantly higher concentrations (ca. 1×10^{17} at/cm^3) as compared to nitrogen.

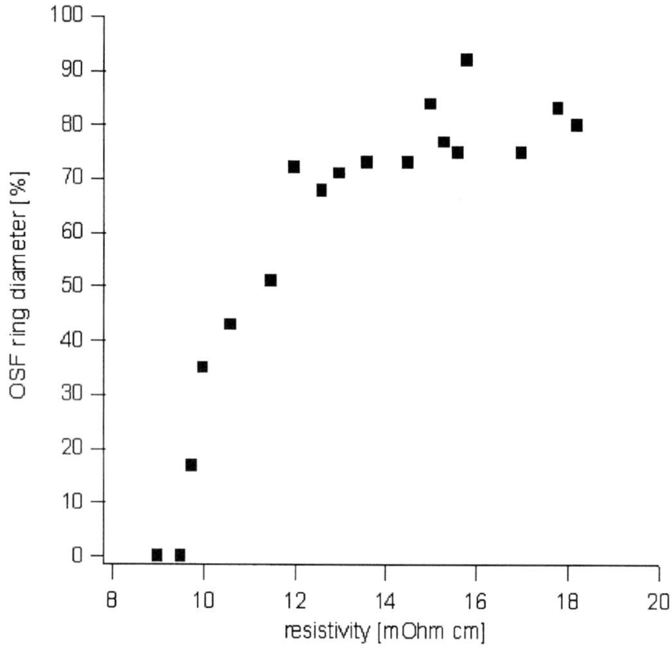

Figure 13. Variation of the OSF ring diameter of crystals grown under the same CZ conditions as a function of resistivity and the boron concentration [91], respectively.

5. FORMATION OF OSF RING

The ringlike distributed OSF's which mark the boundary between the vacancy and the Si interstitial dominated region are generated by oxygen precipitates of a platelet shape which grow particularly large at the edge of the void region and, there, exceed a critical size necessary to form stacking faults in a subsequent wafer oxidation [87]. The critical radius of these grown-in platelets is ca. 70 nm. The reason for this enhanced growth is a local maximum of the residual vacancy concentration after void formation has stopped. The absorption of free vacancies allows the oxygen precipitate, which occupies twice as much volume than the corresponding silicon lattice, to nucleate and grow without building up notable strain energy.

The qualitative explanation of this local maximum is fairly simple. As pointed out in 3.2, the original radial vacancy profile decreases from the crystal center towards the crystal rim. Therefore, the critical supersaturation for void formation is first reached at the crystal center at relatively high temperatures and the free vacancies are quickly consumed in this area (Figure 14). As the cooling of the crystal proceeds, voids are also nucleated in the regions of lower initial vacancy concentration and, thus, vacancy consumption now also occurs further away from the crystal center. As the removal of vacancies works better at higher temperatures where the diffusivity is large, the radial vacancy distribution finally turns into a minimum in

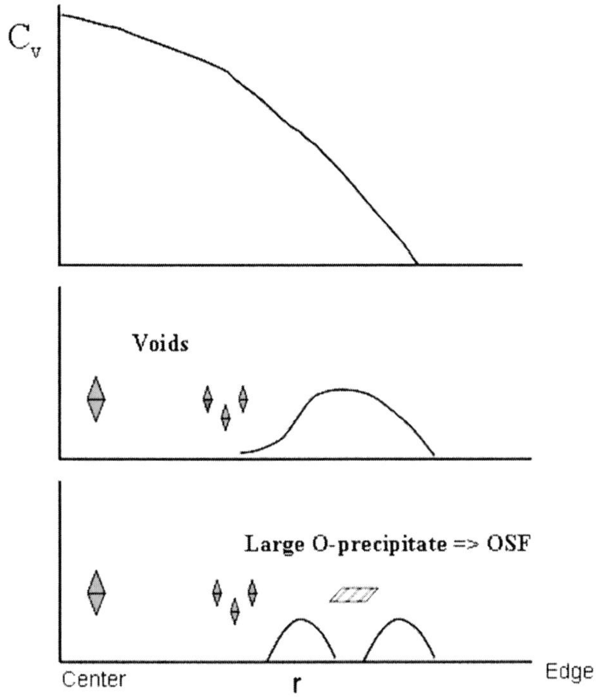

Figure 14. Schematic evolution of the radial vacancy profile and the accompanying void and large O_i- precipitate formation at decreasing temperature [5].

the crystal center and the maximum of the residual vacancies gradually moves toward the V/I boundary upon further cooling. As free vacancies strongly favor the nucleation of oxygen precipitates, the first oxygen aggregates are formed at this local maximum close to the V/I boundary on the vacancy rich side. The relatively high nucleation temperature at this position results in a larger size but lower density of the corresponding grown-in oxygen precipitates as compared to those nucleated at lower temperatures away from the V/I boundary. Consequently, they preferrably reach the critical size for OSF formation in a subsequent wafer oxidation process. The exact position of the OSF ring is therefore slightly on the V-rich side and not at the V/I boundary. At sufficently fast cooling rates of the growing crystal, the oxygen precipitates are nucleated at higher density but smaller size and cannot grow to the critical size. Thus, OSF formation is prevented in agreement with experimental findings. The same density/size effect is obtained if the oxygen content is lowered and, thus, OSF's are suppressed again. The mechanism is the same as it has been describe above for the void formation.

The nucleation of the first oxygen precipitates quickly depletes the residual vacancies in their neighborhood. Hence, the previous vacancy maximum at the V/I boundary is now converted into a pronounced minimum which suppresses further nucleation in this area. As a result, two vacancy peaks with a region of large oxygen precipitates inbetween are frozen in as

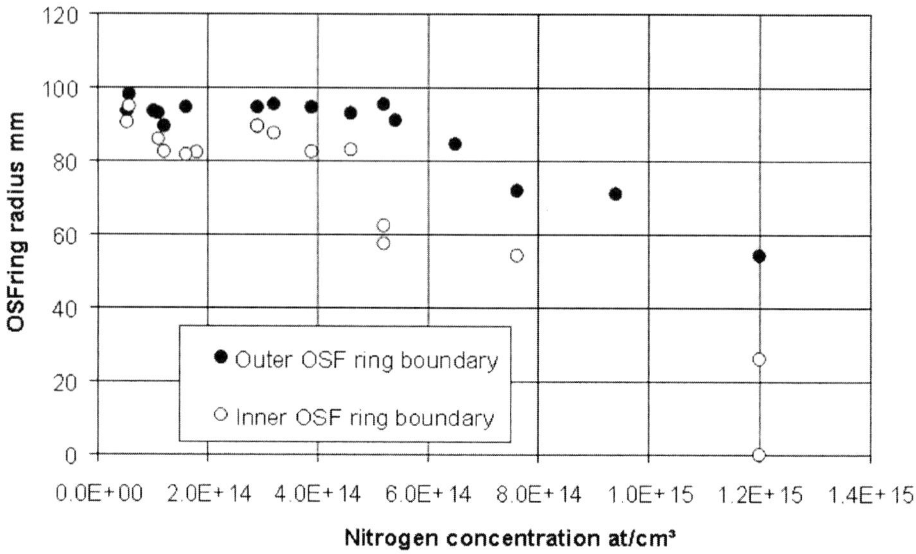

Figure 15. Variation of the outer and inner OSF ring boundary as a function of the nitrogen concentration [88].

the crystal cools down to room temperature. Later on, when the wafers are subjected to further heat treatments, e.g. a nucleation step at 780°C followed by a growth step at 1000°C, new oxygen precipitates with high density are preferrably generated at the radial positions of the two remaining vacancy maxima. This results in a distinct profile of the interstitial oxygen concentration after such heat treatments. In the radial band where the large precipitates and the OSF ring are located, respectively, the intersitial oxygen concentration is only slightly reduced as further nucleation of precipitates during wafer heat treatment is suppressed there. The comparatively low density of precipitates does not consume much of the interstitial oxygen. On the other hand, the high density bands at the adjacent location of the former vacancy peaks coincide with pronounced minima of the interstitial oxygen content.

A considerable change of the width and the radial position of the OSF ring is observed if the crystals are nitrogen doped. With increasing nitrogen content, the outer and the inner OSF ring boundary are shifted towards the crystal center [88]. However, the shift is much larger for the inner OSF ring boundary resulting in a significant widening of the OSF ring width (Figure 15) until it extends over the entire crystal volume. In contrast to nitrogen, carbon tends to suppress OSF formation.

6. CZOCHRALSKI CRYSTAL GROWTH

The method originally developed by Jan Czochralski [89] uses a crucible to hold the melt from which the crystal is grown. The principal setup is shown in Figure 16. The crystals are grown in a flowing argon gas ambient under reduced pressure (8 - 800 mbar). The temperature of the melt must be carefully controlled and adjusted to just above the crystallization temperature before a monocrystalline seed crystal is dipped into the melt.

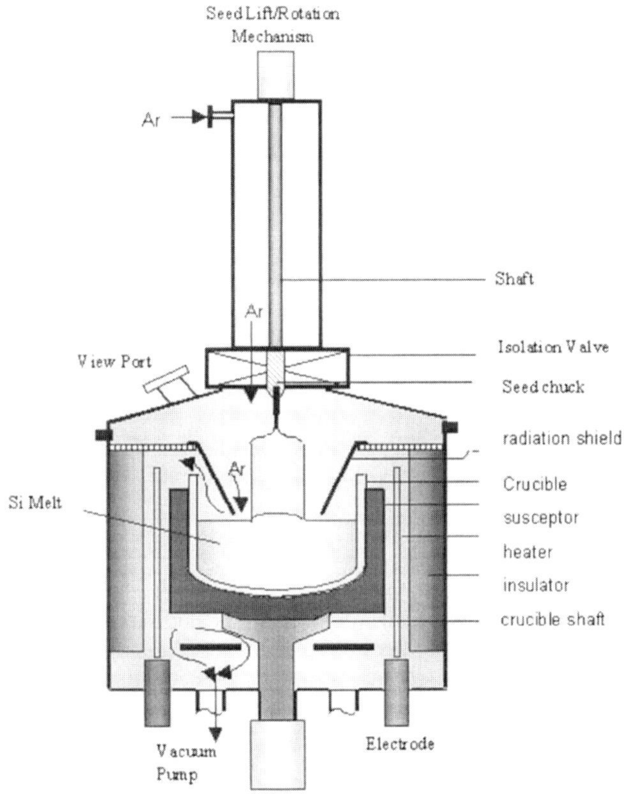

Figure16. Schematic illustration of a CZ puller with a growing crystal [90].

As the seed acts as a heat sink, the temperature at the seed slightly drops below the melting point and crystallization is started. Dislocation free growth is initiated by increasing the pull rate and slimming down the seed neck diameter to a few mm. After a few cm's of neck growth, the socalled Dash neck, all residual dislocations have been diverted towards the surface of the neck and the diameter can now be enlarged without generating new dislocations. As the melt volume shrinks during further crystal growth, the crucible is lifted to keep the level of the free melt surface constant. This allows for an easier process control, in particular the crystal diameter control. The latter is generally based on optical methods which make use of the large emissivity difference between liquid and solid silicon. The respective crucible and silicon charge sizes are obviously related to the crystal diameter (see Table). The currently maximum crystal diameter of 300mm requires charge sizes of 200 - 400 kg in crucibles of 28" - 32". The maximum pull rate decreases with increasing diameter as the dissipation of the latent heat becomes less effective [91,92]. Typical pull rates for 300mm crystals are in the range of 0.5 - 1.2 mm/min. As mentioned in 4.1 and 4.3, the pull rate has a major influence on the defect behavior of the crystal and, thus, cannot always be maximized for economical reasons.

Table
Relation between crystal diameter, charge size and crucible diameter of CZ growth processes.

Crystal diameter [mm]	Charge size [kg]	Crucible diameter [mm]
100	28 - 32	356
125	28 – 50	356 – 457
150	42 – 70	356 – 457
200	65 – 120	457 – 610
300	200 – 400	711 – 914

The only suitable crucible material for silicon crystal growth is silica (SiO_2). The choice of crucible materials is not only limited by the dissolution rate due to the extreme chemical reactivity of the silicon melt, but also by the low solubilities of impurities in solid silicon. The latter does not allow for the growth of long dislocation free crystals as the dissolution of the crucible increases the impurity concentration beyond the solubility limit in the growing crystal. The only exception is silica as the dissolution proceeds via $SiO_2 + Si \rightarrow 2\ SiO$. SiO, however, has a very high evaporation rate which prevents a continuous oxygen enrichment of the melt over time and, instead, an equilibrium between oxygen enrichment by crucible dissolution and oxygen loss by evaporation can be established. About 99% of the dissolved oxygen is removed from the melt by SiO evaporation [93].

The quality of the silica crucible is of utmost importance for the dislocation free yield of the growth process. The corrosive attack of the silicon melt does not proceed via a homogeneous dissolution, but by a pitting corrosion [93]. This locally enhanced corrosion can release small silica particles into the melt which may drift to the crystal/melt interface and stop dislocation free growth. The pitting corrosion is mainly induced by contamination, in particular, surface contaminants. Silica particles may also be washed into the melt from gas bubbles in the silica, which grow over time and, finally, break up. One parameter which enhances bubble growth is the OH content of the silica. These corrosion mechanisms can severely limit the crucible lifetime and, therefore, they are of special concern for large thermal budget processes, e.g. crystal diameter ≥ 200mm and recharge processes.

As silica is rather soft at the melting point of silicon, the silica crucible must be placed in a mechanically stable graphite susceptor. This graphite susceptor is surrounded by a meander shaped cylindrical heater. In case of large crucible diameters as required for crystal diameters ≥ 200mm, the crucible is also often heated from the bottom by an additional heater. An important part of the hotzone setup is the thermal heat shield which surrounds the growing crystal. The heat shield defines the cooling behavior and thermal gradients and, therefore, the defect quality of the growing crystal. Meanwhile, special designs also allow for water cooling of the heat shield without the risk of a steam explosion in case of a malfunction. Consequently, the pull rate could be considerably increased (ca. 50%) while the defect size was substantially reduced.

All graphite parts which are exposed to SiO suffer from surface corrosion which will eventually result in the flake off of particles. The particles can fall into the melt and stop dislocation free growth. Hence, the lifetime of these graphite parts is limited and they must be replaced after a certain number of runs.

One of the major challenges in CZ crystal growth is the adjustment of the oxygen content in the growing crystal according to customer specifications. The technically relevant oxygen concentrations range from 4.5 to 8.5×10^{17} atoms/cm^3. The oxygen incorporation in the growing crystal is essentially determined by the evaporation of SiO from the free melt surface, the dissolution rate of the wetted crucible surface and the melt convection. As the ratio between free melt surface and wetted crucible surface increases with decreasing melt volume, one would principally expect lower oxygen concentrations at the tail of the grown crystal. However, it turns out that the influence of the melt convection is usually dominant which makes the prediction of the oxygen content very difficult. This is a consequence of the fact that the melt convection is turbulent and, therefore, strongly fluctuating with time. This is particularly true for larger charge sizes (>50 kg). Although considerable progress in the numerical simulation of turbulent melt flow has been achieved over the last years, turbulent melt convection is still poorly understood, in particular in near boundary regions. A complicated problem is the calculation of the correct SiO evaporation rate as the argon gas flow and pressure not only influences the partial pressure of SiO above the free melt surface, but also the near surface melt convection which, in turn, impacts the near surface oxygen concentration gradient. A satisfactory solution of this complex situation has not yet been found and, thus, the adjustment of the required oxygen content is still often done by trial and error rather than by theoretical calculations. Nevertheless, the long history and profound experience of CZ silicon crystal growth has enabled the silicon manufacturers to develop processes which provide axially flat profiles within a wide range of oxygen levels. Due to the high evaporation rate, the unknown segregation coefficient of oxygen is technically irrelevant.

In recent years, magnetic fields have been applied to reduce melt fluctuations and stabilize the melt temperature. They also have a strong impact on the oxygen concentration. The effective field direction is usually aligned either horizontally [94] or vertically [95] to the crucible/crystal rotation axis. A special configuration is the socalled cusp field [96] where two parallel coils generate two opposite fields. The resulting overall field has a minimum in the space between the two coils. It is therefore possible to grow the crystal in a nearly field free environment while the crucible walls are exposed to a large field strength. The advantage of this

field configuration is that the melt convection close to the growth interface and at the crucible wall can be controlled by separate parameters which avoids agonizing compromises between adjusting the required oxygen content and optimizing its radial variation. In most cases, static magnetic fields are used which can only damp melt convection. The necessary field strengths are usually high (several thousend Gauss) which makes the magnets rather large and heavy. In order to reduce the high energy costs, many silicon manufacturers have switched to relatively expensive superconducting magnets. In contrast to static magnetic fields, alternating fields are driving the melt convection and, thereby, stabilize the melt flow pattern [97]. This also results in a strong reduction of temperature fluctuations. The necessary field strengths are almost two orders of magnitude smaller as compared to static fields. This fact does not only considerably lower the costs for the magnets, but also avoids safety issues.
The axial concentration profile of impurities which have a negligible evaporation rate from the melt is described by

$$C_s = k_{eff}\, C_l\, (1 - s)^{keff-1} \tag{8}$$

where C_s = impurity concentration in solid silicon, C_l = impurity concentration in the silicon melt and s = crystallized fraction of the melt. The dependence on s implies that the concentration gradient strongly increases towards the tail end of the crystal for small k_{eff}´s. This behavior is rather unfavorable for n-type crystals as the segregation coefficients of all n-type dopants are significantly smaller than 1 - resulting in a substantial axial resistivity variation. Consequently, tremendous efforts were carried out in the past to develop a continuous recharging (CCZ) technology of small and flat crucibles which could compensate for the segregation effect by keeping the melt volume constant. Despite of more than 20 years of worldwide development acitivities, continuous recharging is still not used as a standard production process. One of the reasons is that dislocated crystals cannot be dumped back into the melt and the growth process restarted as the comparatively small crucible cannot accommodate additional melt volume (or the melt volume would be changed, respectively). Thus, the dislocated part of the crystal after structure loss has to be scrabbed. On the other hand, the much higher complexity of the CCZ process notably rises the risk of structure loss. As a rule of thumb, dislocations generated during the growth process propagate back into already dislocation free grown material over a distance of approximately one diameter. Therefore, large diameter processes particularly suffer from structure loss which eventually renders the CCZ technique uneconomical. However, the socalled discontinuous recharging method is widely in use. Here, many short crystals are grown from a standard crucible which is replenished after every crystal to the original level. As only a small fraction of the melt is used up per crystal, the axial resistivity profile is relatively flat and fairly thight resistivity specifications can be economically produced. The number of runs is not only limited by the crucible lifetime, but also by the enrichment of residual impurities like carbon in the melt which are introduced by the recharged poly silicon.

The further growth of the crystal diameter has become a real challenge for the future development of the CZ technology as upscaling will be significantly more difficult than in the past [91]. The next larger diameter will be 450 mm and will require charge and crucible sizes of more than 400kg and 36", respectively. The large thermal budget of such processes may prolongate melting and remelting times (if the crystal has dislocated and must be dumped back into the melt) to the point where the silica crucible is too corroded to restart the process after a structure loss. A structure loss, on the other hand, renders nearly half a meter of already dislocation free grown material useless which means that the whole charge has to be scrabbed in most cases. Consequently, perfect processes with nearly 100% dislocation-free yield upon first dip in would be mandatory to avoid a dramatic cost increase. However, this does not seem to be very realistic for the foreseeable future as the general complexity of the growth processes aggravates with larger charge size. An additional problem arises from the fact, that, as mentioned above, the growth rate decreases with larger crystal diameters. This does not only lower the output, but it may also change the defect behavior to the extend that the crystal is no longer fully vacancy rich. As has been shown [92], G cannot necessarily be reduced in proportion to the pull rate without the risk to destabilize the cylindrical growth of the crystal. Hence, it becomes increasingly difficult to keep the parameter $V/G > \xi_{tr}$ at larger diameters. As a result, the formation of L-pits at the crystal rim may become unavoidable which, by no means, is acceptable to the device industry. The termination of the Japanese SSI project in the year 2000, although a 450mm dislocation free crystal was successfully grown, demonstrates that the cost issue lacks a solution for the time being.

7. FLOATING ZONE CRYSTAL GROWTH

There are principally two different concepts for the floating zone technique [98,99]. One is the pedestal method where a poly silicon rod is inductively melted at its upper end by a high frequency (HF) coil (Figure 17a). Due to the extremely high surface tension of liquid silicon, the resulting melt bath can be safely balanced on top of the poly silicon rod. A seed crystal is then dipped into the melt bath and the growth process is started in the same manner as

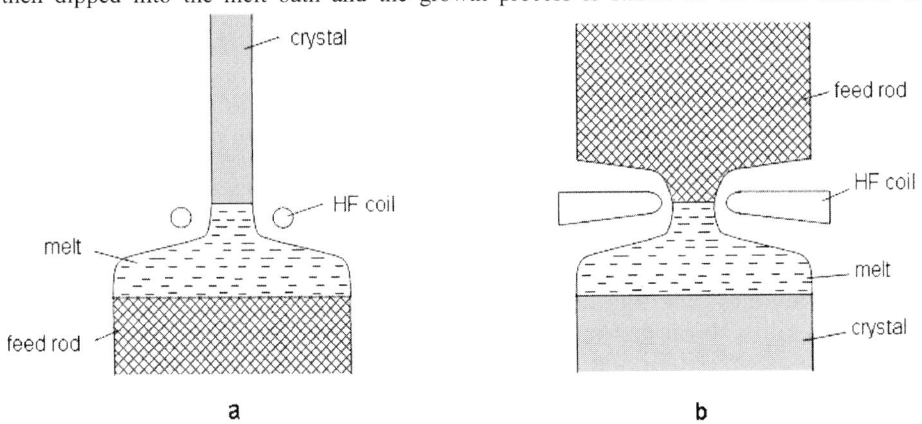

Figure17. Schematic illustration of the FZ growth process. Pedestal method (a), needle-eye method (b).

described for CZ growth. Crystals grown by this method are limited to a maximum diameter of ca. 10-20 mm as the liquid bridge between the poly silicon rod and the growing crystal cannot be maintained for larger diameters. Today, the pedestal method is still used for the production of seed crystals and slim rods. The latter serve as a starting material for the production of poly silicon according to the Siemens process.

The socalled needle-eye technique overcomes the above diameter limitation (Figure 17b). The lower end of a poly silicon rod is inductively melted by a pancake shaped one-turn HF coil. The melt flows through a central whole (needle eye) of the coil down onto the growing crystal below the coil. The process is started by generating a melt drop which hangs down at the lower end of the poly rod. Subsequently, a seed crystal is dipped into the melt drop from below through the central hole of the HF coil. The growth process is then initiated similar to the pedestal and CZ method except that the configuration is upside down. Due to the thin Dash neck, the crystal must be supported by a suitable mechanical device which is pressed against the seed cone end of the growing crystal after a sufficiently large crystal diameter has been reached. In the past, two different concepts of the needle-eye technique have been realized. The first one uses an axially movable coil which allows for a small height of the FZ puller and, thus, low costs. However, at higher HF powers, flexible HF cables which are a necessary prerequiste of this concept are no longer suitable. For a crystal diameter of $\geq$ 100 mm, rigid HF power transmissions have to be used which entail an essentially fixed position of the coil. As a consequence, the poly rod as well as the growing crystal must be axially moved over relatively long distances according to the grown crystal length. FZ pullers for large diameter crystals are therefore rather expensive due to their enormous height.

The most important part of the FZ puller is the HF coil. Although the one-turn coil with its pancake shape and the central hole is still the standard design, the specific details and the precise manufacturing of the coil are of utmost importance. E.g. deviations of a few thenths of a millimeter from the calculated shape at the central hole can result in complete process failures. The most important requirements which a proper coil design has to meet, are a homogeneous melting of the poly rod and a stable cylindrical growth of the crystal. However, the coil geometry also has a strong influence on the radial dopant distribution in the crystal. It is therefore a tedious task to develop a coil design which facilitates an economical production and, at the same time, meets the quality expectations of the device manufacturers. Today, it is possible to simulate the radial dopant profiles for a given coil geometry [100,101] which has tremendously contributed to savings in time and costs of FZ development acitivities.

A severe drawback of the FZ technology is the risk of arcing between the two electrodes of the HF power supply at the coil. Arcing stops the growth process due to the break down of the HF field and often also damages the coil. It is not only promoted by the high voltage between the electrodes but also by the high temperature above the silicon melt bath which partly ionizes the argon gas ambient. The risk of arcing can be lowered by adding small amounts of H_2 or N_2 to the argon ambient. While N_2 has a very favorable effect on the suppression of defects (see section 4.4.1), H_2 doping was found to generate large voids in the crystal at partial pressures above a few mbar [102] and, therefore, is generally not applied.

The doping with electrically active elements is carried out by gas doping through nozzles built into the coil which direct the gas jet onto the melt surface. As a doping gas B_2H_6 is used for p-type and PH_3 for n-type crystals. The continuous doping during the growth process along with the constant melt volume results in an axially flat resistivity profile independent of the segregation coefficient of the dopant. Thus, FZ crystals can be produced within very tight resistivity ranges for p- and n-type.

The fact that the FZ method avoids any contact of the silicon with other materials during the growth process allows for the production of extremely pure crystals with resistivities of several thousand ohmcm's. The multiple application of FZ pulling to further purify the crystals as it was often done in the early days of silicon technology is not necessary anymore owing to the extreme purity of today's poly silicon rods.

The typical pull rates are between 2 and 3 mm/min which is substantially faster as compared to the CZ technique. In addition, the thermal budget of the process is extremely small which dramatically shortens the start up and cool down phase. Consequently, FZ pullers have considerably higher output. Along with the avoidance of consumable costs, one would expect FZ crystals to be be cheaper than CZ crystals. However, this cost advantage is more than offset by the much higher cost for the solid poly silicon feed rods as compared to the poly silicon chunks used for CZ growth. The origin of the latter cost difference is the rather low deposition rate in the production of FZ poly rods.

A severe drawback of the FZ technology is further the higher technical hurdles to increase the crystal diameter. The main challenges are the suppression of arcing and the reduction of the thermal stress in the crystal during growth. The latter is so strong that the crystals abruptly fall into pieces if a certain stress limit is exceeded. The <111> orientation is specifically susceptible to this failure type. A major problem is also the availability of suitable poly rods of a sufficiently large diameter. The poly rods must not have any cracks and have to show a rather smooth surface. Although progress has been achieved in this regard over the last years, there now seems to be a diameter limit of around 160 - 170mm for poly silicon rods which will be very difficult to surmount by the Siemens process. Today, 200mm <100> crystals and 150mm <111> crystals can be produced by the FZ method. Nevertheless, it is clear that a further increase of the diameter will be no longer possible by a further upscaling of the technology.

Up to now, FZ material is predominantly used for those devices where high resistivities and a low oxygen content are indispensable, i.e. high power and, more recently, high frequency devices. This may however change in the foreseeable future. With the development of high efficiency solar cells for which FZ crystals would be the ideal base material due to its high and long time stable minority carrier lifetime a new high volume market would open up, if the costs can be substantially reduced. This can only be achieved if the FZ technique is considerably modified so that much cheaper Si granules can be used as a feed stock material instead of expensive poly rods.

8. SUMMARY/OUTLOOK

The understanding of bulk defects in silicon crystals has greatly advanced during the last decade. One of the important prerequisites for this success was the rather precise determination of the physical properties of vacancies and Si interstitials which had been very uncertain in the past. Today, the bulk properties can be exactly tailored to the needs of the device manufacturers and, in many cases, their behavoir in device processes can be accurately predicted by computer simulations. As the design rule continues to shrink, the trend to smaller defect sizes and, eventually, defect free material will persist. The latter is already indispensable for the layer transfer technology which will play a major role for silicon on insulator (SOI) based devices, where the transferred top layer must be completely free of vacancy aggregates. The focus of defect investigations is currently shifting to intrinsic point defect aggregation under the influence of specific impurities like nitrogen and carbon. The so far presented theoretical models of the observed effects, however, are not yet well established and require additional verification or modifications, respectively.

The next larger diameter step is already fixed and will enlarge the currently 300 mm crystals to 450 mm. This will require charge sizes of about 500 kg in 40" silica crucibles which, up to now, will render the growth process uneconomical. The date of changeover to the next larger diameter will therefore depend on the solution of this cost issue. Furthermore, the inevitably lower pull rates are expected to yield crystals which contain L-pits unless they are heavily doped with boron. Thus, at the current stage, p-p+ epi wafers would be the only viable 450 mm product that meets the requirements of the device industry. However, most device manufacturers will not be able to switch to p-p+ epi wafers due to their specific device design. Consequently, a solution of the L-pit problem in 450 mm crystals is mandatory.

In case of FZ crystals, the 200mm diameter which is currently introduced can probably not be further increased by the classical floating zone technique. The main issues which might turn into show stoppers are the diameter limitation of the feed rod, the high voltage at the HF coils which increases the arcing risk and the excessive thermal stress of the growing crystal. A further diameter step will therefore require a major modification of the FZ method. A necessary prerequisite will be to switch from solid polysilicon feed rods to silicon granules as a feed stock material to overcome the diameter limitation of the solid feed rods.

ACKNOWLEDGEMENT

The author is greatly indebted to Walter Haeckl for many helpful discussions and providing some of the figures.

REFERENCES

1. W.C. Dash, J.Appl.Phys. 30 (1959) 459.
2. W.C. Dash, J.Appl.Phys. 31 (1960) 736.
3. Lark-Horowitz in Semiconductor Materials, Conf.Univ.Reading, Butterworths, London (1951) 47.
4. M.S.Schnoeller, IEEE Trans.Electron.Devices ED-21 (1974) 313.
5. W.v.Ammon, "Point Defects in Silicon" in the Handbook of Electronic and Optoelectronic Materials, ed. by S.Kasap and P. Capper, Springer Science, to be published.
6. P.M.Petroff, A.J.R. deKock, J.Crystal Growth 30 (1975) 117.
7. S.Sadamitsu, S.Umeno, Y.Koike, M.Hourai, S.Sumita, T.Shigematsu, Jpn.J.Appl. Phys. 32 (1993) 3675.
8. L.I.Bernewitz, B.O.Kolbesen, K.R.Mayer, G.E. Schuh, Appl.Phys.Letters 25 (1975) 277.
9. B.O.Kolbesen, A.Mühlbauer, Solid State Electronics (25)(1982)759.
10. W.Bergholz, W.Mohr, W.Drewes, Mat.Science and Engineering B4 (1989) 359.

11. M.Itsumi, H.Akiya, T.Ueki, M.Tomita, M.Yamawaki, J.Appl.Phys. 78 (10) (1995) 5984.
12. M.Miyazaki, S.Miyazaki, Y.Yanase, T.Ochiai, T.Shigematsu, Jpn J.Appl.Phys. 34 (1995) 6303.
13. J.G.Park, J.M.Park, K.C.Cho, G.S.Lee, H.K.Chung, Proc. The 2nd International Symposium on Advanced Science and Materials, JSPS (1996) 519.
14. M.Itsumi, Material Science and Engineering B73 (2000) 184.
15. T.Bearda, M. Houssa, P.Mertens, J.Vanhellemont, M.Heyns, Appl. Phys.Lett. 75 (9) (1999) 1255.
16. U.Lambert, A.Huber, J.Grabmeier, J.Vanhellemont, R.Wahlich, G.Kissinger, Microelectronic Engineering 48 (1999) 127.
17. E.Dornberger, D.Temmler, W.v.Ammon, J.Electrochem. Soc., 149(4) 1 (2002).
18. C.Kupfer, H.Roth, H.Dietrich, Materials Science in Semiconductor Processing 5 (2003) 381.
19. T.Sinno, E.Dornberger, W.v.Ammon, R.A.Brown, F.Dupret, Materials Science and Engineering 28 (2000) 149.
20. R.Falster, Silicon Wafer Symposium, SEMICON West 99, SEMI 1999, E-13.
21. M.Akatsuka, M.Okui, N.Morimoto, K.Sueoka, Jpn.J.Appl.Phys. 40 (2001) Pt.1, No. 5A, p. 3055.
22. T.Sinno, in: Semiconductor Silicon 2002, Vol.2, Proceedings of the 9th Symposium on Silicon Materials Science and Technology, ed. H.R.Huff, L.Fabry, S.Kishino, The Electrochem.Soc., Pennington, NJ, p.212.
23. D.A.Antoniadis, I.Moskowitz, J.Appl.Phys. 53 (10) (1982) 6780.
24. H.-J.Gossmann, C.S.Rafferty, A.M.Vredenberg, H.S.Luftman, F.C.Unterwald, D.J.Eaglesham, D.C.Jacobson, T.Boone, J.M.Poate, Appl.Phys. Lett. 64 (3) (1994) 312.
25. T.Ueki, M.Itsumi, T.Takeda, Jpn. J. Appl. Phys. 37 (1998) Pt. 1, No.4A, p.1669.
26. M.Itsumi, J.Crystal Growth 237 – 239 (2002) 1773.
27. S.Umeno, Y.Yanase, M.Hourai, M.Sano, Y.Shida, H.Tsuya, Jpn.J.Appl.Phys. 38 (1999) 5725.
28. M.Nishimura, Y.Yamaguchi, K.Nakamura, J.Jablonski, M.Watanabe, Electrochem.Soc. Symp.Proc. 98-13 (1998) 188.
29. J.Ryuta, E.Morita, T.Tanaka, Y.Shimanuki, Jap.Appl.Phys. 29, (1990) L1947.
30. H.Yamagishi, I,Fusegawa, N.Fujimaki, M.Katayama, Semicond.Sci.Techn. 7 (1992) A135.
31. P.J.Roksnoer, M.M.B.Van de Boom, J.Crystal Growth 53 (1981) 563.
32. H.Bender, J.Vanhellemont, R.Schmolke, Jpn. J.Appl.Phys. 36 Part2 No9A/B (1997) L1217.
33. K.Nakai, M.Hasebe, K.Ohta, W.Ohashi, J.Crystal Growth 210 (2000) 20.
34. A.J.R.de Kock, Philips Res. Rept. Suppl. No.1 (1973).
35. H.Föll, B.O.Kolbesen, Appl. Phys. 8 (1975) 319.
36. P.M.Petroff, A.J.R. de Kock, J.Crystal Growth 36 (1976) 1822.
37. J.Chikawa, T.Abe, H.Harada, Semiconductor Silicon 1986, Electrochemical Society Proceedings 86-4 (1986) 61.

38. H.Föll, U.Gösele, B.O.Kolbesen, J.Crystal Growth 40 (1977) 90.
39. R.Schmolke, W.Angelberger, W.v.Ammon, H.Bender, Solid State Phenomena 82-84 (2002) 231.
40. V.V.Voronkov, J.Crystal Growth 59 (1982) 625.
41. W.von Ammon, E.Dornberger, H.Oelkrug, H.Weidner, J.Crystal Growth 151 (1995) 273.
42. M.Hourai, E.Kajita, T.Nagashima, H.Fujiwara, S.Umeno, S.Sadamitsu, S.Miki, T.Shigematsu, Mat. Science Forum 196 – 201 (1995) 1713.
43. E.Dornberger, W.v.Ammon, J.Electrochem. Soc. 143 (5) (1996) 1648.
44. E.Dornberger, J.Esfandyari, D.Gräf, J.Vanhellemont, U.Lambert, F.Dupret, W. von Ammon, in Crystalline Defects and Contamination Control: Their Impact and Control in Device Manufacturing II, PV 97-22, The Electrochemical Soc., Inc., Pennington, NJ, 1997, ed. B.O. Kolbesen, P. Stallhofer, C. Claeys, F. Tardiff, p. 40.
45. M.Hasebe, Y.Takeoka, S.Shinoyama, S.Naito, Defect Control in Semiconductors, ed. K.Sumino, (1990) 157.
46. H.Yamagishi, I.Fusegawa, K.Takano, E.Iino, N.Fujimaki, T.Ohta, M.Sakurada, Semiconductor Silicon 1994, ed.H.R.Huff, W.Bergholz, K.Sumino, The Electrochemical Society, Pennington, NJ, (1994) 124.
47. D.Gräf, M.Suhren, U.Lambert, R.Schmolke, A.Ehlert, W.v.Ammon, P.Wagner, J.Electrochem.Soc. 145(1) (1998) 275.
48. E.Dornberger, J.Esfandyari, J.Vanhellemont, D.Gräf, U.Lambert, F.Dupret, W.v.Ammon, The Electrochem. Soc., Proceedings Vol.98-1 (1998) 490.
49. M.Akatsuka, M.Okui, S.Umeno, K.Sueoka, J.Electrochem.Soc. 150 (9) (2003) G587.
50. J.Furukawa, H.Tanaka, Y.Nakada, N.Ono, H.Shiraki, J.Crystal Growth 210 (2000) 26.
51. R.Schmolke, M.Blietz, R.Schauer, D.Zemke, H.Oelkrug, W.v.Ammon, U.Lambert, D.Gräf, The Eletrochem. Soc., PV 2000-17 (2000) 1.
52. E.Dornberger, W.v.Ammon, J.Electrochem Soc. 143 (5) (1996) 1648.
53. A.Natsume, N.Inoue, K.Tanahashi, A.Mori, J.Crystal Growth 225 (2001) 221.
54. V.V.Voronkov, R.Falster, J.Appl.Phys. 86 (11) (1999) 5975.
55. Z.Wang, R.Brown, J.Crystal Growth 231 (2001) 442.
56. E.Dornberger, W.v.Ammon, D.Gräf, U.Lambert, A.Miller, H.Oelkrug, A.Ehlert, The Electrochem. Soc., Proceedings Vol. 96-13 (1996) 140.
57. J.Esfandyari, G.Hobler, S.Senkader, H.Pötzl, B.Murphy, J.Electrochem.Soc. 143 (1996) 995.
58. V.V.Voronkov, Materials Science and Engineering B73 (2000) 69.
59. V.V.Voronkov, R.Falster, J.Crystal Growth 198/199 (1999) 399.
60. T.Sinno, private communication.
61. J.Vanhellemont, C.Claeys, J.Appl.Phys. 62 (9) (1987) 3960.
62. V.V.Voronkov, R.Falster, J.Appl.Phys. 91 (9) (2002) 5802.
63. N.I.Puzanov, A.M.Eidenzon, Semicond. Sci.Technol. 7 (1992) 406.
64. V.V.Voronkov, R.Falster, J.Crystal Growth 194 (1998) 76.
65. V.V.Voronkov, R.Falster, J.Crystal Growth 226 (2001) 192.

66. T.Abe, M.Kimura, Semiconductor Silicon 1990, Eds. H.R.Huff et al., The Electrochem.Soc., Pennington, (1990) 105.
67. D.-R.Yang, Y.-W. Wang, H.-N.Yao, D.-L.Que, Progress in Natural Science, 3(2) (1993) 176.
68. W.v.Ammon, A.Ehlert, U.Lambert, D.Gräf, M.Brohl, P.Wagner, Eds. H.R.Huff, W.Bergholz and K.Sumino, Semiconductor Silicon, , The Electrochem.Soc., PV 94-10 (1994) 136.
69. K.Nakai, Y.Inoue, H.Yokota, A.Ikari, J.Takahashi, A.Tachikawa, K.Kitahara, Y.Ohta, W.Ohashi, J.Appl.Phys. 85(8) (2001) 4301.
70. J.Takahashi, K.Nakai, K.Kawakami, Y.Inoue, H.Yokota, A.Tachikawa, A.Ikari, W.Ohashi, Jpn.J.Appl.Phys. 42 (2003) 363.
71. F.Shimura, R.S.Hockett, Appl.Phys.Lett., 48 (1986) 224.
72. Q.Sun, K.H.Yao, H.C.Gatos, J.Lagowski, J.Appl.Phys. 71 (8) (1992) 3760.
73. K.Aihara, H.Takeno, Y.Hayamizu, M.Tamatsuka, T.Masui, J.Appl.Phys. 88(6) (2000) 3705.
74. K.Nakai, Y.Inoue, H.Yokota, A.Ikari, J.Takahashi, W.Ohashi, Proceedings of the 3rd International Symposium on Advanced Science and Technology of Silicon Materials, JSPS, (2000) 88.
75. D.Gräf, U.Lambert, R.Schmolke, R.Wahlich, W.Siebert, E.Daub, W.v.Ammon, The Electrochem. Soc., Proceedings Volume 2000-17, (2000) 319.
76. W.v.Ammon, D.Gräf, W.Zulehner, R.Schmolke, E.Dornberger, U.Lambert, J.Vanhellemont, W.Hensel, Extended Abstract of the Electrochem.Soc. San Diego Meeting, Vol.98-1, (1998) Abstract No. 512.
77. K.Nakamura, T.Saishoji, S.Togawa, J.Tomioka, Proceedings of the Kazusa Akademia Park Forum on the Science and Technology of Silicon Materials, ed. K.Sumino, (1999) 116.
78. H.Sawada, K.Kawakami, Phys.REv. B62, No 3, (2000) 1851.
79. H.Kageshima, A.Taguchi, K.Wada, Appl. Phys.Lett. 76 (25), (2000) 3718.
80. W.v.Ammon, R.Hölzl, J.Virbulis, E.Dornberger, R.Schmolke, D.Gräf, J.Crystal Growth 226 (1), (2001) 19.
81. P.Wagner, R.Oeder, W.Zulehner, Appl.Phys. A 46, (1988) 73.
82. W.v.Ammon, P.Dreier, W.Hensel, U.Lambert, L.Köster, Materials Science and Engineering, B36 (1996) 33.
83. M.W.Qi, S.S.Tan, B.Zhu, P.X.Cai, W.F.Gu, M.Xu, T.S.Shi, D.L.Que, L.B.Li,J.Appl.Phys. 69, (1991) 3775.
84. M.Suhren, D.Gräf, U.Lambert, P.Wagner, Electrochem.Soc.Proc. 96 (13) (1996) 132.
85. T.Abe, T.Masui,H.Harada, J.Chikawa, VLSI Science and Technolgy, The Electrochem.Soc., eds. W.M.Bullis and S.Broyda, Pennington, NJ (1985) 543.
86. R.Takeda, T.Minami, H.Saito, Y.Hirano,H.Fujimori, K.Kashima, Y.Matsushita, The Electrochem. Soc. PV 2000-17 (2000) 331.
87. K.Sueoka, M.Akatsuka, K.Nishihara, T.Yamamoto, S.Kobayashi, Mat. Science Forum 196-201 (1995) 1737.

88. W.v.Ammon, R.Hoelzl, T.Wetzel, D.Zemke, G.Raming, M.Blietz, Microelectronic Engineering 66 (2003) 234.
89. J.Czochralski, Z.Phys.Chem. 92 (1918) 219.
90. Reprinted by permission of Siltronic AG.
91. W.v.Ammon, in the Proc.2nd Int.Symp. on Advanced Science and Technology of Silicon Materials, Kona, USA, November 1996 (Ed. by M.Umeno, The Japanese Society for the Promotion of Science, 1996), p.233.
92. W.v.Ammon, E.Dornberger, P.O.Hansson, J.Crystal Growth 198/199 (1999) 390.
93. W.Zulehner, D.Huber, Czochralski Grown Silicon, in J. Grabmeier, editor, Crystals, Growth, Properties and Application, Springer, Berlin-Heidelberg-NewYork, 1982.
94. K.Hoshi, N.Isawa, T.Suzuki, Y.Ohkubo, J.Electrochem. Soc., 132 (3), (1985) 693.
95. K.M.Kim, P.Smetena, J.Electrochem.Soc. 133 (8), (1986)1682.
96. H.Hirata, K.Hoshikawa, J.Crystal Growth 98 (1989) 777.
97. Th.Wetzel, A.Muiznieks, A.Mühlbauer, Y.Gelfgat, L.Gorbunov, J.Virbulis, E.Tomzig, W.v.Ammon, J.Crystal Growth 230 (2001) 81.
98. W.G.Pfann, Zone Melting, John Wiley&Sons, Inc., New York, 1958.
99. A.Mühlbauer, W.Keller, Floating Zone Silicon, Wilcox , New York, 1981.
100. A.Mühlbauer, A.Muiznieks, G.Raming, H.Riemann, A.Lüdge, J.Crystal Growth 198/199 (1999) 107.
101. A.Lüdge, H.Riemann, W.Schröder, A.Mühlbauer, A.Muiznieks, G.Raming, Meeting abstracts of the 193rd Meeting of the Electrochem.Soc., Inc, San Diego, 98-1 (1998) 514.
102. E.Wolf, W.Schröder, H.Riemann, B.Lux, Materials Science and Engineering B36 (1996) 209.

Crystal Growth – from Fundamentals to Technology
G. Müller, J.-J. Métois and P. Rudolph (Editors)

Microchannel Epitaxy-
Physics of Lateral and Vertical Growth and its Applications

Tatau Nishinaga

Toyohashi University of Technology
1-1 Hibarigaoka, Tempakucho, Toyohashi, 441-8085 Japan

A new concept of epitaxy named microchannel epitaxy (MCE) is described. In MCE, lattice information of a substrate is transferred through narrow microchannels while the transfer of defect information is prevented by the presence of amorphous film deposited on the substrate. Two types of MCE, vertical and horizontal MCEs are discussed. In the present article a special focus is put on the horizontal MCE which we call simply MCE, while we call the vertical MCE as V-MCE.

MCE is composed of selective area epitaxy in narrow microchannel and successive epitaxial lateral growth. It was shown that flat MCE layers have been successfully grown for Si/Si, GaAs/GaAs, InP/InP, GaP/GaP, GaAs/Si and InP/Si systems. In highly lattice mismatch heteroepitaxy, although dislocations propagate through the microchannel into the grown layer, wide dislocation free regions have been obtained outside of the dislocated area.

MBE was employed to carry out MCE of GaAs. It was found that V-MCE was successfully conducted and the dislocation density was reduced. MCE with high width to thickness ratio was achieved by MBE by sending molecular beams with a low angle to the substrate surface.

1. INTRODUCTION

There are large demands to grow a high quality single crystal film on a substrate with large lattice mismatch. III-V compounds on Si, III-V Nitrides on sapphire, II-VI compounds on various substrates are examples for such highly lattice mismatch heteroepitaxy (HM^2). Among them GaAs on Si and GaN on sapphire are the most exciting topics towards obtaining respectively opto-electronic integrated circuit (OEIC) and blue laser and light emitting diodes. However, the large lattice mismatch brings in a high density of defects in the grown layer such as dislocations, twins and stacking faults. To decrease the density of these defects, two-step growth has been employed [1-4]. In this method, a buffer layer is deposited at relatively low temperature and the epitaxial growth is conducted at higher temperature. However, the reduction of the defect densities is not enough and much more effective way for the defect reduction has been required.

The reduction of dislocation density by employing lateral growth on SiO_2 from seed openings has been firstly demonstrated by Tsaur et al. in 1982 [5]. They found that the dislocation density of GaAs film laterally grown over SiO_2 on GaAs epitaxial layer on Si substrate with Ge buffer was less than 10^4 /cm^2. After this work, lateral growth has been used not for reducing dislocation density but for obtaining a semiconductor on insulator structure in which the oxide mask was used as the insulator film [6-8]. Jastzebski et al. called this technique as epitaxial lateral overgrowth (ELO) [6]. In 1986 Bauser et al. found in Si/Si LPE ELO that dislocation was not observed in TEM cross sectional image of their samples [9]. However, there was no intention in their work to reduce the dislocation density since Si substrate contains no dislocation. In 1988, the present author employed lateral growth for reducing dislocations in homo-epitaxy of GaAs by liquid phase epitaxy (LPE) which is the best technique to realize wide lateral growth [10].

In 1989, Ujiie and Nishinaga demonstrated that even in GaAs on Si hetero-epitaxy, dislocation free region was obtained in laterally grown GaAs layer over SiO_2 outside of the seed opening area, where dislocation density was extremely high [11]. The epitaxial growth from narrow window can be accomplished in both lateral and vertical directions and both techniques are useful for reducing the dislocation density. Hence, the present author has proposed to use the name of micro-channel epitaxy (MCE) instead of ELO [12].

MCE will open a new horizon of epitaxy. Namely, MCE has a basic idea of technique to separate defect information and lattice information. In usual epitaxy one can grow single crystal film on a single crystal substrate. In this growth the grown film takes the lattice information in the substrate so that the film grows in a single crystal. However, if line or plane defects are present in the substrate, the grown film will take this information and the defects will be inherited into the grown film. In MCE one can take the lattice information through the narrow window (micro-channel) while one can eliminate the defect information by the amorphous film present on the substrate. Hence, the new name, MCE contains wider and more basic concept than ELO.

MCE has a wide range of applications to practical devices. Up to the present, we have achieved dislocation free GaAs and InP epitaxial layers on Si substrate [13-15] and made new devices such as AlGaAs/GaAs lasers on Si substrate [16]. Recently, reduction of dislocation density has been accomplished by using epitaxial lateral growth from a ridge structure by LPE in InGaAs/GaAs [17,18] and by metalorganic vapor phase epitaxy (MOVPE) in GaN/SiC and GaN/Sapphire [19-22].

In the present paper the concept of MCE, which includes partly re-proposal of the modernized ELO is described first and second a few examples of homo-epitaxy cases, such as Si/Si, GaAs/GaAs and GaP/GaP are demonstrated. Then, the applications of MCE to highly lattice mismatched heteroepitaxy such as to GaAs/Si and InP/Si are reviewed and finally recent results of MCE by low angle incidence MBE are described.

2. CONCEPT OF MICROCHANNEL EPITAXY

In conventional epitaxy, as shown in Figure1, the defect information of the substrate is transferred

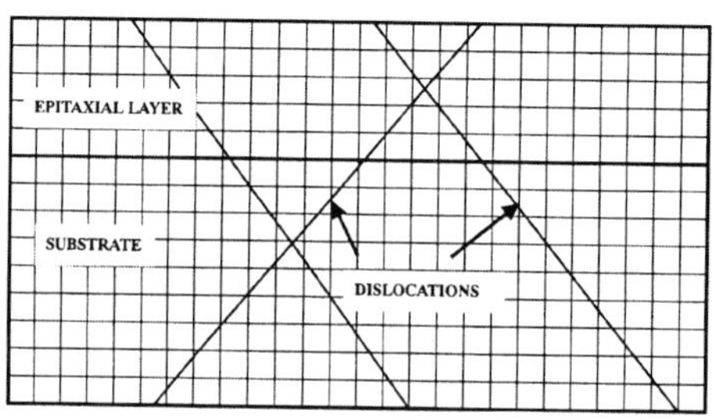

Figure 1. Inheritance of defects in conventional epitaxy

into the epitaxial layer as well as the lattice information. Hence, it is difficult to obtain epitaxial layers without twins, stacking faults and dislocations, if they are present in the substrate. As described before the main idea of MCE is to take the lattice information of the substrate through a narrow microchannel and to elimimate the defect information. There are two methods to realize this idea which are given in Figure 2(a) and (b). Figure 2(a) shows horizontal MCE in which lattice information is transferred through the microchannel while defects are prevented to propagate into the epitaxial layer due to the presence of the amorphous layer. Figure 2(b) shows the vertical MCE in which epitaxial growth is conducted in vertical direction from narrow channels. With choosing proper substrate orientation one can eliminate dislocations from the epitaxial layer by letting them go out from the side surfaces.

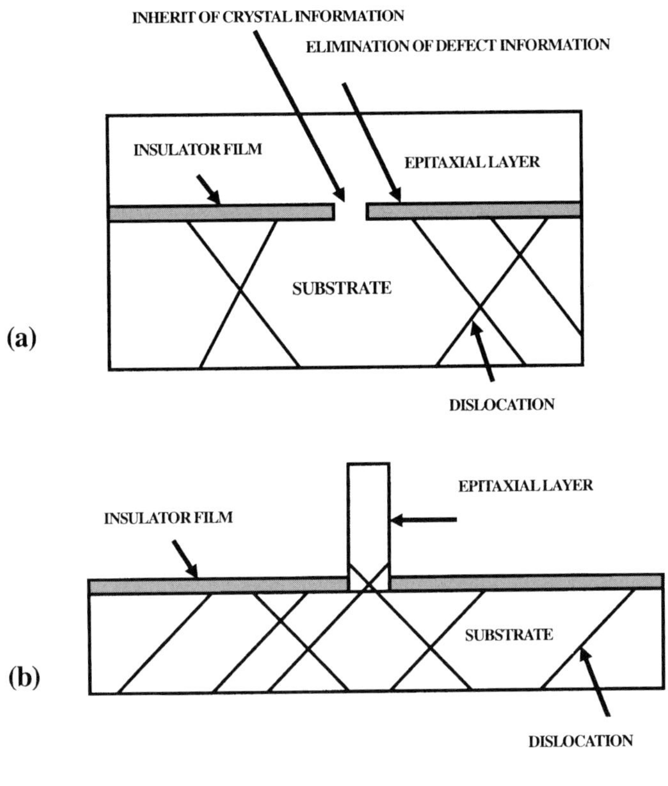

Figure 2. Concept of Microchannel Epitaxy (MCE). Horizontal MCE (a) and vertical MCE (b).

The advantages of creating new name of MCE instead of using old ELO are as follows. First, new concept always gives new ideas. For example, vertical MCE will not be invented by old ELO concept. Second, MCE naturally leads to the idea to decrease the width of microchannel as small as possible because to get lattice information the width of a few nanometers is enough. By decreasing the width to such dimension, one might get even complete dislocation free epitaxial layer on highly lattice mismatched substrate and one can separate epitaxial layer even electrically from the substrate. Hence, it is beneficial for scientific community to give a new name for a new concept.

In the present paper we will focus on the horizontal MCE and review its historical developments. In the following we call horizontal MCE as MCE for simplicity and we call vertical MCE as V-MCE. Real processes of MCE are as follows. First, a selective epitaxy is carried out through narrow microchannels opened in an amorphous film pre-deposited on the substrate and then growth is conducted in lateral direction using the grown film in the window as the seed. Since the area containing the defects generated at the interface between the substrate and the heteroepitaxial layer is restricted within the region over the line seed, one can get a dislocation-free lateral layer grown from the microchannels. So far, defect-free regions have been successfully obtained for GaAs on GaAs substrate [10], Si on Si [23], GaP on GaP [24], GaAs on Si [11, 25], InP on Si [15, 27] and a very strong reduction of dislocation density was reported by Usui et al. for GaN on sapphire [28].

To conduct MCE, one should realize large anisotropy in epitaxial growth. Namely, a high growth rate in the lateral direction should be realized compared with that in the vertical direction. So far, a large anisotropy has been observed by utilizing the difference in the growth rates on a facet and on an atomically rough surface. The growth rate on the atomically rough surface shows a linear increase as interface supersaturation is increased while on the facet, if there is no dislocation with screw component, the growth does not occur before 2D nuclei are generated at relatively high interface supersaturation. Even in the case where a dislocation with screw component exists, the vertical growth rate is much less than that on the rough surface and one can get the condition for a large growth anisotropy to realize MCE.

3. MCE EXPERIMENTS BY LPE

3.1. Si

MCE of Si by LPE with Sn and other metallic solutions has been successfully conducted [9, 23, 29-31]. In LPE of Si at 800 ~ 900 ℃, the facet which appears in LPE with Sn solution is limited only to (111). Hence, one has to use the substrate with (111) surface. The direction of the microchannel should be chosen so that the side surfaces become atomically rough. To find the best orientation for the microchannel we conducted growth employing a pattern in radial arrangement as shown in Figure 3.

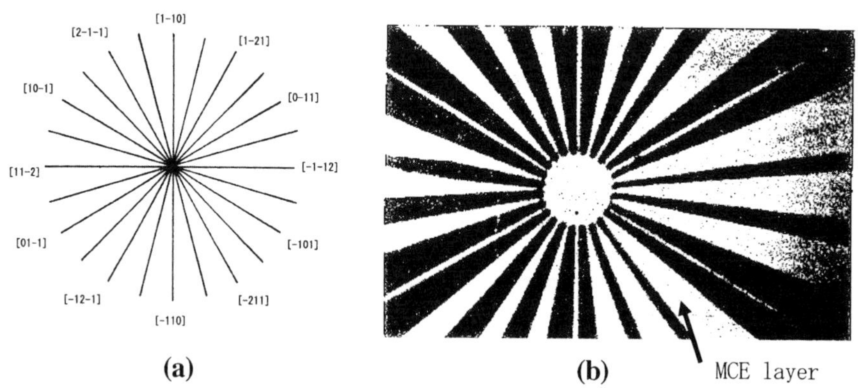

Figure 3. MCE layers of Si from a radial pattern on (111) Si substrate. Figure (a) shows the mask pattern and (b) does the photograph of the grown layer taken by optical microscope.

Figure 3 (a) shows the crystallographic orientations and (b) does the MCE layers grown from a radial microchannel arrangement. As seen in the figure, the orientations off from [-110] and equivalent ones give high lateral growth rate. On the other hand growth fom the microchannels in [-110] or equivalent orientations gives the slowest lateral growth velocity. This is because the lateral growth fronts of the MCE layers are covered with {111} facets. We have chosen [211] as the direction of the microchannel [23]. Another point of interest is that the surface of the MCE layer is extremely flat and no macrosteps are observed although a plenty number of macrosteps are observed on the surface of Si epitaxial layer grown by conventional LPE. In the case of Si epitaxy, the substrate is usually dislocation free. Hence, the reduction of the dislocation density is not important but it is reported that although threading dislocation is not present in Si substrate, there are dislocation loops generated depending on the condition of bulk crystal growth. Such loops may be punched out at the surface and play a role to generate threading dislocations at substrate and epi-layer interface. MCE can be used to eliminate such dislocations.

Typical surface of Si MCE layer grown from parallel microchannel pattern is shown in Figure 4. As seen in the figure, the surface is covered by (111) facet and extremely smooth like optical mirror. In the best case, the width and the thickness so far obtained in our laboratory were 245μm and 3.2 μm, respectively as shown in Figure 5.

The width to the thickness ratio, which is defined as W/T ratio, was around 75. Since the top is covered with (111) facet, the surface becomes completely flat. Instead of using line seed one can employ a ridge-seed, which has the height larger then the SiO_2 thickness. The ridge-seed gives better control for the MCE layer thickness [29].

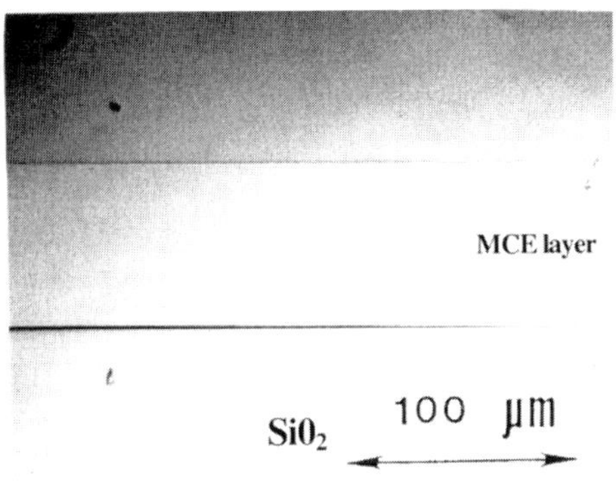

Figure 4. Typical MCE surface photograph taken by optical microscopy.

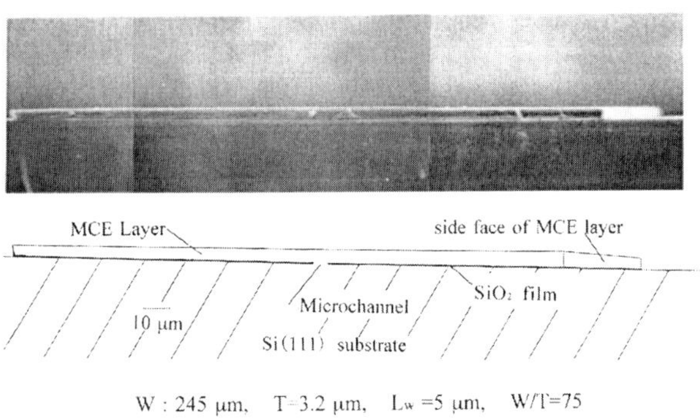

Figure 5. Cross section of Si MCE layer, where W, T and L_W denote width, thickness and microchannel width of the MCE layer, respectively.

3.2 GaAs

3.2.1. GaAs on GaAs

As shown in Figure 6 the growth velocity in lateral direction depends strongly on the orientation of the microchannel [32] namely the microchannel in [-110] and [010] directions on (001) substrate give the lowest lateral velocity while the microchannels with the orientation between these two directions give large lateral growth. This is because the sides of the MCE layer grown from [-110] and [010] microchannels are covered by one of (100), (111)A and (111)B facets as shown in Figure 7. But, the sides of MCE layers grown from other microchannels become atomically rough. The grown layer was chemically etched and the etch pits density was measured. It was found that there are no etch pits in the layer above the oxide layer [10].

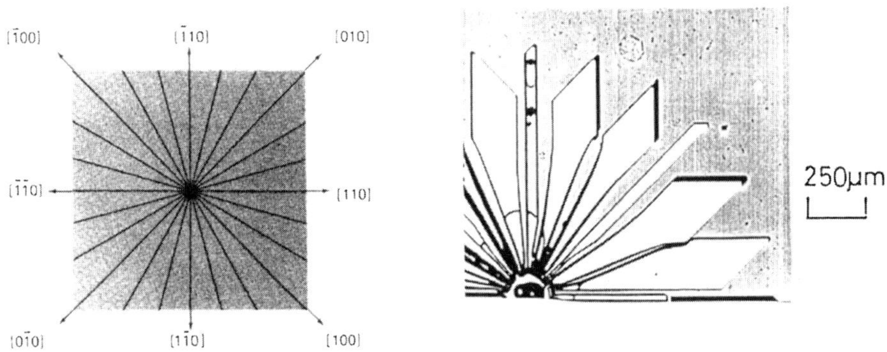

Figure 6. MCE of GaAs on (001) GaAs substrate. Mask pattern (a) and MCE grown layers (b).

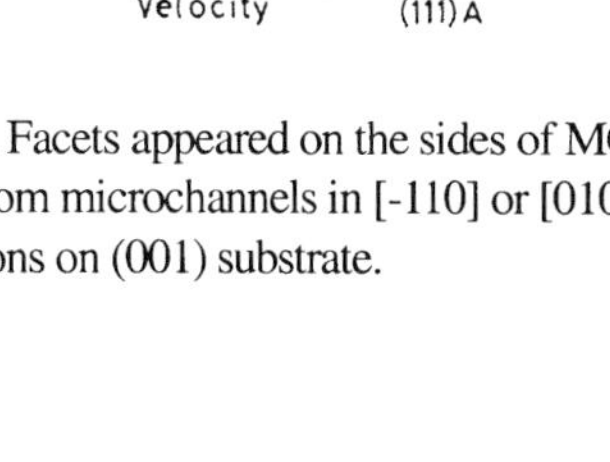

Figure 7. Facets appeared on the sides of MCE layers grown from microchannels in [-110] or [010] orientations on (001) substrate.

In Figure 7 growth velocity in lateral direction is shown as a function of angle from [110] orientation. As seen from the figure lateral growth velocity takes minimum at a special orientation of the microchannel, where the side surface is covered by facets while in other orientations growth velocity becomes large since the side surface is atomically rough.

3.2.2. GaAs on Si

There is a long history of technology to obtain high quality GaAs on Si. However, the lowest dislocation density so far obtained by conventional epitaxy technique is $10^6/cm^2$, which is too high to fabricate a laser diode

(LD) with a reasonable lifetime. We have employed MCE to grow high quality GaAs on Si [11, 13, 14, 25]. There is a difficulty to use LPE directly for the growth of GaAs on Si. In GaAs LPE, Ga is used as the solution. However, when Ga is brought into contact with Si substrate, the Si substrate is attacked by Ga melt strongly so that the direct LPE of GaAs on Si can not be conducted. To solve this problem, a thin GaAs layer of a few μm thickness was epitaxially deposited on Si substrate by MBE or MOCVD.

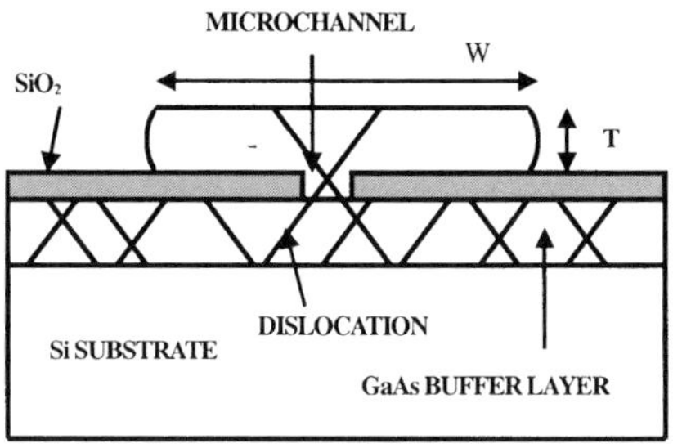

Figure 8. MCE of GaAs on Si substrate coated by thin GaAs buffer layer.

A dislocation-free area on Si (111) substrate by MCE was successfully achieved in 1989 [11]. In this experiment, we have used a Si substrate covered with MBE grown GaAs.

The schematic cross section of MCE of GaAs on Si is given in Figure 8. The surface of MCE layer after etching is shown in Figure 9. As seen in the figure, etch pits appeared in the region above the microchannel and one sees no etch pit appeared outside of this region. Since the thickness of the MCE layer was large, cracks were generated due to the difference in thermal expansion coefficients between GaAs and Si. The widening of the cracks happened by the chemical etching.

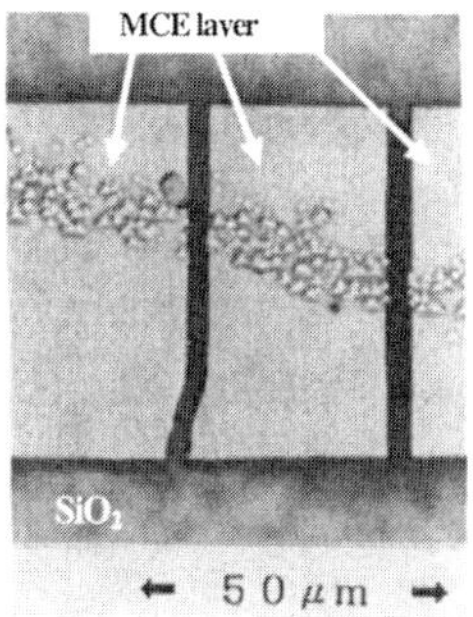

Figure 9. MCE GaAs on (111) Si after etching. Since the thickness is large, cracks were generated due to the difference in the thermal expansion

For device application, growth on (001)Si is more desirable. However, the growth on (001) gives wider dislocated area above the microchannel. This is because the angle between the slip plane and the substrate is 55 degree and dislocations from the substrate propagate on this plane. The configuration of slip planes in the MCE layer, microchannel and substrate is shown in Figure 10.

As seen in Figure10, the dislocated area on the surface of MCE layer is increased as the growth proceeds so that to get wide dislocation-free area, one should grow in lateral direction much more than in the vertical direction. In other words one should get a large anisotropy in the growth.

To increase the growth anisotropy it was found that Si doping in the MCE layer is very effective

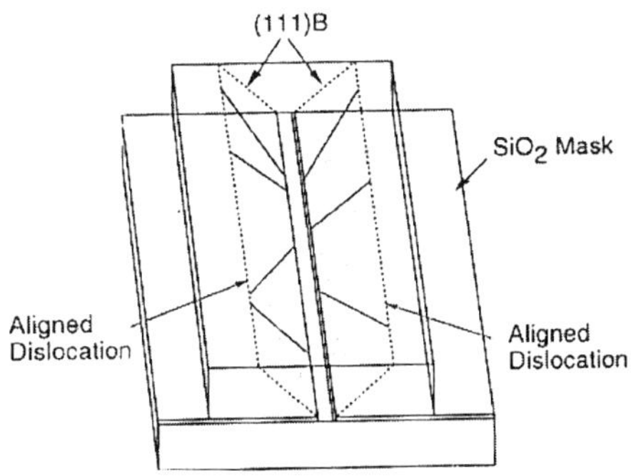

Figure10. Slip planes in MCE layer and microchannel

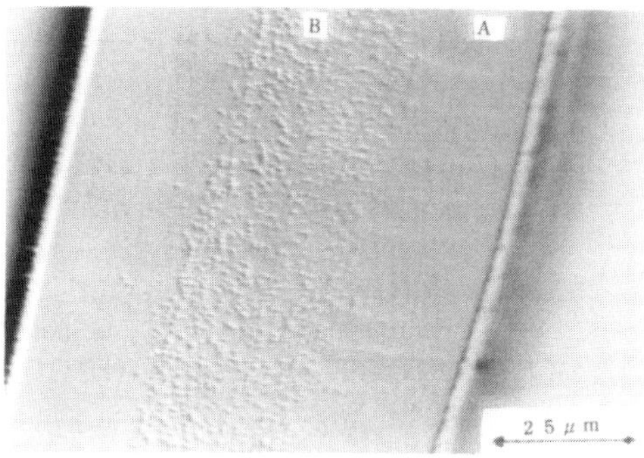

Figure 11. GaAs MCE layer grown on (001) Si substrate after chemical etching. Si is doped in MCE layer to get flat surface and high growth anisotropy. In the figure, A and B denote respectively the regions of the etch pit free and the high etch pit density.

This has been explained as the Si doping in GaAs causes the retardation of step advancement in the growth and hence the suppression of vertical growth [25].

Figure 11 shows GaAs MCE layer doped with Si after molten KOH etching [25]. It is seen that etch pit free regions are present on both sides of the dislocated area existing above the microchannel. Although the channel width was several microns, the width of the dislocated region is nearly 30 μm due to the widening effect explained with Figure 10. In this case, the width of the etch pit free region is around 25μm, which is not enough to be used for device fabrication. To increase the width of the etch pit free region, we repeat many experiments by changing the growth rate, growth temperature and the distance between the microchannels. After optimizing the growth conditions it was possible to get MCE layers with very wide etch pit free region and small thickness. Figure 12 shows the examples.

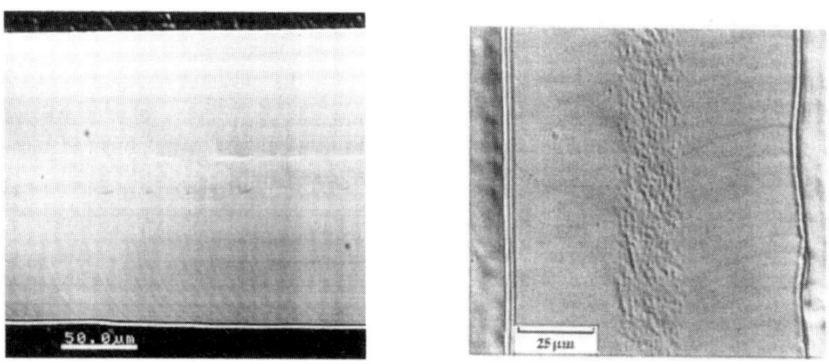

(a) **(b)**

Figure 12. MCE layer grown at optimized growth conditions. As-grown MCE layer (a) and after molten KOH etching (b). The sample of (b) is different from that of (a). The width of the MCE layer in (a) is 195 μm. One sees some structure in the center of (a), which shows the presence of a high dislocation density.

Progress of LPE MCE technique makes it possible to increase the width and to decrease the thickness of MCE layer. Now, we do not see any cracks in the MCE layers. The best data so far we obtained is that the thickness, the width and W/T ratio are 12 μm, 200 μm and 17μm, respectively [14].

Figure 13. TEM cross sectional photograph of GaAs MCE layer on GaAs coated Si substrate.

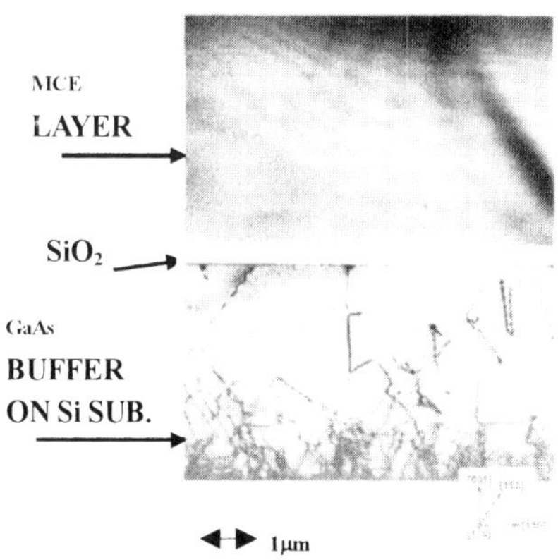

Figure 14. Magnification TEM photograph of Figure 13 near the interface of the SiO_2 and GaAs buffer layer.

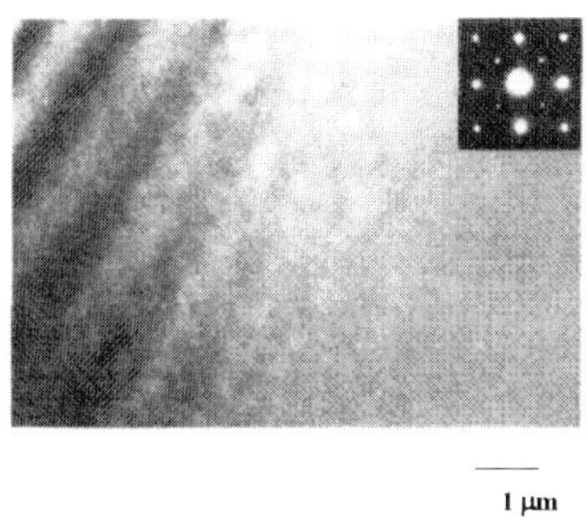

Figure 15. TEM plane view of GaAs MCE layer on Si substrate. From Figure 13 to 15 are taken by Dr. M. Tamura of JRCAT-ATP.

Figure 13 shows the TEM cross sectional photograph of GaAs MCE layer grown on GaAs coated Si substrate. The growth of GaAs on Si by MOCVD was conducted by Dr. H. Mori and Dr. M. Tachikawa of NTT Opto-Electronics Lab. and TEM photographs were taken by Dr. M. Tamura of JRCAT-ATP. As seen in the figure, there are a large amount of dislocations in the GaAs buffer layer on Si substrate. Some dislocations propagate on {111} planes through the microchannel. Thus, the dislocated area expands as the growth proceeds hence to decrease this area one should grow the MCE layer in horizontal direction as large as possible compared with that in vertical direction. Figure 14 shows a part of Figure 13 near the SiO_2 film. One sees clearly that the propagation of the dislocations is stopped at the interface between SiO_2 and GaAs buffer layer and there is no dislocation seen above the SiO_2. Since the sample of TEM is very thin, the cross sectional TEM photograph cannot be the clear evidence to show the layer is dislocation-free. Hence, TEM plane view was taken and is shown in Figure 15, which shows the MCE layer is dislocation-free.

3.3. InP and GaP

3.3.1. InP

MCE of InP was successfully carried out by LPE with In solution. However, since the growth temperature is relatively low compared with GaAs LPE MCE, the surface tension of the In melt takes a higher value which makes it difficult to get a contact of the melt to the surface of the substrate in the microchannel. To solve this problem, In metal was deposited before growth by vacuum evaporation on the InP substrate, after SiO_2 was deposited and the microchannel was opened [33].

Although MCE of homoepitaxy gives an almost dislocation-free layer, sometimes the whole epitaxial layer is grown by steps generated at one single dislocation with screw component. This makes it possible to measure the inter-step distance in a wide area on the MCE layer by AFM and to calculate the interface supersaturation by the equation given by Cabrera and Levine by assuming proper interfacial free energy of the InP and In solution [34]. Figure 16 shows a MCE island and AFM picture of a screw dislocation observed at one corner of the MCE island. It was found that this screw dislocation generates all the growth steps which cover the whole surface of the island.

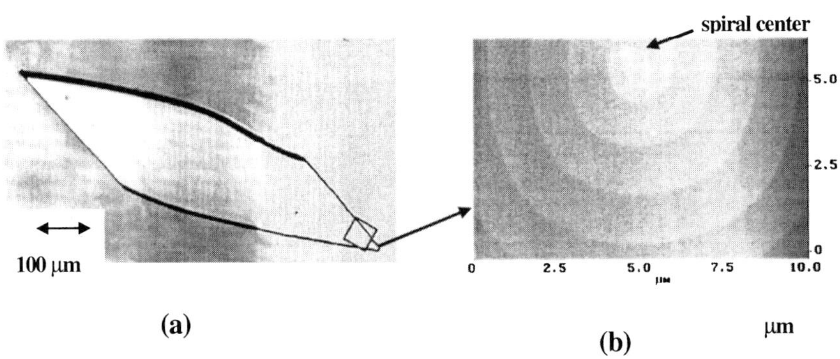

Figure 16. MCE island of InP(a) and AFM image of a screw dislocation observed at the corner(b). The single screw dislocation generates all steps for the growth.

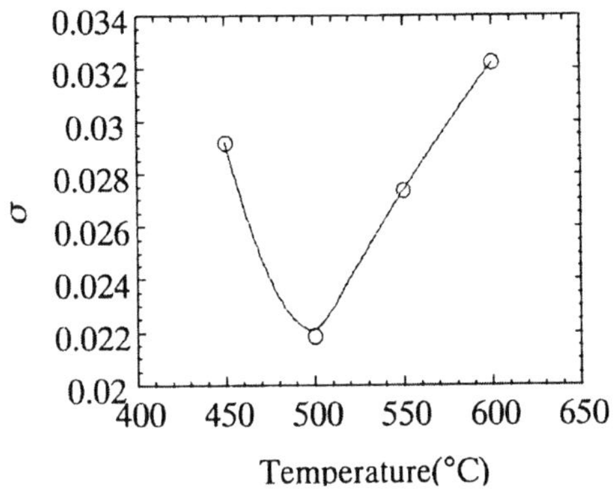

Figure 17 Supersaturation ratio (σ) vs. growth temperature (W/T) in MCE of InP.

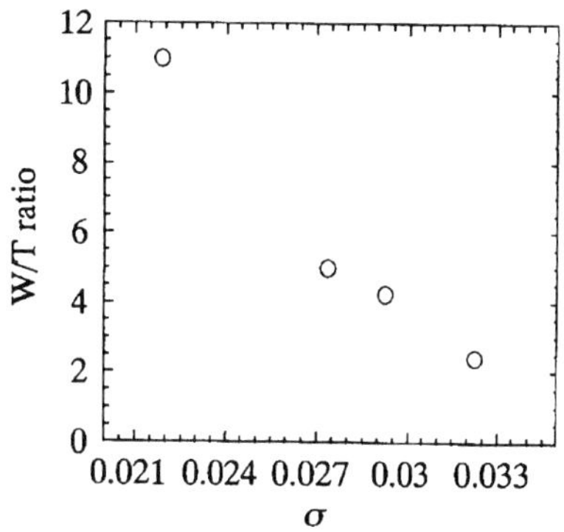

Figure 18. Width to thickness ratio vs. supersaturation ratio (σ).

In Figures 17 and 18 supersaturation ratio vs. growth temperature and width to thickness ratio (W/T) vs. supersaturation ratio are shown, respectively. It is seen in Figure 17 that the supersaturation ratio has a minimum. At the higher temperature side, solute concentration decreases as the growth temperature is decreased while in the low temperature side solute incorporation velocity decreases as the growth temperature decreases and interface supersaturation increases. It is seen in Figure 18 that W/T ratio increases rapidly as the supersaturation ratio is decreased. This is because as the supersaturation ratio decreases, the growth velocity on a facet decreases under-linearly due to the screw dislocation growth mechanism. On the other hand, the lateral growth front is atomically rough, which means the growth velocity decreases linearly as a function of the supersaturation. From this experiment it is shown that lower interface supersaturation is better to obtain higher MCE width to thickness ratio [35-37].

MCE growth of InP on Si substrate is also successfully conducted. Substrate employed for LPE MCE was prepared by Tachikawa and Mori of NTT by MOCVD in which 2μm GaAs was first grown on Si substrate then 13 μm InP was deposited. InP LPE was conducted on the top of the MOCVD-grown InP. MCE InP on Si substrate after chemical etching is shown in Figure 19.

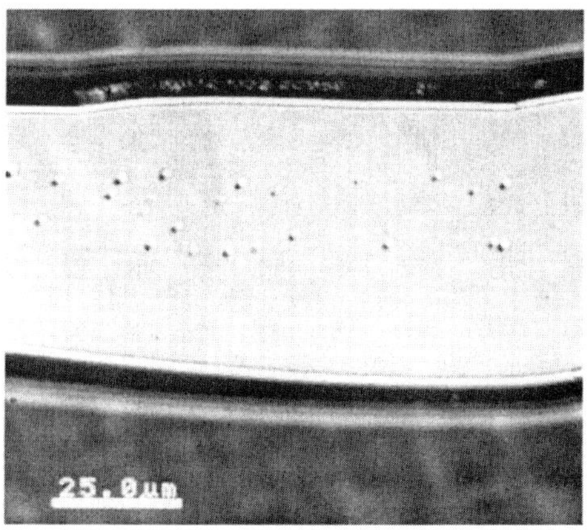

Figure 19. InP MCE on Si substrate after etching. Etch pits appeare above the microchannel.

3.3.1. GaP

MCE of GaP on (111) GaP substrate was conducted [24]. By choosing proper microchannel orientation, a large W/T ratio was also obtained. Lateral growth behavior is quite similar to Si MCE on (111) Si substrate and we have chosen [-211] orientation for the microchannel. The growth temperature was around 1000 ℃ the growth time was typically one to two hours.

Figure 20 shows the MCE layer of GaP after RC etching (AgNO$_3$<mg>:HNO$_3$<cc>: HF<cc>:H$_2$O<cc>=4:2:3:4 at 60 ~ 90 ℃ for 1-3 min.), which is employed to reveal etch pits for GaP. As seen in the figure, etch pits appeared outside of MCE layer where originally SiO$_2$ was present but by etching SiO$_2$ was removed and etch pits of the GaP substrate are revealed. We do not see any etch pits on the laterally grown area although a few etch pits are seen just above the microchannel. Those etch pits are corresponding to the threading dislocations propagated through the microchannel from the substrate.

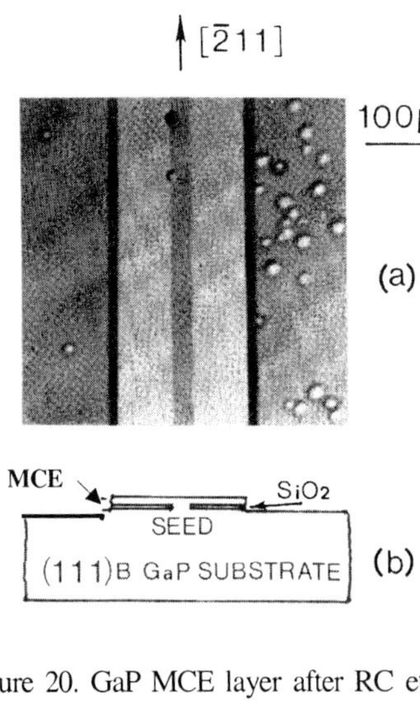

Figure 20. GaP MCE layer after RC etching (see text). Since GaP is transparent one sees the microchannel through the MCE layer.

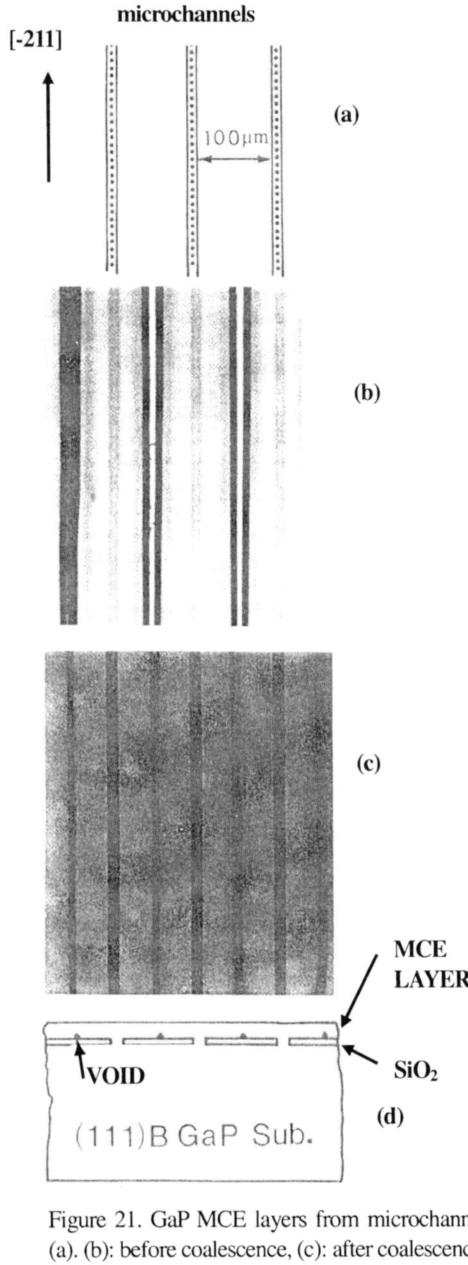

Figure 21. GaP MCE layers from microchannel (a). (b): before coalescence, (c): after coalescence. (d) shows schematic cross section of (c).

Figure 21 shows how coalescence happens during the growth. In this experiment, microchannel separation of 100 μm was employed. The figure (b) shows the MCE layers before the coalescence where the growth is started from 1018 ℃ and is finished at 988 ℃. The growth was conducted for 60 min. To observe the coalescence, higher growth rate and longer growth time were required. Figure (c) shows the MCE layers after coalescence. Here, we employed higher growth starting temperature of 1045 ℃and growth was continued for 2 hours. As seen in the figure, coalescence occurred smoothly. Since GaP is transparent for visible light one can see both of microchannels and linear voids at the MCE layer-substrate interface.

The formation mechanism of the linear void generated at the coalesced front might be as follows. The key issue is the growth velocity difference at the top and the bottom corners of the MCE layer. If the growth rate at the top corner is larger than that of the bottom corner, the coalescence happens at the top corner first, which gives linear void at the bottom of the MCE cross section depends strongly on the growth velocity. If the growth velocity is large, the top corner has the tendency to grow faster than the bottom corner, resulting in the formation of the linear void. On the other hand if the growth rate is very slow, the bottom corner grows faster. In our experiment, by decreasing the growth rate it was possible to get void-free coalescence.

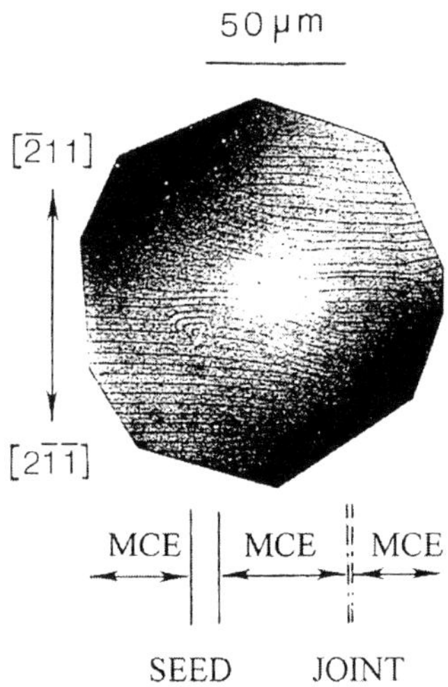

Figure 22. Monomolecular steps observed on MCE layer of GaP near the coalescence point (joint).

Figure 22 shows a photograph taken by Nomarsky differential interference contrast microscope (N-DICM) combined with an image processor. As seen in the figure monomolecular steps are revealed. Evidence for the resolution is given in ref. [24]. On the surface of MCE layer given in the figure, monomolecular steps are seen to propagate in one direction. Even if there is a dislocation coming from the substrate as seen in the figure, this can not be active since the density of the steps sent from strong step source is much larger.

Another interesting fact is that no dislocation spiral is seen at the coalescence front. This means that no dislocation is generated at the coalescence point. It has been reported, that a large number of dislocations is generated when two lateral growth fronts meet in the growth of nitride semiconductors [29]. This problem will be discussed in the next section.

3.4. Coalescence of MCE layers

If one wishes to get a uniform and wide area of the MCE layer, one can continue the growth until two laterally growing MCE layers coalesce. The coalescence of two laterally growing fronts has been already demonstrated in the growth of InP by halogen CVD by Vohl et al. [39]. As shown in the previous section, the coalescence between two MCE layers is possible in the GaP/GaP system and there is no dislocation with screw component present in the coalesced region [24]. Nagel et al. and Banhart et al. studied the coalescence of laterally overgrown layers by LPE on SiO_2 in the Si/Si system [38, 40]. They found that by choosing proper growth conditions a coalescence without the generation of defects is possible.

Recently, we have conducted systematic studies on the coalescence in LPE MCE in InP/InP [41] and GaAs/GaAs systems [42]. In the following we discuss this problem taking GaAs MCE as an example.

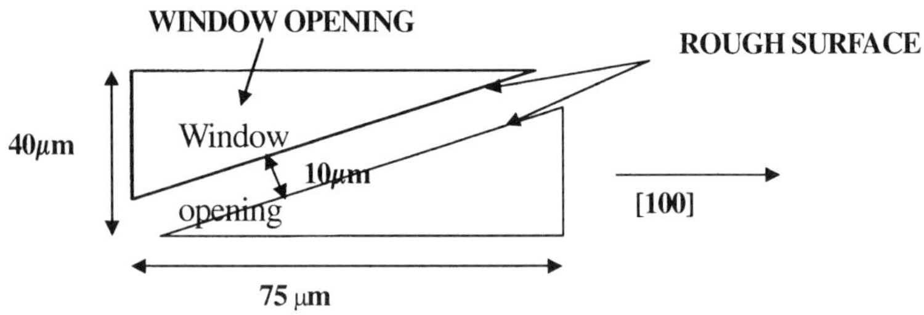

Figure 23. The pattern for window openings which have two parallel sides of rough surfaces.

3.4.1. Lateral coalescence from two parallel seeds

Figure 23 shows the triangle patterns of the window openings, which was designed to have two sides in parallel. The orientation of these parallel sides was chosen to have off-orientation from [100] so that the lateral growth front becomes atomically rough. Figure 24 shows a series of photographs before and after growth with the pattern of Figure 23. Figure 24(a) shows the window openings before growth and (b), (c) and (d) after a growth period of 20, 40 and 80 min, respectively. From figure (b) it can be seen that four {100} and {111} facets appear at the corners of the triangle growth islands. The lateral growth occurred from two rough edges of the triangular window openings. It can be seen that the lateral growth velocity was higher at the ends of the open area between the two islands. This can be easily understood because the local supersaturation in the end regions is

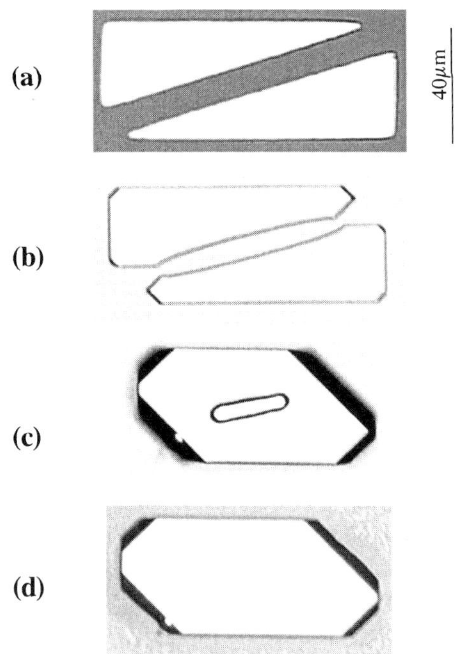

Figure 24. MCE layers grown from the mask pattern given in Figure 23. Before growth: (a), after growth: of 20 (b), 40 (c) and 80 (d) min.

higher than that in the center area. Thus, the lateral coalescence firstly occurred at both ends of the gap between the triangle islands and then proceeded toward the center area as shown in (c). From (d) it can be seen that after 80 min the lateral coalescence was finished and the MCE island is surrounded by {100} and {111} facets. This experiment shows, if the lateral growth occurs from two triangle windows with rough side surfaces in parallel, the coalescence takes place in a "two-zipper mode".

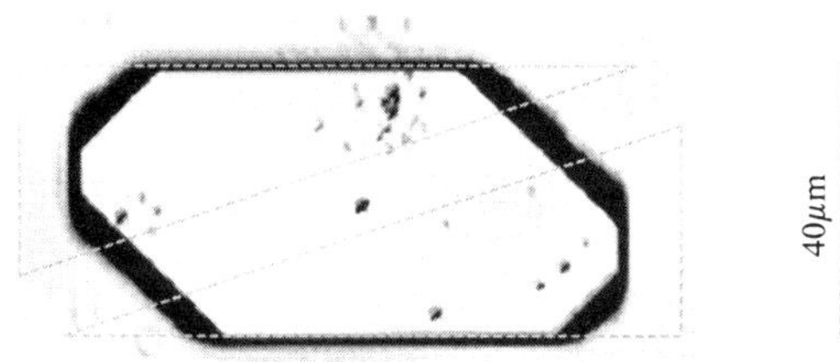

Figure 25. Optical microphoto of the MCE island laterally coalesced from two parallel seeds shown in Figure 24 (a) after molten KOH etching. In the figure the dotted lines indicate the original position of the window openings.

Figure 25 shows a photograph of the MCE island by optical microscope after molten KOH etching. In the figure the dotted lines indicate the original position of the triangle window openings. From Figure 25 it can be seen that a dislocation etch pit appears at the center part of the coalesced area but there is no other etch pit found in the coalesced area. Comparing Figure 25 and Figure 24, one can find that this dislocation etch pit appears at the last coalesced point. Namely, when the coalescence occurs in a "two-zipper mode" the dislocations are generated in the last coalesced point. Among these coalesced islands, nearly 90% islands showed the presence of the etch pits at the last coalesced area. Among the 90%, 60% of them show a single etch pit while the rest show two to three etch pits near the center of the coalesced area. Hence, there are nearly 10% islands which show no etch pit. However, there is no guarantee on the coalescence happened without the generation of any dislocation. We should confirm this by more detailed investigations such as by TEM or X-ray topography.

3.4.2. Lateral coalescence from non-parallel seeds

Figure 26 shows the patterns of the window openings which were designed for the lateral growth to start from seeds which are aligned non-parallel to each other and in off-orientation from [100]. In the figure the closed areas indicate the window openings. Figure 27 (a) shows the GaAs substrate patterned by Figure 26 (a) before the growth. Figure 27 (b), (c), (d) and (e) show the MCE island after the growth

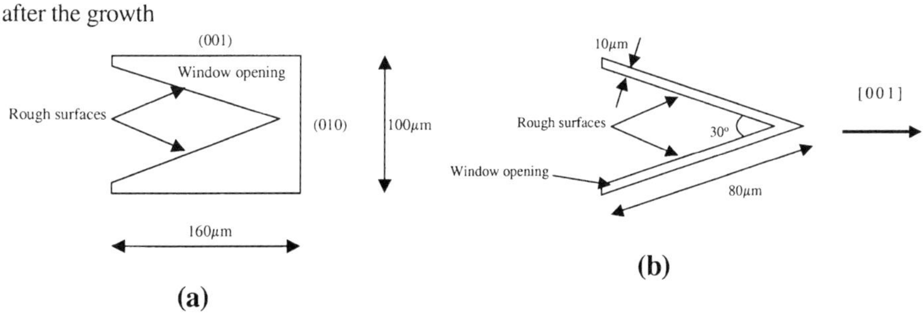

Figure 26. Two patterns of the window opening which were designed to have two non-parallel sides from which the lateral growth occurs.

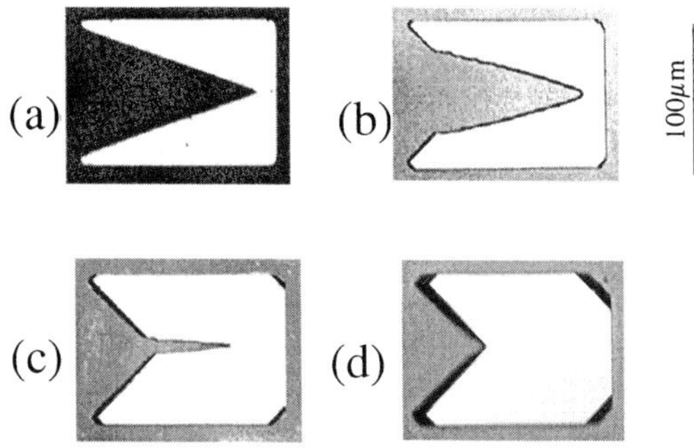

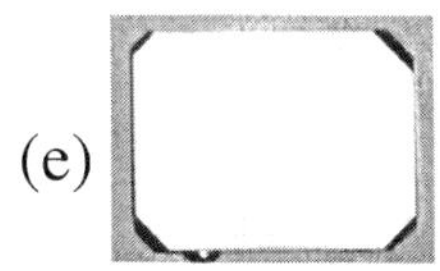

Figure 27. Photographs of MCE layers grown from the seed given in Figure 26(a). Before the growth (a), after the growth of 1 (b), 3 (c), 5 (d) and 7 hours (e).

1, 3, 5 and 7 hours, respectively. From Figure 27 (b) it can be seen that after 1 hour {111} facets appear in the end parts of the two arms of the MCE island but the rest part of V-shape growth front was still surrounded by rough surface. From Figure 27 (c) it can be found that the lateral coalescence began from the bottom of V-shape growth front and then proceeded toward the open area. From Figure 27 (d), one can see that after 5 hours the lateral coalescence is already finished because the MCE island is surround by {100} and {111} facets. However, since the reentrant corner can supply steps, the open triangle area in the left hand side of the MCE island was finally filled to give a nearly rectangular MCE island as shown in Figure 27 (e).

Figure 28. Photographs of MCE layers grown from the seed given in Figure 26(b). Before growth (a), after the growth of 20 (b), 60 (c), 80 (d) and 120 min (e).

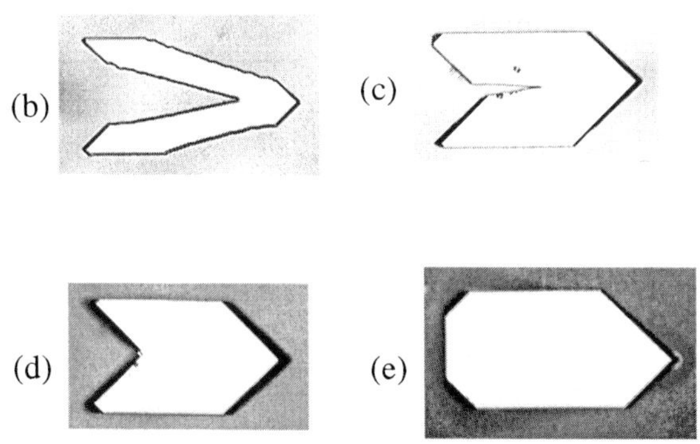

Figure 28 (a) shows the GaAs substrate patterned by Figure 26 (b) before growth. Figure 28 (b), (c), (d) and (e) show the photographs after the growth of 20, 60, 80 and 120 min, respectively. It can be seen that after 20 min, {100} and {111} facets already appeared at all ends of the V-shape pattern but other parts of the sides were surrounded by rough surfaces and thus lateral overgrowth occurred at both inner and outer sides of the V-shape pattern. After 60 min, as seen in Figure 28 (c), all of the outer periphery of the MCE island is covered by facets. It means that the lateral overgrowth at outer periphery was finished. However, a small V-shape inner side was still surrounded by rough surfaces. From Figure 28 (c) it can be found that the lateral coalescence began from the bottom of the V-shape pattern and then continued toward the open area of the V-shape like the case in Figure 27. After 80 min, the coalescence of lateral layers grown from two non-parallel seeds finished except the triangle open area at the left hand side of MCE island. However, as explained in Figure 27, due to the reentrant corner effect, the triangle open area was finally filled. Figure 28 (e) shows the photograph after the completion of the coalescence.

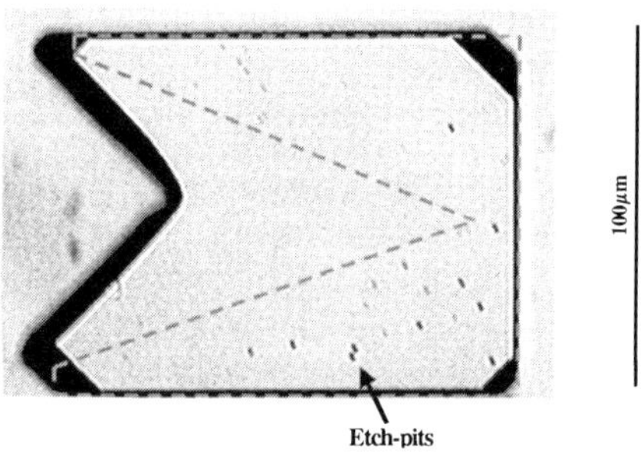

Figure 29. Optical microphotograph of the MCE island laterally coalesced from non-parallel seeds given in Figure 26(a) after molten KOH etching. In the figure the dotted lines indicate the original position of the window openings.

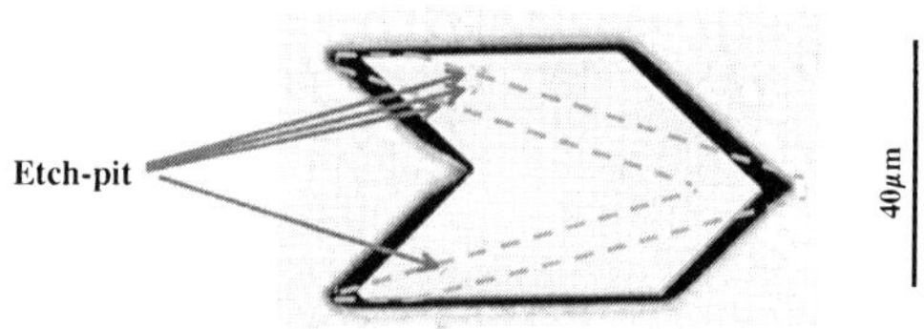

Figure 30. Optical microphotograph of the MCE island lateral coalesced from non-parallel seeds given in Figure 26(b) after molten KOH etching. In the figure the dotted lines indicate the original position of the window openings.

Figures 29 and 30 show the optical microphotographs of the MCE islands laterally coalesced from non-parallel rough surfaces shown in Figures 26 (a) and (b) respectively, after molten KOH etching. It can be found that there was no dislocation etch pit existing in the laterally coalesced areas. In other words, when non-parallel seeds are employed, the lateral coalescence occurs in a "one-zipper mode". In this case, the coalescence starts from one point and no dislocation is generated in the coalesced area. For the etching experiments we have employed the samples at the coalescing stage of (d) in Figures 27 and 28 and not the final stage of (e) in the same figures. This is because if we wait until the final stage of the coalescence, the vertical thickness of the islands becomes large which makes it sometimes difficult to distinguish the etch pits caused by the dislocations generated at the coalescence of those propagated from the windows, since dislocations propagate on {111} plane and not perpendicular to the substrate. On the other hand, to see the generation of the dislocation in one-zipper mode, the etching of the samples at the stage of (d) is enough.

In conclusion it was shown that if we used non-parallel microchannels no etch pit appeared at the coalesced area, but if we used parallel microchannel etch pits appeared at the last point of the coalescence. This can be explained as follows. When non-parallel microchannels are used the coalescence starts from one single point [40], while if a short parallel microchannel is used the coalescence starts from two points from the both ends of the microchannel which leads to the generation of dislocations at the end point of the coalescence [41, 42].

4. MICROCHANNEL EPITAXY OF GaAs BY MBE

4.1. Vertical microchannel epitaxy (V-MCE) of GaAs

As discussed in the section 2, vertical MCE (V-MCE) is also useful to reduce the density of dislocations. For this purpose we have employed MBE and tried to conduct V-MCE of GaAs [43].

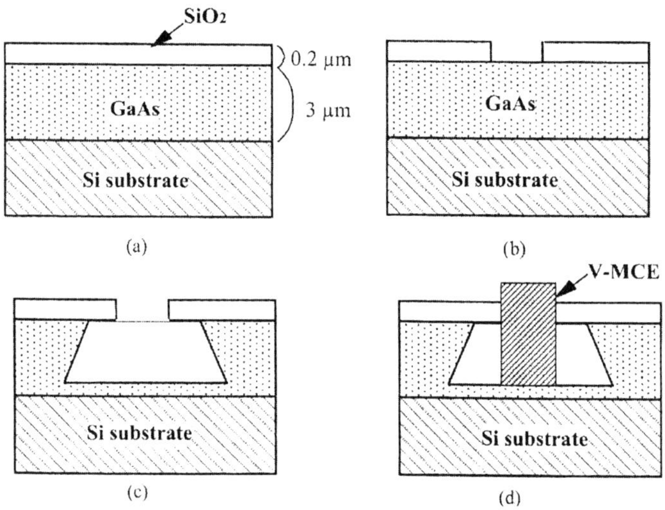

Figure 31. Schematic illustrations of the fabrication process for a V-MCE structure. (a): deposition of SiO_2, (b): patterning of SiO_2, (c): formation of cavity by anisotropic etching of GaAs buffer layer and (d): V-MCE growth by MBE.

The steps of the preparation process of the substrate for V-MCE are illustrated in Figure 31. First, GaAs with a thickness of 3 μm was deposited by MBE on the Si substrate which was 1° misoriented toward [110] orientation. Then, a 0.2 μm thick SiO_2 film was deposited as shown in Figure 31 (a). Microchannels were opened by conventional photolithography as shown in the same figure (b). The width of the microchannel was varied from 2 to 10 μm. The orientation of the microchannel was fixed as [110]. Cavities with SiO_2 film on the top were fabricated by anisotropic etching as shown in (c). At the bottom of the cavity we left a thin layer of GaAs. This layer plays the role of a buffer layer for the heteroepitaxy of GaAs on Si to prevent defects such as antiphase domains. After thermal annealing at 590 ℃ to evaporate oxides, V-MCE was conducted for 740 min at a high growth temperature of 640 ~ 650 ℃ and with a low growth rate of 0.28 μm/h to prevent polycrystalline growth on the SiO_2 mask. A III-V flux ratio of around 8 was employed. Figure 31 (d) shows schematically the V-MCE structure after the growth. The dislocation density was evaluated by counting etch pits after molten KOH etching at 390 ℃ for 2s.

The etch pit density on the top of the V-MCE structure depends on the aspect ratio of V-MCE, which is defined as the ratio of the height to the top width. If the aspect ratio is large, threading dislocation can easily go out from the sides of the V-MCE structure. The largest aspect ratio so far obtained was 1.4 and in this case, the reduction of the dislocation density was 1/5. The reduction ratio was not big compared with horizontal MCE, where it was possible to obtain a dislocation-free area.

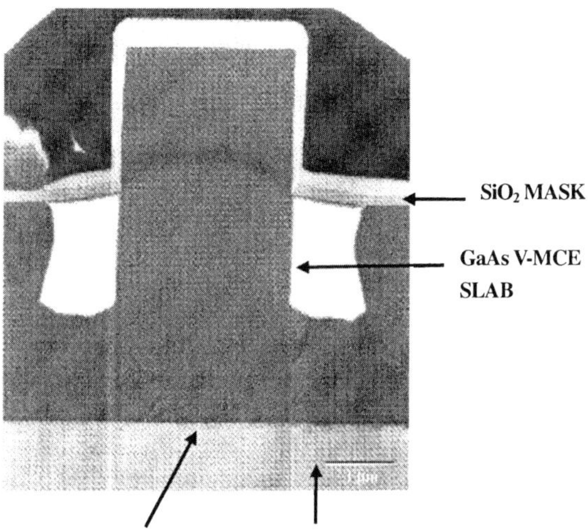

Figure 32. TEM image of V-MCE on Si. The height V-MCE is 3.5 μm and the width of the microchannel is 2.5 μm.

Figure 32 shows a TEM cross sectional photograph of V-MCE. The growth temperature of this sample was 650 ℃. It can be seen that the cross section is that means the top of V-MCE is covered by a (001) face. The width and the height are 2.5 μm and 3.5 μm, respectively, which gives the aspect ratio of 1.41. The TEM photograph indicates no plane defects nor dislocation in the upper part of V-MCE slab, although dislocation images are seen in the bottom part of V-MCE slab, where dislocations with a large density were generated at the GaAs-Si interface. Since the thickness of the TEM sample is very thin such as 0.2 μm, a missing dislocation image can not be the evidence for a dislocation-free material. By etch pit measurement on top of the V-MCE, a dislocation density around 2 x $10^7/cm^2$ was found. Although this number looks large, V-MCE slab can reduce the dislocation density in principle to 1/3 of the original density because there is one more degree of freedom for the dislocation propagation. If one employs the V-MCE shape of square rod instead of vertical slab, it should be possible to get a dislocation-free part in the top of the structure.

4.2. Microchannel epitaxy of GaAs by low angle incidence MBE

It was found that by sending Ga and As_4 fluxes with low angle with respect to the substrate in MBE, one can grow thin layers of GaAs over insulator film [44-46]. This technique is called as LAIMCE (low angle incidence microchannel epitaxy). The schematic illustration of this technique is given in Figure 33. The definitions of various angles are shown in figure (a) and the cross sections of the substrate and Ga/As_4 fluxes before and after growth are given in (b) and (c) respectively. It was also found by sending the fluxes with low angles that the selectivity is improved largely so that the selective area epitaxy becomes easier even by MBE under normal conditions.

In Figure 33 (c), d denotes the microchannel width, W the width for the lateral grown part of the epilayer and T the thickness. In some cases the value of W is difficult to be defined since the fronts of the lateral growth have various shapes. Therefore, we have defined L_m and W_m, respectively, as the maximum bottom width and the maximum width of the lateral grown part of the epilayer. Figures of merit of the epilayers are W_m and the W_m/T ratio. For device fabrication, large W_m is required and a greater value of the W_m/T ratio is preferable to utilize a dislocation-free area when this technique is applied to heteroepitaxy.

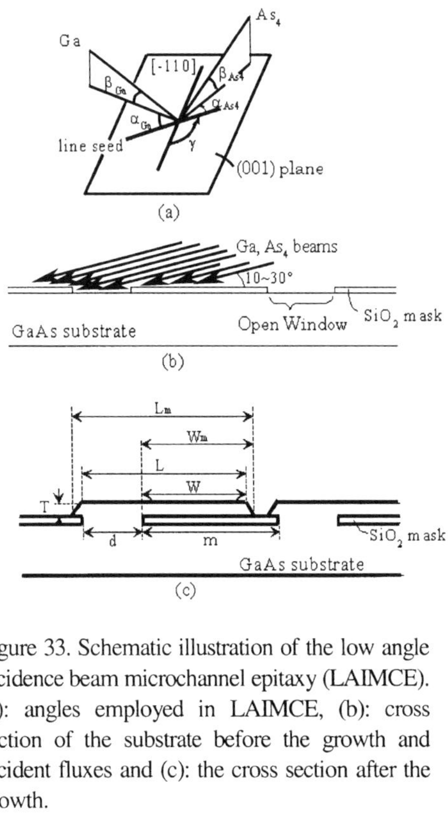

Figure 33. Schematic illustration of the low angle incidence beam microchannel epitaxy (LAIMCE). (a): angles employed in LAIMCE, (b): cross section of the substrate before the growth and incident fluxes and (c): the cross section after the growth.

The growth conditions employed in the experiments were a substrate temperature, an As_4 pressure, and a growth rate of 610℃, 3.8×10^{-5} Torr and 1.08 μm/h, respectively. The LAIMCE angles β_{Ga} and β_{As4} were 11° and 23°, respectively. The 1.08 μm/h growth rate was calculated from the period of the intensity oscillations of the reflection high-energy electron diffraction (RHEED) specular beam and corresponds to the growth rate of GaAs on GaAs (001) when the Ga beam is perpendicular to the substrate. With the angle of the Ga flux employed in LAIMCE, the horizontal and vertical growth rate were calculated as 0.21 μm/h and 1.06 μm/h respectively from simple geometrical consideration. The horizontal growth rate of 1.06 μm/h is calculated for an open window line seed perpendicular to the Ga flux (α_{Ga} = 90° in Figure 33(a)).

The morphology and the shape of the grown layers were investigated by AFM and cross-section analysis. We have employed an ULVAC MBC-300 MBE system for the growth. After organic chemical cleaning and chemical etching by 25%NH_4OH: 40%H_2O_2: H_2O= 4:1:20, a SiO_2 film was deposited on the GaAs (001) substrate by spinning of an organic solution (OCD, Tokyo Ohka Kogyo Co. Ltd.) and baking at 500 ℃. The oxide thickness was estimated by AFM to be 56 nm. The typical values of d and m were 1 and 2.3 μm, respectively. The results are given in Table 1 [46].

When the microchannel is aligned in low index orientations such as [-110], [110] and [010], the fronts of lateral growth are covered by facets such as (111)A, (111)B and (-101), respectively. Once, the front is covered by a facet, the velocity of lateral growth is reduced and one can not get high value of W_m/T ratio. Figure 34 shows the AFM cross sectional analysis for the microchannel orientation of [010]. It is seen that the front is covered by a (-101) facet. As the direction of the microchannel is

Table 1. Results of experiments with d + m = 3.3 μm, β_{Ga} and β_{As4} of 11° and 23°, respectively.

epilayers	LAIMCE parameters			figures of merit	
alignment	α_{Ga}	α_{As4}	γ	W_m (μm)	W_m/T ratio
[-110]	34°	11°	0°	0.62	7.3
[110]	56°	79°	90°	poor morphology	
[010]	79°	56°	-45°	0.73	8.4
[010] + 10°	69°	46°	-35°	1.45	22.3
[100] - 10°	1°	22°	35°	1.29	19.8

misoriented from [010], the side surface of the MCE layer becomes atomically rough and it serves as a sink of Ga adatoms that arrive on the MCE top surface. The intersurface diffusion of Ga adatoms from top surface to the side surface enhances the lateral growth and makes the value of W_m/T ratio high.

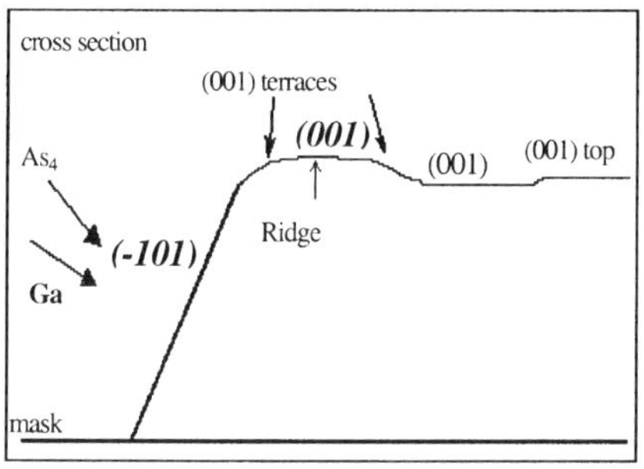

Figure 34. AFM cross section analysis of the epilayer grown on the microchannel aligned along [010] direction. d = 1 μm, d + m = 3.3 μm. T = 87 nm, W_m = 730 nm, W_m/T = 8.4.

[010] + 10° and [100] -10° are equivalent in crystallographic orientation. It is clearly seen from Table 1 that a difference in α_{Ga} and α_{As4} results in a difference in figures of merits between [100] - 10° alignment and [010] + 10° alignment. Since the direction [100] -10° is equivalent crystallographically to the [010] + 10°, differences in W_m and W_m/T ratio between the two alignments should be attributed to the difference in the mass transport to the sidewalls. Table 1 shows that the value of W_m is 12% bigger for the [010] + 10° alignment than that for the [100] - 10° alignment although the direct Ga beam incidence is, from the geometrical calculation, up to 50 times larger on the sidewall of the epilayer of [010] + 10° alignment, than that of [100] - 10° alignment. The Ga atoms directly impinging on the sidewall do not contribute to lateral growth as much as concluded by the geometrical calculation. Therefore, intersurface diffusion is dominant in the growth process of LAIMCE.

Among the various orientations of the microchannel, + 10° off [010] gave the widest layer. When this orientation was employed the side surface becomes atomically rough and is prevented being covered by various kinds of facets. In this case, the width of the MCE layer is of the order of 1.5 μm and the thickness is of the order of 70nm, which gives the aspect ratio, Wm/T, of the order of 20.

5. CONCLUSIONS

The concept of microchannel epitaxy (MCE) was presented. There are two types of MCE. One is horizontal MCE, which is called simply as MCE and the other is vertical MCE which is called V-MCE.

Application of MCE to various homo- and hetero-epitaxy system were described. It was shown that MCE can give dislocation-free area even for highly lattice mismatch heteroepitaxy, such as GaAs and InP on Si. It was also demonstrated that high quality epitaxial layers were obtained for various systems such as Si/Si, GaAs/GaAs, InP/In and GaP/GaP.

V-MCE was carried out by using MBE and the reduction of dislocation density was confirmed. MCE of GaAs on GaAs substrate by MBE was also successfully achieved employing low angle incidence molecular beam epitaxy. The largest ratio of the MCE width to the thickness was more than 20 and the width so far obtained was 1.45 μm. It is suggested that MCE can be applied to materials systems not only of semiconductors but also of oxides and of other crystal materials.

ACKNOWLEDGMENTS

The present work was mostly carried out in The University of Tokyo. The present author would like to thank Dr. S. Naritsuka who is now working for Meijo University and Dr. M. Tanaka of The University of Tokyo for their helps to carry out the present work. He also thanks many graduate students, under graduate students and research students at The University of Tokyo especially Y. Suzuki, T. Nakano, Y. Ujiie, S. Sakawa, S. Zhang, S. Naritsuka, Y. S. Chang, Z. Yan, G. Bacchin, Y. Matsunaga, A. Umeno and W. D. Huang for their collaborations. The author also would like to thank Dr. H.Mori and M.Tachikawa for their support especially for supplying high quality GaAs or InP coated Si substrates and Dr. M.Tamura for the clear TEM photographs of GaAs on Si.

REFERENCES

1. S. Ohnishi, Y. Hrokawa, T. Shiosaki and A. Kawabata, Jpn. J. Appl. Phys., 17 (1978) 773.
2. S. Nishino, Y. Hazuki, H. Matsunami and T. Tanaka, J. Electrochem. Soc., 127 (1980) 2674.
3. M. Ishida, H. Ohyama, S. Sasaki, Y. Yasuda, T. Nishinaga and T. Nakamura, Jpn. J. Appl. Phys., 20 (1981) L541.
4. H. Amano, N. Sawaki, I. Akasaki and Y. Toyoda, Appl. Phys. Lett., 48 (1986) 353.
5. B-Y. Tsauer, R.W. McClelland, J.C.C. Fan, R.P. Gale, J.P. Salemo, B.A. Vojak and C.O. Bozler, Appl. Phys. Lett., 41(1982)347.
6. L. Jastzebski, J.F. Corboy, J.T. McGinn and R. Pogliar, Jr., J. Electrochem. Soc., 130 (1983) 1571.
7. D.R. Brudbury, T.I. Kamins and C.W. Tsao, J. Appl. Phys., 55 (1984) 519.
8. T.S. Jayadev, E. Okazaki, H. Petersen and M. Millman, Electron. Lett., 21 (1985) 327.
9. E. Bauser, D. Kass, M. Warth and H.P. Strunk, Mater. Res. Soc. Symp. Proc., 54 (1986) 267.
10. T. Nishinaga, T. Nakano and S. Zhang, Jpn. J. Appl. Phys., 27 (1988) L964.
11. Y. Ujiie and T. Nishinaga, Jpn. J. Appl. Phys., 28 (1989) L337.
12. T. Nishinaga and H.J. Scheel, in Advances in Superconductivity VIII, eds. H. Hayakawa and Y. Enomoto, Springer-Verlag Tokyo, 1996 p.33.
13. Y.S. Chang, S. Naritsuka and T. Nishinaga, J. Crystal Growth, 174 (1997) 630.
14. Y.S. Chang, S. Naritsuka and T. Nishinaga, J. Crystal Growth, 192 (1998) 18.
15. S. Naritsuka, T. Nishinaga, M. Tachikawa and H. Mori, Jpn. J. Appl. Phys., 34 (1995) L1432.
16. S. Naritsuka, Y. Mochizuki, Y. Motodohi, S. Ohya, N. Ikeda, Y. Sugimoto, K. Asakawa, W.D. Huang and T. Nishinaga, 20th Electronic Materials Symposium. Nara, Japan, (2001) L37.
17. Y. Hayakawa, S. Iida, T. Sakurai, Y. Yanagida, M. Kikuzawa, T. Koyama and M. Kumagawa, J. Crystal Growth, 169 (1996) L613.
18. S. Iida, Y. Hayakawa, T. Koyama and M. Kumagawa, J. Crystal Growth, 200 (1999) L368.

19. T. Gehrke, K.J. Linthicum, P. Rajagopal, A.D. Batchelor and R.F. Davis, MRS Internet Semicond. Res., 4S1 (1999) G3.1.
20. D.B. Thomson, T. Gehrke, K.J. Linthicum, P. Rajagopal, A.D. Batchelor and R.F. Davis, MRS Internet Semicond. Res., 4S1 (1999) G3.37.
21. T.S. Zheleva, S.A. Smith, D.B. Thomson, T. Gehrke, K.J. Linthicum, P. Rajagopal, E.P. Carlson, W. Ashmawi and R.F. Davis, MRS Internet Semicond. Res., 4S1 (1999) G3.38.
22. K.J. Linthicum, T. Gehrke, D.B. Thomson, K.M. Tracy, E.P. Carlson, S.A. Smith, T.S. Zheleva, C.A. Zorman, M. Methrgeny and R.F. Davis, MRS Internet Semicond. Res., 4S1 (1999) G4.9.
23. Y. Suzuki and T. Nishinaga, Jpn. J. Appl. Phys., 28 (1989) 440.
24. S. Zhang and T. Nishinaga, Jpn. J. Appl. Phys., 29 (1990) 545.
25. S. Sakawa and T. Nishinaga, Jpn. J. Appl. Phys., 31 (1992) L359.
26. S.Naritsuka and T.Nishinaga : J.Crystal Growth 146(1995)314.
27. S. Naritsuka, T. Nishinaga, M. Tachikawa and H. Mori, J. Crystal Growth, 211 (2000) 395.
28. A. Usui, H. Sunakawa, A. Sakai and A.A. Yamaguchi, Jpn. J. Appl. Phys., 36 (1997) L899.
29. Y. Suzuki and T. Nishinaga, Jpn. J. Appl. Phys., 29 (1990) 97.
30. Y. Suziki, T. Nishinaga and T. Sanada, J. Crystal Growth, 99 (1990) 229.
31. S. Kinosita, T. Suzuki, T. Nishinaga, J. Crystal Growth, 115 (1991) 561.
32. S. Zhang and T. Nishinaga, J. Crystal Growth, 99 (1990) 292.
33. S. Naritsuka and T. Nishinaga, J. Crystal Growth, 203 (1999) 459.
34. N. Cabrera and M. M. Levine, Phil. Mag., 1 (1956) 450.
35. Z. Yan, S. Naritsuka, T. Nishinaga, J.Crystal Growth, 192 (1998) 11.
36. Z. Yan, S. Naritsuka and T. Nishinaga, J. Crystal Growth, 198/199 (1999) 1077.
37. Z. Yan, S. Naritsuka and T. Nishinaga, J. Crystal Growth, 203 (1999) 25.
38. N. Nagel, F. Barnhart, F. Phillipp, E. Czeck, I. Silier, and E. Bauser, Appl. Phys., A57 (1993).
39. P. Vohl, J. Crystal Growth, 54 (1981) 101.
40. F. Barnhart, N. Nagel, E. Czeck, I. Silier, F. Phillipp and E. Bauser, Appl. Phys. A57(1993) 441.
41. . Yan, Y. Hamaoka, S. Naritsuka and T. Nishinaga, J. Crystal Growth, 212 (2000) 1.
42. W. D. Huang, T. Nishinaga and S. Naritsuka, Jpn. J. Appl. Phys., 40 (2001) 5373.
43. Y. Matsunaga, S. Naritsuka and T. Nishinaga, J. Crystal Growth, 237-239 (2002) 1460.
44. G. Bacchin, T. Nishinaga, J. Cryst. Growth, 208 (2000) 1.
45. G. Bacchin, A. Umeno and T. Nishinaga, Appl. Surf. Science, 159/160 (2000) 360.
46. A. Umeno, G. Bacchin and T. Nishinaga, J. Crystal Growth, 220 (2000) 355.

Crystal Growth – from Fundamentals to Technology
G. Müller, J.-J. Métois and P. Rudolph (Editors)

Epitaxial technologies for short wavelength optoelectronic devices

S. Figge[a], C. Kruse[a], T. Paskova[b] and D. Hommel[a]

[a] University of Bremen, Institute of Solid State Physics, Semiconductor Epitaxy, Otto-Hahn-Allee, D-28359 Bremen, Germany

[b] Department of Physics and Measurement Technology, Linköping University, S-58183 Linköping, Sweden

Modern short wavelength light emitting devices are based on materials like GaN and ZnSe. The most important epitaxial growth methods for these materials will be presented. Molecular beam epitaxy (MBE), metalorganic vapour phase epitaxy (MOVPE) and hydride vapour phase epitaxy (HVPE) will be discussed in detail as well as *in-situ* characterization methods allowing to optimize crystal growth conditions. Exemplarely, conventional edge emitting laser diodes, quantum dot laser and vertical-cavity surface emitting lasers will be discussed.

1. INTRODUCTION

The first epitaxial growth method used to produce optoelectronic devices was the liquid phase epitaxy LPE. It is still important for infrared light emitting diodes (LEDs). Modern epitaxial methods dominating the growth of short-wavelength semiconductor laser diodes are the molecular beam epitaxy (MBE) and the metalorganic vapour phase epitaxy (MOVPE). Both methods are far from the thermodynamical equilibrium in which bulk crystals are typically grown. This is of special importance for wide gap materials were compensation mechanisms by native defects prevent usually n- or p-type doping under equilibrium conditions.

In case of ZnSe-based compounds only MBE allows a sufficient p-doping using a nitrogen radical source. MOVPE can produce high quality material as well but no hole conductivity could be obtained up to now. Therefore, in the first chapter the molecular beam epitaxy will be described for wide gap II-VI compounds. This method requires an ultra high vacuum (UHV) allowing a variety of in-situ characterization methods to be used for crystal growth optimization. It will be shown how MBE can be used to realize different kinds of laser diodes with quantum wells or quantum dots in the active region. The importance of *in-situ* reflectometry measurements will be demonstrated for surface emitting structures. Until now only ZnSe-based laser are able to cover the green spectral region.

In contrast to II-VI compounds, it turns out that the MOVPE is the favored growth method for nitrides, as the equilibrium vapor pressure of nitrogen is high at elevated growth temperatures. The MOVPE growth reactions will be discussed and the nucleation process of GaN presented in more details. In contrast to all other compound semiconductors used for optoelectronic devices,

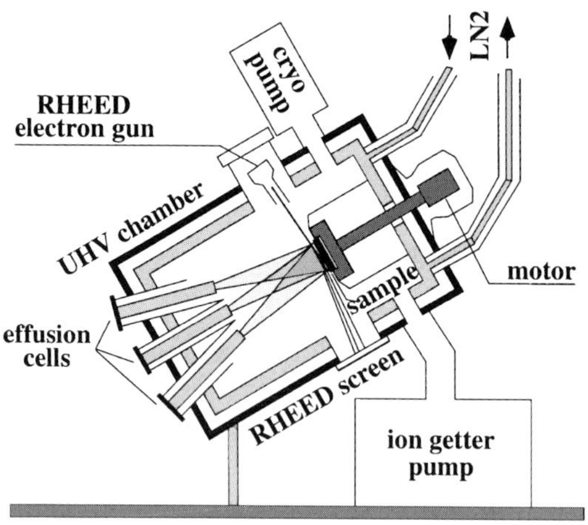

Figure 1. Schematic drawing of a MBE growth chamber. The species are evaporated from the effusion cells on the left hand side and impinge on the substrate on the right hand side. Each source can be covered with a shutter so that the molecular beam is interrupted. Thus the growth of the desired compound can be controlled.

InGaN quantum wells are an efficient light emitter despite the very high defect density of the GaN grown on sapphire or SiC substrates. Nevertheless, especially for laser diodes of high output power a defect density reduction is needed. Since there are no native GaN bulk crystals available which are suitable for wafer production, efforts are made to grow thick GaN layers and to use them as substrates for the epitaxy of the light emitting structures.

In the last 1-2 years it became evident that the best method to realize such GaN templates of thicknesses up to a few millimeter is the hydride vapour phase epitaxy. Therefore, in the 3rd chapter the basics of HVPE growth will be discussed in detail. Special attention will be paid to the microstructure and the morphology of HVPE GaN. Nichia company demonstrated first high power laser diodes grown homoepitaxially on HVPE GaN substrates [1].

2. MOLECULAR BEAM EPITAXY

Molecular beam epitaxy is a technique for crystal growth on substrates with the precision of a single atomic monolayer (ML) reaching growth rates in the range of 1 μm/h or approximately 1 ML/s, respectively. The main aspect is that the deposition occurs under conditions of ultra-high vacuum with backround pressures of about $1 \cdot 10^{-11}$ Torr, which allows the realization of layers with high cleanness. For that purpose, elements of high purity are provided either by thermal evaporation from heated sources (filled with solid or liquid material) or activation of gases using e.g. a plasma. Important types of sources are Knudsen cells, cracker cells and plasma cells. In a Knudsen cell the flux of the atoms or molecules is mainly determined by the temperature of the cell, i.e. the vapour pressure of the species. A variation of the flux for a certain element (necessary e.g. for changing the composition within a compound), is comparatively slow, since the stabilization of temperature inside the source takes several minutes. A cracker cell consists of two separate parts, a volume filled with material (bulk) and a heated tube for cracking larger molecules into smaller ones. The flux of a cracker cell is adjusted by a valve and can be changed over orders of magnitude within seconds. Another advantage is the higher capacity of the source compared to a Knudsen cell. This type of source is available for e.g. As, P, Se and S.

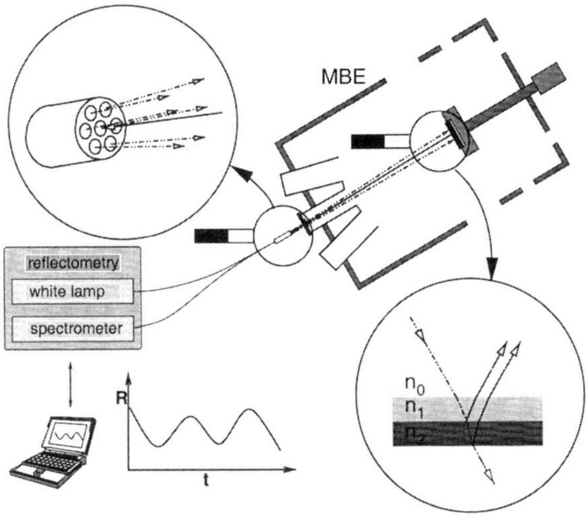

Figure 2. Experimental setup of an *in-situ* reflectometry at a MBE chamber.

An efficient method to utilize gases like nitrogen or hydrogen for the growth process is the activation of these species in a plasma. It is created in a plasma cell by exciting the gas via irradiation with radio frequency (rf) waves or electron cyclotron resonance. Due to the UHV the mean free path of the particles is much larger than the dimensions of the chamber, i.e. the deposition process occurs far away from the thermodynamic equilibrium and is therefore mainly determined by the kinetics of the adsorbed species at the growth front of the crystal.

In Figure 1 a schematic drawing of a MBE system is depicted. The vacuum is typically provided by a cryogenic pump and an ion getter pump, the surrounding cryo shroud filled with liquid nitrogen (LN_2) also improves the pressure by freezing out a certain amount of molecules to the walls of the chamber. A movable shutter is located in front of each source and the substrate, which allows a fast interruption of the molecular beam and therefore the realization of sharp interfaces between two types of compounds. An overview concerning MBE is given in [2].

2.1. *In-situ* characterization methods

Due to the UHV conditions used for the MBE growth technique several surface sensitive characterization methods that utilize electrons like reflection of high-energy electron diffraction (RHEED) or X-ray photoelectron spectroscopy (XPS) can be performed. Especially RHEED is one of the standard experimental setups installed at a MBE chamber, since it gives important information e.g. about surface roughness and reconstructions, as well as the lateral lattice constant of the epitaxial layer. For the optical characterization ellipsometry and reflectometry are widely used. Ellipsometry provides information even for very thin layers below 5 nm by evaluating the phase of the light. On the other hand this method is comparatively sensitive to the overgrowth of viewports and therefore not suitable for growth monitoring during the whole run. The method of spectroscopic reflectometry will be discussed in the following.

The experimental method of reflectance measurement is comparatively simple but nevertheless able to provide many useful informations concerning the optical properties of the epitaxial layers. A schematic drawing of the setup at the MBE chamber is shown in Figure 2. The light

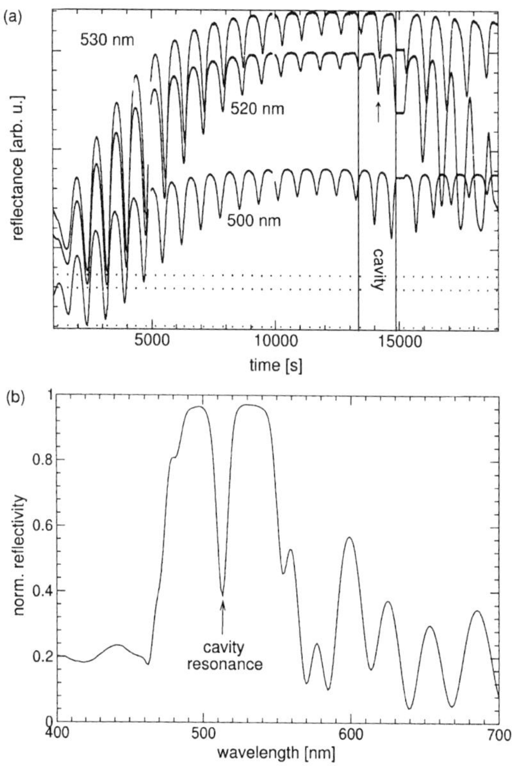

Figure 3. (a) *In-situ* reflectance transients recorded at three different wavelengths for a II-VI-based microresonator consisting of a 17 period bottom DBR, a λ-cavity and a 6-period top DBR. The signal at 520 nm shows a significant decrease, while the other ones remain almost constant. (b) Reflection spectra at room temperature. The lorentzian shaped resonance line at 514 nm is clearly visible. This measurement was performed with an Filmetrics F30 instrument.

of a white lamp passes through the outer fibres of a fiber bundle to the substrate with normal incidence, while the reflected light is focused back into the middle fiber and finally analyzed by a spectrometer. This setup can also be realized with two separate fibres by using a beam splitter. At the right bottom part of Figure 2 the reflection of light at the interface of two materials with different refractive index is depicted. For reasons of clarity an angle of non-perpendicular incidence has been chosen. A wavelength resolved reflectometry measurement gives access to the dispersion $n(\lambda)$ of the refractive index, the extinction coefficient $k(\lambda)$ and layer thickness. Especially for the realization of distributed Bragg reflectors and vertical-cavity surface-emitting lasers (VCSELs) the use of reflectometry is of high value since it measures directly the optical thickness without necessity of knowing the refractive index of a certain compound. The distance between a minimum and a maximum in the time dependant transient signal is given by $d = \lambda / 4n$ (d: thickness, λ: wavelength, n: refractive index).

An example for a reflectance measurement recorded at three different wavelengths during the growth of a II-VI-based microresonator is shown in Figure 3(a). At each minimum and maximum, respectively, the layer type is switched from low index material to high index material and vice versa (quarterwave thickness each). As a consequence, the typical thickness oscillations are superimposed by an overall increase of the reflectance. By adding a sufficient number of Bragg mirror pairs, the normalized reflectivity saturates at a value close to R = 1. After deposition of a λ-thick cavity, the reference signal recorded at a wavelength of 520 nm shows a

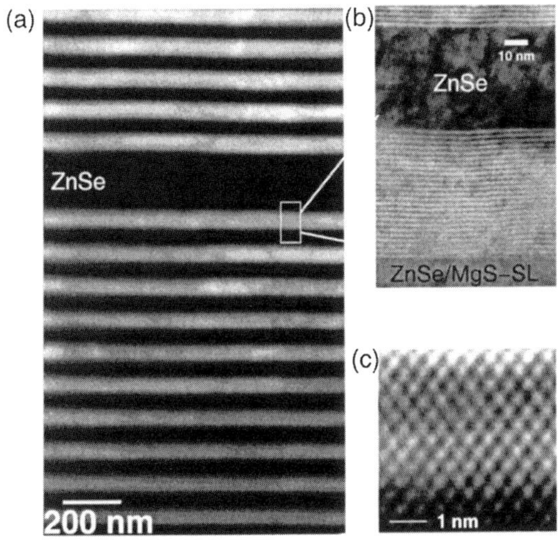

Figure 4. Cross section transmission electron microscopy (TEM) image of a II-VI-based microresonator. The reflectance data of this structure is depicted in Figure 3. (a) Overview: the ZnSe layers appear dark while the MgS/ZnSe superlattices (SLs) appear bright. (b) Magnification of a superlattice. (c) High-resolution TEM image of single ZnSe and MgS layers in a SL.

significant reduction of intensity (see Figure 3(a)). This behavior results from the development of the cavity resonance line in the center of the reflection stopband (Figure 3(b)). The quality factor of the microcavity can be determined by using the equation $Q = \lambda/\Delta\lambda$ ($\Delta\lambda$: linewidth of resonance, λ: spectral position of resonance). A quality factor of $Q = 70$ can be evaluated from Figure 3(b). Microcavities for VCSELs generally have Q-values in the range of several thousands.

Figure 4 shows a transmission electron microscopy (TEM) crossection of this structure. The ZnSe layers (high index) have a dark contrast, while the ZnSe/MgS superlattices (low index) appear bright (Figures 4(a) and (b)). A special feature of this microcavity is that the low index material of the distributed Bragg reflector (DBR) is realized as a short-period super lattice (SL), to stabilize the MgS in the zinkblende structure. In Figure 4(c) single monolayers within the superlattice are resolved. The contrast is inverted compared to Figures 4(a) and (b) in this case.

In order to monitor the spectral position of the DBR stopband more precisely, the use of a reflectometry setup that is able to detect many wavelengths at the same time is very helpful. The data shown in Figure 5 were recorded during MBE growth of an AlGaN/GaN-DBR deposited on a GaN/sapphire MOVPE template layer.

2.2. Growth of ZnSe-based devices

With regard to the realization of II-VI-based optoelectronic devices, MBE is superior to MOVPE in terms of the structural and optical quality. However, the most important fact is that for MOVPE a stable and reproducible p-type doping has not been achieved yet. In the following sections a short overview is given about recent developments concerning II-VI laser diodes in our group.

2.2.1. Edge-emitting laser diodes

In Figure 6 the standard structure of an edge-emitting laser diode (LD) is depicted. The active region consists of a single 2-5 nm thick ZnCdSSe quantum well (QW) which is surrounded by ZnSSe barrier layers with a higher bandgap that act as a waveguide. This waveguide is

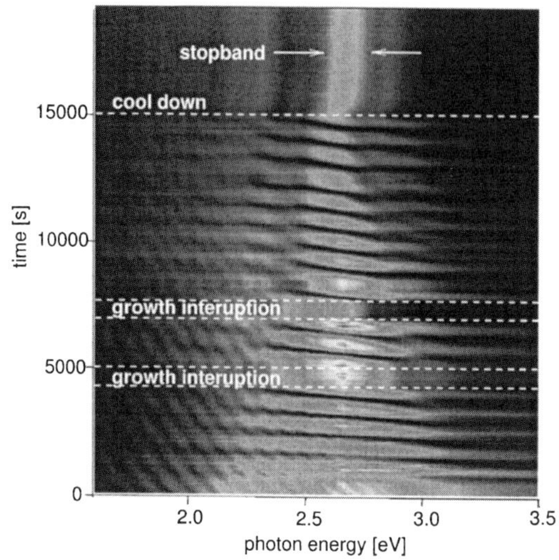

Figure 5. Multi-wavelength reflectometry signal recorded during growth of an AlGaN/GaN distributed Bragg reflector on a GaN/sapphire template layer. This measurement was performed with an EpiR instrument from LayTec.

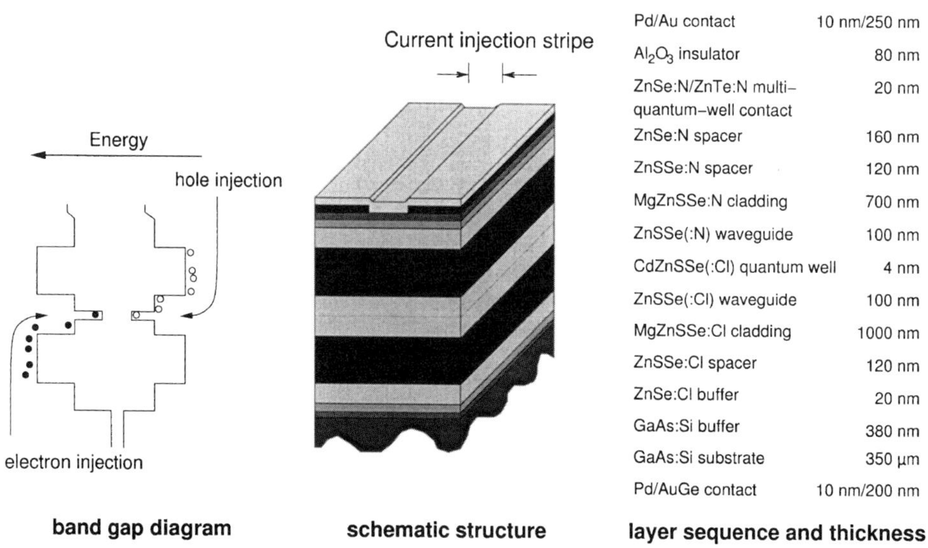

Figure 6. Schematic view of a ZnSe-based edge-emitting laser diode including layer sequence and band gap diagram (bending not taken into account).

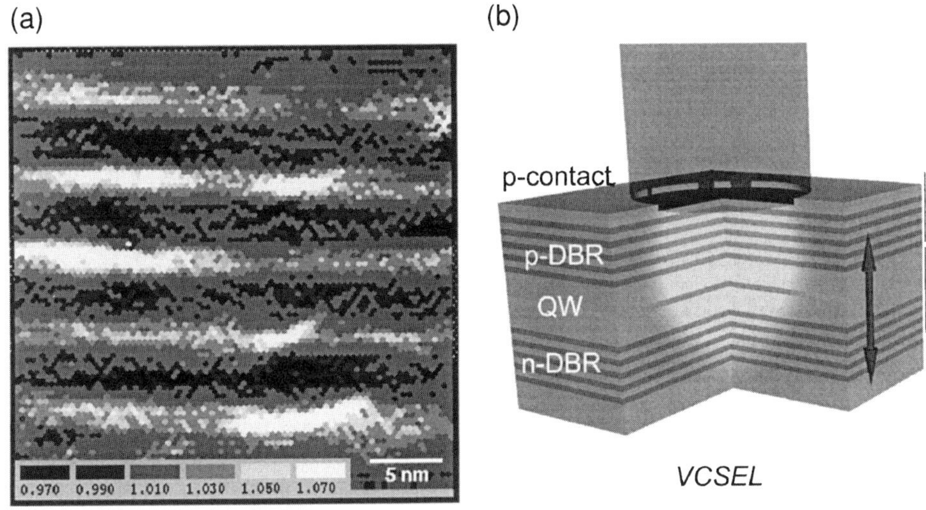

Figure 7. (a) High-resolution cross section TEM micrograph of five CdSe quantum dot sheets inbetween ZnSSe barrier layers. The superimposed gray scale represent the results of a digital analysis of a lattice image (DALI). Bright and dark areas indicate a high Cd an S content, respectively. The key denotes the relative lattice constant (TEM image and DALI analysis by R. Kröger). (b) Schematic drawing of a VCSEL. The light is emitted perpendicular to the wafer surface. Due to the circular aperture the beam has a low divergence.

sandwiched between ZnMgSSe cladding layers with a further raised bandgap and a lowered refractive index. The resulting index step causes an improved guiding of the light in lateral direction and therefore an increased overlap of the optical wave with the active region of the device. The emission of these LDs cover a wavelength range between 500 - 560 nm [3].

2.2.2. CdSe quantum dots

The precision of MBE concerning deposition on a monolayer scale combined with the *in-situ* method RHEED allows detailed growth studies. For example, the transition from two-dimensional to three-dimensional growth which occurs during the deposition of quantum dots (QDs) can be monitored directly. Taking advantage of this together with an **ex-situ** non-destructive analysis using high-resolution X-ray diffraction (HRXRD), CdSe QD-stacks suitable for application as active region in an edge-emitting LD have been realized [4]. A TEM-micrograph of a 5-fold QD-stack evaluated by digital analysis of an lattice image (DALI) is shown in Figure 7(a). For the DALI the brightness maxima of a TEM image are assigned to rows of atoms and from their distances the local lattice constants are determined [6]. By comparison with a reference lattice the local lattice constant can be evaluated and finally the composition of a ternary compound can be determined using Vegard´s law ([7]).

2.2.3. Vertical-cavity surface-emitting laser

Also an optically-pumped monolitic (i.e. fully epitaxially grown) vertical-cavity surface-emitting laser (VCSEL) with a Q-value exceeding 2000 has been achieved [5]. A schematic drawing of such a device is shown in Figure 7(b). The layer sequence is similar to the design depicted in Figure 4, only the ZnSe layers in the SL were replaced by ZnCdSe layers in order to reach lattice matching of the structure to the GaAs substrate. The use of the *in-situ* reflectometry as shown in Figure 3 was of major importance during the optimization process for this complex structure.

3. METALORGANIC VAPOR PHASE EPITAXY

Metal organic vapor phase epitaxy (MOVPE) is one of the latest growth techniques and was employed the first time by Manasevit in the year 1968 [8]. The first growth of GaN and AlN films with this method was carried out in 1971 [9]. In contrast to MBE, growth is taking place at moderate pressures ranging from 50 Torr up to atmospheric pressure. This is the most fundamental advantage for the epitaxy of nitrides, as the equilibrium pressure of nitrogen at the GaN surface rises quickly at temperatures above 800 °C. The MBE growth of GaN is therefore limited to temperatures below 800 °C, as it is not possible to obtain an equilibrium between educts and products in vacuum at higher temperatures. Thus the layers tend rather to thermal etching than to grow. In MOVPE a broader range of growth conditions can be achieved and growth is possible up to temperatures of 1050 °C with growth rates up to 2 μm/h. The reactants can be quickly switched onto and off the reactor, which makes the growth on a scale of single monolayers possible.

3.1. Gas system and precursors

Most of gas flow through the MOVPE-reactor consists of carrier gases, like nitrogen or hydrogen. The main task of the carrier gas is the transport of reactants to the sample surface. As it is not possible to disperse metals in the carrier gas, precursors are needed which have a moderate vapor pressure at room temperature. One possible choice utilized by MOVPE are metal organic compounds. In these compounds the metal is attached to one or more organic radicals. For the epitaxy of group-III nitrides the most commonly used precursors are Trimethylgallium $Ga(CH_3)_3$, Trimethylaluminium $Al(CH_3)_3$, Trimethylindium $In(CH_3)_3$ and Bis(cyclopentadienyl)magnesium $Mg(C_5H_5)_2$ (for doping). These metal organics are either solid or liquid and they exhibit vapor pressures which are ranging from 0.04 Torr ($Mg(C_5H_5)_2$) to 187 Torr ($Ga(CH_3)_3$) at room temperature.

These metal organics are kept in a container, the so called bubbler, and the carrier gas is led through the bubbler in order to form a saturated solution of the metal organic with the carrier gas. The bubbler is stored in a temperature controlled bath to adjust the vapor pressure of the metal organics. The total pressure inside the bubbler is adjusted by a pressure controller at the outlet of the bubbler. The gas flow through the bubbler can be determined by a mass flow controller at the inlet of the bubbler, as it is shown in Figure 8. The precursor flow F_{prec} leaving the bubbler can be easily determined by the ratio of the vapor pressure p_{vap} to the total pressure p_{tot} and the inlet gas flow F_{in}:

$$F_{prec} = F_{in} \times \frac{p_{vap}}{p_{tot}} \tag{1}$$

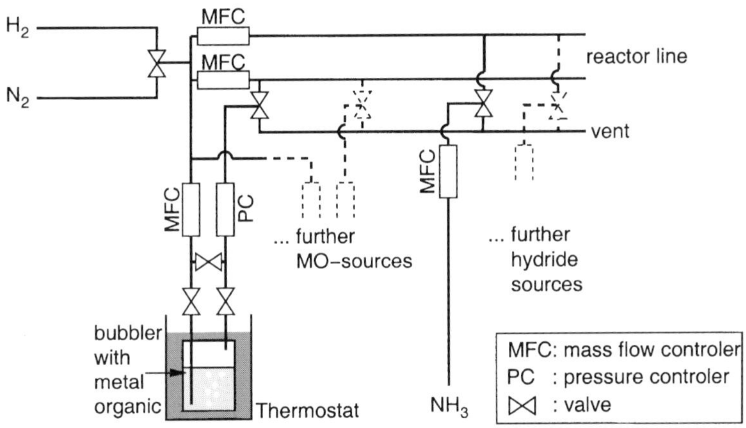

Figure 8. Schematics of the gas system of a MOVPE system.

Besides the metal organics some hydride precursors are used in a MOVPE system. In the case of group-III nitrides a nitrogen precursor is necessary as the bonds in molecular nitrogen (N_2) are to stable to be used for epitaxy. Therefore, Ammonia (NH_3) is used, which decomposes sufficiently at typical growth temperatures between 500 °C to 1050 °C. As there are no suitable silicon organic precursors the hydride Silane (SiH_4) is used for doping. The amount of the hydride precursors are controlled by a mass flow controller.

The precursor flows are added to the carrier gas in two separate reactor lines for the hydride (group-V and group-IV) and metalorganic (group-III and group-II) precursors. This inhibits a reaction between the different precursors inside the tubing system.

3.2. Reaction kinetics

The gas flows of the group-III and group-V precursors are intermixed inside the reactor close to the sample surface. The metal organic precursors of In, Ga and Al are decomposed irreversibly above the sample surface according to the following reactions:

$$R_3M(g) + \frac{3}{2}H_2(g) \longrightarrow M(g) + 3RH(g) \tag{2}$$

where R are the organic radicals like CH_4 or C_2H_5 and M are the metal atoms like In, Ga and Al. The chemical reactions to form the Nitride alloy at the deposition zone are:

$$M(g) + NH_3(g) \longrightarrow MN(s) + \frac{3}{2}H_2 \tag{3}$$

Then, the equilibrium equations for these reactions are as follows:

$$K_M = \frac{a_{MN} \times P_{H_2}^{3/2}}{P_M \times P_{NH_3}} \tag{4}$$

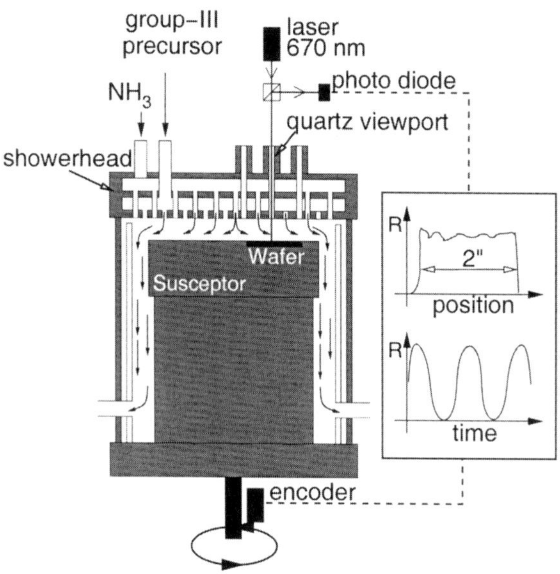

Figure 9. Cross section of a closed coupled showerhead reactor.

where a_{MN} are the activities of the binary compounds in the alloy and the P_i's are equilibrium partial pressures of gaseous species above the surface. These activities and the equilibrium constants K_i are varying in a wide range for the compounds. Therefore the growth conditions have to be carefully chosen for the growth of ternary or quarternary alloys. Detailed thermodynamical calculations on the growth of Nitrides can be found elsewhere [10, 11].

The second factor influencing the growth is the transport of the reactants to the sample surface. As the gas flow inside the reactor is laminar the reactants have to diffuse through a boundary layer to the sample surface. Therefore, the velocity of the gas above the sample surface and the viscosity of the gas are affecting the growth. The higher the gas velocity and the lower the viscosity of the gas the faster is the transport of the reactants to the sample surface.

3.3. Reactor

In principle two different reactor designs are used for MOVPE growth: reactors with *horizontal* and *vertical* gas flow. However, the main difference in the various applied reactor designs is the position of the gas intermixture and the pattern of gas flow. Nevertheless, all reactor designs have in common that a laminar gas flow is obtained and that the intermixture of gases is as close as possible to the specimen to prevent parasitic reactions in the gas phase. Therefore, we will focus exemplarily in this paper on a so called closed coupled showerhead reactor as it is shown in Figure 9.

The design of the close coupled showerhead reactor is of the vertical type. The gases coming from the two reactor lines are introduced into two separate cavities on top of the reactor. They are flowing into the reactor through small nozzles which are covering the whole ceiling of the reactor (showerhead). The wafers are lying plane on a graphite susceptor which is in a distance of approximately 1 cm to the showerhead. The susceptor can be rotated and is heated from the backside.

The rotation of the susceptor has to major tasks: It counterbalances inhomogeneities in the growth which can occur due to non-uniform heating and also influences the gas flow pattern inside the reactor. Due to the high temperature gradient between the sample surface and the reactor ceiling of approximately 1000 °C strong buoyancy forces are tending to drive convection cells inside the reactor head. The rotation of the susceptor is adding addition centrifugal forces which is driving the gas flow to the outer part of the reactor. Besides this, the downward gasflow through the showerhead is pressing the gas onto the susceptor, which leads to a reduction of the available space for convection cells.

3.4. Reflectometry and nucleation scheme

The lack of GaN substrates is a severe handicap in GaN epitaxy. Commonly used substrates are sapphire ($a_{Al_2O_3}$ = 4.759 Å) and SiC (a_{SiC} = 3.081 Å). Due to the lattice mismatch to GaN a_{GaN} = 3.188 Å) the critical thickness of the GaN layer in both cases is low. On sapphire the critical thickness is in the range of only a few nanometer [13]. Therefore, the direct growth of GaN on the substrate normally leads to threading dislocations with densities in the range of 10^{11} cm^{-2}. One of the most important developments in nitride growth is a nucleation scheme from Amano et al. [12] which is nowadays used to obtain lower dislocation densities during the initial growth.

Due to the gas ambient inside a MOVPE reactor it is not possible to implement in-situ characterization methods involving electrons. Therefore, the only applicable in-situ method is optical reflectometry. The reflectometry used in our reactor consists out of a laser diode (λ=670 nm) pointing in normal direction onto the sample and a photo diode detecting the reflected signal. Using an A/D-converter the reflectance transient can be recorded on a computer. Despite the limitation to a single wavelength, this method can give a good insight on the growth mechanism of the initial GaN growth. Therefore, the GaN nucleation on sapphire substrates as seen by normal incidence reflectometry (see Figure 10) will be explained in the following.

At the beginning of growth the reflection intensity is normalized to the bare sapphire substrate as a reference. Before deposition, the sample is heated up to the growth temperature (1050 °C) to accomplish a thermal cleaning of the substrate. Afterwards the wafer is cooled down to approximately 800 °C and is exposed to a ammonia flow to achieve a nitridation of the surface. Due to this, the oxygen of the sapphire is exchanged in the uppermost layers by nitrogen forming a kind of AlN phase at the surface. This procedure does not change the reflectivity of the wafer and the reflectometry signal stays constant.

The nucleation scheme of Amano et al. [12] employs a GaN or AlN nucleation buffer which is grown at low temperatures around 550 °C and has a thickness of 20 to 30 nm. Due to the low growth temperature this buffer grows in a mixed phase consisting out of amorphous, cubic and hexagonal material. In reflectometry this buffer growth can be noticed as an increase of the reflectance (Figure 10 (a)) as the refractive index of GaN is higher than the one of the sapphire.

Then the growth will be interrupted and only ammonia will be supplied to the reactor. In the following the temperature is increased to normal growth temperature of 1050 °C and is kept there for approximately 2 minutes. In this stage a recrystallization of the buffer takes place. The amorphous and cubic GaN have less thermal stability than the hexagonal one and an Oswald ripening of the hexagonal parts of the buffer takes place. At the end of the recrystallization the amorphous and cubic phases vanished and only few hexagonal nucleation sites are left on the surface. The raise of temperature during recrystallization increases the reflectance signal due to an increase of refractive index of the GaN, but as soon as the ripening takes place the signal

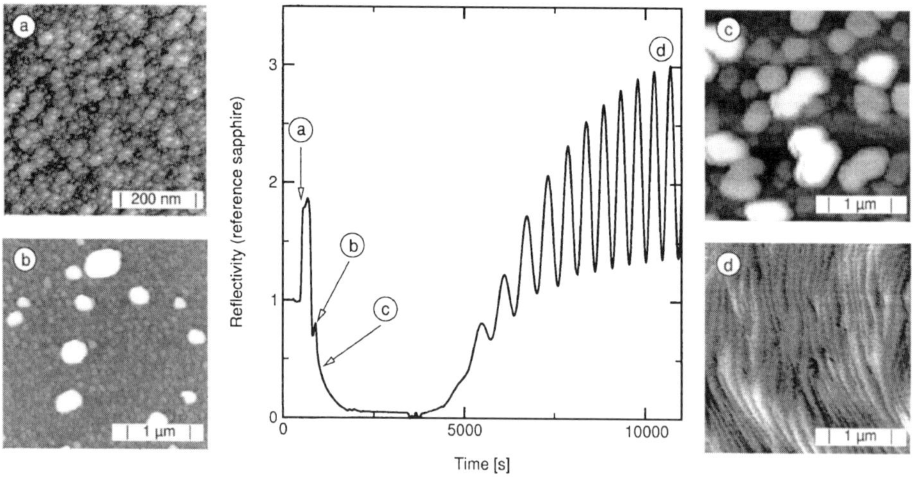

Figure 10. Initial growth steps of GaN as seen by in-situ reflectometry (center) and ex-situ atomic force microscopy. The different stages are (a) low temperature nucleation layer, (b) after recrystallization, (c) initial 3D island growth and (d) coalesced layer [14].

drops due to roughening and thinning of the buffer layer (Figure 10 (b)).

As soon as the reflectivity of the sample is close to 1 the growth is continued by switching the Trimethylgallium to the reactor. The growth starts at the remaining nucleation sites as the sticking coefficient of GaN is low on the sapphire at those temperatures, leading to a 3D-island growth mode. Due to the roughening of the surfaces this causes a further decrease of the reflectivity (Figure 10 (c)) below the value of the sapphire. The lateral to vertical growth rate of the islands can be influenced by the group-III to group-V precursor ratio. A high ammonia flow enhances the lateral growth of the islands and leads to a fast coalescence and smaller grain sizes [14]. During coalescence the surface of the GaN layer gets smoother and the growth can be seen in reflectometry as increasing thickness interference oscillations (Figure 10 (d)).

The individual grains in the GaN layer are twisted and tilted which respect to each other. Therefore, during coalescence of the grains edge type threading dislocations are formed at the grain interfaces to compensate the twist. The twist of the grains does not seem to be effected by the growth parameters, but the size of the grains influences the edge type dislocation density: If the grains are larger, the total length of grain boundary drops, leading to a reciprocal dependence of the edge type threading dislocation density on the grain diameter (see Figure 11). However, the tilt of the grains seems to have a lower impact on the coalescence of the grains and therefore the screw type threading dislocation density stays unaffected by the grain size at a level of 10^8 cm^{-2}.

The size of the grains can be controlled in this nucleation scheme by the variation of the growth parameters. A longer recrystallization and a lower ammonia flow leads to fewer nucleation centers for the initial growth causing larger grain sizes. Furthermore, a slow lateral growth during coalescence caused by high growth pressure and low ammonia flows will give rise to larger grain sizes. But this optimization has two major limitations. On the one hand a low nucleation site density and slow lateral growth leads to a long coalescence phase of the

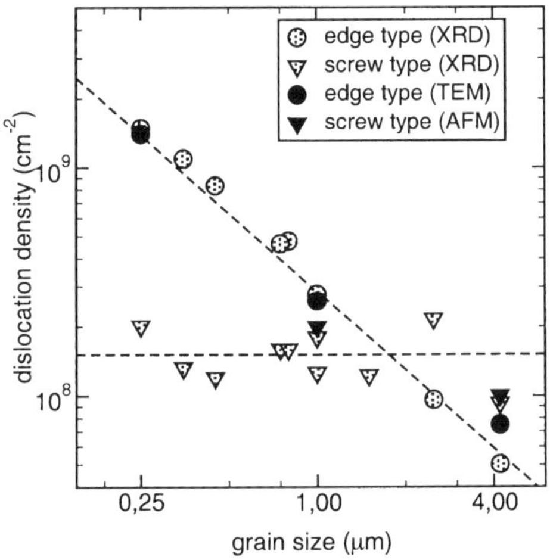

Figure 11. Threading dislocation densities of GaN determined by x-ray diffraction, transmission electron microscopy and atomic force microscopy in dependence of the grain size.

island growth which is unpractical for device growth. On the other hand the risk of the nucleation of GaN on the bare sapphire between the nucleation sites increases and leads to a reduced crystal quality. Therefore threading dislocation densities can be hardly reduced to values below 10^8 cm^{-2}. Thus, other techniques have to be employed in order to obtain lower dislocation densities.

In 1997 a techniques called lateral epitaxial overgrowth (ELOG) involving dielectric stripe masks was developed [15, 16]. In this method a SiO_2 layer is deposited onto a GaN layer and stripes along the {11-20} direction of the GaN are prepared by wet chemical etching (see Figure 12 (a)). The dislocations ending up in the SiO_2 stripes are forming dislocation loops as the SiO_2 shows no epitaxial relation to the GaN. The following growth of GaN starts in the openings of the stripe mask, as the sticking coefficient of GaN is low on the insulator (Figure 12 (a)). The dislocations which are penetrating through the mask openings are bend on the {1-101}-like facets of the growth surface into the lateral growth direction (Figure 12 (c)). During coalescence these dislocations are partially annihilated by the formation of dislocation loops, whereas the remaining defects are accumulated at the coalescence interface (Figure 12 (d)). As a result regions exist above the dielectric stripes which show lower dislocation densities, below 10^6 cm^{-2}. This method is very useful for the production of laser diodes, as the laser stipes can be placed along the {11-20} direction in the regions with the low defect density (see Figure 13) [17].

But even with the help of the ELOG technique the growth on sapphire substrates has some drawbacks. Only small areas are available on the substrate which are sufficient for laser diodes, but not for large LED structures, and the thermal conductance of sapphire ($\kappa_{Al_2O_3}$ = 0.25 W/cm^2 [19]) is much worser compared to GaN (κ_{GaN} = 1.3 W/cm^2 [18]). Thus, GaN substrates are preferable for the production of high power light emitting devices. As the bulk growth of GaN is very complicated [20] currently thick GaN layers are produced by hydride vapor phase epitaxy (HVPE) on sapphire, which can obtain much higher growth rates up to 100 μm/h. Such thick GaN layers can be separated from the sapphire substrate by laser ablation [21, 22] and used as substrates in MOVPE.

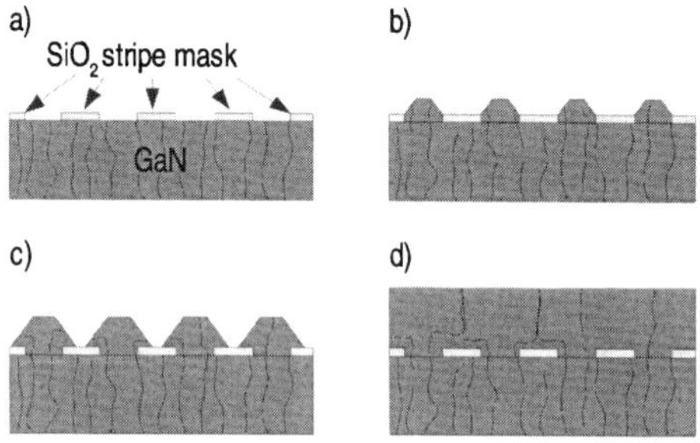

Figure 12. Principle of ELOG growth: (a) patterned SiO_2 mask, (b) initial regrowth, (c) lateral overgrowth and (d) coalescence.

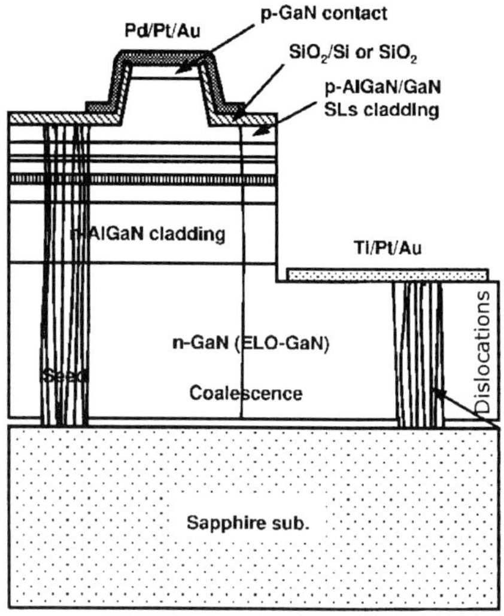

Figure 13. Placement of a laser diode on top of an ELOG structure [17].

4. HYDRIDE VAPOR PHASE EPITAXY

The development of the HVPE technique started in the 1960s firstly for arsenides and phosphides. Ten years later it was successfully demonstrated for growing the three main nitride compounds as well (AlN [23], GaN [24] and InN [25]). The biggest advantage of the technique is its ability to produce high quality material at high growth rates due to a high surface migration of the halide species, motivating the versatility of HVPE as a growth method for both device applications and quasi-substrate application after separating the thick films from the substrate. The latter application has recently attracted significant attention hoping to overcome the major problem in the nitride technology, namely the lack of native substrate. Among all the bulk growth techniques under investigation today for nitrides, the HVPE is the most promising technique since it utilizes a process at more favorable conditions, namely relatively low growth temperature and pressure. This should make the growth process easier to be handled and reduces the operating cost.

4.1. Basic principles of HVPE

4.1.1. Chemistry - reactions and precursors

The HVPE growth process for deposition of GaN proceeds via two steps:

$$Ga(l) + HCl(g) \longrightarrow GaCl(g) + 1/2H_2(g) \tag{5}$$

$$GaCl(g) + NH_3(g) \longrightarrow GaN(s) + HCl(g) + H_2(g) \tag{6}$$

In the first chemical reaction hydrogen chloride (HCl) reacts with pure metallic Ga forming a metal chloride compound GaCl and H_2 at temperature of about 850 °C. The second reaction producing GaN is between the GaCl, as the group III precursor with a hydride of the group V elements (NH_3) at higher temperature typically in the range of 1050-1100 °C. The preferential choice of these precursors is based on a few arguments: (i) The chloride precursor (HCl) is preferable for the first reaction over bromides (HBr) and iodides (HI) due to chloride higher vapor pressure and higher thermodynamic driving force. However, the HCl is a highly corrosive gas and can destroy metal parts of the reactor equipment. An alternative approach used by several groups is the usage of pre-synthesized chlorides ($GaCl_3$) instead of HCl gas and thus, the first chemical reaction of the process can be omitted. Unfortunately, the chlorides are usually available at lower purity and special measures for storage in bubblers and heated tubes are required in the reactor design. (ii) The hydride is not always a preferable group V precursor in the HVPE growth process. For instance, the growth of GaAs or GaP does not employ hydrides (AsH_3 or PH_3) being extremely dangerous gases. Instead halides ($AsCl_3$ or PCl_3) can be used. In the HVPE growth of nitrides it is vice versa, NH_3 should be used as the source of the group V element rather than nitrogen halide (NCl_3), which is highly explosive. However, a possible decomposition of the NH_3 especially at the high growth temperature (with a decomposition factor α) should be considered as a possible reaction reducing the efficiency of the main growth reaction.

$$NH_3(g) \longrightarrow (1-\alpha)NH_3(g) + \alpha/2N_2(g) + 3\alpha/2H_2(g) \tag{7}$$

In addition for a precise analysis of the chemical system in the HVPE growth of GaN, the competing decomposition reactions need to be accounted for as well:

$$GaN(s) + HCl(g) \longrightarrow GaCl(g) + 1/2N_2(g) + 1/2H_2(g) \tag{8}$$

$$GaN(s) \longrightarrow Ga(l) + 1/2N_2(g) \tag{9}$$

4.1.2. Thermodynamics and kinetics of the HVPE GaN growth process

Detailed thermodynamical calculations of the driving forces of the chemical reactions of HVPE growth of GaN was published by several groups [26, 27, 28]. Based on their analysis, some specific features can be pointed out, different from the growth of other III-V materials, and should be taken into account for successful HVPE-GaN growth. (i) The partial pressure of GaCl was estimated to be much higher than that of $GaCl_2$ and $GaCl_3$ and thus, it is the only stable chloride species that is obtained from the first reaction. (ii) The efficiencies of both reactions (5) and (6) were estimated to be higher in an inert ambient atmosphere than in hydrogen atmosphere, which is favorable for other III-V systems. (iii) Although at atmospheric pressure and temperature higher than 250 °C, the NH_3 should be thermodynamically completely dissociated, the decomposition of the NH_3 is estimated to be very slow due to the large kinetic barrier of breaking N-H bonds. Thus, a factor α of about 4% is not expected to have strong negative effect on the deposition process. (iv) The driving force of the competing etching reactions (8) and (9) was estimated to be comparable with the driving force of the deposition reaction (7). Fortunately, the growth is thermodynamically favored because of the high supersaturation of the reactants in the growth zone.

The kinetics of the process and growth mechanisms occurring at the solid/vapor interface during HVPE growth of GaN are still poorly understood. The available models [28, 29] assume a surface process involving the following steps: (i) adsorption of NH_3 molecules, (ii) adsorption of N atoms coming from ammonia decomposition; (iii) adsorption of GaCl molecules on

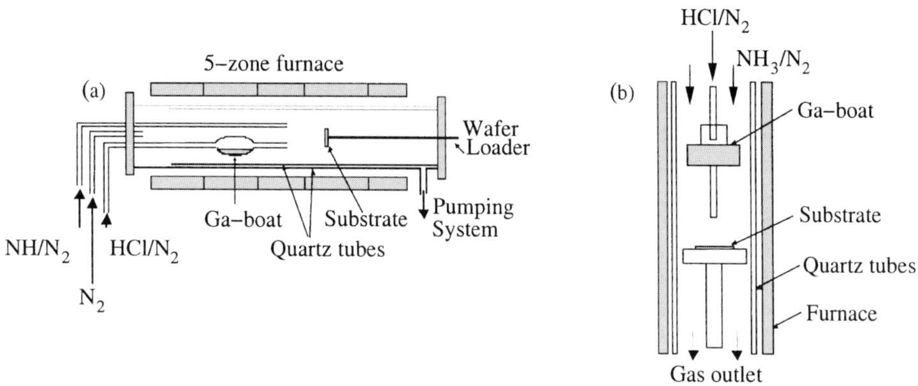

Figure 14. Schematic drawings of horizontal (a) and vertical (b) HVPE reactors.

the N atoms forming NGaCl; (iv) decomposition of the NGaCl via different desorption mechanisms. Two of them were suggested in analogy with the GaAs model: desorption forming HCl and desorption forming $GaCl_3$ [30]. Additionally $GaCl_2$ desorption was suggested to account for the experimental results [31]. Statistical treatment of the dynamic equilibrium between the adsorbed and gas phase species allowed explicit expressions of the growth rates via the different pathways. Although several papers were dealing with the dependence of the growth rate on the growth conditions, the information obtained is not sufficient for a complete picture of determining growth mechanisms in the HVPE-GaN growth.

The 3D nucleation growth is a common growth mechanism for heteroepitaxy of systems with high lattice mismatch between the substrate and the layer, and islanding driven by an increasing strain component in the total system free energy usually leads to relaxation of the system [30]. A study of microstructural evolution in the early stages of HVPE-GaN growth has proved the islanding and revealed a very fast and abrupt full coalescence [31], in contrast to slower island coalescence in MOVPE-GaN growth [14]. A 2D-multilayer growth is a realistic mechanism for later growth stages, ensuring relatively smooth morphology [32, 33] and a contribution of a spiral growth mechanism was also evidenced at growth rates higher than 80 μm/h [33]. So, a simultaneous operation of the three mechanisms: 3D nucleation followed by 2D-multilayer as a base and spiral growth which provides steps and enhances nucleation sites might explain the high growth rates approached (up to 200 μm/h) in the HVPE growth of GaN.

4.1.3. Reactor designs

The conventional HVPE reactor has a hot-wall quartz glas vessel and is operating at atmospheric pressure. The two chemical reactions needed in the HVPE growth require at least two temperature zones. Most of the reactors have more separately controlled zones, either for a better control or for additional metal sources (Al, In, Zn) aiming at alternative nitride growths and/or doping possibilities. Resistive heating is commonly used, ensuring hot wall design to avoid condensation of the metal chloride molecules on unheated surfaces. The active gases are usually delivered to the mixing point through separated parallel quartz glas liners, but coaxial arrangement of the gas inlet tubes is also used.

There are two main modifications of the HVPE reactor: horizontal and vertical reactor design.

A schematic drawing of horizontal reactor is shown in Figure 14 (a). The uniformity of the wafer deposition can be ensured either by tilting the wafer with respect to the horizontal gas flows or by rotation of the substrate holder when the wafer is perpendicular to the flow direction. In the vertical design, the reactants are typically introduced through the top and the substrate rotation is always introduced Figure 14 (b). Additional development of the vertical model is a technique of lowering the substrate holder isothermally into a dump tube [32]. This allows interruption of the growth, and then either change and equilibration of the gas flows in the main tube, or slow cooling of the sample by further lowering the holder at the end of the growth. An alternative modification is an inverted vertical reactor where the process gases are supplied through the bottom inlet flange, while the substrates are placed in the upper part where the gases are mixed. The inverted vertical design provides possibility for raising the substrate holder and minimization of solid particle contamination of the growing surface.

4.1.4. Substrates

To produce free-standing quasi-bulk material the substrate should be removed. There are several approaches, depending mainly on the type of the substrate used. The easiest way for substrate removal is a chemical etching in case of using Si or GaAs as a substrate. They both are inexpensive and industrially established which makes the approach very attractive. Very promising results have already been reported on free-standing GaN layers grown on GaAs substrate combined with ELOG technique [34], while elaborating the Si substrate is still at the initial stage [35].

In case of using SiC substrate, the GaN film can be separated by reactive ion etching (RIE) of the SiC in SF_6 containing gas mixture [36]. However, it is an expensive approach and crack-free GaN on SiC is very difficult to be reproducibly grown. When the films are deposited on sapphire substrates, mechanical polishing or laser-induced liftoff using different pulsed UV lasers processes [37, 38] are suggested. The latter process was proven to work reproducibly for thick films by irradiating the sapphire/GaN interface with intense laser pulses just at the absorption edge of GaN. It leads to a fast and strong heating which causes thermal decomposition in the interface regions of the film, yielding metallic Ga and nitrogen gas effusion. At the moment, sapphire is the most elaborated substrate both for HVPE growth and separation [39, 40], in spite of its large lattice and thermal mismatch with the GaN.

4.1.5. Nucleation schemes

In order to minimize the defects generated due to the highly mismatched heteroepitaxial nature of the HVPE growth of GaN on sapphire, three main types of nucleation schemes have been developed. The first approach is a direct growth on bare substrate using pre-treatments in different atmospheres at growth temperature. The second approach utilizes buffer layers. Reactively sputtered ZnO was the first buffer developed for HVPE-GaN growth, showing very good material characteristics [32].

Low temperature AlN and GaN buffer layers were established to provide a good nucleation surface and thus solved many problems in heteroepitaxial MOVPE growth on sapphire. However, the same approach was generally not successful in HVPE growth. Instead, single crystalline thin layers deposited at high temperature (HT) could be used as template layers, typically deposited by other techniques: AlN layers produced by low-energy ion-assisted reactive sputtering or GaN layers grown by MOVPE [41, 42]. Another approach is a use of more complex structures as templates for HVPE overgrowth of thick films like ELOG [43, 44] and Pendeo [45, 46] structures grown by MOVPE technique.

4.2. Material characterization

In-situ characterization of the HVPE films is generally not easy because of the resistive heating and is not performed. The main characteristics that are of interest for the quasi-bulk applications are typically accessed by ex-situ techniques. Detailed reviews on the characterization of HVPE-GaN grown by different heteronucleation schemes are published by several groups [47, 48, 49, 50, 51, 52, 53].

4.2.1. Structural and optical quality

The structural and optical characteristics of HVPE-GaN quasi-substrates are of significant importance since they are typically inherited in the overgrown device structure. X-ray diffraction, photoluminescence (PL) and Raman scattering (RS) are most commonly used techniques to access these properties. The structural quality is considered comparable to MOVPE grown material in some cases, however the mosaicity is typically increased with increasing film thickness. The optical quality is generally superior, with strong and narrow exciton emission lines and the absence of some defect-related undesirable emission bands as the yellow emission. The average crystal quality always improves with film thickness, which is related to a reduction of the dislocation density with the distance from the film/substrate interface (down to 10^6-10^7 cm^{-2} in a $\sim$300-μm-thick film).

4.2.2. Microstructure

In addition to the strong impact on all the properties, the ensemble of structural defects has a significant influence on crack formation in thicker films. The latter is of critical importance for producing sufficiently thick, large-area, crack-free, free-standing GaN layers. Two techniques can directly visualize the structure and defects in the films: cathodoluminescence (CL) and transmission electron microscopy (TEM). The microstructure of the thick HVPE-GaN films is determined by the nucleation scheme used.

The HVPE films grown on bare sapphire exhibit three sublayers [48, 50, 51]. A panchromatic CL image of a cross-section of such film is shown in Figure 15(a). (i) The main part of the film, extending over several tens of micrometers in thickness, exhibits an uniform emission distribution and low carrier concentration ($\sim 10^{16}$ cm^{-3}). (ii) A bright layer built up from individual columnar structures is clearly visible near the film/substrate interface. The columns have an average height from a few to several tens of micrometers and some of them extend to the top surface of the film and show up as pits of hexagonal shape [48]. Using spatially resolved CL and Raman spectroscopy, the hexagonal pits and bright interface region are found to have high residual free-carrier concentration in the range of 10^{18}-10^{19} cm^{-3}, manifested by broad free electron recombination band emission and yellow hint of the films, due to higher oxygen impurity incorporation [48, 50]. (iii) In addition, a narrow (about 1-μm-thick) region can be identified nearest to the substrate interface showing a greatly reduced radiative efficiency related to a very high degree of structural imperfections [48].

As visualized in both CL (Figure 15(a)) and TEM (Figure 15(b)) images, the columns are associated with planar stacking mismatch defects initiated at the layer/sapphire interface, both in the basal planes and in the 1-100 prismatic planes Figure 15(c). The stacking mismatch

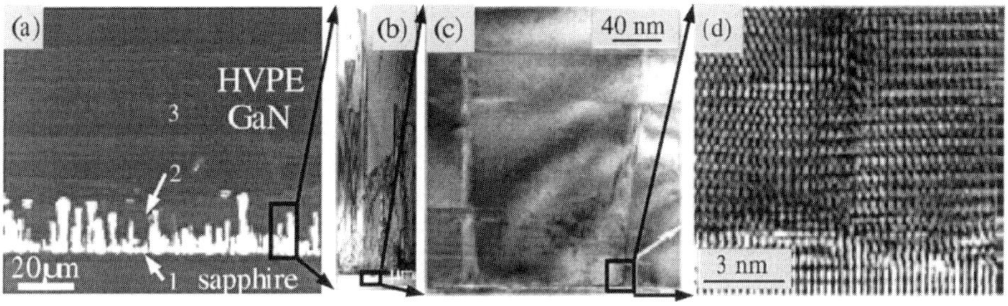

Figure 15. (a) A panchromatic CL image in cross-section of a thick HVPE-GaN layer grown directly on sapphire showing three specific growth zones: (1) the highly disordered interface region; (2) the columnar region and (3) the good quality region. (b) Low magnification TEM image of the layer/substrate interface. (c) TEM image zoomed at the generation region of a column. (d) HRTEM image revealing a stacking mismatch boundary.

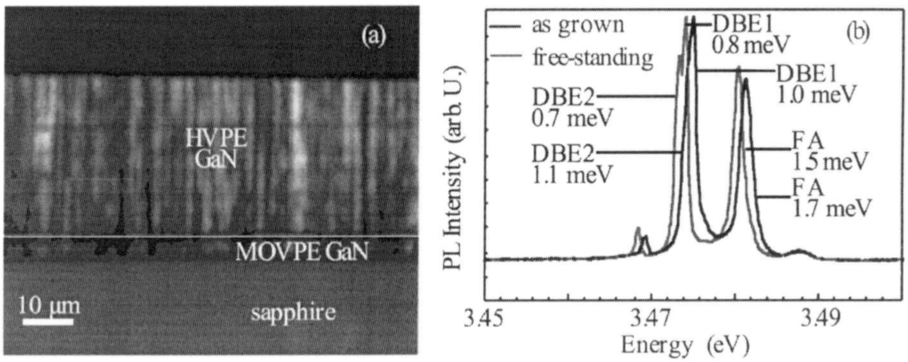

Figure 16. (a) A panchromatic CL image and (b) PL spectra of as-grown and free-standing GaN film grown by HVPE on sapphire with undoped MOVPE-GaN template.

boundary is found to be initiated at a substrate bilayer step (Figure 15(d)) from the coalescence of two islands that grow on two adjacent terraces on both sides of the step. The formation of these domains in the initial stages of the GaN growth accounts for the relaxation of the large mismatch between the GaN and the sapphire. The growth rate was found to have a very strong impact on the columnar formation. A decrease of the growth rate reduces both the nonradiative interface and the columnar regions, however, the defect regions can not be avoided in such highly mismatched systems without using buffers and in addition, a decrease of the growth rate is undesirable for time-efficient bulk growth.

The microstructure of thick GaN films grown on HT buffers has much more uniform characteristics [44, 49]. Employing templates helps to: (i) eliminate the large scale structural defects at the sapphire/GaN interface (Figure 16(a)) and large hexagonal pits on the film surface; (ii) reduce the concentration and nonuniform distribution of residual free carriers; and (iii) relieve partly the compressive biaxial strain in the HVPE-GaN layers (as seen by the shift of the po-

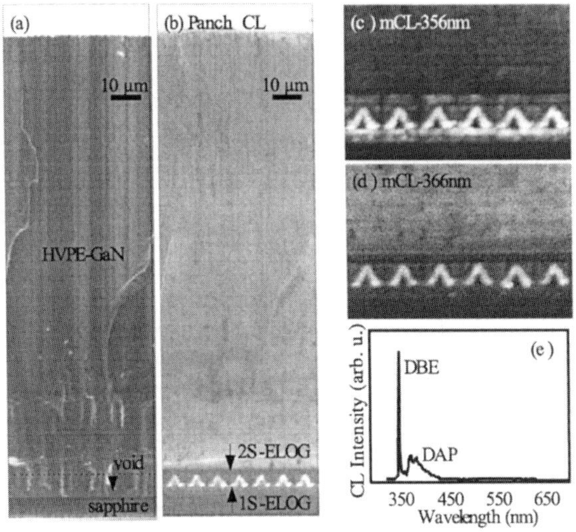

Figure 17. (a) A secondary electron image and (b) a panchromatic CL image in a cross section of an HVPE-GaN layer grown on a two-step ELOG template. (c,d) Monochromatic images of the near interface region recorded at 356 nm and 366 nm, respectively. (e) CL spectrum taken in the HVPE-GaN area, revealing a donor-bound exciton (DBE) and donor-acceptor pair (DAP) emissions.

sition of the excitonic emission in the photo luminescence (PL) spectra of Figure 16(b)). The overall improvement of the quality of the layers grown on all the buffers can be understood in terms of a reduction of extended defect concentrations. Similar improvement of the properties with thickness is observed, although growth of such high-quality, thick GaN films without cracks is more difficult, having not enough defects to compensate the mismatch strain.

The HVPE growth of thick GaN films using complicated ELOG templates was reported to result in a significant reduction of the defect density and was successfully used for the preparation ofdevices with increased lifetime [54]. Pendeo and two-step (2S) ELOG templates were also used with even further reduced defect density [53]. In spite of complicated defect, strain and emission distributions in the early stages of the growth (Figures 17(c,d)), a more uniform main part of the film is followed (as shown in the panchromatic CL image in Figure 17(b)). A common feature for the growth on such templates is the formation of voids (Figure 17(a)) in the coalescence areas, found to partially release the strain. This allows an increase of the critical thickness for crack appearance, and in some cases leads to a self-separation of the film.

4.2.3. Morphology

The morphology of the HVPE films is of determining importance for the next epitaxial overgrowth of the device structure. The macro-morphology of the as-grown thick HVPE-GaN layers is typically a hillock type (Figure 18(a)) although a use of buffers or a decrease of the growth rate at last stages of the growth may result in a surface smoothening (Figure 18(b)). However it is interesting to note that the micro-morphology of the film, as revealed by higher magnification electron microscopy (Figure 18(c)), is very smooth, almost featureless only with surface pits sized from several to hundred micrometers depending on growth rate and using buffers. The micro-morphology of the hillock slopes revealed by atomic force microscopy (AFM) also show the flatness of film surface with roughness down to 5-6 Å(Figure 18(d)). Such a morphology is characteristic for HVPE-GaN films grown at relatively low growth rates (less than 80 μm/h). Further increasing of the growth rate, which is indeed a goal for time efficient bulk growth,

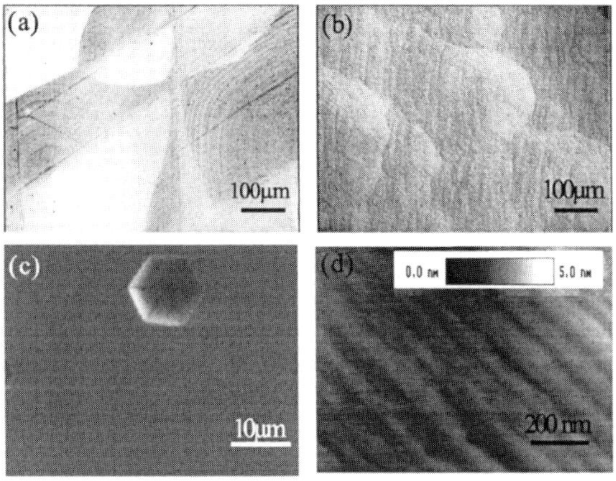

Figure 18. Macro-morphology of thick HVPE-GaN films grown on sapphire without buffer (a) and with MOVPE GaN template (b), revealed by optical microscopy. Micro-morphology revealed by SEM (c) and AFM (d).

leads to a deterioration of the micro-morphology typically manifested by micro spiral-type of hillocks due to increased contribution of a spiral growth mechanism. Clearly, the development of a polishing procedure is needed before further epitaxial proceeding of device structures can be achieved.

5. CONCLUSIONS

For the realization modern short wavelength optoelectronic devices a broad spectrum of epitaxial methods has to be used. The choice of the suitable growth technique depends on the special characteristics of each material system. For example, the growth of thick GaN layers by HVPE is the most promising method in order to obtain high-quality GaN substrates and to overcome one of the basic problems of nitride technology.

REFERENCES

1. S.-I. Nagahama, N. Iwasa, M. Senoh, T. Matsushita, Y. Sugimoto, H. Kiyoku, T. Kozaki, M. Sano, H. Matsumura, H. Umemoto, K. Chocho and T. Mukai, Jpn. J. Appl. Phys. 39 (2000) L647-L650
2. M. A. Herman, H. Sitter, Molecular Beam Epitaxy, Springer Series in Materials Science, Vol. 7, 2nd edition, 1996.
3. M. Klude, G. Alexe, C. Kruse, T. Passow, H. Heinke, D. Hommel, phys. stat. sol. (b) 229 (2002) 935.
4. M. Klude, T. Passow, R. Kröger, D. Hommel, Electr. Lett. 37, (2001) 1119.
5. C. Kruse, S. M. Ulrich, G. Alexe, E. Roventa, R. Kröger, B. Brendemühl, P. Michler, J. Gutowski, D. Hommel, phys. stat. sol. (2004) in press.
6. A. Rosenauer, S. Kaiser, T. Reisinger, J. Zweck, W. Gebhardt, D. Gerthsen, Optik 102, 63 (1996).
7. L. Vegard, Zeitschrift für Physik 5, 17 (1921).

8. H. M. Manasevit, W. I. Simpson, J. Electrochem. Soc., 116 (1969) 1725.
9. H. M. Manasevit, F. M. Erdmann and W. Simpson, J. Electrochem. Soc. 118 (1971) 1864.
10. A. Koukito, Y. Kumagai, and H. Seki, J. Cryst. growth 221 (2000) 743.
11. R. M. Watwe, J. A. Dumesic and T. F. Kuech, J. Cryst. growth 221 (2000) 751.
12. H. Amano, N. Sawaki, I. Akasaki, Y. Toyoda, Appl. Phys. Lett. 48 (1986) 353.
13. C. Kim, I. K. Robinson, J. Myoung, K. Shim, M.-C. Yoob and K. Kim, Appl. Phys. Lett. 69 (1996) 2359.
14. S. Figge, T. Böttcher, S. Einfeldt and D. Hommel, J. Cryst. Growth 221 (2000) 262.
15. A. Usui, H. Sunakawa, A. Sakai, A. Atsushi and Yamaguchi, Jap. J. Appl. Phys. 36 (1997) L899.
16. O. Nam, M. D. Bremser, T. S. Zheleva and R. F. Davis, Appl. Phys. Lett 71 (1997) 2638.
17. T. Tojyo, S. Uchida, T. Mizuno, T. Asano, M. Takeya, T. Hino, S. Kijima, S. Goto, Y. Yabuki, M. Ikeda, J. Appl. Phys. 41 (2002) 1829.
18. E. K. Sichel and J. I. Pankove J. Chem. Sol. 38 (1977) 330.
19. L. M. Belayaev, Ruby and Sapphire, Amerind Publishing Co. New Delhi (1980).
20. I. Grzegory, J. Phys.:Condens. Matter 13 (2001) 6875.
21. W. S. Wong, T. Sands and N. W. Cheung, Appl. Phys. Lett. 72 (1998) 599.
22. M. K. Kelly, R. P. Vaudo, V. M. Phanse, L. Görgens, O. Ambacher and M. Stutzmann, Jap. J. Appl. Phys. 38 (1999) L217.
23. T. L. Chu, D. W. Ing and A. J. Noreika, Solid State Electronics 10 (1967) 1023.
24. H. P. Maruska and J. J. Tietjen, Appl. Phys. Lett. 15 (1969) 327.
25. L. A. Marasina, I. G. Pichugin and M. Tlaczala, Krist. Tech. 12 (1977) 541.
26. V. S. Ban, J. Electrochem. Soc. 119 (1972) 751.
27. A. Koukitu, S. Hama, T. Taki, and H. Seki, Jpn. J. Appl. Phys. 37 (1998) 762.
28. E. Aujol, J. Napierala, A. Trassoudaine, E. Gil-Lafon and R. Cadoret, J. Crystal Growth 222 (2001) 538.
29. A. Trassaudaine, E. Aujol, R. Cadoret, T. Paskova and B. Monemar, Mater. Res. Soc. Symp. Proc. 639 (2001) G.3.2.
30. J. E. Greene, in Multicomponent and multilayered thin films for advanced technologies: techniques, fundamentals,and devices, Ed., O. Auciello, Kluwer Academic Publishers, Boston, 1993, p.39.
31. Y. Golan, X. H. Wu, J. S. Speck, R. P. Vaudo and V. M. Phanse, Appl. Phys. Lett. 73 (1998) 3090.
32. R. J. Molnar, W. Götz, L. T. Romano and N. M. Johnson, J. Crystal Growth 178 (1997) 147.
33. T. Paskova, E. B. Svedberg, L. D. Madsen, R. Yakimova, I. G. Ivanov, A. Henry and B. Monemar, Mater. Res. Soc. Symp. Proc. 537 (1999) G3.16.
34. K. Motoki, T. Okahisa, S. Nakahata, N. Matsumoto, H. Kimura, H. Kasai, K. Takemoto, K. Uematsu, M. Ueno, Y. Kumagai, A. Koukitu and H. Seki, Materials Science and Engineering B93 (2002) 123.
35. S. Gu, L. Zhang, R. Zhang, G. W. Wicks and T. F. Kuech, Proc IWN2000, Nagoya, Jpn, IPAP Conference Series C1 (2000) 23.
36. Y. V. Melnik, A. Nikolaev, I. Nikitina, K. V. Vassilevski and V. A. Dmitriev, Mater. Res. Soc. Symp. Proc. 482 (1998) 269.
37. W. S. Wong, T. Sands and N. W. Cheung, Appl. Phys. Lett. 72 (1998) 599.
38. M. K. Kelly, R. P. Vaudo, V. M. Phanse, L. Görgens, O. Ambacher and M. Stutzmann,

Jpn. J. Apl. Phys. 38 (1999) L217.
39. R. P. Vaudo, G. R. Brandes, J. S. Flynn, X. Xu, M. F. Chriss, C. S. Christos, D. M. Keogh and F. D. Tamweber, Proc IWN2000, Nagoya, Jpn, IPAP Conference Series C1 (2000) 15.
40. S. S. Park, I. W. Park and S. H. Choh, Proc IWN2000, Nagoya, Jpn, IPAP Conference Series C1 (2000) 60.
41. T. Paskova, J. Birch, S. Tungasmita, R. Beccard, M. Heuken, E. B. Svedberg, P. Runeson, E. M. Goldys and B. Monemar, Phys. Stat. Sol. (a) 176 (1999) 415.
42. R. Cadoret and A. Trassaudaine, J. Phys. C 13 (2001) 6893.
43. A. Usui, H. Sunakawa, A. Sakai and A. A. Yamaguchi, Jpn. J. Appl. Phys. 36 (1997) L899.
44. P. Vennégués, B. Beaumont, V. Bousquet and P. Gibart, J. Appl. Phys. 87 (2000) 4175.
45. R. Davis, T. Gehrke, K. J. Linthicum, P. Rajacopal, A. M. Roskowski, T. Zheleva, E. A. Preble, C. A. Zorman, M. Mehregany, U. Schwarz, J. Schuck, R. Grober, MRS Int. J. Nitride Semicond. Res. 6, (2001) 14.
46. T. Paskova, E. Valcheva, P. P. Paskov, B. Monemar, A. M. Roskowski, R. F. Davis, B. Beaumont and P. Gibart, Diamond and Related Materials (2004) in press.
47. R. J. Molnar, in Gallium Nitride (GaN) II, Eds., J. I. Pankove and T. D. Moustakas, Acad Press, San Diego, 1999, p.1.
48. T. Paskova and B. Monemar, in III-Nitride Semiconductors: Growth, Ed., O. Manasreh, Taylor & Francis Group, NY, 2003, p.175.
49. F. Dwikusuma, S. E. Babcock and T. F. Kuech, in Vacuum Science and Technology: Nitrides as seen by the technology, Eds., T. Paskova and B. Monemar, Research Signpost, Trivandrum, 2002, p.79.
50. H. Morkoç, Materials Science and Engineering R33 (2001) 135.
51. W. Zhang and B. K. Meyer, Phys. Stat. Sol. (c) 0 (2003) 1571.
52. K. Hiramatsu, in Vacuum Science and Technology: Nitrides as seen by the technology, Eds., T. Paskova and B. Monemar, Research Signpost, Trivandrum, 2002, p.125.
53. P. Gibart, B. Beaumont and P. L. Vennégués, in Nitride Semiconductors, Handbook on Materials and Devices, Eds., P. Ruterana, M. Albrecht and J. Neugebauer, Wiley Verlag, Weinheim, 2003, p.45.
54. S. Nakamura, Semicond. Sci. Technol. 14 (1999) R27.

Solution growth methods at low and high temperatures

J. Zaccaro[1], B. Menaert[1], D. Balitsky[2] and A. Ibanez[1]

[1]Laboratoire de Cristallographie, CNRS, BP 166, 38042 Grenoble cedex 9, France.

[2]Laboratoire de Physicochimie de la Matiere Condense, Universite Montpellier II, Place E.Bataillon, cc003, 34095 Montpellier cedex 05, France

In this lecture we introduce different types of solution growth methods at low (close to room temperature) and high temperatures, (up to 1000°C), such as slow cooling, temperature gradient, hydrothermal and flux methods. We illustrate all these techniques with examples taken from our studies at CNRS-Grenoble which include growth of hybrid organic inorganic salts and of crystals of the potassium titanyl phosphate family. These materials exhibit non-centrosymmetric crystals structures and high nonlinear optical coefficients. We have carried out their crystal growth to study their optical properties and to develop devices like optical parametric oscillators. Hydrothermal technique will be presented in the case of aluminum and gallium phosphate crystals, which are isomorphous to α-quartz. These crystals, prepared at the University of Montpellier, exhibit promising piezoelectric properties for the development of resonators and physical sensors. In addition, for both low and high temperature ranges, we will give examples of industrial crystal growth processes such as growth of potassium dihydrogen phosphate (KDP), potassium titanyl phosphate (KTP) and α-quartz crystals.

In addition to these standard solution growth techniques, some more particular ones are presented. Indeed, solution crystal growth, carried out under typical conditions at low supersaturations of the solutions (relative supersaturations σ of around 1%), exhibit low growth rates, less than 1mm/day. By applying an overheating and/or an ultrasonic treatment of the solution, it is possible to inhibit the spurious nucleation that takes place at higher supersaturations. Thus, through this type of treatment of the solution during the crystal growth process, higher supersaturations σ can be applied, σ around 10-20%, and thus increase significantly the growth rates, over 1cm/day. This rapid crystal growth is illustrated in the cases of potassium dihydrogen phosphate and hybrid organic-inorganic salts. Also, the liquid phase epitaxy and the traveling solvent zone methods is briefly introduced.

Finally, several difficulties generally associated with crystal growth from solution are mentioned. For instance, the first step of these methods is to obtain large seeds by spontaneous nucleation or by splicing crystal cuts. Crystals grown in solution also exhibit typical defects such as solvent inclusions associated with growth instabilities and problems of hydrodynamics. As other types of crystal growth methods, impurity segregations, twins or bunches of dislocations can be present. All these crystal defects can be well evidenced using X-rays diffraction topography. Several examples of this characterization are presented from experiments carried out in our laboratory with a Lang chamber or at the European Radiation Synchrotron Facility (ESRF) with the section topography method.

Crystal Growth – from Fundamentals to Technology
G. Müller, J.-J. Métois and P. Rudolph (Editors)
2004 Published by Elsevier B.V.

Materials and crystal growth for photovoltaics

Thomas Surek

National Renewable Energy Laboratory, Golden, Colorado 80401, U.S.A.

Photovoltaics (PV) is *solar electric power* – a semiconductor-based technology that converts sunlight to electricity. Three decades of research has led to the discovery of new materials and devices and new processing techniques for low-cost manufacturing. This has resulted in improved sunlight-to-electricity conversion efficiencies, improved outdoor reliability, and lower module and system costs. The manufacture and sale of PV has grown into a $4 billion industry worldwide, with more than 560 megawatts of PV modules shipped in 2002. The key contributions to progress over the years have been made in areas of crystal growth and materials science –from discovering and demonstrating new PV materials and device structures to identifying and developing scalable and potentially low-cost manufacturing approaches. Controlling the chemistry and defect structures in the materials has been a major factor in the ongoing improvements in solar cell efficiencies over the years in all PV technologies (Figure 1).

While the progress in laboratory solar cell efficiencies is impressive, significant differences remain between the best performances and the theoretically predicted values for each solar cell technology.

Furthermore, the efficiencies of commercial (or the best prototype) modules are only about 50% to 65% of the "champion" cells shown in Figure 1. Closing these gaps is the subject of ongoing and future research and will, for the most part, require the talents and ingenuity of the crystal growth and materials science communities in improving the PV materials and in scaling up to large-scale manufacturing. This lecture reviews the most significant advances in PV materials and devices research in the various technologies from the current-generation (crystalline silicon); to the next-generation (thin films); to future-generation PV technologies.

The latter includes innovative materials and device concepts that hold the promise of significantly higher conversion efficiencies and/or much lower costs. The focus of the lecture is on crystal growth advances and remaining technical issues, specifically in developing large-scale, potentially low-cost approaches to module manufacturing for the various PV technologies.

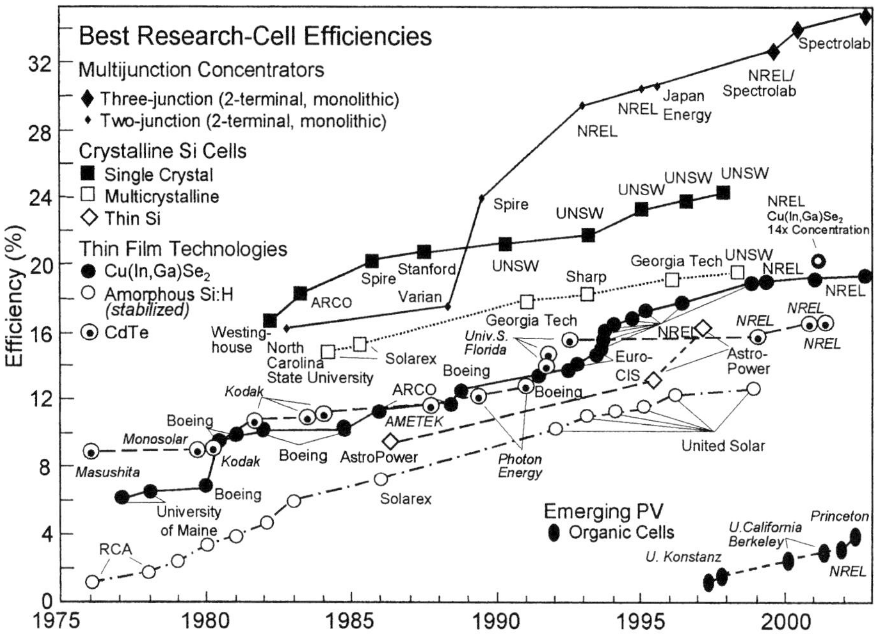

Figure 1: Progress in solar cell efficiencies (1976 to present) for various research or laboratory devices. All these cell efficiencies have been confirmed and were measured under standard reporting conditions.

Crystal Growth – from Fundamentals to Technology
G. Müller, J.-J. Métois and P. Rudolph (Editors)

Point Defects in Compound Semiconductors

D. T. J. Hurle

H.H. Wills Physics, University of Bristol, Tyndall Avenue, Bristol BS8 1TL, UK

Experimental methods used to determine the concentrations and charge states of native point defects in III-V and II-VI semiconductors are outlined. The use of a chemical thermodynamic model to unify the very large amount of experimental data on the III-Vs is described.

It is demonstrated how dopant solubility, self and dopant diffusion, annealing behaviour and phase extent are all coupled to types, concentrations and charge states of the native point defects.

1. INTRODUCTION

In equilibrium at all temperatures above the absolute zero, a crystal contains point defects. These comprise vacant lattice sites, atoms located in the interstices between the lattice sites and, in the case of compound semiconductors, atoms sitting on the wrong lattice site (anti-site defects). Point defects are either native (ie involving only atoms of the pure crystal) or they are comprised of foreign atoms that may have been deliberately added ('dopants') or are residual impurities.

Many of the reviews of point defects in semiconductors are written by physicists for physicists and focus on the identification, site symmetry and the electrical and optical properties of the defects and their complexes. A good example is the review by Skowronski [1]. The present review differs in being directed toward giving crystal growers an insight into how the point defect populations depend on the conditions of growth, annealing and diffusion.

In III-V and II-VI compound semiconductors native point defect concentrations can, at typical growth temperatures, be comparable or even greater than the intrinsic carrier concentration and can therefore influence the position of the Fermi level. This results in a complex interaction between electrically active dopants and the native point defects. The equilibrium concentration of neutral point defects grown into a crystal depends on both the growth temperature and on the composition of the nutrient phase.

Semiconductor crystal growers have always been much concerned with how the incorporation of foreign atoms (both dopants and residual impurities) depends on the conditions of growth. But there has been little understanding of the factors that control the incorporation of native point defects or indeed of which native point defects are dominant. Often able to exist in several electrical charge states, native point defects can influence dopant incorporation and the physical properties of the crystal, all of which can affect the performance of fabricated devices. Point defects can also be generated in non-equilibrium concentrations principally by particle irradiation and also during some low temperature growth and diffusion processes. However these fields are too broad to be covered here.

Techniques for detecting and determining the concentrations of native point defects in semiconductor crystals are expanding and the importance of these defects in determining the properties of the grown crystal is becoming realised. The most studied system to date is probably the role of grown-in vacancies and interstitials in controlling the formation of oxygen induced stacking faults and of so-called 'B' defects in melt-grown silicon crystals. By modelling the process Voronkov [2] and others have inferred that $\sim 10^{15}$ cm^{-3} vacancies and a similar number of interstitials are incorporated, at equilibrium, into the crystal at the melting point. The balance between the number of each of them retained in the cooling crystal controls the defect formation and depends on the quotient of growth rate to temperature gradient. The growth rate dependence arises from a supposed non-equilibrium trapping of the native point defects. Whilst it is believed that vacancies in Si can be multiply charged [3], their concentration is insufficient to influence the Fermi level and therefore their charge state is not generally taken into account in modelling the process.

In the III-V and II-VI compound semiconductors the equilibrium concentrations of native point defects at the melting point are markedly greater than in Si. For example, it can be inferred from the studies of Oda et al [4] and of the group of Bublik and Morizov [5] that, in GaAs, there are of the order of 10^{19} cm^{-3} arsenic vacancies and arsenic self interstitials at the melting point. Most of the arsenic vacancies are positively charged and their concentration just exceeds the number of electron-hole pairs at the melting point so that the Fermi level position lies significantly above its intrinsic position [6].

This fact, together with the additional parameter of component activity in the nutrient phase (or melt), makes the point defect equilibrium situation in compound semiconductors much more complicated than it is in silicon. Furthermore, there being two sub-lattices, the number of native point defect species is three times greater in the binary compounds. Simple counting of unpaired electrons suggests that, in III-V compounds, vacancies on the Group V sub-lattice should exhibit donor-like behaviour whilst those on the group III sub-lattice should be multiply chargeable as acceptors. Both theory [7] and experiment [8] broadly bear this out. (The charge state of a defect is here taken to be the difference between the charge at a lattice site with and without the defect being present). Cation self-interstitials appear to act as multiply ionisable donors in at least some of the III-Vs and II-VIs. The situation with anion self-interstials is less clear. Finally, in III-V semiconductors, an anti-site defect might be expected to behave as a pentavalent impurity on the cation (Group III) site (ie as a double ionisable donor) and as a trivalent impurity on the anion (Group V) site (ie as a doubly ionisable acceptor). Known behaviour of GaAs and GaSb is in accord with this [9,10].

2. SOME EXPERIMENTAL TECHNIQUES FOR THE DETERMINATION OF NATIVE POINT DEFECT CONCENTRATIONS AND THEIR CHARGE STATES

2.1. Coulometric Titration

The most direct and unambiguous way of determining deviation from stoichiometry (δ) of a pure crystal is to count the number of atoms of each type using Faraday's law. For a binary material it is sufficient to titrate for just one component if additionally the total mass of the semiconductor sample is known with sufficient accuracy. δ can be expressed in terms of contributions from the native point defects in each sub-lattice:

$$\delta = \delta As - \delta Ga \tag{1}$$

where:

$$\delta Ga = [Ga_i] - [V_{Ga}] + 2[Ga_{As}] - 2[As_{Ga}]$$

$$\delta As = [As_i\} - [V_{As}] + 2[As_{Ga}] - 2[Ga_{As}]$$

The subscript 'i' denotes an interstitial, 'V' is a vacancy with its subscript denoting its sub-lattice. An anti-site defect is denoted by the symbol for the misplaced atom subscripted with the sub-lattice on which it is sitting. Square brackets indicate concentrations of the defects summed over all their charge states. Interstitial atoms can sit in more than one position in the crystal interstice but we ignore this for the moment.

Jordan et al [11] first demonstrated use of this technique to determine the solidus curve for GaP in the vicinity of its melting point. More recently several workers have applied the technique to melt-grown GaAs [12-14]. Of these, the work of Oda et al appears to be the most careful and extensive [12]. They demonstrated that all melt-grown crystals were As-rich with the As excess depending on melt composition (as expected) having a maximum deviation from stoichiometry of around 3.10^{18} cm^{-3}.

2.2. Density/lattice parameter measurements

The quotient of the density (ρ) and the cube of the lattice parameter (a_o^3) is the mass per unit cell. If we subtract from a measured value of this quantity the mass of an ideal (ie defect-free) unit cell then we obtain the mass excess (or deficiency) per unit cell (δN). This is given by:

$$\delta N = \rho a_0^3 - 4(M_{As}+M_{Ga})/N_a$$

$$= \{M_{As}([As_i] - [V_{As}]) + M_{Ga}([Ga_i] - [V_{Ga}]) + 2(M_{As}-M_{Ga})([As_{Ga}] - [Ga_{As}])\}/N_a \tag{2}$$

where N_a is Avogadro's number. If we know which defects are dominant then equation 2 can be simplified. Thus, for GaAs, the atomic weights of anion and cation are nearly equal ($M_{As} \sim M_{Ga}$) and therefore the terms involving the antisite defects in equation 2 will be negligible. Further, since the mass per unit cell increases linearly with As content in the melt (as it does also for InAs) , we know that defects on the As sub-lattice must dominate. Comparing with equation 1 we see that

$$\delta - (N_a / M)\, \delta N \sim [V_{Ga}] - [Ga_i] \tag{3}$$

which gives a rough measure of the deviation on the Ga sub-lattice. $M = (M_{Ga}+M_{As})/2$ is the mean atomic mass. Bublik, Morizov and co-workers [5,15-19] have made such measurements on melt grown crystals of the six binary systems (Ga,In)-(P,As,Sb). The data of Bublik et al. [14] for GaAs is shown in Figure 1. Achievable experimental precision is set by the density measurements and is of the order of a few times 10^{17} cm^{-3}. (Lattice parameter can be measured to a very high accuracy relatively) [5].

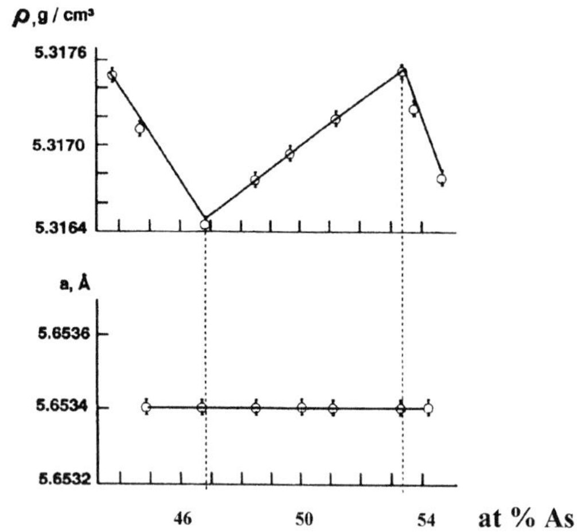

Figure 1. Density and lattice parameter versus melt composition for gallium arsenide. Taken from Bublik et al [14].

2.3. Positron Annihilation

Positrons are trapped by neutral and negatively charged vacancies. Vacancy ionisation energies can be determined by observing the change in positron lifetime as the Fermi level is swept through the ionisation level by changing the sample temperature and/or doping level The lifetime of a positron in a vacancy will depend on the environment of that vacancy as well as on its charge state. It is therefore possible to distinguish between isolated vacancies on each sub-lattice and vacancies bound to dopant atoms.

An estimation of the vacancy concentration can be obtained from the rate at which positrons are trapped. However this does require a knowledge of the proportionality constant (the trapping coefficient) which has to be obtained from a calibration by independent means. There is a recent and comprehensive book on defect studies of semiconductors using positrons by Krause-Rehberg and Leipner [20].

2.4. X-ray quasi-forbidden reflection

For any binary lattice there exist some weak reflections for which scattering from the two atoms are in opposite phase. A small deviation from stoichiometry will, for these reflections, produce a relatively large change in structure factor. Fujimoto [21] calculates that, for GaAs, a difference in atom fraction of Ga and As of 2.5×10^{-5} produces a 0.1% change in reflected intensity. The technique can therefore detect non-stoichiometry to a good accuracy but insufficiently precise knowledge of structure factors prevents it from being an absolute method.

2.5. Diffusion studies

An indirect determination of the charge state of a native point defect which mediates either the self diffusion or the diffusion of an impurity can be obtained from a modelling of the diffusion process. This is a complex subject and has produced some conflicting results. See section 9 below.

2.6. Scanning Tunnelling Microscopy

This relatively recent technique is most applicable to the identification of dopant-vacancy complexes. For example these can be imaged on cleaved {110} surfaces of heavily Si-doped GaAs [22]. A rough estimate of the volume concentration can be obtained by counting the density of near surface defects. The limit of sensitivity is of the order of $10^{17}cm^{-3}$.

It was observed that isolated V_{Ga} formed on the surface after cleavage of n-doped samples and of V_{As} with p-doped samples. This is assumed to be a Fermi level effect. By studying the time evolution of the vacancy density and extrapolating back to the time of cleavage an estimate of the number of vacancies intrinsic to the uncleaved sample could be obtained [22].

2.7. Spectroscopic Techniques

There is a wide range of techniques that can give information on defect structure. Electron paramagnetic resonance, widely used in the identification of point defects in Si and Ge is rather less useful for the III-Vs because the resonances are much broader. Notwithstanding, the technique was used to identify the anti-site defect As_{Ga}. For reviews see [23, 32].

Amongst the optical techniques local vibrational mode spectroscopy (LVM) and Raman spectroscopy can distinguish different lattice configurations of isolated and complexed dopants having an atomic mass lower than that of the matrix atoms. Independent calibration is necessary to obtain quantitative results. It has been extensively used in the study of Si and B in GaAs [24].

Deep level transient spectroscopy (DLTS) is the most powerful method available for the study of the energy levels of deep traps [25]. The technique involves analysis of the temperature dependence of the capacitance transient of a reversed biased Schottky barrier.

Finally we mention photo-luminescence (PL), widely used as a characterisation technique to yield maps of the spatial distribution of alloy composition and of specific non-radiative recombination centres.

2.8. Carrier concentration and mobility measurements

Where charged native point defects or dopant-native point defect complexes are preserved on cooling to room temperature, Hall coefficient and resistivity measurements can be combined to give information on concentrations of these defects. Ionisation energies can be obtained from temperature dependant Hall coefficient studies.

If it is technically possible to make measurements at a temperature above that at which point defect aggregation occurs then, by making measurements at several temperatures, it is possible to obtain values for both the enthalpy and entropy of formation of the ionised dopant and native point defects. This has been employed with some of the II-VI compounds [26].

2.9. Thermodynamic modelling of dopant solubility data

Consider the incorporation during melt or LPE growth into a GaAs crystal of Te as a donor located on an As site. The incorporation reaction is: $Te_l + V_{As}^0 = Te_{As}^+ + e^-$ where the superscript zero denotes the neutral state. e^- denotes an electron. The mass action constant for this reaction is:

$$K_{Te}(T) = [Te_{As}^+]n_{gt}/([V_{As}^0]\,[Te_l]) \tag{4}$$

assuming that Te forms an ideal solution in the Ga/As melt. n_{gt} is the conduction band electron concentration at the growth temperature. Rearranging this equation we obtain for the segregation coefficient of the Te:

$$k_{Te} = [Te_{As}^{+}]/[Te_{l}] = K_{Te}(T)\,[V_{As}^{0}]/n_{gt} \tag{5}$$

n_{gt} is given by the condition that the crystal be electrically neutral:

$$n_{gt} = n_i^2/n_{gt} + [Te_{As}^{+}] + [V_{As}^{+}] \tag{6}$$

where n_i is the intrinsic electron concentration and V_{As}^{+} is an ionised arsenic vacancy present at the growth temperature. From equation 5 we see that k_{Te} is a constant so long as the temperature and n_{gt} are constant. From equation 6 we see that if $[V_{As}^{+}]$ is negligible then n_{gt} is constant and therefore k_{Te} is also constant so long as $[Te_{As}^{+}] \ll n_i$. When this latter condition ceases to be valid the crystal is extrinsic at the growth temperature and k_{Te} progressively falls. Whilst k_{Te} is constant the solubility curve is linear but becomes sub-linear as k_{Te} falls.

However the solubility curve for Te in LPE grown GaAs remains linear up to a much higher dopant concentration than is predicted by this argument. This is due to the fact that, under these growth conditions, $[V_{As}^{+}] \gg n_i$ at the growth temperature and the curve becomes sub-linear only when $[Te_{As}^{+}] > [V_{As}^{+}] \gg n_i$. Fitting to the experimental solubility curve therefore yields a value for the ionised arsenic vacancy concentration. Performing this over a range of temperatures allows one to obtain values for the enthalpy and entropy of the vacancy formation reaction. Hurle [6] has used this approach to make a comprehensive analysis of point defect incorporation in GaAs. The methodology is outlined in sections 3.2 and 3.3 below.

3. THEORETICAL MODELLING OF NATIVE POINT DEFECT CONFIGURATIONS AND THEIR FORMATION AND IONISATION ENERGIES

3.1. Introduction

A proper consideration of this topic is beyond the scope of this presentation. The following is an outline only. The simplest approach is a bond breaking thermodynamic model that gives indications of trends. This is an approach widely considered by van Vechten [27]. An analysis of GaAs behaviour based on this approach has been given by Tan [28]. More fundamental, so-called first principles, approaches have increased in sophistication with advances in understanding and in computing power. Where the atomic position of the defect can be assumed basic approaches to electronic structure include use of pseudo-potentials and molecular orbital theory [29]. However, it is known that for some defects electronic structure and atomic position are closely coupled and here density functional methods are employed [30]. The most famous and much studied case is that of the metastable state of the EL2 defect in GaAs [46]. These approaches essentially neglect entropy and so give information strictly only about the crystal at the absolute zero of temperature. At the high temperature of bulk crystal growth it is evident that the entropy term in the Gibbs free energy cannot usually be neglected. In consequence, first principle calculations frequently do not well describe the experimental data relating to crystal growth [6,31].

To obtain a full understanding of the native point defects in a material one needs to construct a thermodynamic model treating each native point defect as a 'chemical' entity and utilise the methods of chemical thermodynamics pioneered by J. Willard Gibbs [33] and its application to crystals as expounded by Kröger [34]. Then by fitting experimental data obtained from appropriate techniques such as those outlined in section 2, one can obtain values for the enthalpy and entropy of formation of each of the dominant defects. From this one can obtain their concentrations at equilibrium in the crystal under any imposed conditions of temperature, pressure etc. This has recently been done by Fochuk et al [35] for CdTe and by Hurle [6] for GaAs.

In the following an outline of the methodology is given with GaAs as the example.

3.2. Neutral species

The following equilibrium reactions and their mass action constants define the concentrations of neutral point defects at equilibrium in a binary crystal at given temperature and activity of one of the components. The selected component here used is As and the activity is expressed in terms of the partial pressure of the dimer As_2. One could equally well have expressed the activity in terms of any other vapour species ie of As or As_4 or in terms of $[As_l]$, the concentration of As in a melt or solution. These several entities are related to each other through the thermodynamics (ie p, T, x diagram) of the Ga-As system. Having defined an equilibrium state in terms of the arsenic dimer partial pressure, the Ga partial pressure at equilibrium is defined by Gibbs phase rule and is given [36] by:

$$p_{Ga}p_{As2}^{1/2} = \exp(-g_f/kT) \tag{7}$$

where g_f is the free energy of formation of GaAs from gaseous gallium monomers and arsenic dimers at one atmosphere total pressure and temperature T. We can write the appropriate reactions for the individual point defects and their mass action equations as follows:

$$1/2As_2(v) = As_i^0: \qquad K_{AI} = [As_i^0]p_{As2}^{-1/2} \tag{8}$$

$$As_{As} = V_{As}^0 + As_i^0: \qquad K_{FA} = [As_i^0][V_{As}^0] \tag{9}$$

$$0 = V_{As}^0 + V_{Ga}^0: \qquad K_S = [V_{As}^0][V_{Ga}^0] \tag{10}$$

$$Ga_{Ga} = Ga^0_i + V_{Ga}^0: \qquad K_{FG} = [Ga_i^0][V_{Ga}^0] \tag{11}$$

$$As_i^0 + V_{Ga}^0 = As_{Ga}^0: \qquad K_{AG} = [As_{Ga}^0]/([As_i^0][V_{Ga}^0]) \tag{12}$$

$$Ga_i^0 + V_{As}^0 = Ga_{As}^0: \qquad K_{GA} = [Ga_{As}^0]/([Ga_i^0][V_{As}^0]) \tag{13}$$

The Ks are mass action constants. The reactions are here written in a form in which only one of them (equation 8) involves the external phase. Any linear combination of the equations is also a valid representation but only six independent equations are needed to define the system.

3.3. Charged native point defects and electroneutrality

Supposing that we know the energies of ionisation of the defects (if indeed they have ionised states lying within the band gap) then we can determine the fraction in any ionised state from the condition of crystal electroneutrality. In the general case this is:

$$n + \Sigma_a[V_{As}{}^{a-}] + \Sigma_b[V_{Ga}{}^{b-}] + \Sigma_c[Ga_{As}{}^{c-}] + \Sigma_d N_a{}^{d-}$$

$$= p + \Sigma_e[V_{As}{}^{e+}] + \Sigma_f[As_i{}^{f+}] + \Sigma_g[[Ga_i{}^{g+}] + \Sigma_h[As_{Ga}{}^{h+}] + \Sigma_i N_d{}^{i+} \qquad (14)$$

where n is the conduction band electron concentration, p the hole concentration, N_a and N_d are the concentrations of dopant acceptors and donors and the summations are over all the possible degrees of ionisation of the defects (a...i).
The concentrations of the individual ionised defects are related to the neutral concentrations by the Fermi energy as are n and p.

$$n = N_c F_{1/2}(\eta) \text{ and } p = N_v F_{1/2}(-e_g - \eta) \qquad (15)$$

where $\eta = e_f/kT$, e_f being the Fermi energy measured from the conduction band edge and $e_g = E_g/kT$ where E_g is the energy gap. $F_{1/2}$ is the Fermi integral of order 1/2.

For entities forming donor-like states

$$[D^+]/[D^0] = \beta_d \exp(-e_d - \eta) \qquad (16)$$

where β_d is the degeneracy factor for the donor and $e_d = E_d/kT$, E_d being the donor ionisation energy. For acceptors we have

$$[A^-]/[A^0] = \beta_a \exp(e_g - e_a - \eta) \qquad (17)$$

where β_a is the acceptor degeneracy and $e_a = E_a/kT$, E_a being the acceptor ionisation energy.
The electroneutrality condition has to be solved numerically for the Fermi energy from which the concentrations of all the point defects can then be obtained. If the semiconductor is non-degenerate under all experimental conditions, then the Fermi integral can be replaced by the Boltzmann exponential.

4. ISOLATED NATIVE POINT DEFECTS

4.1. Vacancies

The entropy of dissociation of vacancies, interstitials and substitutional dopants means that at the high temperatures of growth they will exist as isolated entities. As the crystal cools the $T\Delta S$ contribution to the Gibbs free energy falls and defect complexes form. Coulombic interactions between charged isolated defects often provide the driving force.
In compound semiconductors, vacancies can usually exist in more than one charge state and are, as we have seen, present at high temperature in concentrations which can influence the Fermi level position. In the II-VI compounds high temperature Hall coefficient measurements

have been used to obtain estimates of their concentrations [8]. In GaAs high temperature Hall and resistivity measurements, made by Nichols et al. [37], have been interpreted by Hurle [6] to give an estimate of the V_{As}^{+} concentration along the As-rich solidus.

4.2. Self interstitials

It is commonly assumed, and first principle calculations support the idea, that the enthalpy of formation of an interstitial is so large that self interstitials will not be formed in high concentrations. In the case of GaAs this is quite contrary to the unambiguous experimental information obtained from titration and from density/lattice parameter measurements. It is an unresolved problem.

Self interstitials can exist in several different configurations. In the zinc-blende lattice they can be located at the centre of the unit cell in a non-bonding configuration having full cubic symmetry. Alternatively they can be located in a bond-centred configuration between neighbouring anion and cation species. Finally they can exist as split interstitials where, for example, two anions sit astride a single anion vacant site [38].

As to ionisation states, in GaAs it is found from diffusion studies (see section 9) that Ga_i are multiply ionisable donors. Diffusion and electrical data (see sections 9 and 6 respectively) would seem to require that As_i are not ionised during self diffusion or melt growth [6, 39].

In CdTe, calculations and fitting to electrical data suggest that Cd_i is a donor present at relatively high concentration [40]. Te_i is assumed to be an acceptor but modelling suggests that it is present only at relatively low levels [41]. It is believed to compensate donor doped material resulting in enhanced solubility.

4.3. Antisite defects

Antisite defects might be expected to be most prevalent in crystals which are dominantly covalent with only a small degree of ionicity, They would therefore be expected to be more important in III-V compounds than in the II-VIs. This appears to be born our in practice. The only reported II-VI antisite defect is Te_{Cd} in HgCdTe [45]. In the III-Vs they can be thought of as doubly ionisable impurities; donors in the case of anions on a cation site and acceptors for cations on an anion site. This simple concept is in accord with the behaviour of GaAs where the As_{Ga} defect is known to have a first ionisation level at 0.75 eV below the conduction band and a second one 0.54 eV above the valence band edge. A defect believed by some authors to be GaAs has acceptor levels at 78 and 203 meV above the valence band. However there is some dispute about the nature of this defect [42].

The As_{Ga} defect would appear not to be stable at the melting point of GaAs but rather to form during cooling of melt-grown material to room temperature. It is also present (at lower concentration) in epitaxial material grown under As-rich conditions [43]. A model of the reactions occurring during cooling has been proposed by Hurle [6].

The P_{Ga} defect has been identified by EPR in LEC grown GaP [44] but has not been as extensively studied as the EL2 defect in GaAs (see [32] for a review). It is not known whether or not the defect is only formed on cooling of the crystal or whether it is grown-in at the melting point.

In GaSb the Ga_{Sb} defect is present in material grown by LPE both from Ga-rich and Sb-rich solutions [10]. It is also present in melt grown material. The fact that all the experimental data points can be fitted onto a single solubility curve suggests that, unlike GaAs, it is incorporated at the growth interface rather than as the result of cooling of the crystal. A concentration in excess of 10^{17} cm^{-3} is found in melt-grown material resulting in undoped material always being p type [47]. First and second ionisation levels are at 34.5 and 102 meV above the valence band respectively.

5. THE COOLING CRYSTAL

As the crystal cools to room temperature the solubility of the point defects will decrease and the room temperature properties reflect this. In the absence of any electrochemical effects (i.e. ionisability) a maximum in the solidus as the temperature is lowered is characteristic of all sparingly soluble entities whether they be native point defects or dopant atoms [48]. Therefore point defects will become supersaturated in the lattice below some sufficiently low temperature. If that temperature is still high enough for diffusion processes to occur within the timescale for which the crystal is held at temperature then some aggregation of point defects will occur. This can take the form of voids, of precipitates or of dislocation climb resulting from the condensation of vacancies or interstitials at jogs in a dislocation [49]. Alternatively isolated point defects can combine with each other to form point defect complexes. These can involve only native point defects (such as a Schottky divacancy $V_{Ga}V_{As}$ or an antisite defect) or can involve a native point defect and a dopant atom. A common example of a complex formed by a native point defect binding to a dopant atom is the donor atom bound to a cation vacancy found both in some III-V and II-VI compounds as we shall see in section 7.

The As_{Ga} defect in GaAs is believed to form as a result of diffusion of arsenic interstials (which are incorporated at a high concentration in melt-grown crystals) to Ga vacancies which are also in supersaturation but at a much smaller concentration leading to a concentration of As_{Ga} of up to $\sim 2\text{x}10^{16}$ cm^{-3} depending on melt composition. The observed dependence on melt composition has been shown to be compatible with the proposed V_{Ga} mechanism [6].

If point defect supersaturation occurs only at temperatures below that for which diffusion is possible then the distribution in the crystal at room temperature of the defects present may represent equilibrium at some higher temperature. However if there is more than one important point defect then the temperatures at which each is 'frozen-in' to the lattice may differ in which event determining the non-equilibrium state of the crystal at room temperature will be more complicated. If diffusion of a point defect in the lattice is very rapid (as with some non-native interstitials in II-V compounds for example) then atoms or native point defects can be lost to the surface of the crystal during cooling. In this case the composition retained at room temperature will depend on the sample size and on the activity of the rapid diffusant in the ambient phase since this will influence the rate of out-diffusion.

A common requirement is to relate the room temperature carrier concentration to the growth conditions. In addition to all the changes that can occur during the cooling process due to rearrangement of point defects produced by supersaturation of one or more entities, there is also to be considered the electron redistribution produced by movement of the Fermi level as the crystal is cooled.

6. PHASE EXTENT

Having determined the concentrations of all the important point defects at two temperatures by employing a sufficient combination of experimental techniques and/or from some first principles modelling, one can determine the enthalpy and entropy of formation of each defect and hence its concentration at any temperature and component activity. Specifically one can determine the deviation of the crystal from its stoichiometric composition at the phase boundary with the liquid phase – its solidus curve. The separation in composition between the anion-rich and cation-rich boundaries at any temperature gives the phase extent at that temperature. At very low temperatures the point defect concentrations become negligibly small and, assuming that there is no low temperature phase change, the phase extent diminishes to zero at the absolute zero with the crystal having its stoichiometric composition.

At the congruent melting point the phase extent will by definition be zero. The composition at which this occurs will not in general be the stoichiometric composition (ie we have $\delta \neq 0$). To grow a stoichiometric crystal will require that the melt is richer in one or other component. Titration and density/lattice parameter measurements provide an unambiguous method for determining the congruent composition and the melt composition that is in equilibrium with a stoichiometric crystal.

To do this we first need to identify the dominant defects. Again we illustrate this with the case of GaAs. Firstly it is known from extensive studies on the deep level EL2, now known to be the As_{Ga} antisite defect, that it is never present in a concentration above a few 10^{16} cm^{-3} and then only after annealing at a temperature well below the melting point [50]. Material quenched from just below the melting point (1511K) has virtually no As_{Ga}. A similar situation exists for a very deep donor level believed to be Ga_{As} [51]. We can therefore neglect antisite defects.

Both titration and density/lattice parameter data show that all melt grown material is As-rich. Its lattice parameter is essentially constant with varying melt composition and so the size of the unit cell remains constant. The density however, increases strongly with As concentration in the melt and so we conclude that the dominant defects must be the Frenkel components on the As-sub-lattice i.e. As_i and V_{As}. If these are treated as the only defects, then unique values for their Gibbs free energies of formation at the melting temperature can be obtained by requiring that their values are in accord with both the absolute value of the density and its dependence on melt composition. Calculated data is shown in Figure 2. To obtain the enthalpies and entropies of formation similar data for material equilibrated at a lower temperature is required. Unfortunately this is not available for GaAs. However, equivalent information can be obtained by fitting to electrical data [6]. This enables one to plot the complete solidus (Figure 3). We see that the congruent melting point is on the As-rich side (~50.023 at% As) and that to grow a stoichiometric crystal requires a melt composition of ~44.8 at% As.

Using the same procedure for the other zinc-blende III-Vs for which the Bublik group have made measurements the following trends can be observed (Hurle, unpublished):

The density/lattice parameter data of the Bublik/Morizov group reveals that both GaP and InP are also dominated by the Group V sub-lattice defects (P_i and V_P). Because of their relatively large size, Ga_i and In_i would be unlikely to be present in large concentrations. Were they to be so then it would be expected that the lattice would be significantly dilated in Group III-rich material and this is not observed. For both materials the congruently grown crystal is nearly stoichiometric being slightly P-rich for GaP (and therefore similar to GaAs) but slightly In-rich for InP.

InAs similarly exhibits a near-constant lattice parameter indicating that significant numbers of In_i are unlikely to be present. Density increases with increasing As content in the melt as with GaAs, indicating again the dominance of the As Frenkel components As_i and V_{As}. However, in contrast to GaAs, the congruent point is very close to the stoichiometric composition.

The situation is markedly different in the antimonides of Ga and In. For GaSb again the lattice parameter is constant [18], suggesting that the relatively large Sb atom is not present interstitially in significant concentration. The density varies only slightly and rather erratically with melt composition. This suggests that the stoichiometric composition is well away from the congruent point. (Note: the interpretation by the authors of this erratic

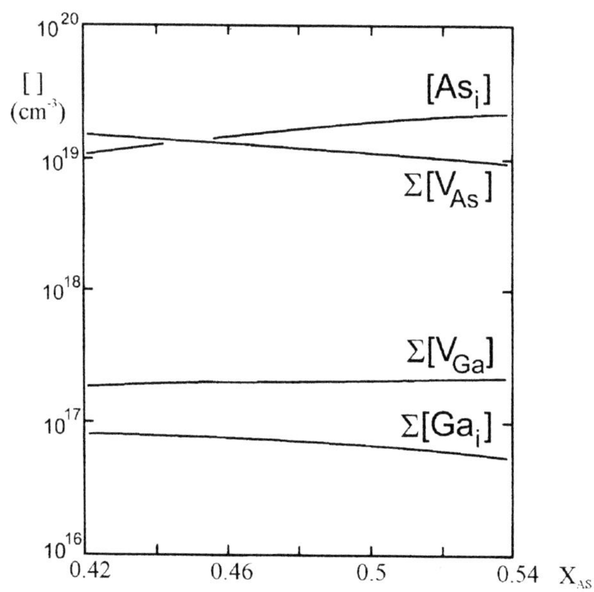

Figure 2. Calculated variation of the concentrations of the principal native point defects in GaAs at its melting point as a function of melt composition. (Hurle [6])

variation is not consistent with thermodynamics). The lack of a coherent trend of density with melt composition suggests that the dominant defects are the Schottky components V_{Ga} and V_{Sb}. The absolute value of the density is also consistent with a total vacancy concentration of a few times 10^{18} cm^{-3}. It is known that all melt grown and LPE GaSb is p-type containing a deep donor identified as the antisite defect Ga_{Sb}. This is present in melt grown material at about one order of magnitude lower concentration than that inferred for $[V_{Ga}]+[V_{As}]$ from the density data. These numbers imply that, at least over the range of melt compositions for which high quality crystals free of the effects of constitutional supercooling can be grown, all GaSb is Ga-rich.

The density data for InSb [52] also requires that the dominant defect(s) is/are vacancies. The weak trend with melt composition is best fitted with a somewhat greater number of Sb vacancies than In vacancies at the congruent point so that the crystal is probably group III-rich as is GaSb.

There are therefore clear trends as one goes from phosphides through arsenides to antimonides. The phosphides have high concentrations of Frenkel components on the group V sublattice. This is also true of the arsenides but with slightly lower concentrations. The antimonides on the other hand are dominated by vacancies on both sublattices. The trend

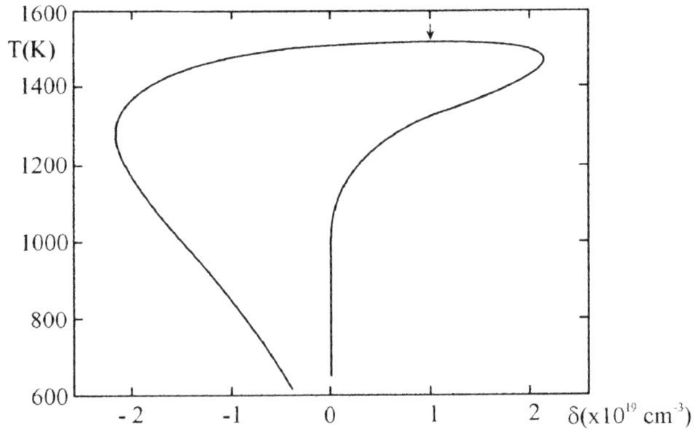

Figure 3. Calculated phase extent of GaAs. Arrow indicates the congruent melting point. Increasing arsenic-richness to the right.
(Hurle [6]).

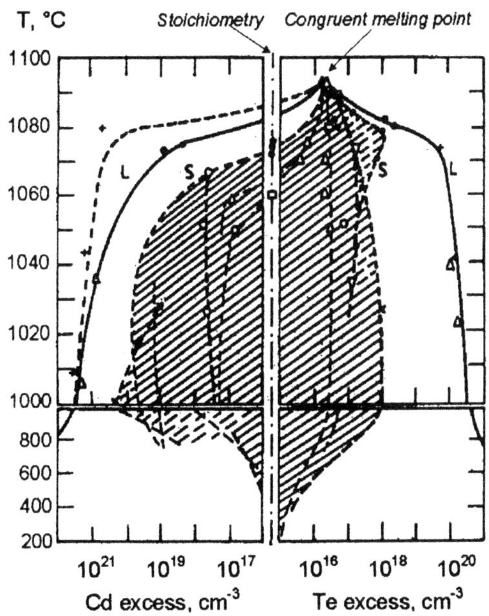

Figure 4. Phase extent of CdTe. (Rudolph et al. [54].

from phosphides to antimonides is for the composition of crystal grown from a congruent melt to move from group V-rich to group III rich.

As one moves from Ga compounds to In compounds having the same group V component the deviation of the composition of a crystal from stoichiometry appears to diminish.

Antisite defects appear to be more prevalent in the Ga compounds (As_{Ga} in GaAs and Ga_{Sb} in GaSb) than in the equivalent In compounds.

Much less is known about the II-VI compounds. CdTe is the most studied binary [53] following pioneering work by Kröger [34]. Hall effect data on quenched samples has been interpreted in terms of charged Cd vacancies and interstitials from which the solidus has been deduced. Rudolph et al [54] have compiled the data for CdTe and this is shown in Figure 4. However recent work by Greenberg [55], in which he deduced deviation from stoichiometry simply from total vapour pressure measurements in a conservative system, indicates a phase extent that is significantly greater than that obtained from Hall data. This suggests either that the high temperature state was not preserved on quenching of the Hall samples or that the dominant native point defects are not charged.

7. DOPING

7.1. The donor-cation vacancy complex

Charged point defects interact through their influence on the Fermi level. For example, the solubility of a donor dopant is enhanced by co-doping to a sufficiently high concentration with an acceptor. Less well appreciated is the fact that dopant solubility can be markedly affected by charged native point defects and, conversely, that doping affects the native point defect solubility if the native defect is charged. The effect of As vacancy concentration on Te solubility in GaAs was illustrated in section 2.9.

Dopant atoms can also form complexes with native point defects both by incorporation as the complex at the growth interface and by reaction in the crystal during cooling or annealing. The most important such defect in the III-V and II-VI compounds appears to be the donor-cation vacancy complex. In II-VI compounds this is known as an 'A'centre.

The phenomenon can be illustrated again with Te doped GaAs. Early on it was found that all donor doped GaAs was compensated with a compensation ratio at relatively light doping levels of about 0.25. This ratio was a) independent of the arsenic activity of the phase from which the crystal was growing (i.e. it was the same for melt, LPE and VPE growth) and b) independent of the specific chemical dopant, whether a group IV or a group VI donor. Hurle [56] showed that this was thermodynamically possible if, and only if, the compensating entity was a singly ionised complex comprised of the donor atom and a gallium vacancy (e.g. $Te_{As}V_{Ga}^-$ or $Si_{Ga}V_{Ga}^-$). Additionally it must be that electroneutrality in the crystal (and therefore the Fermi level position) are controlled by positively charged arsenic vacancies. This can be seen from the following:

The incorporation reaction for the complex and its mass action equation can be written:

$$Te_{As}^{+} + V_{Ga}^{0} = Te_{As}V_{Ga}^{-} + 2h^{+}$$

$$K_{TeV} = [Te_{As}V_{Ga}^{-}]n_i^4/(n_{gt}^2[V_{Ga}^0][Te_{As}^+]) \qquad (18)$$

Similarly, for the ionisation of an arsenic vacancy we have:

$$K_{VA}^{+} = [V_{As}^{+}]n_{gt}/[V_{As}^{0}] \qquad (19)$$

If electroneutrality is dominated by these charged arsenic vacancies then $n_{gt} \sim [V_{As}^{+}]$, so that, substituting this into equation 19 and then re-arranging equation 18 we obtain for the compensation ratio:

$$N_A/N_D = [Te_{As}V_{Ga}^{-}]/[Te_{As}^{+}] = K_{TeV}K_{VA+}[V_{As}^{0}][V_{Ga}^{0}]/n^{4}_{i} \quad (20)$$

where N_A and N_D are respectively the acceptor and donor concentrations. Examining the right hand side of this equation, we note that the mass action constants and n_i are independent of the arsenic activity as is $[V_{As}^{0}][V_{Ga}^{0}]$ the Schottky product. A similar result is obtained if equation 20 is written for a group IV donor species. *No other choice of electroneutrality equation satisfies these conditions.* This autocompensation appears to be present in GaP, GaAs, InAs and possibly GaSb but is not found in InP. Donor doped InP is uncompensated as has been unambiguously shown in radiochemical studies [57].

7.2. Acceptor-anion vacancy complexes

There is only a limited amount of evidence for these complexes in III-V compounds. Zn-doped InP provides the clearest indication. The hole concentration in melt grown Zn-doped crystals depends markedly on the cooling schedule of the crystal. Dlubek and Brummer [58] have found positron annihilation evidence for the formation of a $Zn_{In}V_P$ complex and annealing studies [59] indicate that the defect forms at around 400°C. In GaAs, there is evidence for a $Ge_{As}V_{As}$ acceptor in melt and LPE grown material [6, 60].

7.3. Cation vacancy under-saturation during cooling of n+ crystals

At high doping levels (above a carrier concentration of $\sim 2\times10^{18}$ cm^{-3}) rather dramatic things occur in donor doped GaAs. Annealing a crystal at around 900-1000°C produces a reduction in carrier concentration, the rapid formation of a large number of interstitial dislocation loops and the appearance of a line in the PL spectra which is assigned to the donor-gallium vacancy complex. Additionally, the rate of increase of lattice dilation with doping concentration rises from that predicted by Vegard's law to a value some 15 times greater. This behaviour is also characteristic of donors in GaP, InAs and GaSb but not in InP which, as we saw above, does not form donor-cation vacancy complexes.

All these effects are due to a novel phenomenon: the grown-in gallium vacancies become *undersaturated* as the crystal cools. In general, of course, one expects a super-saturation to develop on cooling a crystal and point defects to be removed from solution by one of a number of possible mechanisms. However once the doping level is sufficiently high for the crystal to be extrinsic at its growth temperature then the solubility of the multiply-charged acceptor V_{Ga} actually increases as the crystal cools so that the fixed, grown-in, gallium vacancy concentration becomes undersaturated. The cooling crystal responds to remove this undersaturation.

To understand this process, we again consider Te-doped GaAs. For simplicity of presentation, we assume Boltzmann statistics apply. The equilibrium concentration of doubly ionised V_{Ga} will be:

$$[V_{Ga}^{2-}] = [V_{Ga}^{0}](n/N_c)^2\exp\{(2Eg-E_1-E_2)/kT\} \quad (21)$$

where E1 and E_2 are the first and second ionisation energies of the gallium vacancy. If the crystal were intrinsic at high temperature, $[V_{Ga}^{2-}]$ would fall with falling temperature because

n/N_c would fall and $[V_{Ga}^{2-}]$ is proportional to $(n/N_c)^2$. However, if the crystal is sufficiently heavily doped for it to be extrinsic at high temperature then n/N_c will rise with falling temperature because N_c falls and the equilibrium concentration will rise. If the grown-in concentration in the crystal remains constant the crystal will become undersaturated.

This undersaturation can be removed in two ways: 1) by the generation of more V_{Ga} and 2) by reduction of n and hence of the *equilibrium* concentration of V_{Ga}. The former can occur by the movement of a Ga atom to an interstitial position (Frenkel reaction):

$$Ga_{Ga} = Ga_i^0 + V_{Ga}^{2-} + 2h^+ \qquad (22)$$

The latter occurs by the formation of more compensating donor-gallium vacancy complexes thereby reducing the carrier concentration n:

$$Te_{As}^+ + Ga_{Ga} = Te_{As}^+ + Ga_i^0 + 2h^+ \qquad (23)$$

This accords with the experimental observation that compensation in the crystal increases along with an increased intensity of the PL line associated with $Te_{As}V_{Ga}$. Both processes generate Ga_i which will be in supersaturation and so the reactions cannot proceed to the right unless this supersaturation is removed. The cooling crystal also has a supersaturated high concentration of As_i and, having relatively rapid interstitial diffusion rates the As_i and Ga_i can combine and aggregate to form the extrinsic (interstitial) dislocation loops mentioned above thereby removing the Ga_i supersaturation:

$$sGa_i^0 + sAs_i^0 = (Ga_iAs_i)_s \qquad (24)$$

The concept of V_{Ga} undersaturation was first introduced by Tan et al [61] and Hurle [6] has used it to provide the above explanation of the annealing behaviour of n+ GaAs. The explanation of the massive lattice dilation is more complex and is beyond the scope of this presentation. It involves further complexing of the $Te_{As}V_{Ga}^-$ [6]. This 'superdilation' is not confined to Te doping; it has also been seen in Se [62] and Sn [63] doped crystals.

Native point defect-related compensation seems to be less prevalent in acceptor doped GaAs. Zinc doped GaAs grown by LPE from both Ga-rich and As-rich solutions seems to be relatively uncompensated even at the very high doping level achievable ($>10^{21}$ cm^{-3}). There is however one report [65] of PL evidence for the complex $Zn_{Ga}V_{As}$ which is the compensating donor analogue of the $Te_{As}V_{Ga}$ acceptor complex.

Thus far only group VI donors and group II acceptors have been considered. Group IV dopants can, in some cases, be amphoteric. There is a strong tendency for the group IV atom to occupy the large vacant site of a missing antimony atom in the antimonides (and therefore to be an acceptor). In the phosphides the group III site is favoured (because of the large strain energy associated with fitting a larger atom into the small phosphorus site). Group IV dopants thus tend to be donors in the phosphides. The arsenides are intermediate with factors other than atomic size seemingly operative. In GaAs the lightest atom, carbon, occupies only the group V site whilst the heaviest, tin, is almost exclusively on the group III site. Si and Ge are intermediate. Ge is the most amphoteric being n type when growth is from the melt and p type when growth is from a Ga-rich LPE solution. Donor-V_{III} complexes are found with Si, Sn and Ge dopants.

8. ANNEALING

Post-growth annealing of a crystal is often carried out in order to optimise the crystal properties for a specific device application. For example slow cooling of n+ GaAs results in the formation of compensating acceptors as was explained in section 7.1. To maximise the carrier concentration (as might be required for making a diode laser for example) one anneals the crystal at a very high temperature not far below its melting point and then quenches it to room temperature. Similarly in CdTe and $Cd_xHg_{1-x}Te$ high temperature growth results in material containing Te micro-inclusions. These are removed by high temperature anneal under Cd-rich conditions.

Conversely one may wish to induce some defect complexing reaction which occurs only below the growth temperature. An example of this is the formation of EL2 in GaAs which can render the crystal semi-insulating (a desired state when fabricating microwave integrated circuits). The EL2 defect is now known to be the As_{Ga} antisite defect and its concentration is maximised by annealing the grown crystal in an As-rich environment at a temperature of around 950°C for several hours [66]. This induces the migration of arsenic interstitials which are in a supersaturated state, to residual grown-in V_{Ga} in the crystal resulting in the reaction: $As_i + V_{Ga} = As_{Ga}$. The number of As_{Ga} defects formed is determined by the number of supersaturated V_{Ga} present in the crystal.

9. SELF DIFFUSION IN GaAs

9.1. Radio-tracer self diffusion measurements

Direct radio-tracer measurements of Ga and As self diffusion have been made by Palfrey [67] following much earlier studies by Goldstein [68]. Palfrey found a similar activation energy (3.0 eV) for each. Much higher values were obtained by Goldstein. Two As diffusion data points obtained by Palfrey for a higher ambient As pressure gave a lower value for the diffusion coefficient and this could be interpreted as indicating that the process was not mediated by As_i. However these measurements are technically very difficult and can depend on the initial stoichiometric state of the crystal. Further confirmation is therefore desirable.

9.2. Gallium sub-lattice diffusion

Being isoelectronic Al diffuses in a near identical manner to Ga in the GaAs lattice producing only a very small elastic strain. It can therefore act as a non-radioactive tracer. This approach has been used both in simple in-diffusion and in studies of the rate of disordering of $GaAs/Ga_xAl_{1-x}As$ superlattices (see the review by Tan [69]).

It has been shown that, under As-rich equilibrium conditions, diffusion is mediated by negatively charged V_{Ga}^{q-}. Under Ga rich conditions Ga_i^{i+} are postulated to be the controlling entities. (The values of q and i remain a partially unresolved issue).

Information on the charge states is obtained by varying the Fermi energy position by heavily doping the sample either n or p type. Superlattice disordering is enhanced in the presence of an n type dopant in the ratio:

$$D_{VGa}(n)/D_{VGa}(n_i) = (n/_{ni})^m \qquad (25)$$

Bracht et al [70], in radio-tracer studies of superlattice disordering of Si and Be doped samples, have elucidated the Ga self diffusion mechanism under their particular conditions. By varying the doping level the V_{Ga} concentration in each charge state could be varied. Fitting the results to a model indicated that V_{Ga}^{2-} was dominant in undoped and n doped material.

9.3. Arsenic sub-lattice diffusion

The iso-electronic dopants P and Sb have been used as tracers of As-self diffusion [70]. Diffusion coefficient values obtained were similar for the two dopants. By varying the ambient As pressure the authors demonstrated that self diffusion was mediated by As_i. (This is contrary to the tentative result of Palfrey mentioned above). Similar results were obtained by Stobvijk et al. using nitrogen as a tracer [72].

By use of heavily n and p doped samples Scholz and Gösele [73] showed that there was no Fermi level dependence of the diffusion rate and hence the As_i must be in a neutral state. This is in accord with the deduction made by Hurle from an analysis of density/lattice parameter and room temperature carrier concentration data [6].

10. DOPANT DIFFUSION IN GaAs

10.1. As-sub-lattice diffusion

It has been shown that a number of dopants diffuse via the so-called 'kick-out' mechanism whereby a mobile interstitial dopant atom kicks out a lattice site atom into an interstitial position and itself occupies the lattice site. Thus, for the case of sulphur diffusion (Uematsui [73]), we have: $S_i = S_{As} + As_i$. The diffusion profile depends on which type of interstitial is the more mobile or, more strictly, which has the higher transport capacity. Denoting diffusion coefficients by "D" and species concentration by "C" with subscript Si indicating a sulphur interstitial and Asi an arsenic interstitial then if $D_{Si}\, C^{eq}_{Si} < D_{Asi}\, C^{eq}_{Asi}$ then the As_i concentration will be in local equilibrium during the diffusion. The superscript "eq" denotes an equilibrium value. In this event the dopant diffusion coefficient bears the following relation to the diffusion coefficient of its interstitial component

$$D^{eq}_S = Di_S\, C^{eq}_{Si}/C^{eq}_S \tag{26}$$

and the dopant distribution will have the usual error function form.

However when $D_{Si}\, C^{eq}_S > D_{Asi}\, C^{eq}_{Asi}$ there will exist locally a *supersaturation* of As_i during dopant in-diffusion. Under these (non-equilibrium) conditions

$$D^{noneq}_S = D_{Asi}\,(C^{eq}_{Asi}/C^{eq}_S)(C^{eq}_S/C_S)^2 \tag{27}$$

This latter condition results in a non-error function in-diffusion profile and so the dopant diffusion coefficient is a function of the dopant concentration and this provides information about the As_i supersaturation.

At high sulphur surface concentration extrinsic dislocation loops are formed in the near-surface region. These probably occur as a result of the V_{Ga} *undersaturation* which occurs as a result of the rise in the equilibrium V_{Ga}^{m-} consequent upon the crystal becoming extrinsic at the diffusion temperature as described in section 7.3.

10.2. Ga sub-lattice diffusion

Studies of silicon diffusion give similar results to those for sulphur diffusion [75] in that auto-compensation due to the formation of donor-V_{Ga} complexes occurs along with extrinsic dislocation loops following a short post-diffusion anneal. This was shown by Herzog et al. [74] to result from V_{Ga} undersaturation.

11. CONCLUSION

Knowledge of native point defects is the link that can provide understanding of electrical, annealing and diffusion behaviour in compound semiconductors. That link is only now being established and that for only a very few of the most important materials. Much remains to be done.

REFERENCES

1. M. Skowronski, in S. Mahajan (ed.), Handbook of Semiconductors, revised edition, Vol. 3, ch. 18, Elsevier, Amsterdam, 1994, p. 1343.
2. V. V. Voronkov, J. Crystal Growth 59 (1982) 625 and 194 (1998) 76.
3. J.A. van Vechten, in S.P. Keller (ed.), Handbook of Semiconductors, first edition, Vol. 3, ch. 1, North Holland Publishing Co., Amsterdam, 1980, p. 1.
4. S. Oda, M. Yamamoto, M. Seiwa, G. Kano, T. Inoue, M. Mori, H. Shimakura and M. Oyake, Semocond. Sci. Technol. 7 (1992) A215.
5. A.N. Morizoov and V.T. Bublik, J. Crystal Growth 75 (1986) 491, 497.
6. D. T. J. Hurle, J. Appl. Phys. 85 (1999) 6957.
7. M.J. Puska, J. Phys. Condens. Matter 1 (1989) 7347.
8. Y. Marfaing, in P. Capper (ed.), Narrow-gap II-VI Compounds for Opto-electronic and Electromagnetic Applications, Chapman and Hall, 1997, ch. 8, p.238.
9. D.E. Holmes, R.T. Chen, K.R. Elliott and C.G. Kirkpatrick, Appl. Phys. Lett. 40 (1982) 46.
10. C. Anayama, T. Tanahashi, H. Kuwatsuka, S. Nishiyama, S. Isozumi and K. Nakajima, Appl. Phys. Lett. 56 (1990) 239.
11. A.S. Jordan, A.R. von Neida, R. Caruso and C.K. Kim, J. Electrochem. Soc. 121 (1974) 153.
12. K. Terashima, S. Washizuka, J. Nishio, A. Okada, S. Yasuami and M. Watanabe, Inst. Phys. Conf. Series 79 (1986) 37.
13. N. Chen, H. He, Y. Yang and L. Lin, J. Crystal Growth 173 (1997) 325.
14. V.T. Bublik, V.V. Kartrataev, R.S. Kulagin, M.G. Mil'vidskii, V.B. Osvenskii, O.G. Stolyarov and L.P. Kholodnyi, Sov. Phys. Crystallogr. 18 (1973) 323.
15. V.T. Bublik, A.N. Blaut-Blachev, V.V. Karataev, M.G. Mil'vidskii, L.N. Perova, O.G. Stolyarov and T. G. Yugova, Sov. Phys. Crystallogr. 22 (1977) 705.
16. A.N. Morizov, V.T. Bublik, I.A. Koval'chuk and O.G. Stolyarov. Sov. Phys. Crystallogr. 31 (1986) 586.
17. A.N. Morizov, V.T. Bublik, V.B. Osvenskii, A.V. Berkova, E.V. Mikryukova, A.Ya. Nashel'skii, S.V. Yakobson and A.D. Popov, Sov. Phys. Crystallogr. 28 (1983) 458.

18. J. Wilke, V.T. Bublik, A.G. Braginskaya, A.N. Popkov, G.P. Kolchina, V.B. Osvenskii, S.P. Grushina and O.G. Stolyarov, Phys. Stat. Sol. (A) 64 (1981) K31.
19. V.T. Bublik, M.N. Kevorkov, V.B. Osvenskii, A.N. Popkov and O.G. Stolyarov, Izvest. Akad. Nauk SSSR, Neorganicheskie Materialy 20 (1994) 364.
20. R. Krause-Rehberg and H.S. Leipner, Positron Annihilation in Semiconductors, Springer, Berlin, 1999.
21. I. Fujimoto, Jap. J. Appl. Phys. 23 (1984) L287.
22. C. Domke, Ph. Ebert, M. Heinrich and K. Urban. Phys. Rev. B 54 (1996) 10288.
23. A.N. Georgobiani and I.M. Tiginyanu, Sov. Phys. Semicond. 22 (1988) 1.
24. R.C. Newman, Semicond. Sci. Technol. 9 (1994) 1749.
25. A.R. Peaker and M R. Brozel, III-Vs Review 12 (1999) 44.
26. M. Wieneke, H. Berger and M. Schenk. Mater. Sci. Eng. B 16 (1993) 219.
27. J.A. van Vechten, Phys. Rev. B 7 (1973) 1479.
28. T.Y. Tan, Mater. Sci. Eng. B 10 (1991) 227.
29. T.A. Kennedy and P.Lin-Chung in Semiconductors and Semi-metals ch.7, p. 357, 1990.
30. A. M. Stoneham, MRS Bulletin December 1991, p22.
31. J.T. Schick, C.G. Morgan and P. Papoulias, Phys. Rev. B 66 (2003) 195302.
32. J.C. Bourgoin and H.J. von Bardeleben, J. Appl. Phys. 64 (1988) R65.
33. J Willard Gibbs, The Scientific Papers Vol 1. Thermodynamics, Dover, NY, 1961.
34. F.A. Kröger, The Chemistry of Imperfect Crystals. North Holland, Amsterdam 1973.
35. P. Fochuk, O. Korovyanko, I. Turkevych, O. Panchuk and P. Siffert, J. Crystal Growth 207 (1999) 273.
36. R.F. Brebrick, in D.T.J. Hurle (ed.), Handbook of Crystal Growth, Vol 1a Elsevier, Amsterdam 1993 p.43.
37. K.H. Nichols, C.M.L. Yee and C.M. Wolfe, Solid State Electron. 23 (1980) 109.
38. J.W. Corbett and J.C. Bourgoin in J. H. Crawford and L. M. Slifkin. (eds.), Point Defects in Solids Plenum, NY Vol 2, p.1.
39. R.F. Scholz and U. Gösele, J. Appl. Phys. 87 (2000) 704.
40. Li Yujie, Ma Guoli and Jie Wanqi, J. Crystal Growth 256 (2003) 266.
41. P. Fochuk, O. Korovyanko and O. Panchuk, J. Crystal Growth 197 (1999) 606.
42. G. Roos, A. Schoer, G. Pensi, J. Wagner, B. K. Meyer and R. C. Newman, J. Appl. Phys. 69 (1991) 1454.
43. N.A. Yakushava, V.Ya. Prinz and Yu.B. Bolkhovityanov, Phys. Stat Solidi A 95 (1986) K43.
44. U. Kaufmann, J. Schneider and A. Rauber, Appl. Phys. Lett. 29 (1976) 312.
45. M. Hanke, D. Hennig and A. Kaschte, J. Vac. Sci. Technol. B8 (1990) 1148.
46. D.J. Chadi and K.J. Chang, Phys. Rev. Lett. 60 (1988) 2187.
47. W.L. Sunder, R.L. Barns, T.Y. Kometani, J.M. Parsey and R.A. Laudise, J. Crystal Growth 78 (1986) 9.
48. R.A. Swalin, Thermodynamics of Solids. J. Wiley NY 1961.
49. H. Klapper in B.K. Tanner and D. K. Bowen.(eds.), Characterisation of Crystal Growth Defects ch.6, NATO Adv. Study Series B, Plenum New York, 1979, p.156.
50. Y. Otoki, M. Sahara, M. Shibata, S. Kuma and M. Kashiwa, Mater. Sci. Forum 196-201 (1986) 1431.
51. K.R. Elliott, J. Appl. Phys. 55 (1984) 3856.
52. A.N. Morizov, T.V. Abaeva and V.T. Bublik, Crystal Res. Technol. 21 (1986) 613.

53. P. Capper (ed), Narrow-gap II-VI Compounds for Optoelectronic and Electromagnetic Applications, Chapman & Hall London, 1997.
54. P. Rudolph, U. Rinas and K. Jacobs, J. Crystal Growth 138 (1994) 249.
55. J.H. Greenberg, J. Crystal Growth 161 (1996) 1.
56. D.T.J. Hurle, Inst. Phys. Conf. Series 33A (1977) 113.
57. M.R. Brozel, M.J. Foulkes, I.R. Grant, L. Lui, D.T.J. Hurle and R.M. Ware, J. Crystal Growth 70 (1984) 191.
58. G. Dlubek and O. Brummer, Appl. Phys. Lett. 46 (1985) 1136.
59. R. Hirana, T. Kanazawa and T. Inoue, J. Appl. Phys. 71 (1992) 659.
60. E.W. Williams and C.T. Elliott, Brit. J. Appl. Phys. J. Phys. D2 (1969) 1657.
61. T.Y. Tan, H-M Yu and U. Gösele, Appl. Phys. A: Solids Surf. A56 (1993) 249.
62. G.M. Kuznetsov, A.D. Barsukov, O.V. Pelevin, V.V. Olenin and N.B. Berdavtseva, Izv. Akad. Nauk SSSR, Neorg. Mater. 9 (1973) 1053.
63. N.A.Anastes'eva, V.T. Bublik, V.B. Ovsenski, M.G. Mil'vidskii, O.G. Stolyarov and L.P. Kholodnyi, Sov. Phys. Crystallogr. 23 (1978) 174.
64. K. Köhler, P. Ganser and M. Maier, J. Crystal Growth 127 (1993) 720.
65. C.J. Hwang, Phys. Rev. 180 (1969) 827.
66. D.J. Stirland, P. Kidd, G.R. Booker, S. Clark, D.T.J. Hurle, M.R. Brozel and I. Grant, Inst. Phys. Conf. Ser. 100 (1989) 373.
67. H.D. Palfrey, M. Brown and A.F.W. Willoughby, J. Electron. Mater. 12 (1983) 863.
68. B. Goldstein, Phys. Rev. 121 (1961) 1305.
69. T.Y. Tan, Mater. Chem. Phys. 40 (1995) 245.
70. H. Bracht, M. Norseng, E.E. Haller, K. Eberl and M. Cardona, Sol. State Comm. 112 (1999) 301.
71. R.F. Scholz and U. Gösele, J. Appl. Phys. 87 (2000)704.
72. N.A. Stobvijk, G. Bösker, J.V. Thordson, U. Södevall, T.G. Andersson, Ch. Jäger and W. Jäger, Physica B273-274 (1999) 685.
73. M. Uematsu, P. Werner, M. Schultz, T. Tan and U. Gösele, Appl. Phys. Lett. 48 (1995) 2863.
74. L. Herzog, U. Egger, O. Breitenstein and H-G. Hettwer, Mater. Sci. Eng. B30 (1995) 43.

Crystal Growth – from Fundamentals to Technology
G. Müller, J.-J. Métois and P. Rudolph (Editors)

Synchrotron radiation x-ray imaging: a tool for crystal growth

José Baruchel

European Synchrotron Radiation Facility, Grenoble, France

A major use of X-rays since their discovery is the visualisation of the inner part of systems that are not transparent for visible light. The X-ray images associated "experimental result" is the contrast that results from the inhomogeneous response of the sample to the beam. The main topics of this lecture on hard X-ray (6-100 keV) imaging are the physical origins of the contrast, and its interpretation, with special emphasis on synchrotron radiation based techniques and applications to crystal growth. We consider more particularly three-dimensional imaging, with spatial resolution in the μm range (microtomography), the qualitative and quantitative use (holotomography) of phase contrast imaging, microbeam based imaging, and Bragg-diffraction imaging revealing inhomogeneities (defects, domains, phases, ...) in single crystals.

1. INTRODUCTION

X-ray imaging started over a century ago, with the discovery of X-rays, allowing the visualization of the volume of systems opaque for other probes. For several decades the only form of X-ray imaging was absorption radiography. About forty years ago Bragg diffraction imaging (X-ray topography) developed into practical use. The "topographic" techniques apply to single crystals and basically consist into mapping the amplitude and direction of the locally Bragg diffracted beams. They reveal crystal defects in the bulk of single crystals, and helped producing large, practically perfect, crystals for the microelectronics industry. These techniques extended beyond semiconductor crystals, and are in particular useful for investigating, induced or growth defects and distortions. Their range of applications grew when synchrotron radiation (SR) was available for condensed matter scientists, including then topics like in-situ crystal growth, defect nucleation and movement, phase transitions.

All the X-ray imaging methods benefited from the use of SR, the improvements being even more pronounced when considering the new features of modern, "third generation", SR machines. Real time and high spatial resolution experiments are routinely performed, and the high coherence of the beam allows exploiting a novel form of radiography, in which contrast arises from phase variations across the transmitted beam, through Fresnel diffraction. Phase microradiography and its three-dimensional companion, phase microtomography, provide useful information on crystal growth-related features (porosities, inclusions, ...). The spatial resolution of these two or three-dimensional imaging techniques is now in about 1 μm.

The highly brilliant sources of the modern SR facilities, coupled with new X-ray optics, allowed achieving micro-beams in the μm or sub-μm range. This leads to another type of images through the use of X-ray microbeam-based scanning imaging approaches, probing for instance the fluorescence or the absorption near the absorption edges.

Whatever the imaging process used, it implies an inhomogeneous response of the sample to the beam. The "experimental data" obtained when using X-ray imaging is the *contrast* that results from this inhomogeneous response. The main topics of this lecture on hard X-ray (6-100 keV) transmission, scattering and diffraction imaging will be the technical aspects of these imaging methods, the physical origin of the contrast and its interpretation.

The initial part of this paper deals with the developments of SR based X-ray imaging techniques that can be of interest for crystal growth. These include firstly transmission (absorption and phase) imaging, in its two or three-dimensional ("microtomographic") version, and secondly a brief report on microbeam-based techniques (fluorescence, diffraction). The third, more extended, part, describes diffraction topography, i.e. a set of techniques that are specific for the investigation of single crystal inhomogeneities. Lastly, we give examples showing the possibilities of these techniques with emphasis on aspects of interest for crystal growth.

2. ABSORPTION AND PHASE IMAGING

2.1. Absorption radiography

The interaction of X-rays with matter can be expressed through the complex refractive index

$$n = 1-\delta + i\beta \qquad (1)$$

that describes both absorption (β) and the phase change (δ) with respect to a path in vacuum. The absorption of X-rays is usually described by the absorption coefficient μ, such that the number of photons N after transmission through the sample writes:

$$N(y,z) = N_0(y,z)\ \exp\left[- \int_{path} \mu(x,y,z)\, dx \right] \qquad (2)$$

where N_0 is the incident beam number of photons. The absorption coefficient μ is related to β through the simple expression $\beta = \lambda\mu/4\pi$, where λ is the X-ray wavelength.

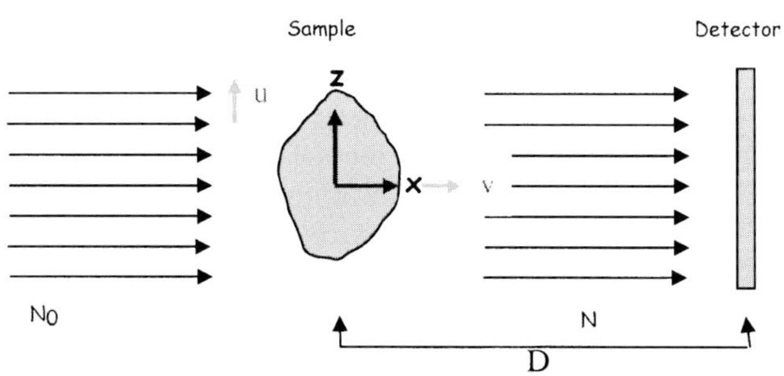

Figure 1. Transmission imaging (radiography): the contrast mechanism can be absorption or, when dealing with a coherent beam, phase. A 3D image (tomography) is obtained by rotating the sample in the beam, recording a series of radiographs, and using reconstruction software.

X-ray absorption imaging consists in detecting the photons transmitted through the investigated object (figure 1). When a two dimensional detector (film, CCD camera) is used, this directly produces a map of the integral of $\mu(x,y,z)$, the linear attenuation coefficient, along the X-ray path, as indicated on expression (2).

The behaviour of the linear attenuation coefficient μ is dominated by the photoelectric effect for X-ray energies E below 200 keV. μ is in this case proportional to $\rho Z^4 E^{-3}$ where ρ is the material density and Z is the atomic number. The separation between Z and ρ implies to perform experiments using at least two energies. Above 200 keV, the Compton effect is predominant and the mentioned proportionality does not hold anymore.

The absorption-related coefficient β changes abruptly with the energy of the photons at the absorption edges, corresponding to the energy required to eject an electron from a deep atomic level. The absorption edges are element-specific, and the logarithmic subtraction of images recorded above and below the edge enhances the visibility of the distribution of the considered element. It is also possible, by tuning the energy of the probing photons in the high absorption side of the edge, to take advantage of the oscillations of the absorption curve (XANES). These oscillations reflect the molecular environment of a given atom and provide the basic mechanism for imaging with chemical sensitivity.

2.2. Microtomography

Computer-assisted tomography provides the three-dimensional (3D) absorption image of a bulky object from the two-dimensional (2D) images (radiographs) recorded at various angular positions of the object, processed using appropriate algorithms and software. From the reconstructed 3D data any virtual slice, or perspective renditions, of the object can be obtained.

Microtomography, with a spatial resolution better than 20 μm, was developed over the last years [1]. Several laboratory instruments using micro-sources providing a cone, polychromatic, beam are now commercially available, and reach spatial resolutions of ≈10-20 μm. However, the best quality images, in terms of signal-to-noise ratio and spatial resolution, are obtained on instruments located in SR facilities. We will here restrict to this SR based technique.

Figure 1 shows the principle of the most commonly used SR-based tomographic acquisition. The beam is monochromatized by using Bragg reflections either from crystals ($\Delta E/E \approx 10^{-4}$) or from multilayers ($\Delta E/E \approx 10^{-2}$). The monochromaticity of the beam allows a quantitative evaluation of the images. When using conventional polychromatic X-ray sources, "beam hardening" artefacts are producing in the reconstructed images, due to the more important attenuation of soft X-rays in the sample. The beam can be considered as parallel, the source being situated far from the sample (150 meters at ID19, one of the imaging beamlines of the ESRF). This "parallel beam" approach implies that no magnification is obtained, and the spatial resolution mainly results from the effective pixel size of the detector. But this setup exhibits the advantage that all the radiographs, recorded on the 2D detector, obtained by rotation of the sample along **z,** are such that each slice, at a given height z, is independent of the other slices. A radiograph (or « projection ») corresponds to a given angle θ, formed (figure1) by the axes **v** (direction of the X-ray beam) and **x** (linked to the sample).

Equation (2) can be rewritten, in this case, as a series of integrals (the projections $p_\theta(u,z)$)

$$p_\theta(u,z) = \ln [N_0(u,z) / N_\theta(u,z)] = \int_{path} \mu[x(u,v,\theta), y(u,v,\theta), z]\, dv \qquad (3)$$

In parallel geometry, $p_\theta(u,z) = p_{\theta+\pi}(-u,z)$, meaning that the acquisition of projections over a range $[0,\pi]$ is sufficient. For a given z the ensemble $\{p_\theta(u,z), \theta \in [0,\pi]\}$ constitutes the Radon transform (or « sinogram ») of the 2D slice. The inverse transform provides the $\mu(x,y)$ map of the slice. The 3D problem reduces, in parallel geometry, to a sequence of 2D calculations[1].

Several reconstruction algorithms, like the filtered back-projection (FBP) or the algebraic reconstruction techniques (ART), have been developed. Their mathematical description is beyond the scope of the present paper, but can be found in [2,3]. The commonly used reconstruction algorithm is the FBP one, because it provides the better ratio "image quality / calculation time". It was shown that the inverse problem can be solved by firstly correcting the projections by a filter $H(f) = f$, f being the spatial frequency[2], and then by performing a back-projection procedure.

The experimental principle of SR microtomography is simple, but a series of requirements are compulsory to obtain a high quality 3D image:

- The best energy is such that $N/N_0 \cong 0.1$ [4]
- The sample has to be totally immersed in the beam at any time of the rotation; this reduces the maximum sample dimension (perpendicular to the rotation axis) to the field of view of the detector[3].
- An accurate alignment is required before the acquisition. The rotation axis has to be parallel to a column of the CCD detector, with a precision of a few 10^{-4} radian. It has, in addition, to be well centered in the image or, at least, its position should be known with an accuracy of a pixel size, to be able to correct the data.
- The number of 2D images necessary is of the order of the number of pixel columns used by the image on the detector (typically 100 – 1000). In addition reference images (images without sample, and without beam) are recorded for flat-field correction. The 2D images recording-time is a function of the source and the experimental setup (going from 1-2 minutes to 1 hour when working at the ESRF).
- Data acquisition for one sample typically represents 2 Gigabytes. Reconstruction programs directly handling the radiographic images acquired during experiment were developed. The reconstruction time is a function of the computer used, being, for a $(1024)^3$ volume, 1-2 hour when using a set of modern PCs working in parallel.
- Care is required to reduce the artefacts on the reconstructed images. They can correspond to a centering error, or to a pixel having a very different efficiency than the others. In both cases spurious rings are observed. On the other hand, as the detector samples the signal with a non-zero step, high spatial frequencies present in the object (steep edges) can generate streaks.

---

[1] The slices are independent in parallel geometry, but not when using a cone beam

[2] In practice the projection is sampled with a pixel size s_p, and the maximum frequency is $1/(2s_p)$. The filter $H(f)$ is consequently set to zero outside this band.

[3] It is possible to visualize only an inner part of the sample (« local » tomography), at the cost of high frequency artefacts being generated on the 3D image by the regions of the sample that are on the X-ray path only during part of the rotation. The obtained "absorption" cannot be used quantitatively, but the structural information is retained.

2.3. Phase imaging

The X-ray beams produced at modern synchrotron radiation facilities exhibit a high degree of coherence. This is due to the very small angular size α of the source as seen from a point in the specimen (α is in the microradian range at the ESRF), leading to a large lateral coherence L_C of the X-ray beam with wavelength λ ($L_C = \lambda / 2\alpha$). L_C is about 100 µm (λ = 1 Å) at the 150 m long ESRF ID19 beamline.

The refractive index of materials (equation (1)) deviates slightly from unity through a decrement δ, which is in the 10^{-5}-10^{-6} range. δ is proportional to the electronic density in the material, hence to its mass density; it involves a dispersion correction when the photon energy is near an absorption edge. Far from absorption edges it can be written as

$$\delta = (\lambda^2 r_0 / 2\pi V) \sum N_j Z_j \qquad (4)$$

where r_0 (= 2.82 fm) is the classical electron radius, and N_j is the number of atoms of atomic number Z_j present in the volume V.

The passage of an X-ray beam through a sample entails a phase shift

$$\phi(y, z) = (2\pi/\lambda) \int_{path} \delta(x, y, z)\, dx \qquad (5)$$

This phase is variable over the cross-section of the beam if the beam has passed through a sample with inhomogeneous thickness and/or structure.

The phase is not directly measurable, and variations in phase do not affect the intensity of the beam as it leaves the specimen. Phase images were obtained in the past using rather complex setups (systems involving several crystals). The qualitatively new aspect of modern SR beams is that they make it possible to reveal, through the "propagation" technique, with great instrumental simplicity [5-7], the phase variations of hard X-rays, be it in simple imagery or in tomography [8]. From the optical point of view, the effect used to change the local phase variations into intensity variations is the interference, at finite distances, of parts of the beam that suffered different phase shifts, but are mutually coherent. It can be described in terms of Fresnel diffraction, or of in-line holography.

The great advantage of this new type of imagery is the increased sensitivity it provides. Figure 2 shows that $\delta >> \beta$ in the hard X-ray range (E > 6 keV). This increased sensitivity allows visualizing tiny objects like cracks [8], micro-pipes [9], µm sized pores [10], or regions exhibiting even a weak (1%) density variation with respect to the matrix (see, for instance, [10,11]).

In practice, obtaining a phase-sensitive image just involves setting the detector at a distance D from the specimen (Figure 1) of the order of 10^{-1} m. An absorption image is obtained if D is small (10^{-3} m range). This corresponds to the fact that the region in the sample that effectively determines the image at a point of the detector, the first Fresnel zone, has a size $r_F = (\lambda D)^{1/2}$. When D is a few mm, the separation of the fringes originating from an area r_F is below the resolution of the detector (µm range), and no interference will be observed, hence the image will be purely due to absorption. For D larger, but such that r_F remains smaller than the size τ of the object to be imaged, the edges of the image behave independently, and provide the only phase contribution to the image. For D even larger, interference fringes built up. The best sensitivity for an object with size τ is obtained for $D \approx \tau^2/2\lambda$.

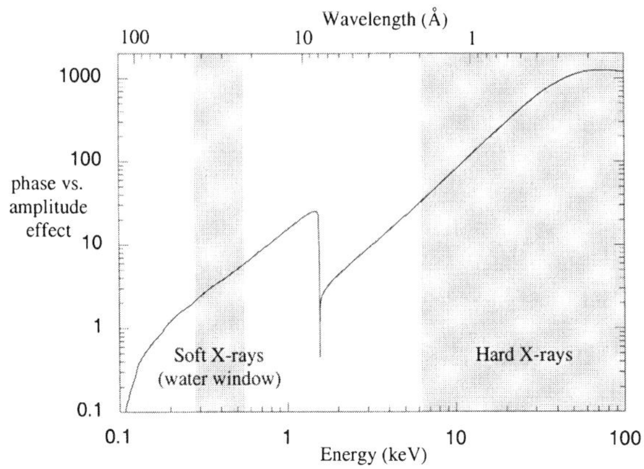

Figure 2. Ratio δ / β as a function of the X-ray energy for Al: $\delta / \beta > 10$ for hard X-rays

Phase variations across the beam exiting the sample lead to variations in intensity, hence to contrast, provided the phase has a two-dimension Laplacian $\nabla^2\phi(y,z) \neq 0$. They show up, for increasing distance D, first through the appearance of a black-white line at the phase discontinuities ("edge enhancement"), then a more and more obtrusive Fresnel fringe system ("holography") which progressively changes towards the Fraunhofer regime. Figures 3 and 4 illustrate the previous statements. Dodecahedral pores present in the bulk of an Al-Pd-Mn quasicrystal grain, which are hardly detectable when D is small, i.e. when absorption is predominant (fig. 3a), are clearly visible for a larger D, and new, smaller pores, can be observed in this case (figs. 3b and 3c).

Figure 4 shows images of an annealed Al-Pd-Mn quasicrystal grain: in addition to the pores, which evolved but remain visible at both distances, inclusions of an actually crystalline phase ("approximant" of the quasicrystalline structure), with a slightly different density, are observed for D = 50 cm.

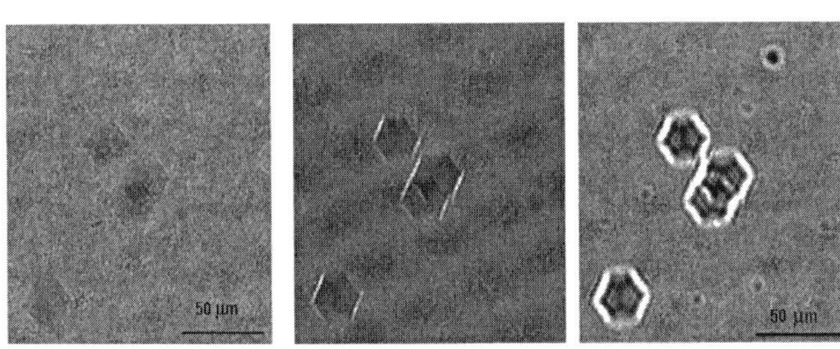

Figure 3. Phase images (λ = 35 pm) of faceted dodecahedral holes in a Al-Pd-Mn quasicrystal, recorded at various distances: a) 13 mm: absorption is predominant b) 100 mm, the edges of the holes are outlined c) 500 mm, a set of interference fringes built up around the holes.

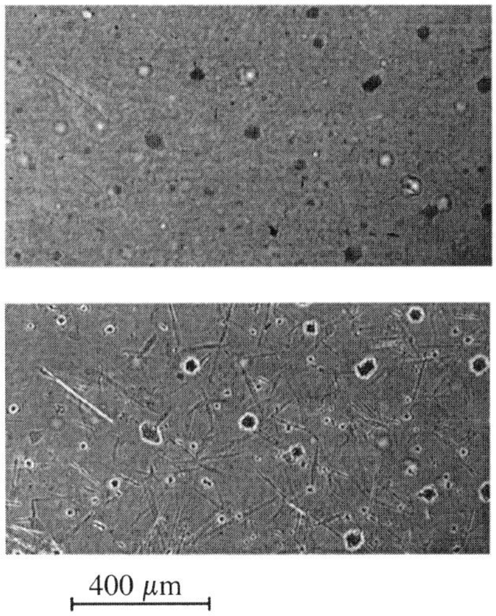

Figure 4. Phase radiographs, obtained at two distances D = 2 cm (above) and 50 cm (below), of porosities and crystalline "approximant" inclusions in a Al-Pd-Mn quasicrystal (courtesy L. Mancini and J. Gastaldi)

In the case of *phase contrast microtomography*, it has been shown that, for phase images, the reconstruction procedure based on the FBP algorithm provides an acceptable approximation for those distances D where the effect of phase discontinuities (edge enhancement) is predominant [8]. Figure 5 shows the result of the application of such a microtomographic procedure to the porosities in a quasicrystalline sample of Al-Pd-Mn: the distribution of the porosities can be clearly visualized.

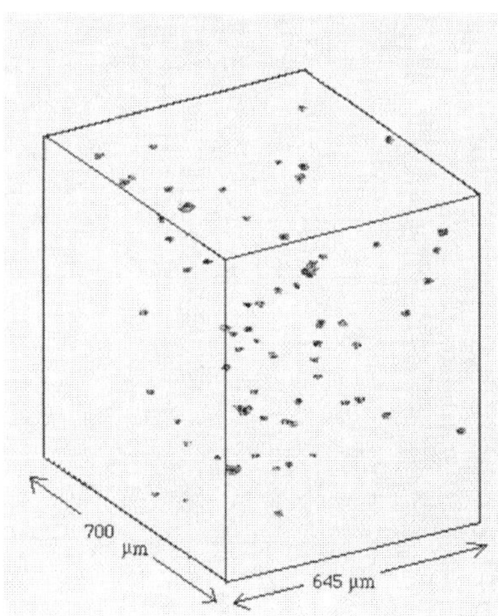

Figure 5. 3D rendering of the phase contrast tomographic image of porosities in a Al-Pd-Mn quasicrystal (courtesy W. Ludwig).

Phase imaging based on the visualization of the edges is not a quantitative technique, and its spatial resolution is limited by the occurrence of the fringes used to visualize the borders. Quantitative phase determination implied developing a procedure for the « holographic » reconstruction of the phase maps. It is based on the combination of several images recorded at different distances. An algorithm, initially developed for electron microscopy by the Antwerp group, was adapted to the X-ray case, and allows the "holographic" reconstruction of the local phase [12]. Once these phase maps are obtained, there is no conceptual difficulty in bringing together many maps corresponding to different orientations of the sample, and in producing the tomographic, 3D reconstruction.

This combined procedure was applied to an Al-Si alloy, quenched from a temperature where a "mainly Al" solid phase is surrounded by a Al-Si liquid one. The density difference between the two phases is in the 1-2% range. For each of 700 angular positions of the sample 4 images were recorded at different distances, yielding 3D quantitative images that show, with a voxel edge size of 1 μm, the distribution of electron density, hence of mass density, in the sample. The spatial resolution is the one of the detector (~1-2 μm). Figure 6 compares the possibilities of absorption microtomography (a), which only shows iron-rich metallic inclusions (b) edge-enhancement, where the boundaries of the phases are outlined by a black-white line (c) holotomography, resulting from the phase retrieval processing: the two phases are clearly distinguishable through their grey level (d) 3D rendering of what was the "liquid" phase. Quantitative phase mapping and tomography ("holotomography") are now operational, and provide a new approach to the characterization of materials on the micrometer scale

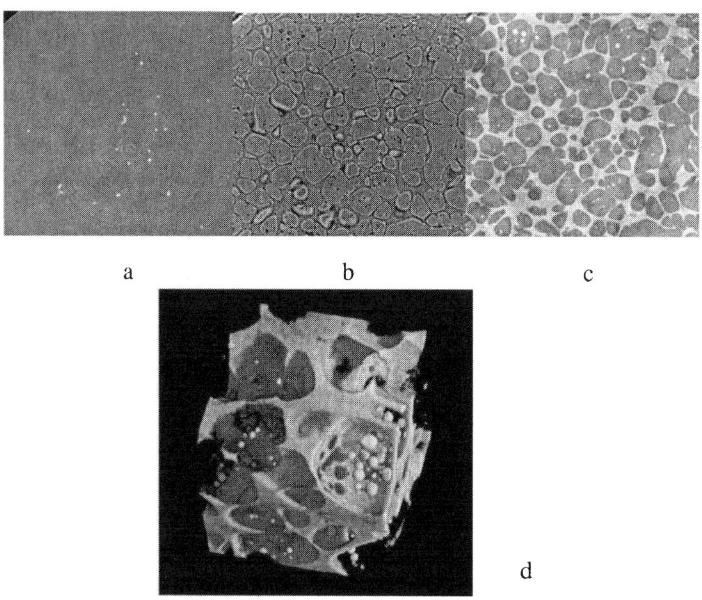

a b c

d

Figure 6. Tomographic images of an Al-Si alloy, quenched from the "semi-solid" state a) Detector to sample distance D = 0.7 cm b) D = 60 cm c) "holotomographic" image, d) 3D rendering showing the former "liquid" phase (courtesy L. Salvo and P. Cloetens).

3. MICROBEAM-BASED X-RAY IMAGING

The absorption or phase-based imaging techniques described above cannot provide information on the local elemental composition and/or structure. This implies combinating scattering measurements with scanning of the sample. The X-ray beams produced at modern SR sources can, because they originate from a very small source, be focused into a very fine spot (0.1-5 μm range) with extremely high intensity. Various focusing elements are currently used to produce small section beams: they include curved crystals, Fresnel zone plates, tapered capillaries, compound refractive lenses, waveguides, and, recently, "Kirkpatrick-Baez" focusing mirrors or multilayers ([13], and references therein). It is then possible to scan the sample and collect quantitative information. This information, displayed in the form of a (y,z) map corresponding to the position of the beam, is the "image" we are concerned with.

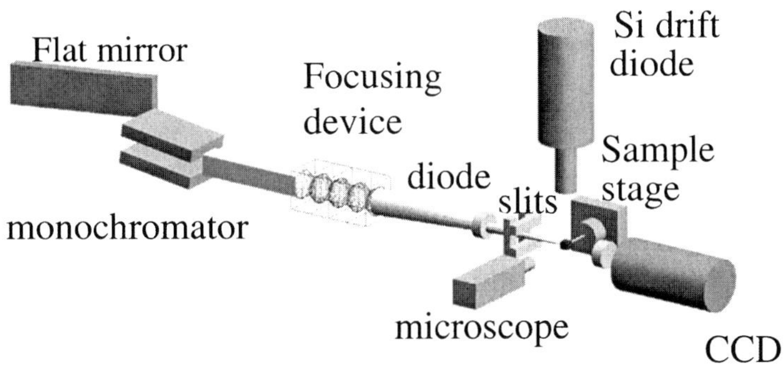

Figure 7. Setup for micro-beam based imaging, which allows producing fluorescence, diffraction or absorption maps (courtesy A. Simionovici)

X-ray microfluorescence allows mapping the presence and concentrations of elements on the sample [14]. A beam of energy E_0 is produced ($\Delta E/E \approx 10^{-4}$), a mirror eliminating the hard ($E > E_0$) radiation (Figure 7), and focused on a small area of the sample. A solid-state detector is set at 90° with respect to the incident beam. A given element produces a fluorescence beam with a characteristic energy E_f. The spatial resolution is, in the hard X-ray range where the depth of focus is usually large (≈ 100 μm), mainly determined by the spot size.

Rindby and coworkers [15] performed an experiment at the ID13 beamline of the ESRF that illustrates the possibilities of this technique. A fly ash particle collected from a lignite-fired power station was analyzed using a tapered capillary that produced a 2 μm beam. Two-dimensional images were produced from the data (fluorescence and diffraction) recorded during the scans. Figure 8 shows microfluorescence maps providing the distribution of trace elements (ppm level) in the particle.

Fluorescence is not the only quantity that can be mapped: wide angle diffracted X-rays, or small-angle scattered X-rays, absorption, or secondary electrons, provide additional information. Scanning imaging simultaneously exhibits high spatial resolution (μm scale in

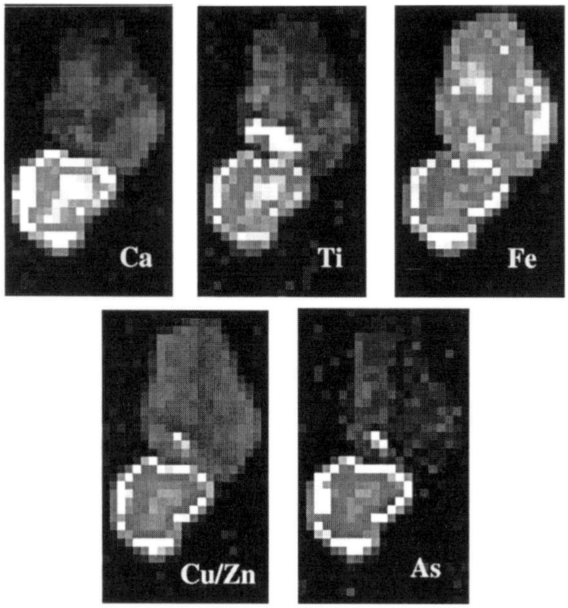

Figure 8. Micro-fluorescence maps of a fly-ash particle showing the distribution of several elements (courtesy A. Rindby and P. Engström)

the hard X-ray range) and elevated chemical selectivity. Local information on different chemical states within systems having the same elemental composition is therefore possible. In addition combined transmission, fluorescence and Compton microtomography was recently implemented ("Integrated Tomographic technique" ITT [16]). It allows a quantitative reconstruction of the internal elemental composition and structure of the measured slice. These new techniques of course require appropriate corrections, based on simplifications and assumptions reducing the complexity of the problem. The problem is substantially simplified when the element of interest is « heavy » with respect to the matrix (like Uranium in soils, or Chromium in organic materials).

This type of techniques and their applications exhibit a rapid expansion, in connection with the availability of modern SR sources.

4. BRAGG DIFFRACTION IMAGING ("X-RAY TOPOGRAPHY")

4.1. Basic principles of X-ray diffraction topography

X-ray topography is an imaging technique for single crystals based on Bragg diffraction. The local variations of the amplitude or direction of the diffracted beam provide information on crystal inhomogeneities. The inhomogeneities that can be visualized using this technique include defects like dislocations, twins, domain walls, inclusions, impurity distribution, and

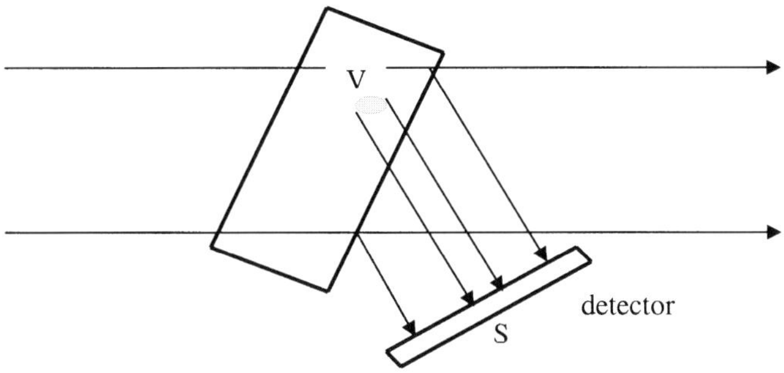

Figure 9. Principle of the topographic methods (see text)

macroscopic deformations, like bending or acoustic waves, present within the single crystal sample.

Figure 9 schematically represents the main idea of all the diffraction imaging methods. The single crystal sample C is set to diffract a fraction of the incoming beam, according to Bragg's law $2\ d\ \sin\theta_B = \lambda$. The diffracted beam is recorded on a position sensitive detector (usually X-ray films, nuclear plates or, increasingly, CCD cameras equipped with a high resolution scintillator). If a small volume V within C behaves differently from the matrix, intensity variations ("contrast") can produce on an area S of the two-dimensional image produced by the diffracted beam. This contrast only occurs if the platelet shaped crystal displays inhomogeneities.

Figure 10 is an example of such a topographic image. It shows the topograph recorded using the 444 reflection from a flux-grown platelet shaped Gallium substituted Yttrium Iron Garnet crystal ($Y_3Ga_xFe_{5-x}O_{12}$, with x~1, Ga-YIG) using $MoK\alpha_1$ radiation. In addition to a distorted region on the upper corner, associated with the crystal fixation, various features are observable on this topograph: dislocations, growth striations and growth sectors, magnetic domains, dissolution bands and the initial stage of a crack. These features are observed because they entail a distortion of the crystalline lattice. These distortions include a lattice rotation $\delta\theta$ and/or a lattice parameter variation $\Delta d/d$. They are, for the growth-striations that are the main observable features, in the 10^{-4} range, whereas the variation of distortion associated with the magnetic domains ("magnetostriction"), as well the distortion between growth-sectors, are in the 10^{-6} one. For dislocations the distortion is a function of the distance to the core; at the level of the image border it is in the 10^{-5} range. A large contrast can therefore correspond to a rather weak variation of distortion.

4.2. Some results of dynamical diffraction theory

A few results of dynamical theory [17], necessary to understand the contrast mechanisms, are given below. We restrict our discussion to symmetrical transmission geometry as shown in Figure 9.

In the case of a *monochromatic plane wave* beam, and a perfect crystal set in Bragg diffraction position, two "wavefields" propagate within the crystal if the direction of the incident beam propagation vector lies within an angle ω_h around the exact Bragg position (ω_h, the intrinsic width of the diffraction curve, often called Darwin width, being of the order of 10^{-5}-10^{-6} radians)

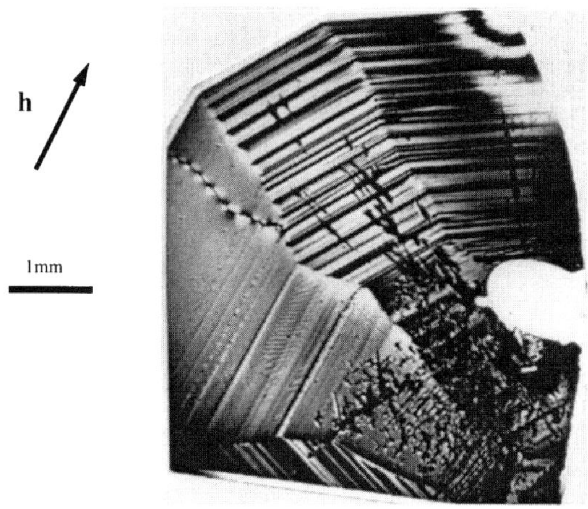

Figure 10. Topograph of a flux grown Ga-YIG platelet shaped crystal; **h** is the reciprocal lattice vector corresponding to the used Bragg reflection

$$\omega_{\mathbf{h}} = (2\lambda^2 C_p |F_{\mathbf{h}}| r_0) / (\pi V_c \sin 2\theta_B) \tag{6}$$

where C_p is the polarization factor ($C_\pi = \cos 2\theta_B$ for π polarisation, i.e. for the electric field vector **E** in the scattering plane, and $C_\sigma = 1$ for σ polarisation, i.e. for **E** perpendicular to this plane), r_0 is the "classical electron radius", V_c is the unit cell volume and F_h is the structure factor corresponding to the Bragg reflection used. A wavefield corresponds to the superposition of one diffracted and one forward-diffracted wave, coupled because the perfect crystal is set for Bragg diffraction. These waves decouple outside the crystal, producing two beams, parallel to the incident and diffracted directions respectively (Figure 11).

Figure 11. Propagation of wavefields within the "Borrmann" triangle ABC in the case of a perfect crystal

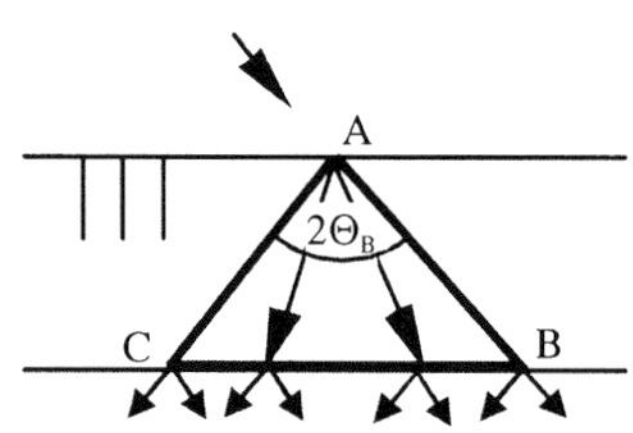

In the case of a monochromatic *spherical wave*, wavefields propagate and interfere within the whole "Borrmann" fan, or Borrmann triangle ABC (Figure 11). The characteristic coupling length between wavefields is Λ_0 (extinction length, or Pendellösung period, which corresponds in the case we are considering to about twice the distance necessary for most of the energy participating to the diffraction process to move from the transmitted to the diffracted direction). Λ_0 typically is in the 1 to 100 µm range.

$$\Lambda_0 = (\pi V_c \cos\theta_B) / (\lambda C_p | F_h | r_0) \tag{7}$$

If the crystal is rotated in a monochromatic beam, or if the beam is polychromatic, the diffracted power is a function of this "variable" (angle, or wavelength), leading to the integrated reflectivity ρ, defined as the area under the diffraction curve. Figure 12 shows ρ for a non-absorbing perfect crystal, as a function of the reduced thickness t/Λ_0, t being the thickness of the platelet shaped crystal, for both the transmission ("Laue") and reflection ("Bragg") cases. These curves show that ρ is proportional to t for $t \ll \Lambda_0$ ("kinematical" limit of the dynamical theory), and that a plateau is reached for $t \gg \Lambda_0$, with an added oscillatory behaviour in the Laue case.

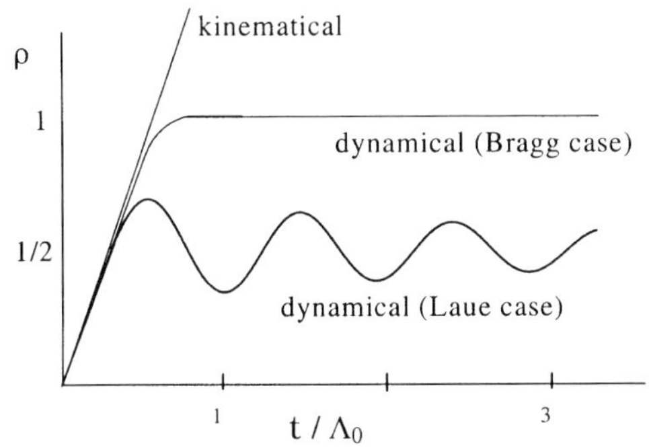

Figure 12. Reflecting power ρ of a non-absorbing perfect crystal as a function of the reduced thickness t / Λ_0 for the transmission, reflection, and the "kinematical diffraction" limit cases.

4.3. Effect of imperfections: contrast mechanisms

For a non-absorbing plate-shaped crystal illuminated by a parallel and polychromatic beam (a good approximation for the simplest and presently most usual topographic technique, white beam synchrotron radiation topography), we can say, within a very simplified approach, that the direction and intensity of the locally diffracted intensity depends upon

- the structure factor F_h
- the angle θ formed by the considered lattice planes, and the incident beam
- Y, a parameter that incorporates the modifications introduced by the crystal inhomogeneities on the perfect crystal dynamical theory results.

The beams diffracted by two neighbouring regions I and II of the sample produce contrast on the topograph if

1) $(F_h)_I \neq (F_h)_{II}$, where the differences can either reside in the modulus or the phase or the structure factor. This *structure factor contrast* is an important mechanism when dealing with regions related by a missing centre of symmetry (twins or domains) in non-centrosymmetrical crystals. In this case either the module (Figure 13) or the phase of F_h (see section 4.4), or both, can be different for these I and II regions.
2) $\theta_I \neq \theta_{II}$, i.e. regions I and II are misoriented -subgrains, domains, ...- one with respect to the other: this is *orientation contrast*. Figure 14 shows, for instance, white or black thick lines, corresponding to the separation, or superposition, of the images of subgrains on a white beam topograph.

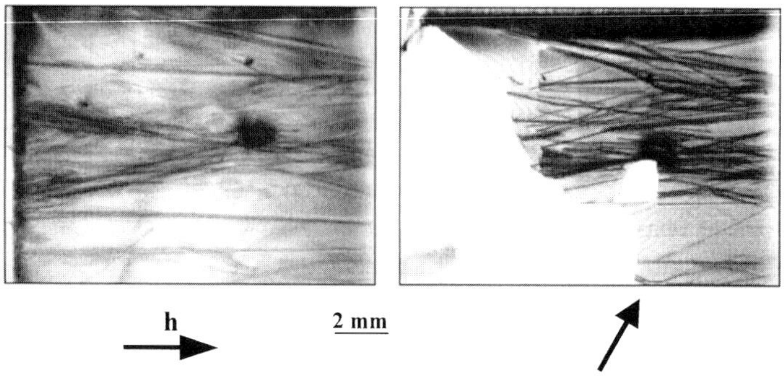

Figure 13. Twins in quartz observed through a "structure factor contrast" mechanism a) 00.6 reflection: the twins are not visible –same modulus of the structure factor- b) 60.2 reflection; the twinned regions structure factors correspond to the 60.2 and 60. -2 reflections, which have very different modulus (2 and 21): this is why one twinned region is hardly visible (white area) on the topograph (courtesy J. Härtwig)

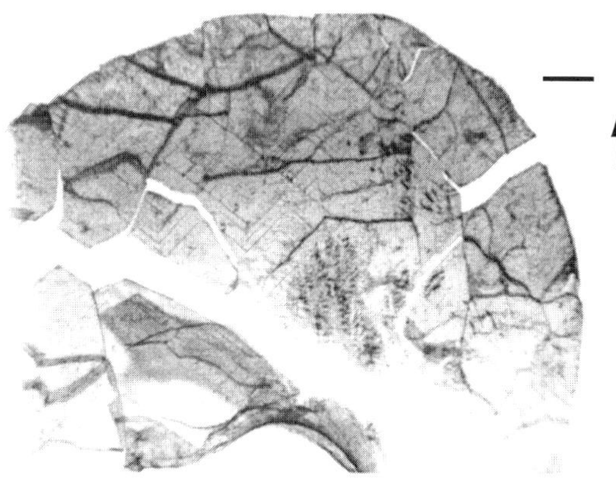

Figure 14. White beam topograph of a Fe-3%Si crystal, $E \approx 25$ keV, showing orientation contrast associated with subgrains. A fir-tree magnetic domains pattern is also visible.

3) $Y_I \neq Y_{II}$, the two regions display *extinction contrast.*

Imperfections modify the perfect crystal behaviour described by dynamical theory. They can hide the interference fringes, or lead to locally enhanced or reduced diffracted intensity, depending on the absorption conditions. New images can occur through the creation of new wavefields at a defect [18].

In the low absorption case the predominant contrast mechanism is the so-called *direct image* process. Let us assume again (Figure 9) that a polychromatic, extended beam impinges on a crystal such that a small crystal volume V contains a defect (inclusion, dislocation,...) associated with a distortion field, which decreases with growing distance from the defect core. A wavelength range $\Delta\lambda/\lambda = \omega_h / (\text{tg}\, \theta_B)$ ($\sim 10^{-4}$) participates in the diffraction by the "perfect" matrix crystal, whereas regions around the defect are at the Bragg position for components of the incoming beam which are outside this spectral range. The defect thus leads to *additional diffracted intensity* on the detector.

The regions that produce the direct image of a given defect are some distance away from the core of the defect. This distance depends not only on the nature of the defect, but also on the diffraction process itself. The lattice distortion acts on diffraction through an angle, the effective misorientation $\delta\theta$, which reflects the change in the departure from Bragg angle that is associated with the existence of the defect:

$$\delta\theta\, (\mathbf{r}) = - (\lambda / \sin 2\theta_B)\, \partial(\mathbf{h}\cdot\mathbf{u}(\mathbf{r})) / \partial(s_h) \qquad (8)$$

where $\mathbf{h}$ is the undistorted reciprocal lattice vector, $\mathbf{u}(\mathbf{r})$ is the displacement vector, and $\partial / \partial(s_h)$ means differentiation along the reflected beam direction. It was shown that the width of the direct image corresponds to the contribution of regions such that $\delta\theta$ is of the order of ω_h. This can be understood when considering that the regions where the distortion is $< \omega_h$ are within the diffraction range of the perfect crystal, and that those where the distortion is $>> \omega_h$ have a negligible contribution for the considered Bragg diffraction spot.

This leads, in the low absorption case we are considering, to an *"intrinsic" image width* of a defect, for a given reflection and wavelength. As an example, for an edge dislocation of Burgers vector $\mathbf{b}$ this width is $\sim \Lambda_0\, |\mathbf{b}\cdot\mathbf{h}| / 4$, ranging in the 1-100 μm scale, and therefore implies a limitation on the density of defects that can be resolved on a topograph (about 10^4-10^5 cm/cm^3 for dislocations).

This also implies that direct images are only seen in crystals that are thick enough. Indeed dynamical theory predicts a width ω_h for thin ($t < \Lambda_0/\pi$) crystals inversely proportional to the thickness. If the crystal is thin only very distorted regions with a strongly reduced volume – and consequently reduced diffracted intensity- can contribute to the direct image, which can therefore not be observable.

The effective misorientation $\delta\theta$ allows, in addition, determining on which reflections a given defect can produce an image. Equation (8) shows that $\delta\theta = 0$ if $\mathbf{u}$ is perpendicular to $\mathbf{h}$. The defect is not visible on the corresponding topographs. A dislocation, for instance, is usually not visible when $\mathbf{h}\cdot\mathbf{b} = 0$. This invisibility criterion is a powerful tool to characterize the defects.

The effective misorientation can also be written, as a function of the local variation in rotation $\delta\varphi$ of the concerned lattice planes and the local relative variation in lattice parameter $\Delta d/d$, as

$$\delta\theta = - (\Delta d/d)\, \text{tg}\, \theta_B \pm \delta\varphi \qquad (9)$$

When taking *absorption* into account, the situation can change drastically. Direct images do not occur in the high absorption case ($\mu t > 10$), and are replaced by dips in the intensity corresponding to the disruption of a dynamical-theory-related effect, the anomalous transmission (also called Borrmann effect). Both types of images ("additional" intensity and "dips") are simultaneously present on the topographs, as well as "dynamical" and "intermediary" images [18, 19] for samples with intermediate absorption ($\mu t \sim 2$-5).

4.4. Diffraction topographic techniques

Bragg diffraction imaging may be used either in *transmission* or in *reflection*. In addition the beam can be *extended* or *restricted*, and the image can be an *integrated* or a *non-integrated* one. Let us consider these different cases.

Figure 8 shows an extended, divergent and/or non-monochromatic beam illuminating the whole sample, or a large area. This results in an image which can, to a first approximation corresponding to predominance of the direct images, be considered as the projection of the defect distribution along the diffracted beam direction.

The incident beam can also be restricted to a small width (typically 10-20 μm) in the scattering plane (Figure 15). In the same first approximation this produces an image of those defects intercepted by the blade-shaped incident beam, hence the name "section" topograph.

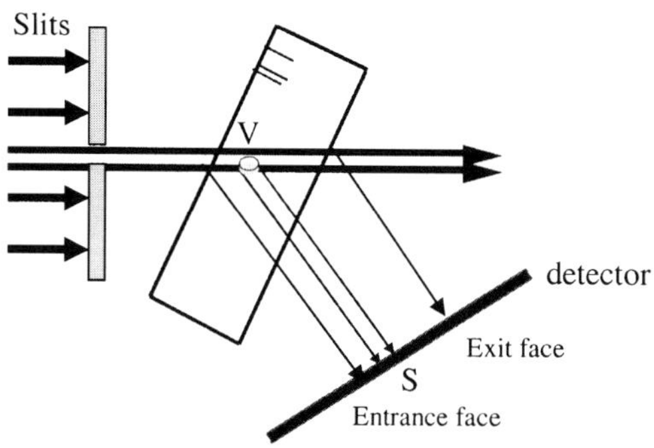

Figure 15. Principle of "section" diffraction topography

In section topography, the position of the direct images within the image can be used to extract the depth of the related defects in the crystal, one edge of the image corresponding to the entrance surface of the sample, the other to the exit one. In very good crystals, with small or moderate absorption, the same interference effect that leads to the oscillations shown on Figure 11, produces interference fringes on the section image, called Kato's fringes. These fringes are very sensitive to crystal distortion, and their modification is one of the first indications of a departure from the perfect crystal situation. The use of a *polychromatic beam* with a sufficiently wide bandwidth often offers advantages. Several reflections are recorded simultaneously, which is very helpful when characterizing defects, and misoriented regions within a sample diffract at once, each region finding a wavelength λ that satisfies the local Bragg condition. This "white beam" version of diffraction topography is identical to the well-known Laue technique (Figure 16), except that the incident beam is broad. Each Bragg spot is now a topograph. This is an *integrated wave diffraction image*, because it results from the superposition of contributions from a range of wavelengths. It is mainly sensitive to the variations of lattice plane orientation $\delta\varphi$, which produce a change in the diffracted beam direction, and not to $\Delta d/d$. "White beam" topography is widely used at synchrotron radiation facilities. The exposure time is often of the order of a fraction of second, and it is possible to follow the evolution of any one of the Bragg spots in "real time", using a CCD camera equipped with a scintillator.

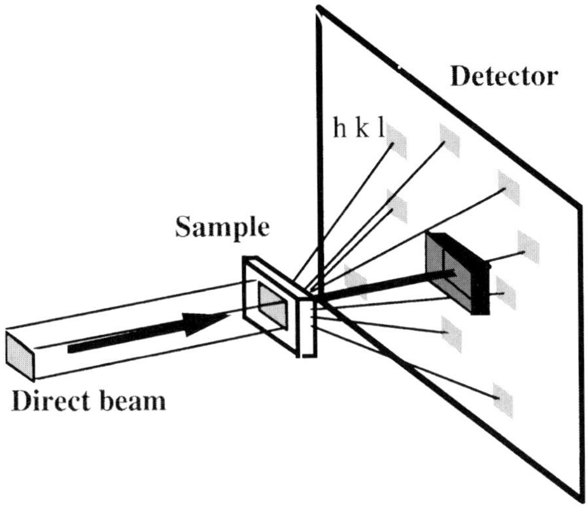

Figure 16. Topography using a "parallel" polychromatic beam: each Bragg spot is a topograph

The incident beam on the sample can also be *monochromatized.* This "monochromatic wave" nevertheless displays a range of wavelengths ($\Delta\lambda/\lambda$) as well as an angular divergence ω. Let ω_c be the diffraction width of the crystal (which reduces to ω_h in the case of a "perfect" crystal). If the divergence ω [and/or ($\Delta\lambda/\lambda$) tg θ_B] is > ω_c the image is still an integrated wave topograph.

If ω [and ($\Delta\lambda/\lambda$) tg θ_B] << ω_h the technique is called *plane wave topography.* It is not only suited to detect weak deformations (10^{-6}-10^{-8} range), but also to obtain images of defects like dislocations (see Figure 18a) with much more details than in the case of direct images. Its use allows a quantitative analysis, because the local diffracted intensity is closely related to the effective misorientation $\delta\theta$. An example of the use of "plane wave" topography is shown on figure 17: the seed of the investigated quartz crystal and the various growth sectors exhibit different impurity contents, which result into small variations of the lattice parameter (in the 10^{-5}-10^{-7} range).

If an image is recorded at an angular position very far away (typically a few ω_c) from the center of the diffraction curve, the main contribution to Bragg diffraction originates from the distorted regions, and not from the perfect crystal matrix. This allows reducing the "intrinsic" width of the images and is known, by analogy with electron microscopy, as *weak beam* topography (Figure 18b).

The weak beam technique can be used to visualize defects not producing a "direct image". This is for instance the case of biological crystals, which are usually thin (t<<1 mm), and, being composed by light elements, exhibit weak structure factors. This leads to large Λ_0, and to ratios t/Λ_0 too small, as pointed out in section 3.3, to allow the "direct image" contrast mechanism to be efficient. Figure 19 shows a monochromatic beam image of a thaumatin crystal. The upper part of the crystal, in Bragg position, produces a black, saturated image,

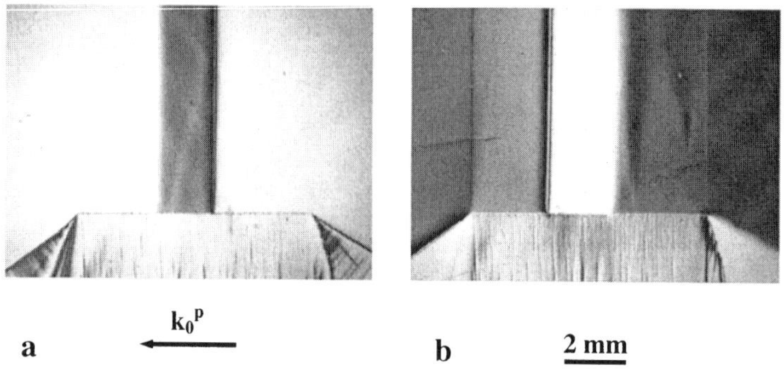

Figure 17. Plane wave $40\bar{4}0$ topographs of a quartz plate with different amount of impurities in the seed plate (rectangle in the middle upper part) and the various growth sectors; k_o^P is the projection of the incident wave vector on the crystal surface; the topographs were recorded a) at about 40% of the maximum intensity on the low angle side, and b) at about 20% of the maximum intensity on the high angle side of the rocking curve (courtesy J. Härtwig)

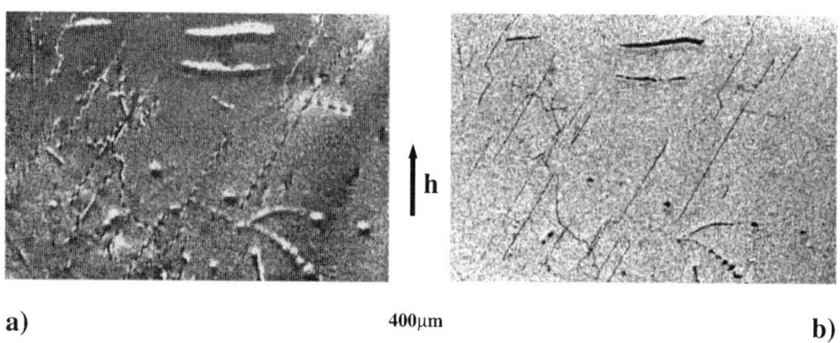

Figure 18. 111 topographs of a Ge crystal recorded using the 35 keV beam produced by a Si 111 monochromator a) top and b) high angle tail ("weak beam" image) of the rocking curve.

whereas the lower one produces a "weak beam" image where dislocation images can be observed. This shows that even in these flexible and large lattice parameter (and large Burgers vector) crystals, dislocations occur.

4.5. Simulation of X-ray topographs

The information provided by the topographs allows extracting *quantitative* data. This is performed through the direct inspection of the images, or through simulations. The simulation process requires introducing a deformation field with free parameters that will be refined by comparing the simulated image to the experimental one. Depending on how rapidly the distortions (expressed by the effective misorientation $\delta\theta(\mathbf{r})$) change locally within the crystal, this may be done using diffraction theories with various complexities. In the case of Figure 17, Bragg case plane wave topographs of a quartz plate with non-homogeneous distributions

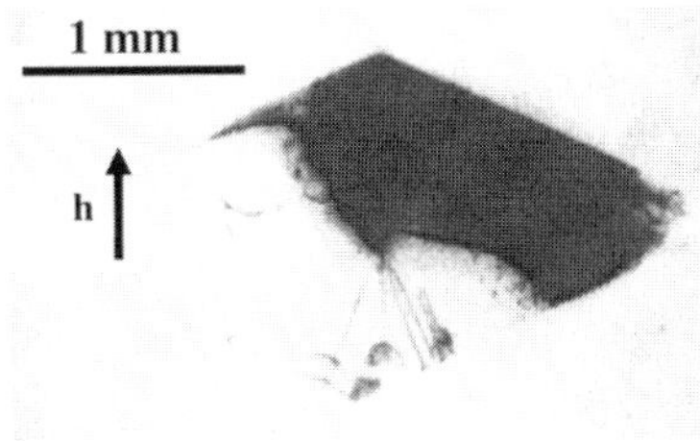

Figure 19. 17.0 topograph of a thaumatin crystal showing dislocations in the "weak beam" lower part of the image (courtesy F. Otalora, V. Stojanoff, B. Capelle and J. Härtwig)

of impurity atoms, already the *local* application of the dynamical theory for perfect crystals provides very good results. In this case the diffracted intensity can be approximated by the simple formula $\mathbf{I_h = I_M R_N(\theta_A - \delta\theta)}$, where R_N is the rocking curve of the set-up, I_M is a normalisation factor, and θ_A is the Bragg angle in a crystal part assumed as the perfect reference one. This means, the contrast depends on the position of the "working point", which is determined, through the parameter $\delta\theta(\mathbf{r})$, by the local Bragg angle on the perfect reference rocking curve. For faster varying deformations (Figure 20) a geometrical-optical approximation may be chosen. The most general case requires a wave-optical approach, which is necessary to calculate, for instance, the contrast of a dislocation. The algorithms give very good results for the plane wave case.

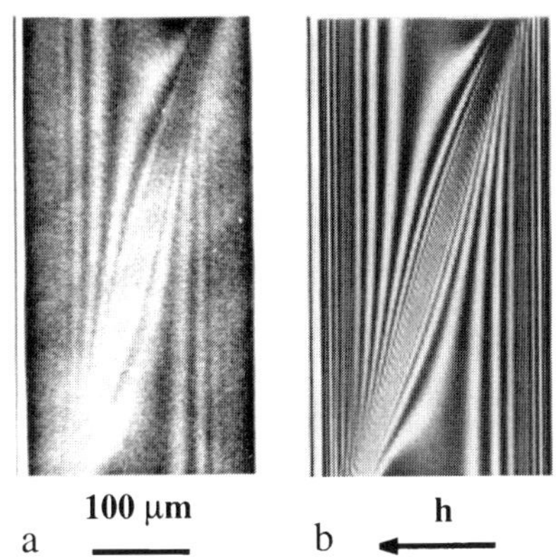

Figure 20. Measured (a) and calculated (b) 422 section topographs of a 565 μm thick silicon crystal with the edge of a 150 nm silicon oxyde film, E = 17.48 keV (courtesy J. Härtwig)

5. EXAMPLES OF APPLICATION OF SYNCHROTRON RADIATION IMAGING TECHNIQUES TO CRYSTAL GROWTH

Synchrotron radiation is now widely used for X-ray imaging. The reduced exposure time (fraction of a second for a topograph, a few ms for absorption or phase radiography) allows investigating evolving phenomena like movements of defects or domains, and first order phase transitions. Specific sample environments (cryostat, furnace, electric or magnetic field, strain device) are added for each particular experiment. Image subtraction can emphasize changes with respect to a reference state. We will give, in this last section, some examples of application of Synchrotron Radiation to topics related to crystal growth, with special emphasis on diffraction topography.

5.1. Propagation of defects from the seed to the growing crystal

Large flux grown KTP crystals produced by "Cristal Laser" were characterized by synchrotron radiation topography. Figure 21b topograph (location indicated on the schematic drawing, Figure 21a), shows that many dislocations propagate from the seed (Figure 21b) and that far from the seed the number of defects is reduced (Figure 21c). It was shown that these dislocations do not originate from the bulk of the seed, but nucleate at the level of irregularities of its surface. Preparing the surface of the seed prior to the crystal growth process could eliminate these dislocations. This work, in addition to the observation of dislocations, growth bands and growth sectors, allowed to visualize twin boundaries and a decrease in the cell parameter c as a function of the distance from the crystal seed [20].

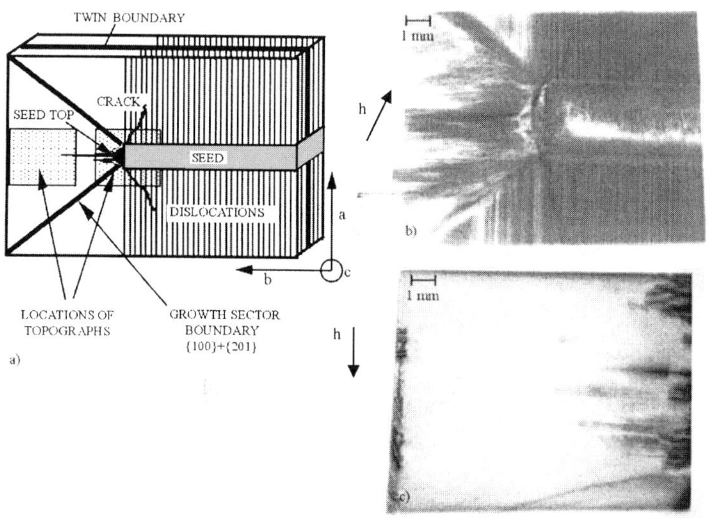

Figure 21. Scheme and topographs of a KTP plate containing the initial growth directions (courtesy P. Rejmankova-Pernot).

5.2 Simultaneous phase and diffraction imaging of porosity in quasicrystals

The combination of phase radiography with X-ray topography provides simultaneous information about local strain fields and growth inhomogeneities. Figure 22 shows the same region of an Al-Pd-Mn quasicrystal grain exhibiting porosity. There is a direct correspondence between the "black-white" contrast observed on the topograph (Figure 22a, and insert) and the pore images observed on the phase radiograph (Figure 22b and insert). This experiment shows that the pores are associated with an "inclusion-like" distortion [10].

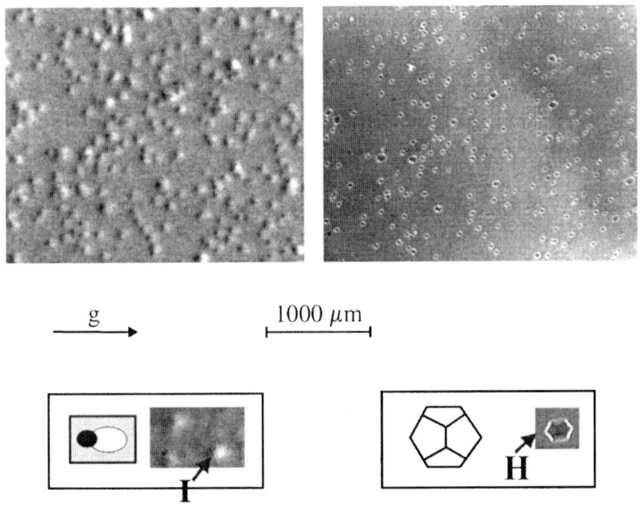

Figure 22. a) topograph and b) phase radiograph of the same volume of an Al-Pd-Mn quasicrystalline grain (courtesy L. Mancini).

Nguyen Thi and co-workers [11] apply the same kind of technique to investigate, in real time, the solidification of metallic alloys.

5.3. Real time investigation of the growth of metallic alloys

White beam topography is a unique tool to investigate crystal growth from the melt, where the orientation of the growing crystal is not known in advance. It allows, more in particular, to visualize the morphology of the solidification front, which determines the microstructure of the solidified material and hence influences its mechanical properties. The directional solidification in ultra-high vacuum of an 0.2-0.3 mm thick Al-(0.73 wt%) Cu binary alloy was investigated by varying parameters like the thermal gradient and the growth rate. Figure 23 shows the transition from a planar interface to a cellular one, when imposing a growth rate higher than the one compatible with a stable planar interface (Figure 23a). The precursor instabilities, the occurrence of defects and the cellular region are clearly seen on Figures 23b and 23c [21].

5.4. Bragg diffraction imaging using a coherent beam

The use of *highly coherent X-ray beams* adds new possibilities to diffraction topography. In this case surface inhomogeneities, or porosities, can be imaged even when they do not produce a strain field, through their Fresnel diffraction image. Recording images at various sample-to-detector distances allows retrieving the involved phase jump.

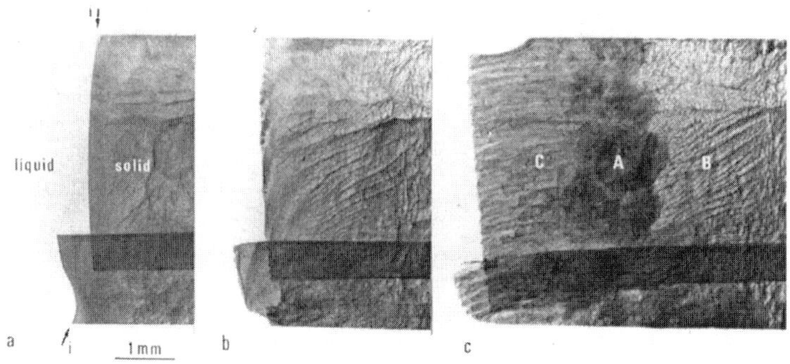

Figure 23. Synchrotron white beam X-ray topographs during melt growth of an Al-Cu alloy, showing a) the planar solid-liquid interface "i" b) the occurrence of instabilities that lead to c) a distorted region A and a cell structure C (courtesy J. Gastaldi, G. Grange and C. Jourdan)

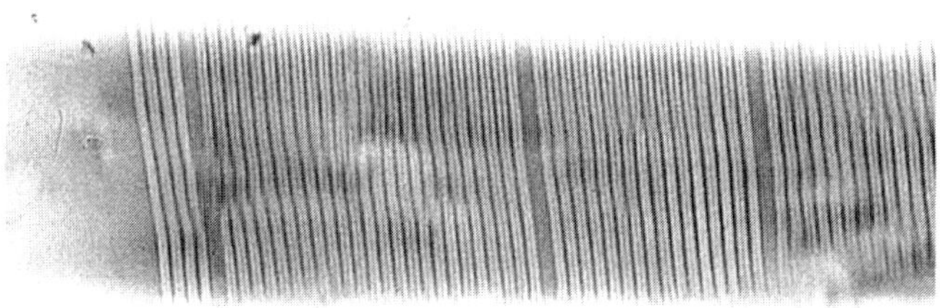

Figure 24. "Coherent beam" section of a periodically poled (9 μm) KTP crystal, showing the ferroelectric domain distribution (D = 40 cm)

The use of Bragg diffraction imaging with a coherent beam allows visualizing 180° ferroelectric domains. Figure 24 shows a coherent beam section topograph of a periodically poled KTP crystal$_3$ where the domains are visualized through a contrast mechanism that mainly originates from the phase shift between the structure factors of oppositely polarized regions [22]. In special cases this technique also allows obtaining quantitative information about atom displacements even though the spatial resolution of the images is on the much larger scale of microns: the extraction of the phase difference from the images gives access to the way ferroelectric domains are connected across the domain wall [23].

6. CONCLUSION

X-ray imaging techniques, and more particularly diffraction topography, are very useful tools to characterize single crystals and growth related features. When coupled with synchrotron radiation, in-situ and real-time investigations are possible and allow studying crystal physics phenomena like the production or movement of defects, and phase transitions.

These techniques are in constant evolution, and spatial resolution improvements, new tools for image processing, or extension to new topics, will surely produce in the near future.

REFERENCES

1. B.P. Flannery, H.W. Deckman, W.G. Roberge and K.L. D'Amico, Science 237 (1987) 1439.
2. A.C. Kak and M. Slaney, Principles of computerized tomographic imaging, IEEE Press, New York, 1988.
3. F. Peyrin, L. Garnero and I. Magnin, Traitement du Signal 13 (1996) 381.
4. W. Graeff and K. Engelke, in Handbook of Synchrotron Radiation, S. Ebashi et al. (eds), North Holland-Elsevier 4 (1991) 361.
5. A.R. Lang, G. Kowalski, A.P.W. Makepeace, M. Moore and S.G. Clackson, J. Phys. D: Appl. Phys. 20 (1987) 541.
6. A. Snigirev, I. Snigireva, V. Kohn, S. Kuznetsov and I. Schelokov, Rev. Sci. Instrum. 66 (1995) 5486.
7. P. Cloetens, R. Barrett, J. Baruchel, J.P. Guigay and M. Schlenker, J. Phys. D: Appl. Phys. 29 (1996) 133.
8. P. Cloetens, M. Salomé, J.Y. Buffière, G. Peix, J. Baruchel, F. Peyrin and M. Schlenker, J. Appl. Phys. 81 (1997) 5878.
9. S. Milita, R. Madar, J. Baruchel, M. Anikin and T. Argunova, Materials Science and Engineering B61 (1999) 63.
10. L. Mancini, E. Reinier, P. Cloetens, J. Gastaldi, J. Hartwig, M. Schlenker and J. Baruchel, Philosophical Magazine A 78 (1998) 1175.
11. Thi H. Nguyen, H. Jamgotchian, Y. Dabo, B. Billia, J. Gastaldi, J. Härtwig, J. Baruchel, T. Schenk and H. Klein, J. Phys. D: Appl. Phys. 36 (2003) A83.
12. P. Cloetens, W. Ludwig, J. Baruchel, D. Van Dyck, J. Van Landuyt, J.P. Guigay and M. Schlenker, Appl. Phys. Lett. **75** (1999) 2912.
13. P. Dhez, P. Chevallier, T.B. Lucatorto and C. Tarrio, Rev. Sci. Instrum. 70 (1999) 1907.
14. F. Adams, K. Janssens and A. Snigirev, J. Analytical Atomic Spectrometry 13 (1998) 319.
15. A. Rindby, P. Engström, K. Janssens and J. Osan, J. Nucl. Instr. Meth. B 124 (1997) 591.
16. B. Golosio, A. Simionovici, A. Somogyi, M. Lemelle, M. Chukalina and A. Brunetti, J. Appl. Phys 94 (2003) 145.
17. A. Authier, Dynamical theory of X-ray diffraction, Oxford Univ. Press (2001).
18. A. Authier, S. Lagomarsino and B.K. Tanner (eds), X-Ray and neutron dynamical diffraction, theory and applications, Plenum, New York, 1996.
19. B.K. Tanner and D.K. Bowen (eds), Characterization of crystal growth defects by X-ray methods, Plenum Press, New York, 1980.
20. P. Rejmankova, J. Baruchel, P. Villeval and C. Saunal, J. Crystal Growth 180 (1997) 85.
21. G. Grange, J. Gastaldi, C. Jourdan and B. Billia, J. Crystal Growth 151 (1995) 192.
22. P. Rejmankova-Pernot, P. Cloetens, J. Baruchel, J.P. Guigay and P. Moretti, Phys. Rev. Letters 81 (1998) 3435.
23. P. Pernot-Rejmankova, P.A. Thomas, P. Cloetens, T. Lyford and J. Baruchel, J. Phys: Condens. Matter 15 (2003) 1613.

Crystal Growth – from Fundamentals to Technology
G. Müller, J.-J. Métois and P. Rudolph (Editors)

Macromolecular Crystals - Growth and Characterization

J. M. García-Ruiz and F. Otálora

Instituto Andaluz de Ciencias de la Tierra, CSIC-Universidad de Granada, Av. Fuentenueva s/n, 18002-Granada, Spain.

Macromolecular crystals are growing according to classical crystal growth mechanisms. They also contain all the crystalline defects found in crystals of small molecules. However, macromolecular crystals and their crystallizing solutions have some peculiarities that affect drastically their quality. This paper reviews the techniques currently used to characterize macromolecular crystals and the current discussion on the relationship between growth conditions and crystal quality.

1. INTRODUCTION

Biological macromolecules such as proteins, carbohydrates, nucleic acids and viruses form crystals. The crystallization of these large molecules takes place by the same mechanisms that account for the growth of crystals of small molecules [1,2], namely normal growth and tangential growth either by two-dimensional nucleation or screw dislocation. However, there are some features of the macromolecular crystals and of the mother solutions from which they grow that are rather specific and are important to understand the growth behaviour, the mechanical and thermal stability of these crystals and therefore their characterization.

For crystallization purposes, a biological macromolecule can be basically considered as a non-penetrable sphere of up to 3 million Daltons[1] in molecular weight, with a diameter of 2 to 20 nanometres, with a non-zero net charge distributed along a surface spotted with hydrophilic and hydrophobic patches. Macromolecules form crystals through intermolecular contacts driven by unspecific attractive interactions which are stabilized by hydrogen bonding, hydrophobic interactions, ion pairing and other weak forces [3].

One of the main differences between macromolecular crystals and crystals of small molecules is the role of water in preserving crystalline order. Water molecules help to maintain the crystal edifice of macromolecular crystals, and sometimes they account for even more then 60% of the crystal mass [1]. Therefore, the characterization of macromolecular crystals is possible only with techniques that do not require an environment for sample preparation or observation that provokes water evaporation.

Another interesting feature of macromolecular crystallization is the purity of the crystallizing solution. Macromolecular solutions are always intrinsically impure and in most cases far more impure than solutions of small molecules. The influence of impurities on crystal quality is still a matter of debate and depends on the type and amount of impurity. It

[1] Unit of mass that equals the weight of a hydrogen atom, or 1.657×10^{-24} grams.

has been shown that unusually large amounts of homologous impurities (for instance of ovoalbumin trapped in lysozyme [4] or of turkey egg white lysozyme in hen egg white lysozyme [5,6]) affects clearly the quality of the crystal. At lower degrees of incorporation or impurity concentration values normal in the real world, the effect is not clear, and we will discuss this subject in some detail.

Today, crystals of biological macromolecules are produced almost exclusively as X-ray samples to be used in the determination of the three-dimensional atomic structure of proteins, nucleic acids or other macromolecules or complexes. Therefore, the characterization is in practice reduced to the assessment of the quality of crystals as samples for diffraction studies, which makes X-ray diffraction characterization techniques the most widely used in macromolecular crystal growth. Very often, the statistical quality of the diffraction data set collected for structural analysis (limit of resolution, R_{sym}, etc...) is misleadingly called "crystal quality". Sample characterization in terms of such statistical indicators of the diffraction data set is extremely useful in practice because they describe the "information content" of the data to be used for solving the structure and, therefore, the quality of the refined structure. But in most cases their interpretation in terms of true crystal quality (i.e. crystalline order) is far from trivial, so we should differentiate both approaches. In fact, the last part of this presentation will focus on the relationship between crystal order and crystal growth conditions. Ex-situ and in-situ experiments performed with different characterization techniques will be discussed to correlate crystal ordering with growth kinetics. Finally, the effects of post-crystallization handling of crystals (such as drying, annealing, freezing, etc) on crystal quality will be considered.

Biochemical characterization of macromolecular crystals is out of the scope of this introductory review. However, the reader should be aware that protein crystals are found to have higher inherent stability than amorphous powders and in some cases enzymes have also been found to have higher activity in their crystalline form as compared with amorphous preparations [7-9]. The application of crystalline proteins to industrial processes has been extended by the demonstration of the stabilization of protein crystals by cross-linking [10,11] and silica gel [12] and also because their topological properties as microporous materials make them a plausible alternative to zeolites [13]. For the same limitations in scope, neither Mach-Zehnder nor Michelson interferometry will be described in this review. These techniques are mainly used to characterize the fluid dynamics around protein single crystals and the kinetics of crystal growth [14,15].

Finally, a terminological clarification. Up to now, macromolecular crystallization has been focused mostly on proteins, either water soluble proteins or membrane proteins (which are proteins that have parts of their surfaces with a strong hydrophobic character). This chapter will not discuss in detail the biochemical aspects of macromolecular crystals. Therefore, we will use the terms protein and macromolecule without distinction. Unless explicitly said, both terms mean in this chapter any protein, nucleic acid, carbohydrate, virus, or any macromolecular complex which is crystallized with negligible consumption of precipitating agent.

2. CRYSTALLIZATION TECHNIQUES OF BIOLOGICAL MACROMOLECULES

Most macromolecules are water soluble and they are usually crystallised from their aqueous solutions. Therefore, protein crystallization techniques make use of the classical ways to create supersaturation from water solutions: (a) water removal, either by evaporation or dialysis (b) solubility change driven by temperature, pH, dielectric constant, ionic strength or polymers.

The solubility of many proteins varies either positively or negatively with temperature. In some cases the dependence of solubility on temperature is large enough to allow crystallization by cooling or heating methods. This thermal method is not currently used, probably because of the concerns of most biochemists with thermal denaturation of proteins. However, several groups have demonstrated the crystallization of several proteins by careful variation of temperature inside microreactors [16,17].

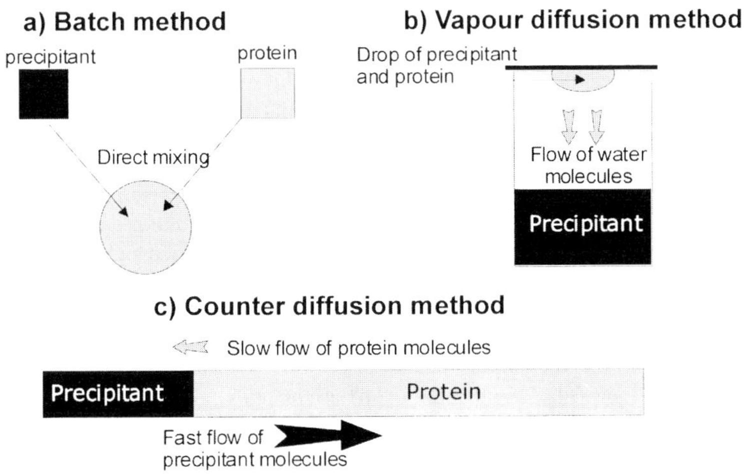

Figure 1. The three main protein crystallization techniques. See text for explanation.

The solubility of proteins varies with ionic strength. It follows that increasing the strength of a protein solution will eventually precipitate the protein. The same also occurs with some polymers such as polyethylene glycols (PEGs). Therefore, when a macromolecular water solution and a solution of a precipitating agent (be it either a salt or a polymer) are thoroughly mixed this will provoke eventually the precipitation of the macromolecule either as a crystal or as an amorphous phase. This mixing method is termed the batch method, and it is the simplest technique used in macromolecular crystallization.

The evaporation technique in protein crystallization is currently performed making use of a simple but elegant method where evaporation is controlled to avoid desiccation of the protein solution. The technique called vapour diffusion has several implementations known as hanging/sitting/sandwiched drop techniques (see Figure 1) [1,18]. In fact crystallization occurs in a drop composed of an undersaturated protein solution which is forced to evaporate until it becomes isotonic with the salt solution contained in an independent reservoir (the well), following Raoult's law. The concentration of salt in the solution within the well is typically twice the concentration of salt in the drop. Therefore, assuming the volume of the drop much smaller than the volume of the solution in the well, the concentration of protein in the drop increases until double the initial concentration. If eventually the critical supersaturation for nucleation is achieved, a precipitation occurs hopefully in the form of crystals. Unlike the batch technique, supersaturation in the drop technique changes at a finite rate. It has been demonstrated that this rate of change of supersaturation depends on the geometry of the system, in particular on the well and drop aspect ratio and the separation distance between them. Indeed, the crucial point is not how far from equilibrium is the solution, but at what rate equilibrium is approached. Note that in the vapour diffusion technique the concentration of protein increases by evaporation as also does the concentration of precipitating agent (being H^+, salts, PEGs, or alcohols). Thus in the vapour diffusion method the driving force is created by a combination of both evaporation and reduction of solubility.

A third technique is the counter-diffusion method, which has also different implementations [19]. Basically the method is based on the diffusion of molecules of protein and precipitant when they are arranged to counter flow one again the other. The precipitating agent and protein solutions are placed together either in direct contact or separated by an intermediate chamber working as a physical buffer, either a gel or a membrane dialysis. The small molecules of the precipitant diffuse faster than the larger molecules of the protein. Therefore, the molecules of precipitant invade the protein solution and supersaturation is created by reduction of solubility according to the solubility dependence of the protein on precipitating agent concentration. The starting concentration of precipitant must be high enough to provoke immediately a precipitation far from equilibrium. Thus a dynamic out of equilibrium precipitation system is created that moves spontaneously towards equilibrium. As a result, the precipitation phenomena taking place form a record of the supersaturation varying in time at different locations within the protein chamber, so that precipitation phenomena take place consecutively with decreasing supersaturation across the protein chamber. Thus, the technique self searches the best crystallization conditions. The main constraint of the technique is that the mass transport must be diffusive. To achieve this, the experiments are performed in capillaries of small diameter ($\leq$ 0.1 mm) or alternatively, after gelling or increasing the viscosity of the protein solution.

Whatever the technique used for protein crystallization the output is crystals of tens or hundreds of microns or in exceptional cases millimetres in size. We have to say, that not much effort has been devoted to increasing the crystal size to the scale of centimetres or even some millimetres with a rational approach. Thus, those interested in characterization of macromolecular crystals have to deal in most cases with the handling of small crystals of few hundreds of microns in size. An additional constriction in the characterization of protein crystals is that they are made up of a large percentage (some cases up to 70%, typically 40-50%) of water. Therefore, protein crystals must always be kept in a water-rich atmosphere. In practice, this means that during characterization experiments, the crystals must be enclosed along with a small drop of the mother solution inside capillaries transparent to X-rays.

3. X-RAY CHARACTERIZATION TECHNIQUES

Many statistical indicators, especially signal to noise ratio, degrade as the disorder of scatterers (atoms) within the crystal increases, giving rise to weaker and/or wider Bragg peaks over a stronger background. This degradation of the diffraction pattern can be due to lattice (or intermolecular) disorder or to intramolecular[2] disorder. The first term refers to the quality of the packing of molecules within the crystal lattice while the second includes thermal motion, presence of impurities and (in macromolecular crystals) flexible domains, alternative configurations, molecular fragments, etc. Depending on the relative contribution of each of these disorder terms, the scattering properties of the solid (hopefully crystalline) phase can be rather different as illustrated in a schematic and qualitative way in Figure 2.

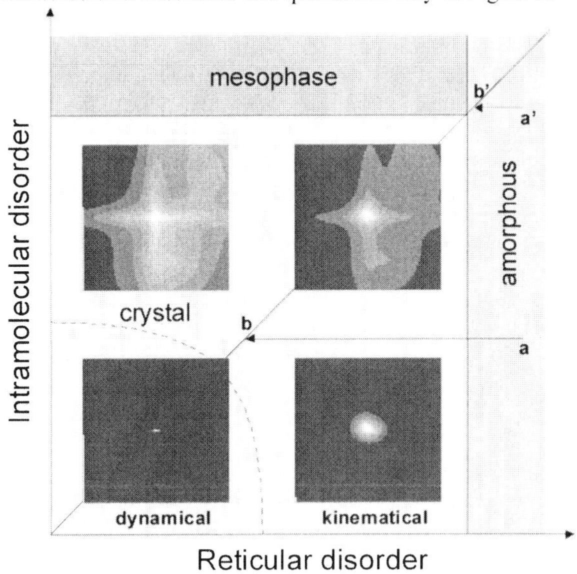

Figure 2. Scattering behaviour as a function of the relative importance of lattice and intramolecular disorder. Diffraction by macromolecular crystals is limited to samples having low or medium levels of disorder; further increases in disorder leading to amorphous or mesophase scattering. Within the "crystal" area of the chart, scattering also changes as a function of the relative contribution of lattice and intramolecular disorder. When both are small, diffraction occurs according to the dynamical theory and the spots in reciprocal space (insets in the chart) are very narrow peaks at the Bragg positions, whose width is close to the "intrinsic rocking width" of the perfect non-absorbing crystal. Higher lattice disorder produces a widening of the diffraction spot and some increase of the background level. High intramolecular (or more strictly "intracell") disorder, on the other hand, increases the background level due to diffuse scattering at directions out of the Bragg peaks. Finally when both lattice and intramolecular disorder are important, wider peaks over an increased background are observed. The *a-b* and *a'-b'* paths show two possible crystal quality enhancements obtained by crystal growth methods (i.e. enhancements of the lattice order) without any enhancement of the intramolecular order. The final crystal quality obtained (corresponding to *b* and *b'* respectively) is strongly dependent on the molecular disorder that cannot be enhanced by a better packing.

2 This really should be called "intracell" or "intraunit" as it refers to whatever is at a crystallographic level smaller than the unit cell (or the asymmetric unit).

If both terms are small, scattering from a "perfect crystal" arises, according to the dynamical theory of X-ray diffraction. This is not a common situation in macromolecular crystallography, although dynamical effects have been shown [20] and even proposed as a means for solving the phase problem [21]. Allowing for some disorder in both components, we enter the "usual" kinematical diffraction regime in which diffraction patterns containing Bragg spots over a diffuse scattering background are routinely collected nowadays for structural analysis. A further increment in intramolecular disorder leads to the "mesophase scattering" region where small angle spots or stripes can be collected from micelles, fibres or other partially ordered structures. Whatever the degree of intramolecular disorder, high levels of lattice disorder imply an amorphous phase, producing exclusively diffuse scattering, which is the starting point for much macromolecular crystallography.

From Figure 2 it is clear that any attempt to improve the structural data quality must look for a reduction of disorder levels, either intramolecular, lattice or both. But there is a practical limit to this improvement: there is no point in having an excellent lattice order with a high intramolecular disorder (and probably it is not possible anyway). The same is true the other way around too; as soon as one of the terms is bad, diffraction data get spoiled. This limitation is illustrated by the solid diagonal line in figure 2. Crystal growth techniques can only improve lattice order, which means that by mastering the growth conditions we can follow (for the best case) the horizontal paths indicated, from either *a* or *a'* to *b* or *b'* in the diagonal. In terms of what can be done by crystal growth methods, both paths are successful, but in terms of diffraction data quality, they are not really the same thing. Intramolecular disorder must be reduced within the feasible limits before performing extensive crystal growth trials and, in any case, must be relatively well known in order to anticipate the improvement one can reasonably expect from crystal growth methods.

Of course this "interplay of the two terms" is not the whole history; diffraction data quality is also highly dependent on experimental aspects related to the X-ray source, the sample handling, the cryo-cooling processing, the amount and composition of amorphous phases in the X-ray path (air, water, glass...), the detector, and even the data processing software. All these aspects are completely beyond the scope and the extension of this chapter; some starting points for diving into them can be found in [22]. The rest of this section contains a review of the X-ray methods for the characterisation of lattice disorder, which is the quantity of interest in macromolecular crystal growth.

3.1. Rocking curves

Lattice defects, including growth sectors, growth striations, impurities, strain fields, etc. cause inhomogeneities leading both to the misorientation of the crystal lattice and to the misspacing of lattice planes. For a given plane *hkl* of the crystal lattice, having an average d-spacing d_{hkl}, if the crystal is exposed to an incident beam s_0 and oriented so that the average Bragg angle θ_{hkl} is such that the Bragg law holds for the wavelength used, a diffracted beam will be scattered in the direction s making an angle $2\theta_{hkl}$ with s_0. Figure 3 (top) shows a sketch of an ideally imperfect (mosaic) crystal within which three crystal volumes *a*, *b* and *c* have been highlighted *a* and *b* share exactly the same d-spacing, but they are slightly misoriented; *a* and *c* have exactly the same orientation, but the d-spacing of the *hkl* lattice plains is slightly different. The whole crystal is assumed to be composed of a number of such volumes showing discrete or continuous variations in orientation ($\theta_{hkl} + \Delta\theta$)

or d-spacing ($d_{hkl} + \Delta d$) in such a way that these properties are no longer constant values but have a definite distribution within the crystal volume.

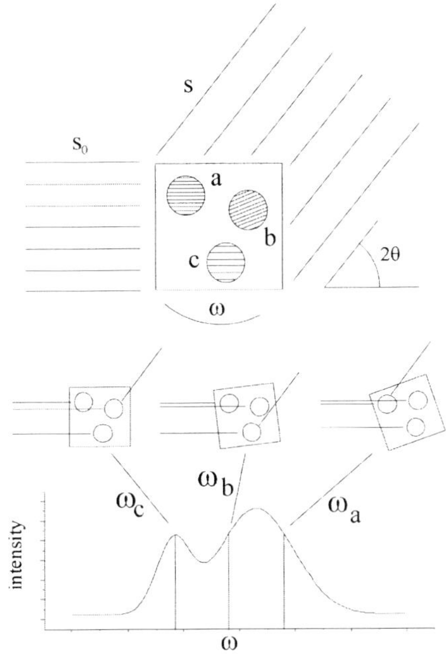

Figure 3. Diffraction from an ideally imperfect (mosaic) crystal. The geometry of the scattering process (top) shows the contribution of three different volumes to the peak profile (bottom). A crystal made up of a number of inhomogeneous domains is exposed to an incident X-ray beam s_0; a diffracted beam is expected in the direction s, making an angle 2θ with the incident beam. Within the crystal, three subvolumes (*a-c*) are highlighted. Subvolume *b* has exactly the same d-spacing as *a*, but is slightly[3] misoriented while *c* has exactly the same orientation as *a*, but a slightly different d-spacing. Obviously, the three subdomains will fulfil the diffraction condition (Bragg Law) at slightly different θ angles, and will contribute an intensity at this angle roughly proportional to their volume. The contributions from all subvolumes in the crystal add up to make the rocking curve.

Rocking curve measurement is the simplest way to characterize this lattice disorder distribution. The intensity profile of a diffraction peak is collected as a function of the angle while rocking the crystal around ω. The collected profile (Figure 3, bottom) contains quantitative information on the statistical distribution of both misorientation and d-spacing. Each of the subvolumes making up the whole crystal will be in the diffraction condition when the Bragg law

$$\lambda = 2(d_{hkl} + \Delta d_i)\sin(\theta_{hkl} + \Delta\theta_i) \qquad (1)$$

[3] Angular deviations and differences in d-spacing are enormously exaggerated in all figures for the clarity of illustrations. See figure 7 for an order of magnitude of these quantities in real crystals.

is locally fulfilled at the ith subvolume for their respective values of misorientation $\Delta\theta_i$ and deviation in d-spacing Δd_i. In particular, the crystal volumes *a*, *b* and *c* in Figure 3 are in the diffraction condition when the rocking angle is equal to ω_a, ω_b and ω_c respectively; other volumes of the crystal make the rest of the distribution illustrated at the bottom, each of them contributing an intensity proportional to its volume. The most often used parameter to quantitatively characterize the mosaic spread is the Full Width at Half Maximum (FWHM) of the distribution. Although this is a good estimate to present statistical information on several measurements, its meaning is only well defined for gaussian-like peaks, being much less useful when the shape of the peak deviates from a single gaussian, as the one illustrated in Figure 3.

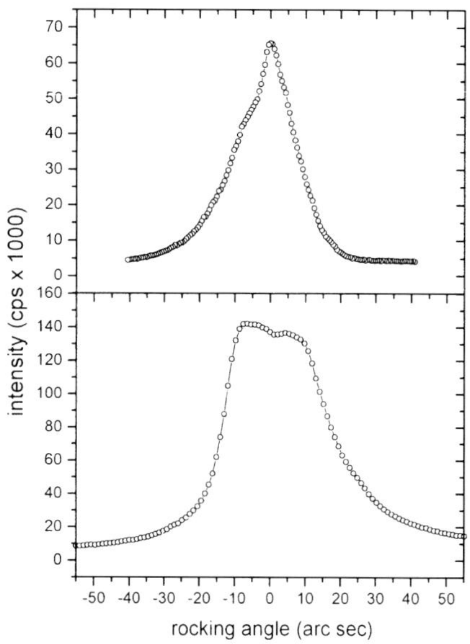

Figure 4. Rocking curves from a tetragonal Hen Egg White Lysozyme crystal rotating around the 2-fold (top) and the 4-fold (bottom) axes. The shape of the peak is clearly different because the contributions from misaligned domains are anisotropic. In this crystal, most of the block misalignment corresponds to rotations around axes close to the 4-fold axis of the structure, making the peak wider when rocking around this direction. The crystal volumes contributing to double peak in the bottom plot are shown in Figure 8.

Rocking curves have been extensively used to characterize protein crystals in trying to understand problems like the distribution of defects within the crystal [23], the relation between growth history and defect distribution [24], or the crystal quality of crystals grown using a given growth method [25,26]. Two typical rocking curves for tetragonal Hen Egg White Lysozyme are shown in Figure 4. The difference between them is the rocking axis that corresponds to the 2-fold axis of the structure in the top plot and the 4-fold axis in the bottom one. The anisotropy is evident, and the fact that the bottom peak is split in two maxima, but not the top one indicates a twist of the crystal lattice around the 4-fold axis (see section 3.3 on topography).

3.2. Reciprocal Space Mapping

In the reciprocal space, the fact that the crystal is made up of a set of crystal subvolumes having slightly different orientation or d-spacing corresponds to the reciprocal nodes not being discrete points but extended probability distributions. Therefore we can visualize reciprocal space spots as reciprocal volumes that have a definite size and cross the Ewald sphere as the crystal rocks around ω. Subvolumes having exactly the same misorientation lie on straight lines along a given q vector. Subvolumes having exactly the same d-spacing lie on a circular arc centred at the origin O of the reciprocal space (Figure 5).

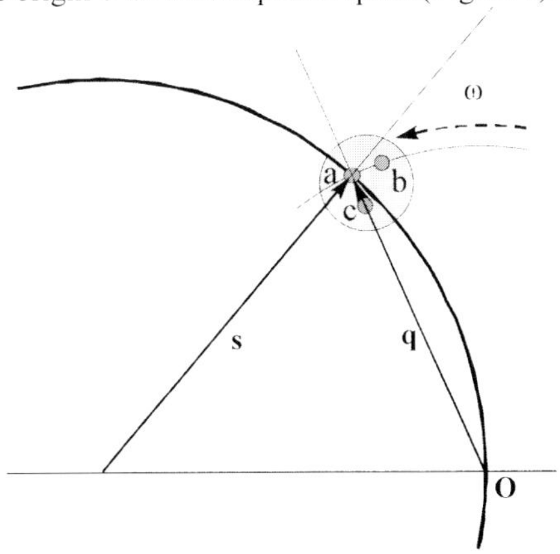

Figure 5. The structure of a single reciprocal space node from a crystal made of subvolumes having different misorientation and d-spacing. The contributions of the three subvolumes (a, b and c) illustrated in Figure 3 is sketched. O is the origin of the reciprocal space, s is a vector in the direction of the diffracted beam, q is the scattering vector. At the rocking angle illustrated, only the a subvolume fulfils the Bragg equation.

It is clear that all reciprocal space points at the intersection of the reciprocal space node and the Ewald sphere correspond to crystal volumes that are simultaneously in the diffraction condition for the given ω angle (Figure 5). This means that the intensity recorded at a given point of the rocking curve contains information on volumes having heterogeneous values for both the misorientation and d-spacing, which is the main drawback of rocking curves. To overcome this problem, reciprocal space maps are used. The key point to note is that these points will diffract at slightly different angles (see Figure 6), so the diffracted beam is to some extent divergent and can be analysed by setting a third crystal (the analyser) in the direction of the diffracted beam. If the mosaic spread of this analyser crystal is small compared to that of the sample, the diffracted beam can be quantitatively decomposed in its components as only part of them will fulfil the Bragg equation at the analyser for a given ω' angle. A two dimensional map of the reciprocal space can therefore be composed in which each point represents a unique combination of ω and ω'. Further geometrical correction of these data produces a map in terms of the parallel $q_{=}$ and perpendicular $q_{\perp}$ components of the scattering vector that, in turn, correspond to the variations in d-spacing and misorientation respectively.

An example of a reciprocal space mapping from a protein crystal (tetragonal thaumatin) is shown in Figure 7. This map corresponds to a crystal made up of four domains having large differences in both misorientation and d-spacing. Reciprocal space mapping has been used for the characterization of biomacromolecular crystals facing problems like the effect of radiation damage [27], impurities [28,29], and growth conditions [30] in lattice disorder. An interesting application of reciprocal space mapping in the field of biological macromolecules, although not directly related to the characterization of single crystals is the characterization of macromolecular interaction in two-dimensional layers [31,32].

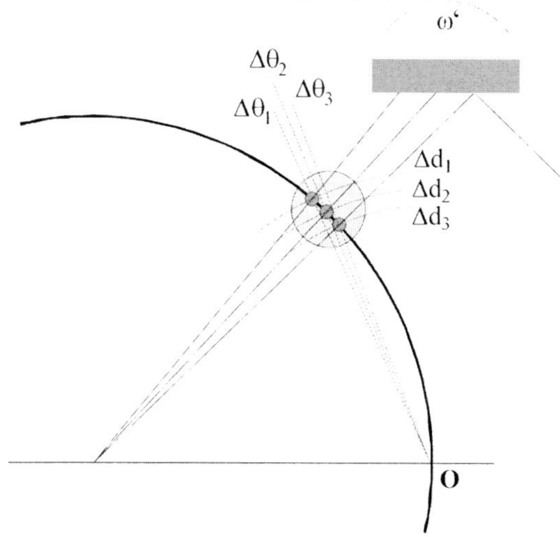

Figure 6. Three different points within a single reciprocal space node having misorientations $\Delta\theta_i$ and d-spacing variations Δd_i and simultaneously fulfilling the Bragg condition. All of them will diffract in the directions indicated from the centre of the Ewald sphere. Therefore, a slightly divergent diffracted beam is produced by the sample due to the extended nature of the reciprocal space nodes. This divergence can be used to separate the contribution of misalignment and changes in d-spacing by using an analyser crystal having a very narrow rocking curve. This is done by scanning the ω' angle around the Bragg angle of a strong reflection from the analyser crystal. This is the operational principle of reciprocal space mapping using a three crystal spectrometer.

3.3. Topography

Reciprocal Space Mapping gives us accurate quantitative information on the distribution of crystal subvolumes having a given misorientation and d-spacing within a single crystal, but due to the geometry of the scattering process, the relation between the reciprocal and the direct space is missing in these data. This means that we know the amount of crystal having some given properties, but we don't know where these volumes are. This complementary information is obtained using X-ray topography.

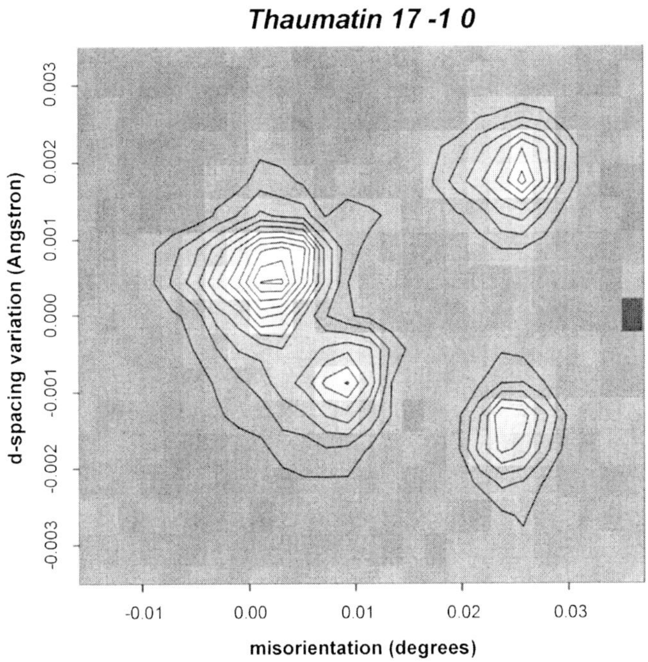

Figure 7. Reciprocal space map of the $17\,\bar{1}\,0$ reciprocal space node from a thaumatin single crystal. The axes have been transformed to show $\Delta\theta$ and Δd values. A double peak is observed close to (0,0) accounting for most of the recorded intensity. A minor peak close to it comes from a subvolume misoriented by 0.007° and having a d-spacing 0.001Å smaller. Even more prominent is the existence of two crystal subvolumes having the same misorientation but a 0.003Å difference in d-spacing (two peaks to the right).

Up to now we have considered the diffracting crystal as a point source for the diffracted beam either by assuming the whole crystal to be a point or by considering that only a point-like volume is diffracting at a time. In real diffraction experiments, different volumes within the crystal are simultaneously contributing to the diffracted intensity, and these volumes can be separated from each other. If the incident beam is a plane wave (i.e. monochromatic and parallel), the diffracted beam will be a projection of all these volumes in the direction of the diffracted beam, i.e. the spot collected into a two-dimensional detector having a spatial resolution high enough will be a projected picture of this density of volume in diffraction condition (Figure 8). This projection is a X-ray diffraction topograph.

Figure 8 illustrates only the mechanism of formation of "orientation contrast" which is typical of mosaic crystals and which can be understood only in terms of diffraction geometry. Large, high quality crystals can also display what is called "extinction contrast" which comes from the differences in diffraction properties between a highly distorted area (like the one close to a dislocation) and the surrounding almost perfect lattice. This kind of contrast can only be understood under the dynamical theory of X-ray diffraction, which is beyond the scope of this text, but is interesting as it is the type of contrast that allows the observation of single defects like dislocations. Figure 9 shows a topograph from a thaumatin crystal in which both types of contrast can be observed. The dark/white contrast between the upper-right part and the lower-left part of the crystal is orientation contrast due to a crystal lattice slightly bent over long distances. The dark straight lines at the bottom are three dislocation pairs going from the crystal nucleus to the crystal face while remaining perpendicular to the growth surface. More dislocations are present in this crystal, but only these show a large enough extinction contrast.

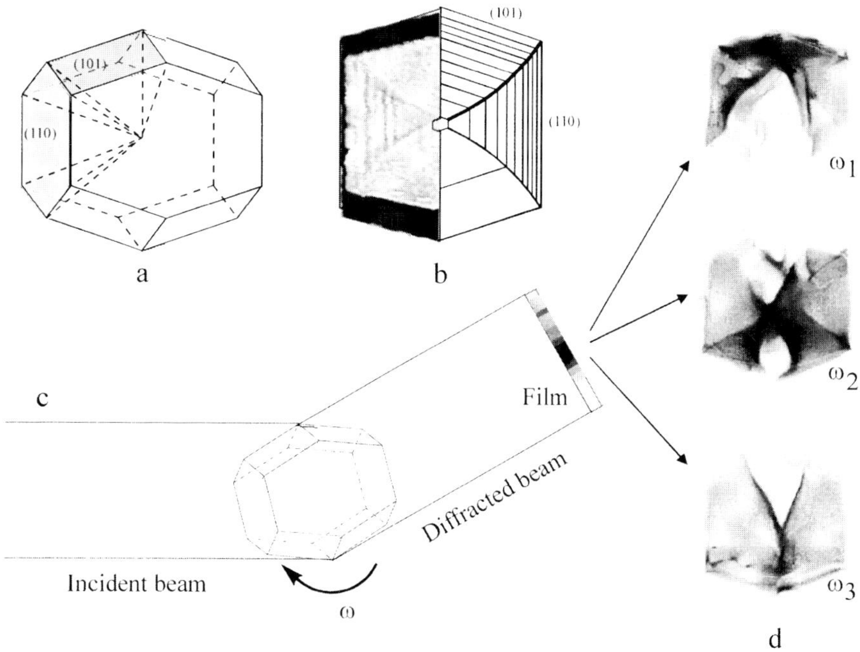

Figure 8. Basis of topography diffraction imaging in kinematical diffraction. The crystal is assumed to be composed of a number of subvolumes having misorientations and d-spacings changing either smoothly or suddenly. Typically, these volumes will be related to the growth sectors, as illustrated in the case of tetragonal lysozyme (a-b) where two types of growth sectors are observed: the prismatic (110) and the pyramidal (101) sectors. For a given ω angle close to the Bragg angle of difraction, some of the subvolumes making up the crystal will be in diffraction condition producing some intensity in the corresponding film area while other subvolumes will not contribute to the diffracted beam (c). Therefore, an image is formed which is the projection in the direction of the diffracted beam of the crystal volumes simultaneously diffracting at this ω angle. Three such images (x-ray diffraction topographs) are shown in (d); the contribution of the different growth sectors is clearly identified as well as the growth sector limits (particularly at ω_2). The topographs shown in (d) were collected from the same crystal producing the rocking curves in Figure 5 while rocking around the 4-fold axis (Figure 5 bottom) at the rocking angles of the left maximum (top sector), the centre of the peak and the right maximum respectively (bottom sector).

Topography is the right tool to use when looking for localized disorder like zonal or sectorial structures or individual defects like dislocations, or to complement other x-ray diffraction techniques. It has been used in biomacromolecular crystallography, for example to study the influence of growth conditions [26,33], the presence of impurities [28], or to characterize single defects in protein crystals [34,35] ([34] uses the Laue diffraction method not described here).

3.4. Combining methods

From the previous discussion, it is evident that rocking curves, reciprocal space maps and topography are complementary techniques. Rocking curves provide a fast method for the characterization of systems that are being followed on line during growth [20] or to get statistical information (by analysing many crystals) on a given problem. Reciprocal space maps are more demanding in terms of equipment, data collection, and processing time, but offer a more detailed and clear view of the problem, especially in situations where simultaneous changes in both misorientation and d-spacing are involved. Finally, topography adds the real space information on defect distribution within the crystal volume that is missing in the other two methods. A nice example of this combination of methods is the assignment of growth sectors responsible for each of the observed features in a rocking curve using topography. The double peak illustrated in Figure 4 (bottom) corresponds to the superimposition of two main peaks from the two opposite 101 sectors of lysozyme crystals whose lattice is twisted around the 4-fold axis of the structure. This fact was clearly demonstrated using a series of topographs collected along the rocking curve (Figure 8d shows part of this series).

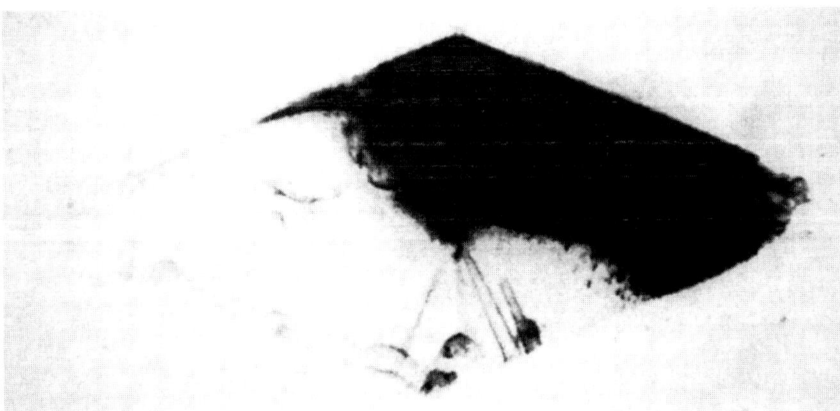

Figure 9. X-ray topograph collected from a thaumatin crystal having the *c* and *a* axes parallel to the paper (*c* axis is horizontal). The overall rhomb shape of the spot is a projection of the crystal form. The large dark/clear contrast between the right-top and the left-bottom parts corresponds to orientational contrast produced by a continuous bending of the crystal lattice over the whole crystal volume. More interesting are the thin dark straight lines at bottom, which correspond to dynamical (extinction) contrast from three pairs of dislocations nucleated close to the crystal nucleus and developing perpendicular to the crystal faces as the crystal grows.

The main drawback of these experiments using multiple complementary characterization methods is the high-end experimental requirements imposed on the beamline, as a large and flexible set of accurate instruments needs to be operated simultaneously or iteratively on the crystal at the diffraction position. In this direction, we must comment briefly on the experimental set-up that has been implemented at the ID19 beamline of the ESRF in the framework of the LS1860 / SC1220 Long Term Project. The set-up consists of an extremely accurate four circle goniometer equipped with a selectable 2 crystal, 3 crystal spectrometer featuring a rich set of on-line detectors (X-ray video camera, photo diode, film holder, scintillator and a MAR345 image plate) and mounted on a beamline providing a quasi-plane wave beam of very high brilliance. Using this set-up, it is possible to collect screenless oscillation datasheets, rocking curves, reciprocal space maps as well as 2 crystal, 3 crystal and section topographs from the same oriented, indexed and undisturbed crystal.

4. CRYSTAL QUALITY FOR STRUCTURAL ANALYSIS

The key parameter defining the data quality of a X-ray diffraction datasheet for structural analysis is the limit of resolution, defined as the inverse value d_{min} of the largest value of the scattering vector modulus $S = 2\sin\theta / \lambda$ for which useful information can be extracted from integrated intensities of Bragg reflections [36]. In practice this limit is defined by fixing some value for the average magnitude of signal to noise ($I / \sigma(I) < 2$) or the disagreement between the intensity of symmetry related reflections ($R_{merge} > 0.2$) in a given resolution shell. This progressive loss of information is due to two main effects: a) the decrease of the atomic scattering factor at increasing S and b) the decrease of electronic density correlation for shorter correlation lengths due to disorder in the crystal. The first contribution is unavoidable as it is a property of X-ray scattering; reducing the second contribution is the main objective in crystal growth and characterization studies. Notice that a perfect crystal is not the best solution because it will diffract in the dynamic regime where the intensity of the diffracted beam scales with the structure factor. Thus a slightly imperfect crystal diffracting in the kinematic regime (where the intensity of the diffracted beam scales with the second power of the value of the structure factor) is a better partner for getting a good diffraction dataset.

All the characterization methods described in section 3 concentrate on Bragg peaks, their intensity distribution and shape, but lattice disorder also contributes (along with intermolecular disorder) to the diffuse scattering, and probably this is the main link between lattice disorder (and consequently crystal growth) and the statistical quality of structural X-ray data. The amplitude scattered by a single atom depends exclusively on the properties of the incident radiation and the number and distribution of electrons in the atom. The scattering process at the atomic level is, in particular, independent of the order of matter, i.e, the fact that the substance being exposed to X-rays is a crystal, an amorphous solid, a liquid, etc. The diffraction process that allows the collection of structural data is a property of the crystal lattice, the ordering of the atoms in the lattice being the only reason for the existence of constructive and destructive interference of the scattered wavelets that produces diffracted beams in discrete directions. Obviously, the existence of defects in the lattice –considering as defects any deviation from the perfect lattice due to either intramolecular or lattice disorder– will introduce deviations in the relative positions of the atoms in the crystal and, therefore will partially destroy the coherence of the interference process. When this occurs, the amplitude of the Bragg peaks decrease because some of the atoms are not contributing to the constructive interference and, for the same reason, the amplitude of the background increases because destructive interference is also incomplete. This increased background is what is called "diffuse scattering" and can be structured or unstructured depending on the properties of the disorder causing it. Therefore, the detrimental effect of crystal disorder is two-fold in the signal to noise statistics and, consequently in the very statistical quality of the structural data. It should be remembered here that, although both lattice and intramolecular disorder contribute to diffuse scattering and, therefore, produce lower $I/\sigma(I)$ values, the efficiency of crystal growth methods in improving these values via improvements in the lattice order is always limited by the presence of intramolecular disorder as shown in Figure 2.

In terms of lattice packing, the main contribution to the disorder in biomacromolecular crystals is their zonal and sectorial structures. Zonal structures correspond to features parallel to the growing interface and therefore they are a decoration of the growth history; these defects are commonly referred as "growth striations" and form due to fluctuations (either externally driven or self-organized) in the growth rate, the composition of the solution, the impurity concentration, or all of them. Impurity content during the growth of the crystal also changes in a continuous way in the absence of growth striation because impurities usually have a different affinity for the crystal phase than the macromolecule being crystallized. This affinity is defined by the partition coefficient

$$K = (C_{is}/C_{ps})/(C_{il}/C_{pl}) \tag{2}$$

where i stands for impurity, p for protein, s for solid (crystal) and l for liquid (solution). This value can be higher than one for homologous impurities (which therefore tend to accumulate into the crystal) and is in general lower than one for foreign impurities (which tend to concentrate in the solution). In any case, the progressive accumulation of impurities into the crystal or the solution creates a gradient in the impurity content within the crystal [40].

The sectorial structure includes all features that develop in the growth direction, in particular the boundaries between growth sectors. A growth sector is simply the crystal volume left behind a single crystal face during the advance of this face (see Figure 8). The boundaries between these sectors are a common place for the accumulation of impurities, dislocations and other types of defects produced during the growth of the crystal or transported to the boundary after the growth. For this reason, they are usually observed in X-ray topographs (see Figure 9). Both growth striations and sector boundaries give rise to stress in the crystal that is elastically accommodated by long range bending or twisting of the crystal lattice. This long range continuous misorientation is the main contribution to the observed mosaic spread in protein crystals [20,37]. The existence of well defined, relatively ordered growth sectors immediately suggests some new directions to explore: on one hand, the use of microfocus beams to get diffraction patters from crystal volumes belonging to a single sector is a possible way of getting better structural data from high mosaicity crystals, avoiding most of the disorder associated with boundary sectors. Other possibilities, less developed at present, could include the use of these volumes (cut from the crystal in some way) as seeds for producing large high quality crystals, as is routinely done with materials like quartz. In this direction, probably the message from the crystal characterization techniques used today in the field of biomacromolecular crystallography is that these crystals are so similar to small molecule crystals that we can still get a lot of benefits from the previous results of materials science and small molecule crystallography.

The fluid dynamics scenario where a crystal grows affects substantially its crystal quality for structural studies. A convection-free environment such as the one provided by gels [38] or microgravity may offer a better scenario for growing crystals of high quality. The reason is the following: imagine a scenario for protein crystallization in which there are a number of growth units in the solution surrounding a growing crystal (see Figure 10a). These growth units may be single molecules, dimers or oligomers or they can be clusters of these units. Unfortunately we have not yet a clear understanding of the distribution of growth units in a crystallizing solution [39]. Because of the flow of growth units towards the crystal face where they become incorporated, a region around the crystal is created, termed the concentration depletion zone (CDZ) where the solute concentration changes continuously from the concentration at the crystal/solution interface C_i to the concentration in the bulk of the solution C_∞. This region develops only when crystals grow in the diffusion controlled or in the mixed regime. In the diffusion controlled regime $C_i = C_e$ (C_e being the equilibrium concentration) while in the mixed regime $C_i > C_e$. No CDZ develops in the kinetic growth regime even if the mass transport is diffusive. In the CDZ, the concentration profile varies with time and is controlled by the relative rates of the two simultaneous processes operating: the diffusive flow of molecules towards the crystal face and the flow of incorporation of the molecules into the crystal lattice. This non-linear evolution of the CDZ is more important in the mixed regime. In the steady state, the diffusive flow towards the crystal face is given by the kinetic equation

$$\frac{dm}{dt} = \frac{D}{\delta} A(C_\infty - C_i) \tag{3}$$

where A is the surface area of the face and the ratio between the diffusion coefficient and the thickness of the CDZ (D/δ) is called the transport coefficient k_d. Once the molecules reach the crystal face after diffusion across the CDZ, there is a reaction step in which the molecules have to find the proper orientation and binding location to fit the arrangement of the crystal structure. This process of surface integration has its own kinetics and can be properly represented by a similar equation

$$\frac{dm}{dt} = \beta A(C_i - C_e) \tag{4}$$

where β is the kinetic coefficient for surface integration. Values of β (surface) for tetragonal hen egg white lysozyme crystals grown from solution have been measured to be $3\text{-}30 \times 10^{-6}$ cm/s [40], i.e. similar to the diffusion coefficient of lysozyme molecules in water $D = 1.16\times10^{-6}$ cm^2/s. The competition between surface kinetics and diffusion transport is measured by the relation

$$\beta L/D \tag{5}$$

where L is the size of the crystal. For $\beta L/D \gg 1$, diffusive mass transport is the rate controlling step, while for $\beta L/D \ll 1$ the crystal grows controlled by the processes taking place at its surface. Avoiding convective mixing in the growth environment will increase the possibility of growing the crystal under slow diffusive mass transport provided that the surface interaction kinetics is faster than the characteristic diffusive flow of macromolecules. However, if we consider reliable the values of β and D for lysozyme, it is clear that a lysozyme crystal will not grow under pure mass transport control if its size is smaller than several millimetres. An exception to this argument can be envisaged if the β step coefficient, defined as β surface divided by the vicinal slope (typically 10^{-2}) is taken as the appropriate parameter constraining the surface interaction, since then the mass transport would control the whole growth process. In fact, it has been reported that, at low supersaturation, lysozyme crystals grow layer by layer, the step energy being the main parameter governing surface kinetics [41]. In addition, the existence of a larger contribution of bulk diffusion with respect to surface kinetics has been measured under microgravity conditions [15].

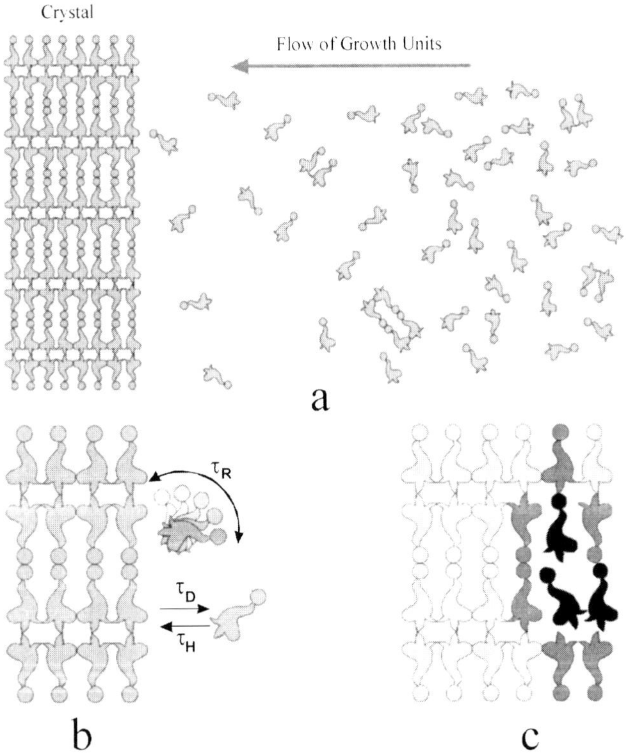

Figure 10. Schematic view of the different steps in the incorporation of a growth unit into a crystal lattice and the effect on crystal quality.

The existence of this strong contribution of bulk diffusion is interesting for crystal quality. Growth units flow towards the crystal and they hit the crystal surface with a hitting frequency $1/\tau_H$ (Figure 10b) and hitting distribution which depends on the concentration C_i of protein in the vicinity of the crystal face i.e., on the overall driving force and on local fluid dynamics. Once on the surface of the crystal, the growth units are weakly bound through unspecific attractions. Therefore they may rotate to find the proper orientation into the crystal geometry and to minimize the crystal energy. However, when a growth unit is immobilized in the crystal structure before it finds the proper orientation, an orientational defect is created, and this can propagate and deteriorate the quality of the lattice (Figure 10c). The time τ_R that a molecule has to find the proper orientation is constrained by a) the frequency of detachment ($1/\tau_D$) and, b) the probability of a new molecule hitting a location close to itself, which depends on τ_H. For macromolecules, τ_R is much longer than for small molecules due to their large volume and to the lower anisotropy of the binding configuration [42]. The value of the parameter τ_R is difficult to decrease except by manipulating the macromolecule to create preferential binding spots on its surface. However, τ_H can be reduced by lowering the concentration of the growth units close to the crystal face. Even in the mixed regime, when mass transport in the bulk of the solution is governed by diffusion, the flow of units towards the crystal is reduced, creating a wider and deeper depletion zone around it. Lower supersaturation values close to the crystal face help to decrease τ_H and to reduce the probability of formation of small pre-critical clusters close to the crystal face, which have a lower probability to fit the geometrical lattice with the proper orientation than single growth units do. Certainly, when the crystal grows beyond a size L, making $\beta L/D >> 1$, the conditions are optimum but the quality of the accretion process and therefore the quality of the diffraction dataset is also enhanced for crystals grown at low supersaturation in the mixed regime.

The above discussion is only valid when the mass transport is governed by diffusion, and it must be emphasized that this observed direct relationship between supersaturation and crystal quality only applies for crystals grown under a similar bulk transport scenario. When convective transport is present, as it is in typical ungelled terrestrial experiments, as well as in gravitationally noisy space experiments, the aggregation of growth units at the crystal faces is disturbed. Crystal sedimentation, as well as the very crystal growth process, provoke convective motion that homogenizes the concentration in the bulk of the solution. As soon as buoyancy enters the scene, convective flow destroys the symmetry of the system and makes thinner and shallower the concentration depleted zone around the crystal.

5. OTHER CHARACTERIZATION TECHNIQUES

It is conventional to define crystal characterization as the assessment of the morphology, structure, composition, physical properties and quality of crystals. Although as discussed above X-ray diffraction techniques are the most informative and most used techniques, many other characterization techniques are also used to understand the growth behaviour and the intimate structure of the protein crystals. We will review briefly some of them and we refer the reader to other works dealing with the subject in detail.

5.1. Optical microscopy

Most protein crystals are transparent to visible light and their size is appropriate for visualization with optical microscopy. Morphology, faceting, and some internal features such as fluid inclusions, sector zoning and growth fronts are currently described and analysed with the help of optical microscopy [47]. Growth rates at the resolution of microns have been measured with the help of time lapse microscopy, both on ground and in space [43,44].

High-resolution confocal laser scanning microscopy has been recently used for protein crystallization studies. This non-invasive technique allows in situ 3-D observation and because of the high resolution achieved it is possible to visualize step displacement and could be applied for growth kinetics studies [45]. Fluorescence microscopy has been used to measure the concentration of phosphor-labelled (ovoalbumin) impurities inside a tetragonal lysozyme crystal [46].

5.2. Atomic Force Microscopy

Among different scanning probe microscopies, atomic force microscopy has revealed itself to be the most powerful tool to investigate the crystal quality of protein crystals at microscopic and nanoscopic scales. The main advantage of this technique is that it can be used to study protein crystals in an aqueous medium similar to physiological conditions. The preparation of the sample is rather tricky. The use of AFM requires fixing a seed crystal beneath carbon fibres on a cover glass and equilibrating it with a solution of protein and salt. Then the glass slip is transferred to the fluid cell of the microscope designed for in situ studies of crystal growth. Fluid cells are available for most commercial models of AFM. The transfer of crystals from one solution to another one of different ionic strength must be carefully performed. Osmotic pressure and charge balance account for some reversible and irreversible cracking of the crystals. To avoid irreversible cracking it is recommended to transfer the crystal from a solution to another one of higher ionic strength [47].

Contact mode or tapping mode can be used. AFM studies have revealed the existence in protein crystals of all kinds of defects found in crystals of small molecules. Thus stacking faults, screw dislocations, vacancies, clusters, and impurities, have been revealed by different groups, mostly by McPherson and coworkers [48-53]. The images are very impressive as they offer faithful information at molecular and in a few cases (of huge molecules) submolecular information [54,55]. Two- and three-dimensional nuclei, steps, terraces, kinks, and other growth features have been observed with very high resolution. The use of the AFM at different scan sizes and rates and using appropriate time lapses between images makes possible the recording of growth kinetics for quantitative studies. It has been claimed that the use of the scanning process does not affect either the surface structure or the dynamics of crystallization. However we have to be cautious about quantitative use of AFM studies of growth kinetics. In fact, it has been demonstrated (see for instance reference [56]) that AFM can be used to manipulate defects very efficiently.

5.3. Electron microscopy and electron diffraction

Electron microscopy and electron diffraction have been used to explore macromolecular crystals to some extent. A review of structural studies of soluble proteins performed with the help of electron microscopy can be read in reference [57]. The main handicap for the use of electron microscopy in the characterization of protein crystals is sample preparation. Nevertheless, transmission cryoelectron microscopy is used today for getting low-resolution phase information for viruses and for complex assemblies. Also electron crystallography of two-dimensional protein crystals is currently used for structural studies at high (about 3 Å) resolution [58,59]. Freeze-etched and heavy-metal decoration techniques have been used by the group of S. Weinkauf to obtain beautiful and informative images of crystal growth faces with transmission electron microscopy [60,61] (Figure 11). Using replicas of crystal surfaces of lumazine synthase single crystals along with image analysis, they have been able to show clearly the existence of rotational disorder. The crystals of these huge molecules (approximately one megadalton) were well-faceted, rotational disorder defects and mismatches in stacking faults being the explanation for their poor diffraction quality.

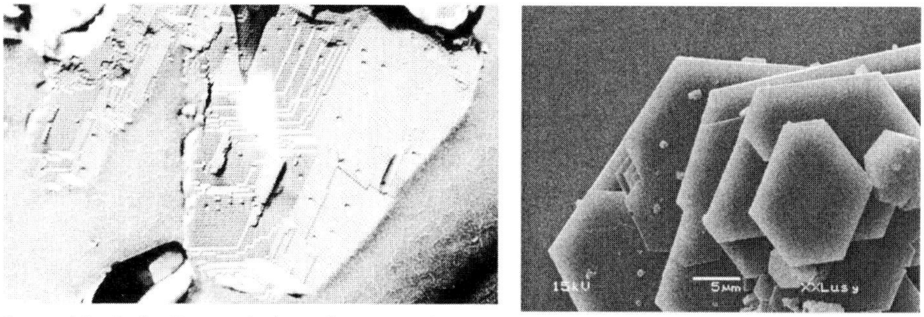

Figure 11: Left: Transmission electron micrograph of a silver replica of a 3-D lumazine crystal showing different defects and crystal features. Right: Scanning electron micrograph showing mismatch between crystal subvolume stacked on 001 basal plane of a lumazine crystal. Photographs kindly supplied by S. Weinkauf.

ACKNOWLEDGMENTS

We acknowledge the financial support of the Spanish Ministerio de Ciencia y Tecnología project numbers ESP2002-03397 and ESP2003-04759. Some of the results shown in section 3 have been obtained from the ESRF long term project LS1860 / SC1220. We also acknowledge Julyan Cartwright for revision of the manuscript.

REFERENCES

1. A. McPherson, Crystallisation of Biological Macromolecules, Cold Spring Harbor Laboratory Press 1999.
2. A.A.Chernov and H. Komatsu, Principles of crystal growth in protein crystallization, in: J.P. Van der Eerden and O.S. Bruinsma, (eds.), Kluwer Academic, Dordrecht 1995.
3. S. Phinet, D. Vivarès, F. Bonneté and A. Tardieu, Methods Enzimol. 368 (2003) 105.
4. C. L. Caylor, I. Dobryanov, S.G. Leamy, C. Kimmer, S. Kriminski, K.D. Finkelstein, W. Zipfel, W. W. Webb, B.R. Thomas and A. A. Chernov, Protein Struct. Funct. Genet. 36 (1999) 270.
5. J. Hirschler and J.C. Fontecilla-Camps, J. Crystal Growth 171 (1997) 559.
6. K. Provost and M.C. Robert, J. Crystal Growth 156 (1995) 112.
7. A. Jen and H. P. Merkle, Pharm.Res. 18 (2001) 1483.
8. A. L. Margolin and M. A. Navia, Angew. Chem. Int. Edit. 40 (2001) 2205.
9. B. Shenoy, Y. Wang, W. Z. Shan and A. L. Margolin, Biotechnol. Bioeng. 73 (2001) 358.
10. M. Ayala, E. Horjales, M. A. Pickard and R. Vazquez-Duhalt, Biochem. Biophys. Res. Commun. 295 (2002) 828.
11. M. S. Doscher and F. M. Richards, J. Biol. Chem. 238 (1963) 2399.
12. J. M. García-Ruiz, J.A. Gavira, F. Otálora, A. Guasch and M. Coll. Mat. Res. Bull. 33 (1998) 1593.
13. L. Z. Vilenchik, J. P. Griffith, N. St Clair, M. A. Navia and A. L. Margolin, J. Am. Chem. Soc. 120 (1998) 4290.
14. P. G. Vekilov and A. A. Chernov, Solid State Phys. 57 (2002) 1.
15. F. Otálora, J. M. García-Ruiz, L. Carotenuto, D. Castagnolo, M. L. Novella and A. A. Chernov, Acta Cryst. D**58** (2002) 1681.
16. W. F. Jones, J. M. Wiencek and P.A. Darcy, J. Crystal Growth 232 (2001) 221.

17. S. Veesler, S. Lafont, S. Marcq, J. P. Astier and R. Boistelle, J. Crystal Growth 168 (1996) 124.
18. A. Ducruix and R. Giege, in "Crystallisation of Nucleic Acids and Proteins: A Practical Approach" (A. Ducruix and R. Giegé, eds), p. 121, IRL Press at Oxford University 1999
19. J. M. García-Ruiz, Methods Enzimol. 368 (2003) 130.
20. F. Otálora, J. M. Garcia-Ruiz, J. A. Gavira and B. Capelle, J. Crystal Growth 196 (1999) 546.
21. E. Weckert, K. Hölzer, K. Schroer, J. Zellner and K. Hümmer, Acta Cryst. D55 (1999) 1320.
22. CCP4 Study Weekend 1999. Special issue of Acta Cryst. D55 (1999)
23. R. Fourme, A Ducruix, M. Riès-Kautt and B. Capelle, J. Synch. Rad. 2 (1995) 136.
24. F. Otálora, J. A. Gavira, B. Capelle and J. M. García-Ruiz, Acta Cryst. D55 (1999) 650.
25. F. Otálora, M. L. Novella, D. Rondón and J. M. García-Ruiz, J. Crystal Growth 196 (1999) 649.
26. O. Vidal, M. C. Robert, B. Arnoux and B. Capelle, J. Crystal Growth 196 (1999) 559.
27. H. M. Volz and R. J. Matyi, J. Crystal Growth 232 (2001) 502.
28. Z. W. Hu, B. R. Thomas and A. A. Chernov, Acta Cryst. D57 (2001) 840.
29. M. C. Robert, B. Capelle, B. Lorber and R. Giegé, J. Crystal Growth 232 (2001) 489.
30. T. J. Boggon, J. R. Helliwell, R. A. Judge, A. Olczak, D. P. Siddons, E. H. Snell and V. Stojanoff, Acta Cryst. D56 (2000) 868.
31. T. Salditt, Current Opinion Struct. Biol. 13 (2003) 467.
32. C. Münster, A. Spaar, B. Bechinger and T. Salditt, Biochim. et Biophys. Acta (BBA) – Biomembranes 1562 (2002) 37.
33. I. Dobrianov, C. Caylor, S. G. Lemay, K. D. Finkelstein and R. E. Thorne, J. Crystal Growth 196 (1999) 511.
34. K. Izumi, K. Taguchi, Y. Kobayashi, M. Tachibana, K. Kojima and M. Ataka, J. Crystal Growth 206 (1999) 155.
35. B. Capelle, Y. Epelboin, J. Hartwig, F. Otálora and V. Stojanoff, J. Appl. Cryst. 37 (2004) 67.
36. R. Fourme, A. Ducruix, M. Riès-Kautt and B. Capelle, J. Crystal Growth 196 (1999) 535.
37. M-C. Robert, B. Capelle and B. Lorber, Methods Enzimol. 368 (2003) 154.
38. J.M. Garcia-Ruiz, M.L. Novella, R. Moreno, J.A. Gavira, J. Crystal Growth 232 (2001) 165
39. J. M. García-Ruiz, J. Struct. Biol. 142 (2003) 22.
40. A.A. Chernov, J. Struct. Biol. 142 (2003) 3.
41. R. F. P. Grimbergen, E. S. Boek, H. Meekes and P. Bennema, J. Crystal Growth 207 (1999) 112.
42. A. A. Chernov, J. Crystal Growth 174 (1997) 354.
43. S. Koszelak, D. Martin, J. D. Ng, and A. McPherson, J. Crystal Growth 110 (1991) 117.
44. J. M. García-Ruiz and F. Otálora, J. Crystal Growth 182 (1997) 155.
45. P. Mühlig, Th. Klupsch, U. Kaulmann and R. Hilgenfeld, J. Struct. Biol. 142 (2003) 47.
46. K. Kurihara, S. Miyashita, G. Sazaki, T. Nakada, S. D. Durbin, H. Komatsu, T. Ohba and K. Ohki, J. Crystal Growth 196 (1999) 285.
47. F.J. López-Jaramillo, A.B. Moraleda, L.A. González-Ramirez, A. Carazo and J.M. García-Ruiz, Acta Cryst. D58 (2002) 209.
48. S. D. Durbin, W. E. Carlson and M. T. Saros, J. Phys. D: Appl. Phys. 26 (1993) 812.
49. S. D. Durbin and G. Feber, Annu. Rev. Phys. Chem. 47 (1996) 171.

50. A. J. Malkin, Yu. G. Kuznetsov and A. McPherson, J. Crystal Growth 232 (2001) 173.
51. P. G. Vekilov, Methods Enzimol. 368 (2003) 170.
52. A. J. Malkin, Yu. G. Kuznetsov, T.A. Land, J.J. De Yoreo and A. McPherson, Nat. Struct. Biol. 2 (1996) 956.
53. J. D. Ng, Y.G. Kuznetsov, A. J. Malkin, G. Keith, R. Giegé and A. McPherson, Nucleic Acid Res. 25 (1997) 2582.
54. A. J. Malkin, Yu. G. Kuznetsov, R.W. Lucas and A. McPherson, J. Struct. Biol. 127 (1999) 35.
55. C. Möller, M. Allen, V. Elings, A. Engel and D. J. Müller, Biophys J. 77 (1999) 1150.
56. S.-T. Yau, B. R. Thomas and P. G. Vekilov, J. Crystal Growth 232 (2001) 188.
57. M.J. Ellis and H. Hebert, Micron 32 (2001) 541.
58. E. Nogales, S.G. Wolf and K.H. Downing, Nature 391 (1998) 199.
59. A. Brisson, W. Bergsma-Schutter, F. OLing, O. Lambert and I. Reviakine, J. Crystal Growth 196 (1999) 456.
60. N. Braun, J.Tack, M. Fischer, A. Bacher, L. Bachmann and S. Weinkauf, J. Crystal Growth 212 (1996) 703.
61. N. Braun, J.Tack, M. Fischer, L. Bachmann and S. Weinkauf, Thin Solid Films 284/285 (2000) 270.

Crystal Growth – from Fundamentals to Technology
G. Müller, J.-J. Métois and P. Rudolph (Editors)

In-situ analysis of thin film growth using STM

U. Köhler, V. Dorna, C. Jensen, M. Kneppe, G. Piaszenski, K. Reshöft, C. Wolf

Experimentalphysik / Oberflächenphysik, Ruhr-Universität Bochum,
D-44780 Bochum, Germany

After a short survey of different microscopical methods for the in-situ analysis of thin film growth the use of scanning probe microscopy is discussed. Attention is mainly paid to scanning tunneling microscopy which is well suited to study the crystalline growth of metallic and semiconducting films with atomic resolution in a wide range of temperatures. The experimental details and limitations of such studies will also be discussed. The application of scanning tunneling microscopy for the in-situ observation of samples during MBE- and CVD-deposition is shown. The experiments show the development of a specific location on the surface during all stages of thin film growth.

Examples are shown for the key stages of epitaxial growth. For the simple case of homoepitaxial growth the STM-data are complemented by kinetic Monte-Carlo simulations which mimic the atomic diffusion behavior during growth in the computer.

STM-examples will be shown for different diffusion events of atoms and small clusters on the surface. Homogeneous nucleation of islands with different shapes determined by the growth temperature and the local bonding is visualized and compared to inhomogeneous nucleation at surface defects or at specific sites of the substrate reconstruction.

The influence of a step edge barrier on the roughness during growth will be shown (kinetic roughening). After the supply of material is stopped, lateral and vertical ripening processes which lead to a new equilibrium of the systems island size distribution and layer roughness are imaged. For two-component systems the intermixing of the deposited material with the substrate resulting in the growth of an alloy is shown.

1. INTRODUCTION

Various microscopical methods are routinely used to analyze the structure of thin films grown on single crystal substrates. Usually only static states of the film are imaged after the actual growth process is finished ("post mortem"). Thin film growth, on the other hand, is a dynamic process quite often far from thermodynamic equilibrium. Therefore, "post mortem" experiments can only provide information on the kinetic processes which govern the growth only in a quite indirect way. By measuring characteristic features like island densities at systematically varied growth conditions (temperature or growth rate) and applying theoretical models, kinetic data can be extracted [1]. For any theoretical analysis it has to be assumed that the state of the thin film which is imaged at room temperature after growth is representative for the state directly during growth. Contrary to this in many systems the structure of the film changes immediately after the flux is stopped due to relaxation processes. Therefore, for many systems "post mortem" experiments do not yield an accurate image of the state of the actual growth processes, especially if the deposition is done at a very high or very low temperature (below room temperature).

The kinetics of thin film growth – and we will limit ourselves to epitaxial growth in a dilute environment – is usually studied with diffraction experiments. Powerful methods are electron diffraction with slow electrons (low energy electron diffraction, LEED), [2, 3] or fast electrons using grazing incidence (reflection high energy electron diffraction, RHEED) [4]. Especially RHEED can be used for in-situ analysis during growth because the grazing incidence of the electrons allows direct access to the substrate surface for simultaneous deposition. Monolayer oscillations in the intensity of the reflected electron beam are often used to determine the thickness of the deposited film. Diffraction experiments deliver reliable statistical information on the atomic periodicity on the surface, the roughness and average values and the distribution of lateral structure sizes. To extract information on the kinetic behavior models of the surface structure have to be set up and compared to the diffraction data.

Microscopy, when applied in-situ directly during growth, can give a much more direct view of the kinetic processes in thin film growth. One necessary prerequisite to study these processes is surface sensitivity of the microscopical method. Some methods which can be applied in-situ during growth are described in the following:

Field ion microscopy (FIM) is very successfully used to study the diffusion behavior of individual atoms or small clusters [5]. Because the substrate has to be prepared in form of an atomically sharp needle the substrate surfaces are restricted to very small areas. Only very stable materials and crystal faces are accessible by FIM studies because of the high electric field which is applied during imaging.

Using high energy electrons surface sensitivity is only possible in grazing incidence. Reflection electron microscopy (REM) is used (like RHEED) for in-situ growth studies [6]. It is capable to resolve atomic steps or lattice distortions but atomic resolution is not possible. The grazing incidence, on the other hand, leads to a strong distortion of the image. In the direction of the electron beam the image is typically compressed by a factor of 100.

Low energy electrons, which are intrinsically surface sensitive, are very difficult to handle in electron optics. A very elegant way around this problem is implemented in the LEEM, the low energy electron microscope [7]. High energy electrons are decelerated to an energy below 100eV just before they hit the sample and accelerated again before they enter the optics for image formation. This way, electron optics is done with high energy electrons but the interaction with the sample with low energy electrons. The normal incidence results in a non-distorted image. Because LEEM is a non-scanning microscopical method the speed of image acquisition is fast, usually video-rate. By using interference effects mono-atomic steps can be detected. Therefore LEEM is very well suited to study the nucleation and the kinetics of thin film growth as long as the lateral resolution of a few nm is sufficient.

Except for FIM the microscopy methods described above do not provide information on the kinetics of growth with atomic resolution. In the last 20 years a variety of scanning probe microscopy methods was developed which use different interactions of a scanning probe with the sample to scan the surface and create an image [8]. The most prominent ones are the scanning tunneling microscopy (STM) [9] and the scanning force microscopy or atomic force microscopy (AFM) [10]. STM uses the tunneling current as sensor for the distance between a sharp metal tip and the sample and therefore is limited to conducting surfaces. Atomic resolution is routinely achieved (at least on semiconductor surfaces) because of the very steep exponential decrease of the tunneling current with the tip-sample distance. AFM utilizes the force between a sharp tip and the sample as a sensor and is therefore also applicable on non-conducting surfaces. Atomic resolution is harder to achieve but is possible on a variety of materials. In this publication we intend to concentrate on scanning tunneling microscopy. Usually STM, like transmission electron microscopy (TEM) or scanning electron microscopy (SEM), is used to study thin film growth processes in the "post mortem" mode which means that the actual deposition is interrupted by stopping the flux of material and the substrate temperature is quenched to room temperature before the surface is imaged. In this paper we will show how STM can be applied in-situ, i.e. directly during growth, to obtain information on thin film growth kinetics. Material is deposited using molecular beam epitaxy (MBE) or deposition from the gas phase (chemical vapor deposition - CVD), while the surface is imaged in the STM to follow one specific location through different stages of epitaxial growth. Sequences of images directly show the kinetic processes of epitaxial growth. After a discussion of experimental details of in-situ applied STM we will show examples illustrating the different stages of epitaxial thin film growth in homo- and heteroepitaxy. A very detailed description of similar in-situ STM experiments on Si and Ge growth on silicon surfaces is given by B. Voigtländer [11].

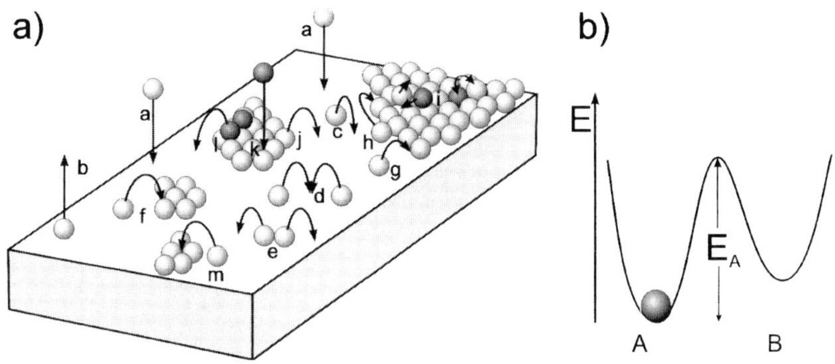

Figure 1. (a) Basic steps of epitaxial growth on an atomic level. a: adsorption on the bare substrate, b: desorption, c: diffusion, d: nucleation, e: decay of an instable nucleus, f: attachment to a stable nucleus, g: attachment to a step edge, h: edge diffusion and attachment to a kink site, j: detachment from an island, i: exchange process, k: adsorption on top of a nucleus, l: downward jump over a step edge, m: upward jump over a step edge. (b) For each jump event from site A to B an activation barrier E_A has to be overcome.

In case of deposition in a dilute environment like the direct beam of atoms in MBE and in low pressure CVD, growth is governed by only a few elementary steps taking place on an atomic scale [1] (see Figure 1). Species deposited onto the substrate (a) may diffuse on the surface (c) if the temperature is sufficiently high. At very high substrate temperature, there is a certain probability for the mobile species to desorb again (b) before being accommodated on the surface. The mobile species may reach an already existing step edge (g). Via edge diffusion (h) the shape of the step edge may rearrange. On the other hand, the diffusing species may also attach to an island nucleated in an earlier stage of the growth (f). If the density of mobile species on the substrate increases, two or more of them may meet to form a new nucleus (d). Below a critical size such a nucleus may be unstable and will decay before further atoms can attach (e). Also from bigger islands fragments may detach again (j). In this way the whole structure of the epitaxial layers can rearrange after the deposition has stopped. Larger islands and step edges can grow at the expense of smaller ones to minimize the total length of step edges on the surface. Atoms landing on top of an island can cross the edge of this island (l) and attach to the lower step edge. An extra energetic barrier (Ehrlich-Schwoebel-barrier) may inhibit this down-flow. Of course, also the reverse process may happen, in which a species moves across a step edge from a lower to a higher level (m). There may also exist a certain possibility for deposited atoms to exchange with substrate atoms. In the case of heteroepitaxial deposition this will lead to the formation of an alloy.

The probability for each of these processes is governed by an energetic barrier E_A to jump form site A to B (see Figure 1b). A precise determination of the energetics of these elementary steps in epitaxial growth is necessary to understand and control thin film growth on an atomic level. The examples shown in section 3 will illustrate that it is possible to image the kinetics of most of these elementary steps of thin film growth directly with the STM.

2. EXPERIMENTAL

Scanning tunneling microscopy is (like any scanning probe microscopy) a method where an image is acquired by scanning a solid object in very close proximity to the sample (see Figure 2). The distance between the surface imaged and the foremost atom of the tip (the tunneling gap) is typically 0.5 –1 nm. This means that the area which is actually imaged is not accessible from on top but only from the side. In the case of MBE from an effusion cell in-situ deposition of material is therefore only possible under an angle of up to 45° to the surface plane (see Figure 2).

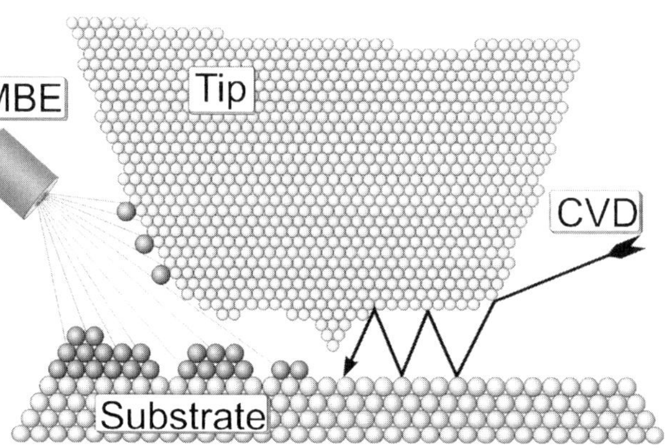

Figure 2. Model of the tip-substrate region during in-situ deposition. In MBE (left) material is deposited only in a direct line of sight. The region which is actually imaged is shaded by the scanning tip. When a gas is used as deposition source (CVD, right), material can reach the area underneath the tip via multiple collisions at tip and substrate.

Figure 3 shows the arrangement of a Knudsen-cell evaporator used in our experiments relative to the scanning tunneling microscope. The microscope which is especially constructed for in-situ growth experiments has an open design which allows a direct line of sight between the tip-substrate area and up to three MBE sources. Here the flux from the effusion cell reaches the substrate surface under an angle of 20° to the surface plane. Even for this geometry the area, which is actually scanned, is shielded effectively by the STM-tip. The deposition rate inside the field of view of the STM is lower than outside. If the tip is in a fixed position above the surface in tunneling distance the deposition rate drops to zero "behind" the tip. In this case a sharp tip shadow in the local coverage is found, which actually can be used to get information on the shape of the tip [11].

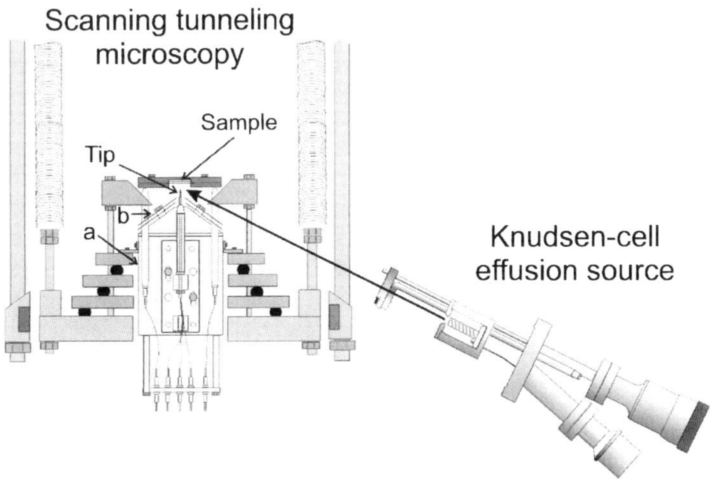

Figure 3. Geometrical arrangement of scanning tunneling microscope and evaporator for in-situ growth studies. A direct line of sight between evaporation cell and STM-tip is necessary. An invar housing (a) and additional heat shields (b) protect the piezo-scanner.

Figure 4 shows a zoom-out from an area where iron was deposited on a W (110) surface while the area in the dashed rectangle was scanned by the STM. In this case the moving tip created a blurred shadow image of the STM-tip with a penumbra where the coverage gradually increases. In the region which is scanned during deposition (dashed rectangle) the coverage is inhomogeneously distributed due to these shading effects. This complicates a later analysis of the images to obtain kinetic data on the epitaxial growth.

On the other hand, the inhomogeneous coverage distribution in the penumbra region can also be utilized to create a wedge-like coverage distribution on a nano scale. An STM image of this region can give an easy overview of behavior of a thin film system with increasing coverage starting from the bare substrate surface in the central shadow region. Figure 5 shows an example of the different stages of Nb-growth on Fe (110) at 300K. The image shows the evolution of the Nb-layer from the phase of nucleation in the submonolayer range in the upper right up to three layers of Nb in the lower left.

One way to avoid the problem of an inhomogeneous coverage distribution is to withdraw the STM-tip a few hundred nanometer between the acquisition of images (by the z-piezo-element of the scanner) to open a direct line of sight to the MBE-source, and to close the shutter of the source when an image is acquired. This way, the STM-images show "snapshots" of a homogeneous coverage which increases from image to image.

Gaseous species used in CVD (precursor gases) usually have a sticking probability smaller than unity and therefore can reach the region directly underneath the tip via multiple collisions. Nevertheless, also in CVD the coverage is inhomogeneously distributed due to the decrease of the local pressure of the precursor-gas underneath the tip. Therefore, the retraction of the tip between the acquisition of images is also advisable in the case of CVD.

The retraction of the STM-tip from the growing film while the material is actually deposited will also help to reduce another problem related to the presence of the STM-tip during growth. Due to the proximity of the tip to the surface an extremely high local electric field (typically 10^9 V/m) and current density (10^{10} A/m^2) are present. These extreme conditions may influence many steps in epitaxial growth like surface diffusion, nucleation or ripening processes. Such an influence of the tip on the surface, which is undesirable in the case of growth studies, can actually be used intentionally for lithographic purposes creating nanostructures with the STM [12].

Although the retraction of the STM-tip during the actual deposition will minimize the influence of the tip on the growth process, it has to be carefully checked for each experiment if the growth behavior found in the in-situ STM study is identical to the undisturbed one.

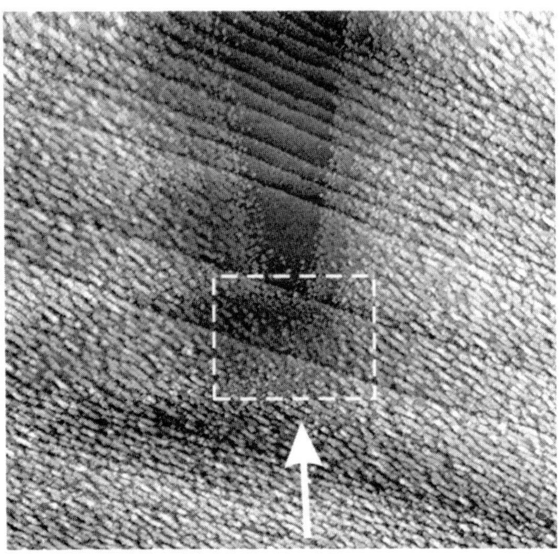

Figure 4. Shadow image of the STM-tip in a Fe-layer on W (110) (image size: 400nm x 400nm). The dashed rectangle marks the area where the tip was scanning during the deposition, the arrow marks the direction of deposition. The movement of the tip causes a penumbra with a gradual increase of the coverage from a completely shielded area in the middle to the edge of the shaded region.

To obtain epitaxial thin films of good crystalline quality an appropriate substrate temperature has to be chosen. For semiconductors a few hundred centigrade above room temperature are necessary, for metals like iron or copper room temperature is already sufficient. Therefore a microscope for in-situ thin film growth studies has to be capable to image surfaces at variable temperatures. Taking STM-images from a sample which is a few hundred centigrade hot requires a special design of the instrument.

The samples have to be heated without disturbing the measurement of the small tunneling current. For semiconductors this can be done using resistive heating by direct current flow through the sample [11]. Often batteries are used to minimize electrical noise. The heating current induces a voltage drop across the sample which has to be compensated to provide the correct bias voltage (voltage between tip and sample). For details, see [11]. For metal surfaces a radiative heating using a tungsten filament behind the sample can be used.

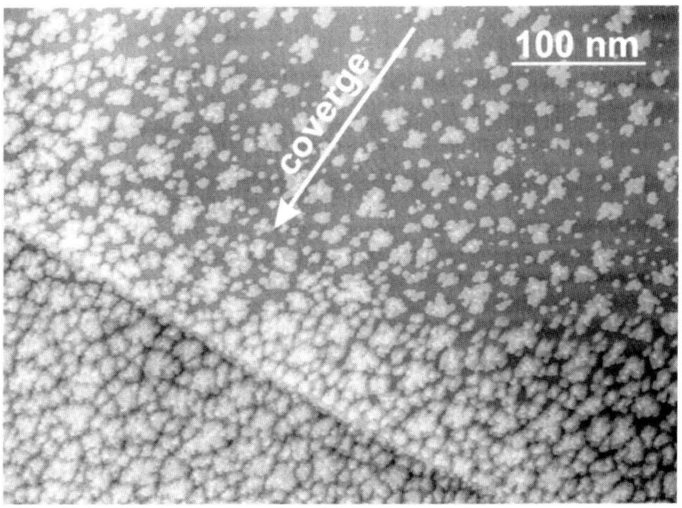

Figure 5. Coverage "nano-wedge" in the penumbra of the tip-shadow of Nb-growth on Fe (110) at room temperature

A sample at elevated temperature will heat the STM assembly radiatively as does the MBE-source pointing at the STM. The piezo-elements of the STM, on the other hand, have to stay at a temperature below 200°C to avoid depolarization. Therefore the scanning device has to be designed in a way that protects the piezo-scanner from this unwanted radiative heating. In the STM design used in our group this is done by encapsulating the scanning unit by an invar-housing and additional heat shields between the hot sample and the scanner (see Figure 3).

A second problem caused by the radiative heating of the scanning unit is thermal drift in the STM-images. A thermal gradient due to the heating will cause different parts of the sample holder and the scanning unit to expand differently leading to a lateral and/or vertical displacement of the STM-tip relative to the sample. This effect can be minimized by a symmetric construction of the whole STM assembly and the use of materials with similar thermal expansion coefficients or by special, expansion compensated designs [13,14]. In our experiments any residual thermal drift in the microscope at elevated sample temperature is corrected on-line during the scan process and numerically afterwards in the data.

The atomic resolution of STM comes along with a relatively slow image acquisition speed. While scanning the surface the scanner has to move the tip vertically to follow the surface contours. The speed of this movement is limited to an up-down frequency of 1000 Hz up to 10 kHz. This results in an acquisition time between five seconds for smaller images and several minutes for larger ones. Therefore, in-situ growth experiments have to be designed to allow following of the kinetic processes by choosing a low deposition rate (typically a few atomic layers per hour) of by lowering the substrate temperature to a scale where diffusion processes are effectively slowed down. In our experiments, we have the capability to perform in-situ STM experiments in a temperature range between 10K and 1100K. The temperature range below room temperature allows e.g. the study of the diffusion of individual metal atoms or small clusters.

On the other hand, there exist several designs in the literature for high speed STMs [13,15]. For small scanning areas and flat samples rates up to 20 images/second have been reached. Since during thin film growth the surface usually is not atomically flat this option is not available for the experiments presented here.

An alternative approach to study faster diffusion events with the STM is to avoid the acquisition of complete images. Instead the STM-tip is positioned above a diffusing species. When this species performs a diffusion jump a feedback loop is used to dynamically follow with the STM-tip. Using this "atom tracking" method dynamic events up to a 1000 times faster than with conventional STM can be measured [16]. The disadvantage is the continuous interaction of the STM-tip with the diffusing atom in contrast to the short time interaction in the case of conventional scanning. A possible influence of the microscope on the kinetic events in the growing layer is, therefore, a much more serious problem.

The experiments shown in this paper were all performed in an ultrahigh vacuum environment at a base pressure $p < 10^{-10}$ mbar necessary for preparation of clean substrate surfaces. The tunneling current was set to a value $\approx$1nA in all images shown in this article.

3. EXAMPLES ILLUSTRATING EPITAXIAL GROWTH

In the following sections we like to present illustrative examples of thin film growth processes studied in-situ with the STM. By taking sequences of STM images a specific location of the substrate surface is followed through the different stages of growth. The acquisition of these sequences takes between 30 minutes up to 24 hours, a sequence may contain a few hundred individual STM frames. When displayed as a time lapse movie (a factor of 100-1000 faster than during acquisition) the observer can gain a quite direct impression of the epitaxial growth behavior. In this publication bond to paper only the parallel display of few selected frames from these longer sequences is possible to show characteristic kinetic changes from image to image.

3.1. Surface diffusion

The mobility of deposited atoms on the surface is an essential prerequisite for the growth of crystalline ordered thin films. The smallest unit to diffuse on the surface, a monomer, is usually an individual atom (see Figure 1). The diffusion of monomers is the basic form of mass transport on the surface. The energetics of their diffusion rules the island density and form and therefore the whole following growth process.

At temperatures where growth proceeds in a crystalline form monomer diffusion is too fast to be imaged by the direct acquisition of sequences of images. To image hopping events of individual atoms the mobility has to be slowed down drastically to fit the slow acquisition speed of STM. The hopping rate h is connected to the activation barrier E_A according to:

$$h = \nu \cdot e^{-E_A/kT} \tag{1}$$

where ν is an attempt frequency for the event of the order of 10^{13} Hz, k the Boltzmann-constant and T the temperature.

For the growth of semiconductors films E_A is of the order of 1eV; therefore a deposition temperature close to room temperature is suitable to slow down the diffusion events to a time scale of a jump in a few seconds which is slow enough for STM. For most metals the diffusion barriers (for quantitative data see [5]) are such that temperatures below room temperature are necessary to slow down the diffusion events sufficiently.

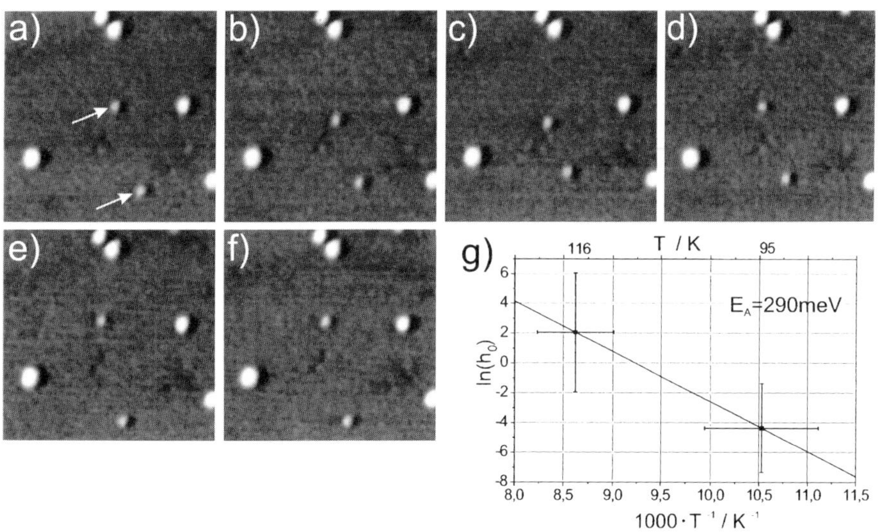

Figure 6. Diffusion of iron atoms on Fe(110) at 95K. The two atoms marked in (a) are mobile (see b - f) whereas larger clusters containing more than two atoms are immobile at 95K. The Arrhenius-plot in (g) for the hopping rates at different temperatures gives an estimate for the diffusion energy E_A.

The homoepitaxial growth of iron on Fe (110) is a prototypical system for the growth of bcc (110) planes. The sequence of STM images in Figure 6 shows the diffusion of two individual iron atoms (marked in Figure 6a) at 95K. From (a) – (f) these atoms perform a random walk motion. The hopping rate h can be determined from these experiments. Larger clusters of more than one iron atom are immobile at 95K. Experiments at different temperatures yield, according to equation (1) an estimate for the diffusion barrier when plotted in form of an Arrhenius-plot (Figure 6g). In the case of the monomer diffusion on Fe (110) an activation energy E_A of 290meV is found.

When two or more monomers meet during their diffusion on the surface at a random location homogeneous nucleation can take place. Stable clusters are formed, which, depending on temperature, may stay fixed on the substrate or perform a random walk motion by themselves. In most theoretical descriptions [1] even a mobility of dimers is neglected. Figure 7 shows on the other hand, that on Fe (110) clusters containing a few iron atoms are mobile at 116K. In the figure different events are displayed. In the sequence a) two mobile clusters meet and form a bigger cluster which afterwards is immobile at 116K. During further deposition this cluster can represent a basis for the growth of a larger island (see section 3.2).

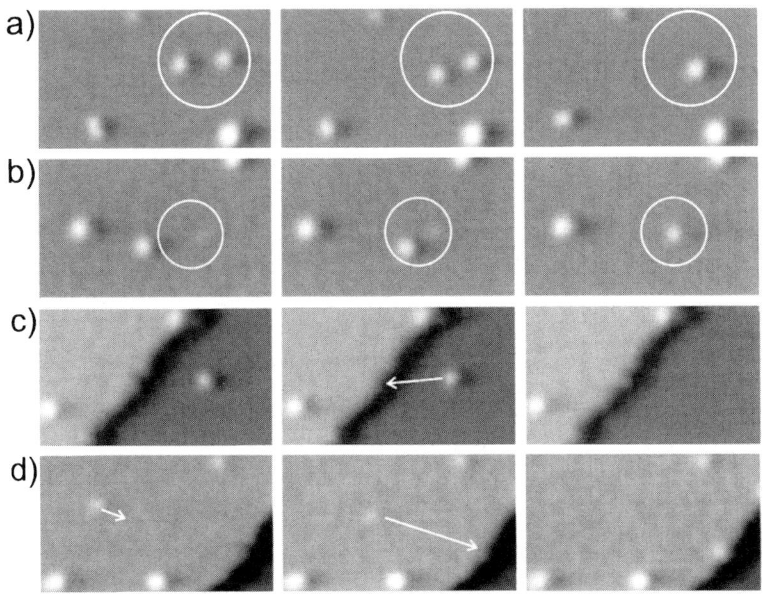

Figure 7. Diffusion of iron clusters on Fe (110) at 116K. The frames are selected to show characteristic elementary steps. (a) Coalescence of two mobile clusters to form a bigger cluster which is immobile at 116K. (b) Trapping of a mobile cluster at an impurity of the substrate. (c) Attachment to the lower side of a step edge. (d) Attachment to the upper side of a step edge.

In Figure 7b a mobile iron cluster is trapped by a defect on the substrate, probably an impurity due to incomplete substrate cleaning. Also this cluster can grow during further deposition forming the basis for inhomogeneous nucleation (see section 3.4). Figure 7c and (d) show two other inhomogeneous nucleation events. In (c) a mobile cluster is the attached to the lower side of a step edge (event (g) in Figure 1). At 116K there is no further mobility along the step edge. In Figure 7d an analogous attachment to the upper side of a step edge is shown. Due to the presence of a energetic barrier the material can not cross the step edge to reach the more stable sites at the lower side of the step edge (see section 3.3).

Systematic experiments as a function of temperature can provide at least qualitative information on these basic steps in epitaxial growth. In simple cases, as shown above for the monomer diffusion, also quantitative information on the underlying energetics can be obtained [17].

3.2. Nucleation and island growth

Surface diffusion has a crucial influence on the density of stable nuclei [1]. The mean diffusion length before a stable nucleus is reached increases with temperature, leading to a decrease in island density. In Figure 8, again for the system Fe / Fe (110), the development of the island density and shape is shown for increasing temperature of deposition.

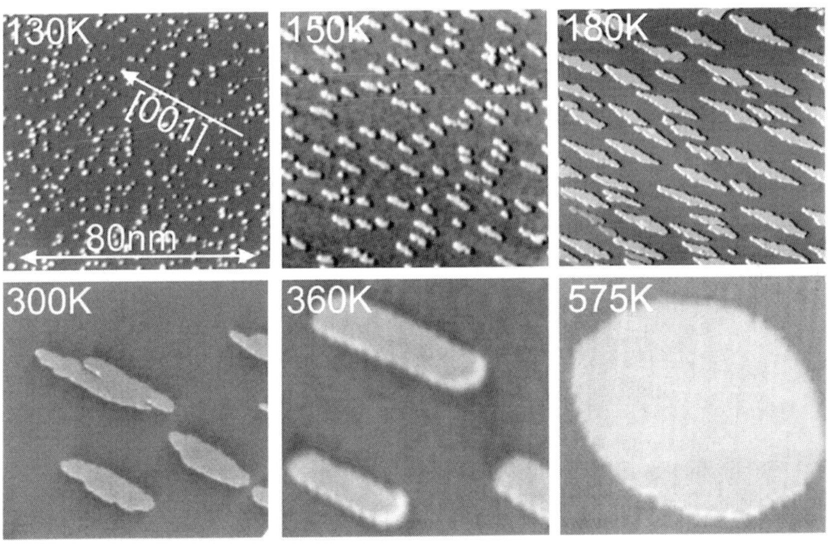

Figure 8. Shapes of Fe islands on Fe (110) for various growth temperatures. The island density decreases with increasing temperature. Ordered island growth starts above 130K.

At 130K stable islands consist of only a few atoms with random shapes. At 150K the islands elongate anisotropically along the [001] direction. This elongation becomes more characteristic at 180K and 300K. The shape of the islands is induced by the crystallographic symmetry of the (110) substrate via anisotropic bonding of atoms to the islands and via anisotropic diffusion [18,19]. In the 180K and the 300K image the edges of the islands are rugged whereas at 360K the islands develop straight edges parallel to the [001] direction and shorter perpendicular ones. This points to an insufficient edge diffusion (see event (h) in Figure 1) up to 300K (see further discussion below), hindering kinetically the development of an equilibrium island shape, whereas at 360K thermodynamically stable edges of the islands can develop. The stability of island edges during growth is determined by the step edge energy which is strongly dependent on the crystallographic direction. A Fe (110)-island grown at

575K (see Figure 8), on the other hand, develops a more round appearance. At this high growth temperature the differences in the thermodynamic stability of the different step edges are negligible compared to the thermal activation. The shape of the island is determined by its desire to minimize its length of perimeter which is smallest for a circle.

Two extreme cases of the influence of edge diffusion on island shape are shown in Figure 9. In (a) the homoepitaxial growth of copper islands on Cu (111) at room temperature is displayed. Here the diffusion of atoms along the step edge is sufficient in every moment for the island to maintain its equilibrium hexagonal shape. When two islands coalesce (see markers in Figure 9a) the resulting larger islands immediately rearrange to a new hexagonal shape. When displayed in form of a movie sequence the behavior of the copper islands reminds of the behavior of water droplets. The copper islands grown on W (110) at room temperature shown in Figure 9b, on the other hand, have a strongly hindered edge diffusion [20]. When a Cu-monomer attaches to the island edge it is immobilized and basically stays at the position of arrival. This way fractal island shapes develop (diffusion limited aggregation). In the case of Cu growth on W (110) the fractal behavior of the island is additionally influenced by the strain in the pseudomorphically grown Cu layer due to the misfit in lattice constant and crystal structure between the film and the substrate [20].

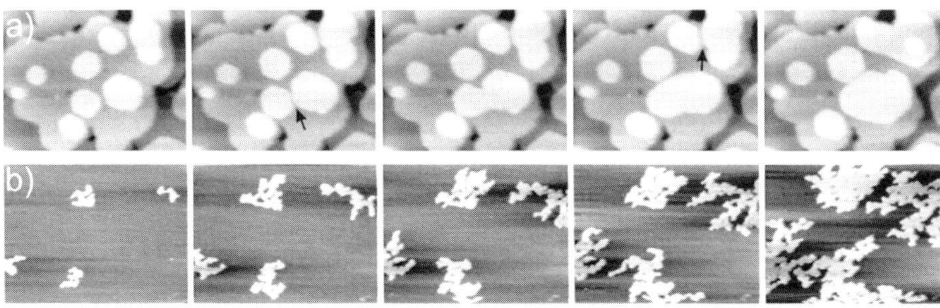

Figure 9. The influence of edge diffusion on the island shape. In the case of homoepitaxial growth on Cu (111) at room temperature (a) a fast diffusion of atoms along the edges of islands leads to a hexagonal rearrangement of the island shape immediately after coalescence (see markers). A hindered edge diffusion, as in the case of copper growth on W (110) at room temperature (b), leads to fractal islands.

3.3. Layer-by-layer-growth and kinetic roughening

For technological purposes it quite often is desirable to grow smooth layers. In the layer-by-layer or Frank-van der Merwe [21] growth mode the lower layer is completed before nucleation of the next layer starts. When the growth is close to thermodynamic equilibrium (low deposition rate, high temperature), the growth mode is determined by the free energy of the surfaces. The thin film wets the substrate completely when the surface free energy of the film material is lower than the one of the substrate, in the opposite case the deposited material forms 3D-islands, Vollmer-Weber growth mode [22].

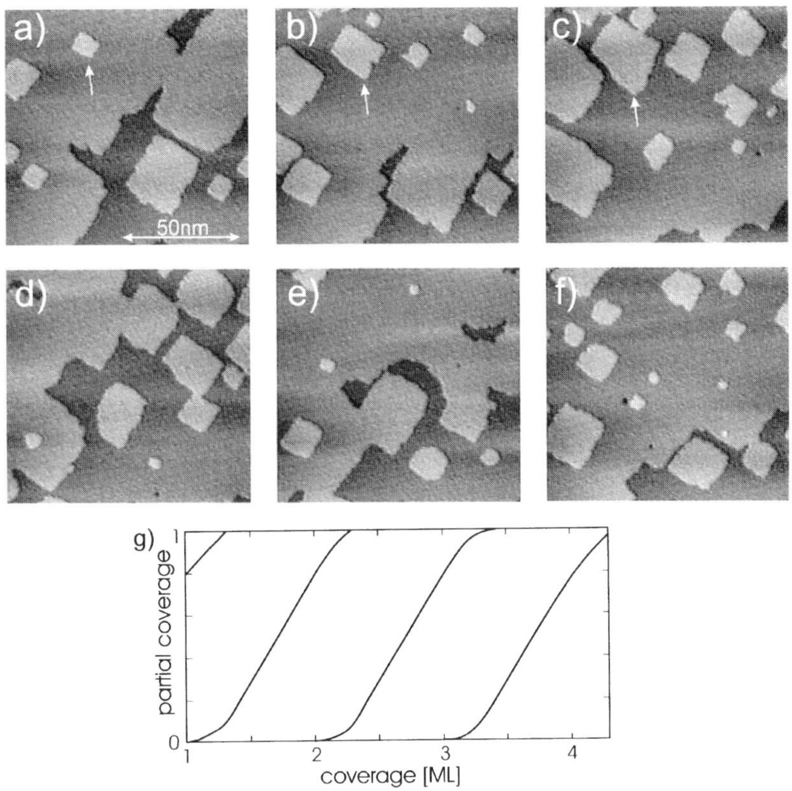

Figure 10. Layer-by-layer growth of iron on Fe (100) at 690K. (a) - (f) shows frames from a sequence of STM-images during the deposition of one monolayer of Fe. The island which nucleates in (a) (see arrow) grows in (b) and (c) and connects to neighboring islands to complete the layer. In (f) the layer is closed and nucleation of the next layer starts. (g) shows a statistical analysis of the layer distribution of the sequence. The partial coverage of one layer increases up to 95% before the next layer starts growing, indicating nearly perfect layer-by-layer growth.

Figure 10 shows as an example for nearly perfect layer-by-layer growth the homoepitaxial growth of iron on Fe (100) at 690K. The island shape is square according to the four-fold symmetry of the bcc (100) substrate. In Figure 10a the marker points to a monolayer island whose lateral spreading is seen in (b) and (c). In (d) this island coalesces with the neighboring islands to complete the layer (e). A massive nucleation of the next layer sets in only after the underlying layer is nearly completely filled. The quality of the layer-by-layer growth can be quantified by plotting the partial coverage for each layer as extracted from the STM data versus the total coverage, see Figure 10g. The graph shows that the lower layer is completed to $\approx$ 95% before the next layer starts. A growth mode like the one shown in Figure 10 results in strong oscillations of the reflected intensity in RHEED experiments.

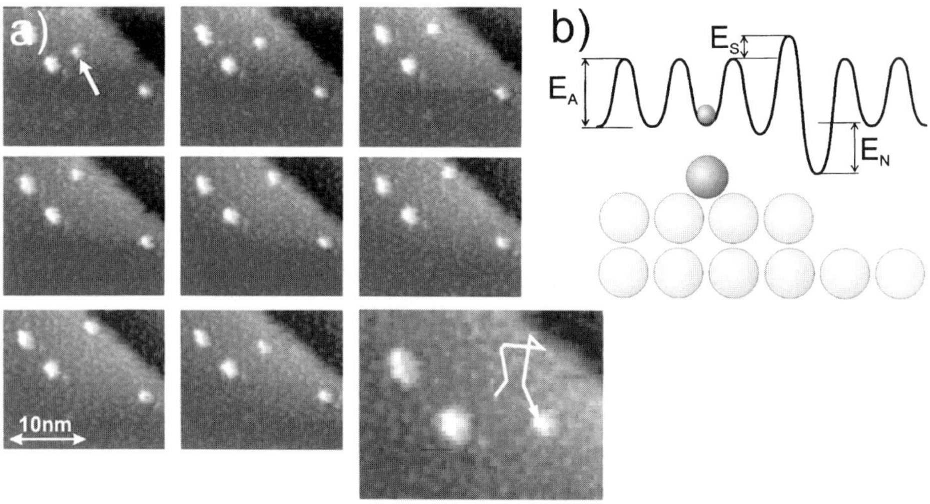

Figure 11. The influence of the Schwoebel-Ehrlich energy barrier on the diffusion in the vicinity of a downward step edge. (a) Diffusion of a holmium clusters on W (110). When the mobile cluster approaches the step edge it is trapped for a while and then repelled. In the enlarged image at the end of the sequence the previous path of the cluster is sketched. (b) Energy contour for an atom when crossing a step edge. Additionally to the normal diffusion barrier E_A an extra barrier, the Schwoebel-Ehrlich E_S barrier is present. E_N represents the bonding to the step edge on the lower side.

Even if thermodynamics of a thin film system is in favor of layer-by-layer growth, kinetics may prevent the formation of a smooth layer. An essential prerequisite for the layer-by-layer growth mode is that atoms landing on top of an island can cross the edge of the island to contribute to the growth of the lower level. In many systems, on the other hand, there exists an energetic step edge barrier additionally to the usual diffusion barrier which prevents atoms from crossing the step edge, the Schwoebel-Ehrlich barrier [23]. Figure 11b shows a schematic sketch of the potential energy curve in the vicinity of a step edge. When a diffusing atom approaches the step edge from the lower side it senses an extra bonding to the step edge E_N. When it approaches the step edge from the upper side the Schwoebel-Ehrlich barrier provides an extra obstacle for the atom to cross the edge. Figure 11a shows the effect of this barrier on a diffusing holmium cluster on W (110) at room temperature. In the first image of the sequence a mobile cluster is marked. When the cluster approaches the step edge during its random walk diffusion it is repelled. In the last (larger) image of the sequence the path of the mobile clusters is sketched.

Whether a Schwoebel-Ehrlich barrier hinders the development of a smooth film depends on the height of this barrier relative to the normal diffusion barrier and the thermal energy. For the homoepitaxial growth of iron on Fe (110) at room temperature the step edge barrier suppresses very effectively the downward flux of atoms. The sequence in Figure 12 shows that the iron layer on Fe (110) roughens at room temperature although the thermodynamic conditions would promise layer-by-layer growth. In this case the roughening is kinetically induced.

A quantitative analysis shows that the layer distribution is roughly described by a Poisson-distribution [18], which mathematically describes the case where an interlayer mass transport is completely forbidden. The islands in higher layers become increasingly anisotropic (see Figure 12d – f) up to a ratio of length to width of ≈10. With increasing total coverage the side walls of the islands convert into regular step trains. Further growth reduces the average step separation until characteristic facets are formed (for details see [18,19]) which remain stable during further growth.

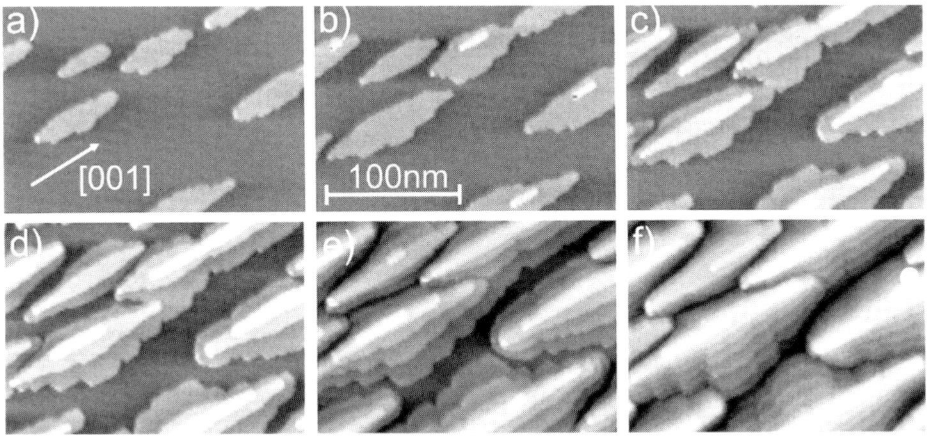

Figure 12. Statistical growth of iron on Fe (110) at room temperature. In-situ STM sequence showing the nucleation of islands in higher levels before the lower level has closed. In this stage of growth the film increasingly roughens.

One way to better understand the atomistics of the thin film growth is to model the behavior of the diffusing atoms by a kinetic Monte Carlo simulation. For the case of the growth of iron on Fe (110) the diffusion barriers for all possible events were calculated using simple models (for details also see [18,19]). The STM sequences showing directly the development of a specific region of the thin film with increasing coverage offer the unique opportunity to directly compare the results of a kinetic Monte-Carlo model with reality. In Figure 13a - c the development of a set of iron islands during growth at room temperature is shown. A digitized version of the first image of the STM sequence is taken as the starting situation of a kinetic Monte Carlo simulation (Figure 13d). Figure 13e and (f) show how these islands would grow within the limits of the theoretical model. By a comparison of the characteristic features of the grown layer in the STM sequence and in the kinetic Monte Carlo model the validity of the underlying energetics of the model can be judged.

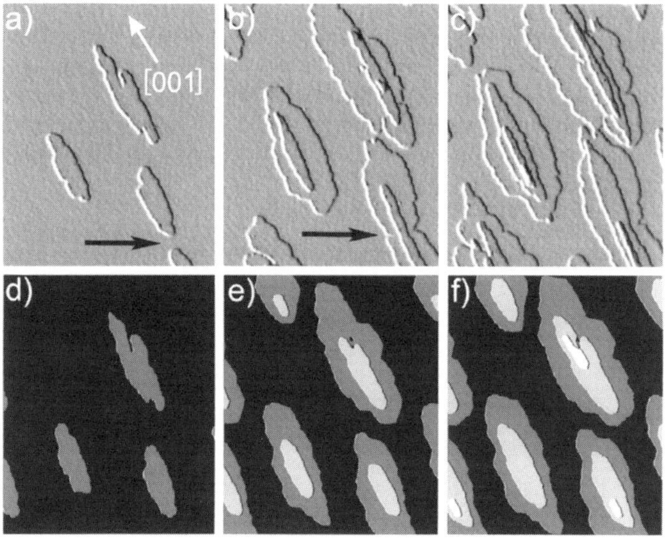

Figure 13. Comparison of a STM-sequence ((a) – (c)) showing the growth of iron islands on Fe (110) at room temperature with results of a kinetic Monte Carlo simulation ((d) – (f)). A digitized version (d) of the first image of the STM sequence (a) is used as the starting situation for the computer simulation.

3.4. Inhomogeneous nucleation

Often nucleation of islands on the surface does not take place homogeneously at random sites but at preferred sites of the surface. Every real substrate surface contains imperfections which provide such sites for inhomogeneous nucleation.

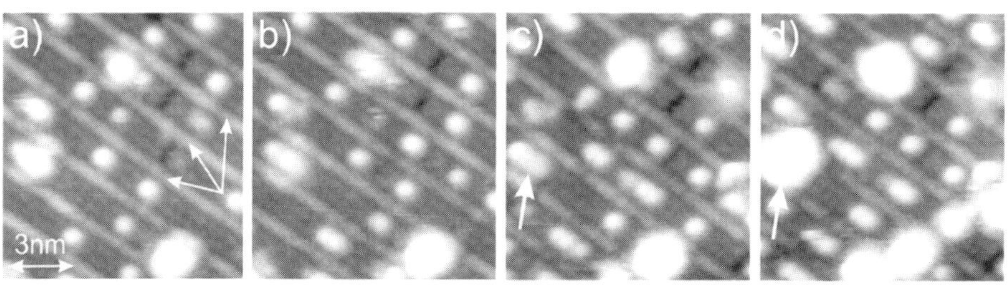

Figure 14. Nucleation of iron on InAs (001) at room temperature. The space between the indium rows of the (2x4)-reconstruction of the substrate (see marker in (a)) provides sites for inhomogeneous nucleation of 2D-islands. In later stages of growth (see marker in (c) and (d)) some of these 2D-islands convert into 3D-islands, the final form in which the iron layer grows on InAs (001).

If the density of nuclei which would be present due to the intrinsic homogeneous nucleation is lower than the density of these special sites the nucleation behavior is dominated by inhomogeneous nucleation.

A reconstruction of the substrate surface itself may provide sites for inhomogeneous nucleation. Figure 14 shows the growth of iron on InAs (100) in the submonolayer range at room temperature. Nucleation takes place only in the open space between the indium rows of the reconstruction [24]. Therefore, the lateral separation of the islands in the submonolayer range is not determined by kinetics, but fixed due to the reconstruction. In later stages of the growth of iron on InAs (100) 3D-islands emerge from the submonolayer nuclei which are also located only in between the In-rows of the substrate.

An imperfection which is unavoidably present on every substrate surface are step edges. Around each step edge a zone exists where no islands nucleate and instead the deposited material is attached to the step edge. The width of this denuded zone is of the order of the average separation of the islands in the case of homogeneous nucleation. If the separation of the edges is lower than the width of this denuded zone the surface grows solely by an attachment of material to the step edges.

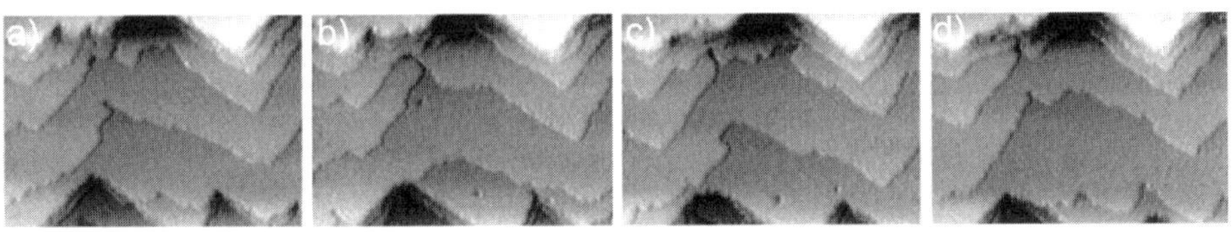

Figure 15. Successive images during step flow growth of iron on Fe (100) at 690K (Image width 100nm)

Figure 15 shows homoepitaxial growth of iron of a Fe (100) substrate at 690K in a region with a high density of monolayer steps. No islands nucleate, just the step edges move in the downward direction. The growth proceeds by step propagation (step flow). This mode of growth is very hard to analyze by diffraction experiments because the average state of the surface does not change. RHEED would not show intensity oscillations which could be utilized as a monolayer counter as in the case of island nucleation.

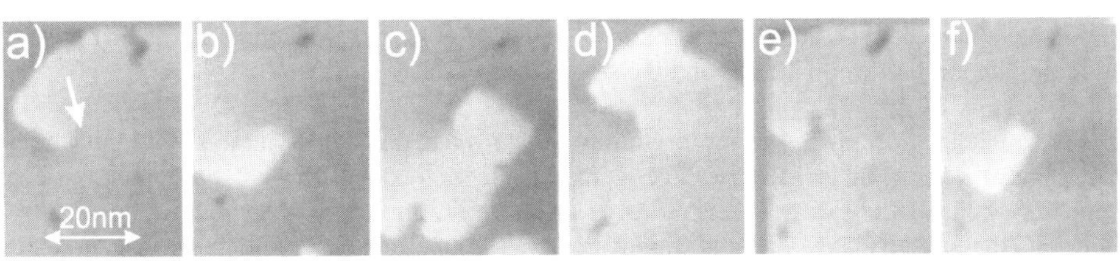

Figure 16. Iron growth at a screw dislocation (see marker in (a)) on Fe (100) at 740K. The coverage increases by one monolayer from (a) – (f).

Dislocations which intersect the substrate surface provide another type of sites for inhomogeneous nucleation. In Figure 16 the growth of iron at a screw dislocation on a Fe (100) substrate is shown. The thickness of the iron layer increases by one monolayer from (a) – (f). As in the case of step flow growth no islands nucleate in the vicinity of the dislocation. Growth is accomplished by a circular motion of the step edge originating at the intersection point of the dislocation line with the surface.

3.5. Relaxation processes after growth

During growth the continuous deposition of material drives the system to a non-equilibrium situation. When this deposition is stopped but the substrate is still kept at elevated temperature the system proceeds to a new equilibrium state on a time scale dependent on temperature. Lateral and vertical relaxation processes generally enlarge the average size of the structures in the thin film and flatten the film when a kinetically induced roughness during growth has developed.

Figure 17 shows the nucleation of islands and the subsequent Ostwald ripening [25] of a Si(111)-film. The deposition was done from the gas phase using disilane (Si_2H_6) as precursor. The high noise in the images is induced by the mobility in the layer and the presence of co-diffusing hydrogen. In Figure 17b three silicon islands nucleate from which two dissolve in (c) and (d). The material which was contained in these islands contributes to the remaining island and to the step edge visible in the image. If a larger assembly of islands is present the average island size increases with annealing time this way. The physical reason for this behavior is the difference in the detachment rate of atoms (see event (j) in Figure 1) from a small island with a large edge to volume ratio compared to a larger island with a smaller edge to volume ratio. For very long annealing times bigger islands will also decay and all material will contribute to the growth at step edges.

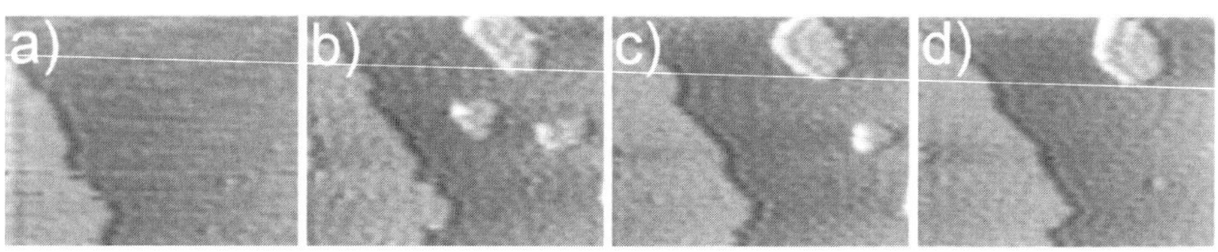

Figure 17. Island coarsening of silicon islands on Si (111) at 700K. Two of the three island which have nucleated during deposition in (b) decay afterwards. Their material feeds the large island in the top part of the image and the step edges.

Vertical relaxation processes require an interlayer mass transport which is only possible if the thermal activation is sufficient to overcome the Schwoebel-Ehrlich barrier at the step edges. In Figure 18 the vertical relaxation of an iron film on Fe (110) is shown. The temperature was increased by 25K after the flux of iron was stopped in order to enhance the tendency of the layer to relax vertically. In Figure 18a the iron film extends vertically across four atomic layers. During the period of annealing the islands in the uppermost layer dissolve. Simultaneously the islands in the second layer grow, showing that indeed a downward flux is present and the Schwoebel-Ehrlich barrier, which very effectively prevents an interlayer diffusion in the system iron on Fe (110) at room temperature (see Figure 12), can be overcome at

425K. The islands in the second layer, which grow from the material supply of the decaying uppermost layer, coalesce and rearrange. Step diffusion additionally leads to an annihilation of kink sites and to rounding of the island shapes. A longer annealing period than the one shown in Figure 18 results in a flat surface consisting of only two layers.

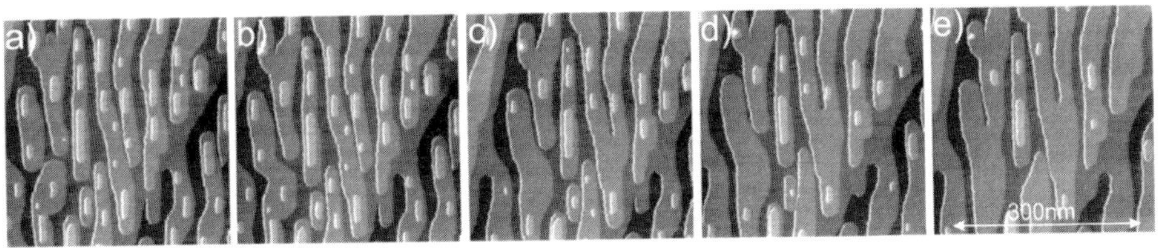

Figure 18. Coarsening processes in a Fe-layer homoepitaxially grown on Fe (110) at 400K and subsequently annealed at 425K. The flattening of the film shows that the Ehrlich-Schwoebel barrier can be overcome at higher temperature.

The example shown in Figure 19 illustrates the relaxation process in the more complex situation of an iron silicide layer on Si (111) when the temperature is gradually increased after the actual growth process has been finished. Here lateral and vertical relaxation of the layer is present simultaneously. The average size of the silicide islands increases and a distinct island shape develops. A detailed analysis shows [26] that the crystalline quality increases in combination with the changes in the geometrical arrangement of the thin film. Additionally, the chemical composition of the silicide changes due to further alloying with the silicon substrate.

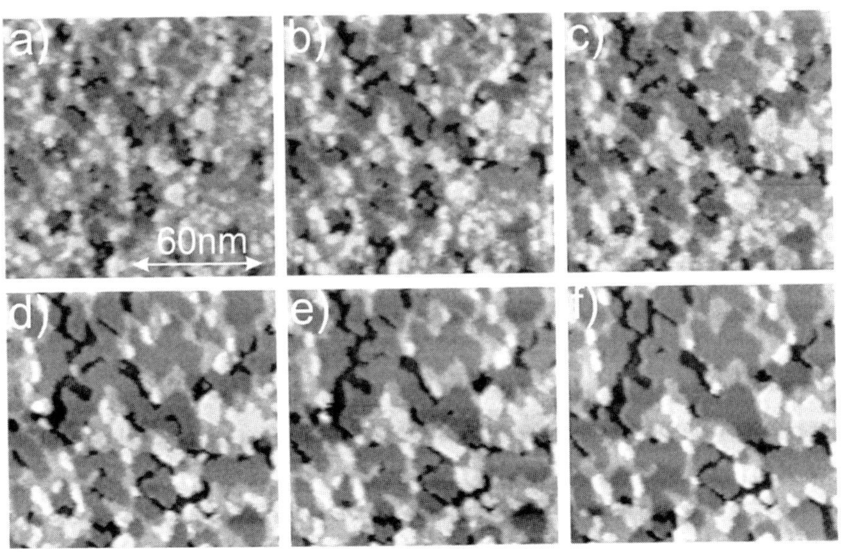

Figure 19. Annealing of a thin iron silicide layer (containing 1 monolayer of iron) at increasing temperature. (a) 576K, (b) 642K, (c) 667K, (d) 682K, (e) 690K, (f) 707K.

3.6. Alloy formation

As a last example for in-situ growth studies with the STM, we like to show sequences of an alloy formation in the growing layer. In many cases an intermixing of deposited material with substrate material represents an unwanted effect (especially for semiconductor MBE). An exception is the formation of silicides during the deposition of metals on silicon substrates. Figure 20 shows the formation of an iron silicide island on Si (111) - (7x7) at 525K on the atomic scale. Before the deposition starts, we have positioned the field of view of the STM to an area of the substrate which shows a perfectly clean (7x7) reconstructed surface to exclude inhomogeneous nucleation. When iron is deposited a triangular silicide island with a (2x2)-reconstruction on-top forms. This island locally consumes atoms from the silicon substrate around the island to form its alloy. Apart from the local disruption the long range structure of the substrate, especially the (7x7) reconstruction, stays intact. A quantitative analysis of the consumed silicon material in relation to the deposited iron results in a composition close to FeSi for silicide islands grown at 525K [26,27].

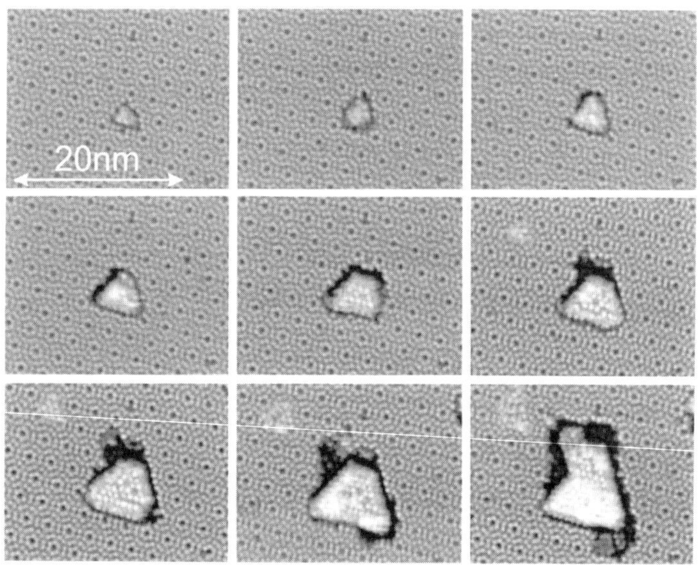

Figure 20. The formation of an iron silicide island on Si(111)-(7x7) during iron deposition at 525K. The local consumption of silicon substrate atoms is visible as a black rim around the island.

When iron is deposited at higher temperature onto Si (111) - (7x7) (see Figure 21) iron silicide is formed by the consumption of silicon which is supplied by a monolayer-wise removal of the substrate. In the sequence shown in Figure 21 from (a) - (f) one monolayer of substrate silicon is consumed in a step retraction mode (inverse process to step flow growth) while triangular silicide islands grow in height. At a temperature of 725K the diffusion length of silicon is sufficiently high to feed growing iron silicide islands with silicon atoms detaching from step edges somewhere on the surface. A local disruption as in the case of the low temperature silicide formation shown in Figure 20 is not found. Again the composition of the silicide islands is determined via the ratio of consumed silicon to deposited iron [27]. At 725K a composition of the silicide close to $FeSi_2$ is found.

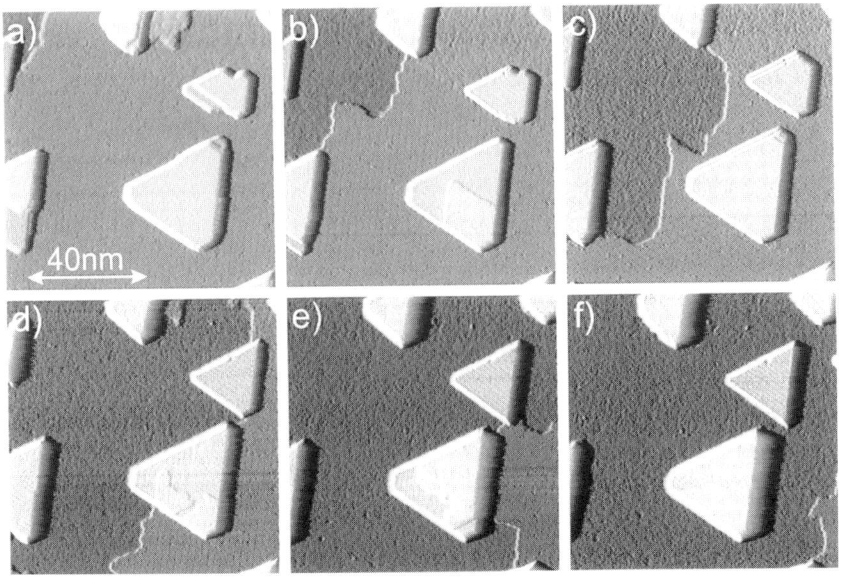

Figure 21. The formation of iron silicide islands on Si (111) - (7x7) during iron deposition at 725K. The higher temperature (compared to Figure 20) leads to a silicon consumption in a step retraction mode. Iron is supplied from the gas phase using iron pentacarbonyle ($Fe(CO)_5$) as precursor.

4. CONCLUSION

The above examples show that it is possible to study the kinetics of thin film growth with scanning tunneling microscopy by a direct sequential imaging of the growing surface with atomic resolution at actual growth temperatures. Especially, most of the basic steps in epitaxial growth sketched in Figure 1 can be directly observed. The conditions of deposition for the direct imaging of the growing thin film have to be specially chosen to meet the requirements of the microscopical method. These conditions are usually far away from the ones used in the technological deposition of thin films (especially with respect to the deposition rate). Therefore the kinetic parameters obtained at the more "prototypical" growth conditions shown here have to be extrapolated to technologically relevant growth conditions. Kinetic Monte-Carlo simulations with parameters obtained from the in-situ STM experiments are one promising way to achieve this transfer. Also a combination with other analytical tools like diffraction experiments is necessary to obtain reliable statistical information in the regimes of growth where STM is not capable to image the growing film directly.

REFERENCES

1. J.A. Venables, Surf. Sci., 299/300 (1994) 798.
2. M.A. van Hove, W.H. Weinberg, C.M. Chan, *Low-energy electron diffraction. Experiment, theory and surface structure determination*, Springer, Berlin (1986).
3. M. Horn-von-Hoegen, Zeitschrift für Physik, 214 (1999) 591(part I), 684 (part II).
4. W. Braun, *Applied RHEED. Reflection high-energy electron diffraction during crystal growth,* Springer, Berlin (1999).
5. G.L. Kellogg, Surf. Sci. Rep., 21 (1994) 1.
6. K. Yagi, Surf. Sci. Rep., 17 (1993) 305.
7. E. Bauer, Rep. Prog. Phys., 57 (1994) 895.
8. G. Binnig, H. Rohrer, Rev. Mod. Phys., 71 (1999) 324.
9. G. Binnig, H. Rohrer, Ch. Gerber, E. Weibel, Phys. Rev. Lett., 49 (1982) 57.
10. G. Binnig, C.F. Quate, Ch. Gerber, Phys. Rev. Lett., 56 (1986) 930.
11. B.Voigtländer, Surf. Sci. Rev., 43 (2001) 127.
12. J.A. Stroscio, D.M. Eigler, Science, 254 (1991) 1319.
13. L. Kuipers, R.W.N. Loos, H. Neerings, J. ter-Horst, G.J. Ruwiel, A.P. de-Jongh, J.W.M Frencken, Rev. Sci. Instrum., 66 (1995) 4557.
14. J.W. Lyding, S. Skala, J.S. Hubacek, R. Brockenbrough, G. Gammie, Rev. Sci. Instrum., 59 (1988) 1897.
15. R. Curtis, T. Mitsui, E. Ganz, Rev. Sci. Instrum., 68 (1997) 2790.
16. B.S. Swartzentruber, Phys. Rev. Lett., 76 (1996) 459.
17. G. Piaszenski, Ph.D. thesis Bochum, Germany (2002) (in German).
18. U. Köhler, C. Jensen, A.C. Schindler, L. Brendel, D.E. Wolf, Phil.Mag., B80 (2) (2000) 283.
19. U. Köhler, C. Jensen, C. Wolf, A.C. Schindler, L. Brendel, D.E. Wolf, Surf. Sci., 454-456 (2000) 667.
20. K. Reshöft, C. Jensen, U. Köhler, Surf. Sci., 421 (1999) 320.
21. F.C. Frank, J.H. van der Merwe, Proc. Royal Soc., London, A198 (1949) 205.
22. M. Vollmer, A. Weber, Z. Phys. Chem., 119 (1926) 277.
23. R.L. Schwoebel, E.J. Shimpsey, J. Appl. Phys., 37 (1966) 3682.
    G. Ehrlich and F. G. Hudda, J. Chem. Phys., 44 (1966) 1039.
24. M. Kneppe, M. Berse, U. Köhler, Appl. Phys. A, DOI: 10.1007/s00339-003-2096-6 (2003).
25. M. Zinke-Allmang, Thin Solid Films, 346 (1999) 1.
26. K. Reshöft, Ph.D. thesis Kiel, Germany (2001), (in German).
27. V. Dorna, Ph.D. thesis Bochum, Germany, Fortschritt-Bericht VDI, 301, Reihe 9 (Elektronik) (1999), (in German).

INDEX